Engineering Haptic Devices

Springer Series on Touch and Haptic Systems

More information about this series at http://www.springer.com/series/8786

Christian Hatzfeld · Thorsten A. Kern
Editors

Engineering Haptic Devices

A Beginner's Guide

Second Edition

Editors
Dr.-Ing. Christian Hatzfeld
Institute of Electromechanical Design
Technische Universität Darmstadt
Darmstadt
Germany

Dr.-Ing. Thorsten A. Kern
Continental Automotive GmbH
Babenhausen
Germany

ISSN 2192-2977 ISSN 2192-2985 (electronic)
Springer Series on Touch and Haptic Systems
ISBN 978-1-4471-6859-1 ISBN 978-1-4471-6518-7 (eBook)
DOI 10.1007/978-1-4471-6518-7

Springer London Heidelberg New York Dordrecht

Softcover reprint of the hardcover 2nd edition 2014

Printed on acid-free paper

Springer is part of Springer Science+Business Media (www.springer.com)

Series Editors' Foreword

This is the 10th volume of the 'Springer Series on Touch and Haptic Systems', which is published in collaboration between **Springer** and the **EuroHaptics Society**.

Engineering Haptic Devices is focused on topics related to the design of effective haptic devices from an engineering point of view. The book is divided into two parts with 15 chapters. Part I is titled 'Basics', which is an introduction to general topics in haptics, such as the haptic interaction modality, the role of users in closed-loop haptic systems and several areas of applications. In Part I, you will also find a taxonomy related to haptic perception, the requirements of the user, and the performance of haptic devices. Furthermore, this part of the volume highlights several methodologies for designing haptic systems, and represents the application of general principles of engineering design to the development of new haptic interfaces. Part II of this volume is focused on studying the main components of haptic systems. The chapters in Part II focus on control, kinematics, actuators, sensors, interfaces, and software development for haptic systems. Finally, the book provides a guide to haptic evaluation. It also shows some examples for the development of haptic systems.

This volume is the second edition of a previous publication from Springer. It has been substantially expanded and several new authors have contributed to this updated edition. The content has been thoroughly revised and more in-depth material has been added to the description of most topics. Some new themes have been included such as the role of the user, the evaluation of haptic systems, and more examples of applications.

This book is of great value to the 'Springer Series on Touch and Haptic Systems' since it covers all engineering aspects for the design of haptic interfaces in a comprehensive manner. We therefore expect it to be well received by the Haptic Community and interesting for teaching these topics. Students will find most subjects studied in master-level courses on haptics, while researchers and engineers will also find it a useful and authoritative source of information for their work in haptics.

May 2014

Manuel Ferre
Marc Ernst
Alan Wing

Preface

The term "haptics" unlike the terms "optics" and "acoustics" is not so well known to the majority of people, at least not in the meaning used in the scientific community. The words "haptics" and "haptic" refer to everything concerning the sense of touch. "Haptics" is everything and everything is "haptic," because it not only describes pure mechanical interaction, but also includes thermal- and pain- (nociception) perception. The sense of touch makes it possible for humans and other living beings to perceive the "borders of their physical being," i.e., to identify where their own body begins and where it ends. While vision and hearing will make us aware of our greater surroundings, the sense of touch covers our immediate vicinity: In the heat of a basketball match a light touch on our back immediately makes us aware of an attacking player we do not see. We notice the intensity of contact, the direction of the movement by a shear on our skin, or a breeze moving our body hairs—all without catching a glimpse of the opponent.

"Haptic systems" are divided into two classes.[1] There are *time-invariant systems* (the keys of my keyboard), which generate a more or less unchanging haptic effect whether being pressed today or in a year's time. Structures like surfaces, e.g., the wooden surface of my table, are also part of this group. These haptically interesting surfaces are often named "haptic textures." Furthermore, there are *active, reconfigurable systems,* which change their haptic properties partly or totally depending on a preselection—e.g., from a menu, or based on an interaction with real or virtual environments.

---

[1] In engineering there are three terms that are often used but do not have definite meaning: System, Device, and Component. Systems are—depending on the task of the designer—either a device or a component. A motor is a component of a car, but for the developer of the motor it is a device, which is assembled from components (spark-plug, cocks, knocking-sensor). It can be helpful when reading a technological text to replace each term with the word "thing." Although this suggestion is not completely serious, it surprisingly increases the comprehensibility of technical texts.

The focus of this book is on the technological design criteria for active, reconfigurable systems, providing a haptic coupling of user and object in a mainly mechanical understanding. Thermal and nociceptive perceptions are mentioned according to their significance but are not thoroughly discussed. This is also the case with regard to passive haptic systems. For active haptic systems, research and industry developed a large number of different universal haptic systems that can be used for different purposes. Because of the large variability of these devices, they sometimes fall short of requirements for certain applications or are—in short, just too expensive. We therefore believe that there is a need for a structured approach to the design of *task-specific haptic systems* on the one hand and a necessity to know about the different approaches for the components and structures of haptic systems on the other hand.

The fact that you have bought this book suggests that you are interested in haptics and its application in human–machine interaction. You might have already tried to sketch a technical system meant to provide a haptic human–machine interaction. Maybe, you are just planning a project as part of your studies or as a commercial product aimed at improving a certain manual control or introducing a new control concept. Maybe, you are a member of the increasing group of surgeons actively using haptics in medical technology and training to improve patients' safety and trying to apply the current progresses to other interventions.

Despite of, or even because of, this great variety of projects in industry and research working with haptic systems, the common understanding of "haptics" and the terms directly referring to it, like "kinaesthetic" and "tactile," is by no means as unambiguous and indisputable as it should be. In this book, we intend to offer a help to act more safely in the area of designing haptic devices. We consider this book as a starting point for engineers and students new to haptics and the design of haptic interfaces as well as a reference work for more experienced professionals. To make the book more usable and practical in this sense, we added recommendations for further insight into most chapters.

It begins with a presentation of the different areas that can benefit from the integration of haptics, including communication, interaction with virtual environments, and the most challenging applications in telepresence and teleoperation. Next, as a basis for the design of such systems, haptics is discussed as an interaction modality. This includes several concepts of haptic perception and haptic interaction and the most relevant results from psychophysical studies that can and have to be applied during the design process of a task-specific haptic system. Please note that this book has been written by and is addressed to engineers from several disciplines. This means that especially psychophysical content is sometimes simplified and shortened in favor of a fundamental basic insight into these topics for engineers working on a haptic device. Next, the role of the user as a (mechanical) part of the haptic system is discussed in detail, since this modeling has a large impact on system properties like stability and perceived haptic quality.

Part I of the book concludes with an extension of the commonly known development models of mechatronic systems to the special design of haptic systems. This chapter lays special focus on the integration of perception properties and ergonomic aspects in this process. The authors believe that the systematic consideration of perception properties and features of the sensory apparatus based on the intended interaction can reduce critical requirements on haptic systems, such as lowering the efforts and costs of development as well as leading to systems with higher perceived quality.

In Part II of the book, an overview of technological solutions, like the designs of actuators, kinematics, or complete systems including software and rendering solutions and the interfaces to simulation and virtual reality systems, is given. This is done with two aspects in mind. First, the reader should be able to find the most important and widely used solutions for recurring problems like actuation or sensing including the necessary technical basis for own designs and developments. Second, we wanted to give an overview of the large number of different principles used in haptic systems that are maybe a good solution for a new task-specific haptic system—or a noteworthy experience of which solution not to try.

The first idea for this book was born in 2003. Originally intended as an addition to the dissertation of Thorsten A. Kern, it was soon thought of as filling a gap: The regrettably small number of comprehensive recapitulating publications on haptics available for, e.g., a technically interested person, confronted with the task of designing a haptic device for the first time. In 2004, inspite of a considerable number of conference proceedings, journals, and Ph.D. theses, no document was available giving a summary of the major findings of this challenging subject.

The support of several colleagues, especially Prof. Dr.-Ing. Dr. med. Ronald Blechschmidt-Trapp and Dr.-Ing. Christoph Doerrer, helped to make the idea of this book clearer in the following years—and showed that this book would have to be much more extensive than originally expected. With the encouragement of Prof. Dr.-Ing. habil. Roland Werthschützky, the first edition was edited by Thorsten A. Kern during a post-doc time. It was funded by the Deutsche Forschungsgemeinschaft (DFG, grant KE1456/1-1) with special regard to the consolidation of the design methodology for haptic devices. Due to this funding the financial basis of this task was guaranteed. The structure of the topic made clear that the book would be considerably improved by contributions from specialists in several areas. In 2008, the German version *Entwicklung Haptischer Geräte* and in 2009 the English version *Engineering Haptic Devices* were published by Springer. Both books sold about 500 copies in total up till now.

In 2010, the idea of a second edition of the book was born. With the change of Dr. Kern from university to industrial employer, the attention also shifted from mainly kinaesthetic to tactile devices. This made severe gaps in the first edition eminent. In parallel, science made great progress in understanding the individual tactile modalities, blurring the borders between different old concepts of the same perception, offering now an opportunity to find an engineering approach to more than the pure vibrotactile perception. It took however until the year 2013 for the work on the second edition to start. In that year, Christian Hatzfeld finished his dissertation dealing with the perception of vibrotactile forces. Also, encouraged by Prof. Dr.-Ing. habil. Roland Werthschützky, he took the lead in editing this second edition. In addition to the first edition, this work was also funded by the DFG (grant HA7164/1-1), pointing out the importance of an adapted design approach for haptic systems.

With the cooperation of Springer and the series editors, the second edition of this book was integrated in the *Springer Series on Touch and Haptic Systems*, as we felt that the design of task-specific haptic interfaces would be complemented well by other works in this series. We wish to thank all the authors who contributed to this book as well as all colleagues, students, and scientists from the haptics community who supported us with fruitful discussions, examples, and permissions to include them in this book. On behalf of many, we would like to point out Lukas Braisz, who was a great support in preparing the figures, especially in the new chapters of the second edition. Special thanks go to our mentor and advisor Prof. Dr.-Ing. habil. Roland Werthschützky, who encouraged and supported the work on both editions of this book.

Since a book is a quite static format compared to the dynamic progress of haptics in general, we set up an accompanying homepage with regular updates on the books topics at http://www.hapticdevices.eu. We hope that this work will alleviate the work of students and engineers new to the exciting and challenging development of haptic systems and serve as a useful resource for all developers.

Darmstadt, April 2014

Christian Hatzfeld
Thorsten A. Kern

Contents

Part I Basics

1 **Motivation and Application of Haptic Systems** 3
Christian Hatzfeld and Thorsten A. Kern
1.1 Philosophical and Social Aspects 4
1.1.1 Haptics as a Physical Being's Boundary 4
1.1.2 Formation of the Sense of Touch 5
1.1.3 Touchable Art and Haptic Aesthetics 6
1.2 Technical Definitions of Haptics 8
1.2.1 Definitions of Haptic Interactions 9
1.2.2 Taxonomy of Haptic Perception 11
1.3 Application Areas of Haptic Systems 13
1.3.1 Telepresence, Teleaction, and Assistive Systems 15
1.3.2 Virtual Environments 17
1.3.3 Noninvasive Medical Applications 20
1.3.4 Communication 21
1.3.5 Why Use a Haptic System? 22
1.4 Conclusions 23
References 23

2 **Haptics as an Interaction Modality** 29
Christian Hatzfeld
2.1 Haptic Perception 29
2.1.1 Physiological Basis 30
2.1.2 Psychophysical Description of Perception 38
2.1.3 Characteristic Values of Haptic Perception 53
2.1.4 Further Aspects of Haptic Perception 65
2.2 Concepts of Interaction 69
2.2.1 Haptic Exploration of Objects 69
2.2.2 Active and Passive Touch 69

2.2.3 Gestures 71
2.2.4 Human Movement Capabilities 72
2.3 Interaction Using Haptic Systems 73
2.3.1 Haptic Displays and General Input Devices 74
2.3.2 Assistive Systems 76
2.3.3 Haptic Interfaces 76
2.3.4 Manipulators 79
2.3.5 Teleoperators 79
2.3.6 Comanipulators 80
2.3.7 Haptic System Control 81
2.4 Engineering Conclusions 81
2.4.1 A Frequency-Dependent Model of Haptic Properties 81
2.4.2 Stiffnesses 84
2.4.3 One Kilohertz: Significance for the Mechanical Design 84
2.4.4 Perception-Inspired Concepts for Haptic System Design 87
References 89

3 The User's Role in Haptic System Design 101
Thorsten A. Kern and Christian Hatzfeld
3.1 The User as Mechanical Load 101
3.1.1 Mapping of Frequency Ranges onto the User's Mechanical Model 101
3.1.2 Modeling the Mechanical Impedance 104
3.1.3 Grips and Grasps 105
3.1.4 Measurement Setup and Equipment 107
3.1.5 Models 108
3.1.6 Modeling Parameters 110
3.1.7 Comparison with Existing Models 118
3.1.8 Final Remarks on Impedances 120
3.2 The User as a Measure of Quality 120
3.2.1 Resolution of Haptic Systems 120
3.2.2 Errors and Reproducibility 121
3.2.3 Quality of Haptic Interaction 121
References 122

4 Development of Haptic Systems 125
Christian Hatzfeld and Thorsten A. Kern
4.1 Application of Mechatronic Design Principles to Haptic Systems 125
4.1.1 Stage 1: System Requirements 127
4.1.2 Stage 2: System Design 128

4.1.3 Stage 3: Modeling and Design of Components 129
4.1.4 Stage 4: Realization and Verification of Components and System 130
4.1.5 Stage 5: Validation of the Haptic System 131
4.2 General Design Goals 131
4.3 Technical Descriptions of Parts and System Components 132
4.3.1 Single Input, Single Output Descriptions 133
4.3.2 Network Parameter Description 134
4.3.3 Finite Element Methods 136
4.3.4 Description of Kinematic Structures 137
References 140

Part II Designing Haptic Systems

5 Identification of Requirements 145
Thorsten A. Kern and Christian Hatzfeld
5.1 Definition of Application: The Right Questions to Ask 145
5.1.1 Experiments with the Customer 146
5.1.2 General Design Guidelines 148
5.2 Interaction Analysis 149
5.3 Technical Solution Clusters 153
5.3.1 Cluster ①: Kinaesthetic 155
5.3.2 Cluster ②: Surface-Tactile 156
5.3.3 Cluster ③: Vibro-Tactile 156
5.3.4 Cluster ④: Vibro-Directional 157
5.3.5 Cluster ⑤: Omnidirectional 158
5.3.6 General Requirement Sources 158
5.4 Safety Requirements 159
5.4.1 Safety Standards 159
5.4.2 Definition of Safety Requirements from Risk Analysis 160
5.5 Requirement Specifications of a Haptic System 166
References 166

6 General System Structures 169
Thorsten A. Kern
6.1 Open-Loop Impedance Controlled 170
6.2 Closed-Loop Impedance Controlled 171
6.3 Open-Loop Admittance Controlled 173
6.4 Closed-Loop Admittance Controlled Devices 173
6.5 Qualitative Comparison of the Internal Structures of Haptic Systems 176
6.5.1 Tactile Devices 177

6.5.2 Kinaesthetic Devices . . . 177
6.6 How to Choose a Suitable System Structure . . . 178

7 Control of Haptic Systems . . . 181
Thomas Opitz and Oliver Meckel
7.1 System Description . . . 182
7.1.1 Linear State Space Description . . . 183
7.1.2 Nonlinear System Description . . . 184
7.2 System Stability . . . 186
7.2.1 Analysis of Linear System Stability . . . 187
7.2.2 Analysis of Nonlinear System Stability . . . 190
7.3 Control Law Design for Haptic Systems . . . 197
7.3.1 Structuring of Control Design . . . 197
7.3.2 Requirement Definition . . . 199
7.3.3 General Control Law Design . . . 201
7.3.4 Example: Cascade Control of a Linear Drive . . . 206
7.4 Control of Teleoperation Systems . . . 208
7.4.1 Two-Port Representation . . . 209
7.4.2 Transparency . . . 210
7.4.3 General Control Model for Teleoperators . . . 214
7.4.4 Stability Analysis of Teleoperators . . . 217
7.4.5 Effects of Time Delay . . . 219
7.5 Conclusion . . . 222
References . . . 223

8 Kinematic Design . . . 227
Sebastian Kassner
8.1 Introduction and Classification . . . 227
8.1.1 Classification of Mechanisms . . . 229
8.2 Design Step 1: Topological Synthesis—Defining the Mechanism's Structure . . . 231
8.2.1 Synthesis of Serial Mechanisms . . . 231
8.2.2 Synthesis of Parallel Mechanisms . . . 232
8.2.3 Special Case: Parallel Mechanisms with Pure Translational Motion . . . 233
8.2.4 Example: The DELTA Mechanism . . . 235
8.3 Design Step 2: Kinematic Equations . . . 237
8.3.1 Kinematics: Basic Equations for Design and Operation . . . 239
8.3.2 Example: The DELTA Mechanism . . . 241
8.4 Design Step 3: Dimensioning . . . 244
8.4.1 Isotropy and Singular Positions . . . 245
8.4.2 Example: The DELTA Mechanism . . . 250
References . . . 251

9 Actuator Design 253
Henry Haus, Thorsten A. Kern, Marc Matysek and Stephanie Sindlinger
9.1 General Facts About Actuator Design 254
9.1.1 Overview of Actuator Principles. 254
9.1.2 Actuator Selection Aid Based on Its Dynamics 257
9.1.3 Gears 258
9.2 Electrodynamic Actuators 261
9.2.1 The Electrodynamic Effect and Its Influencing Variables. 262
9.2.2 Actual Actuator Design. 276
9.2.3 Actuator Electronics 281
9.2.4 Examples for Electrodynamic Actuators in Haptic Devices. 286
9.2.5 Conclusion About the Design of Electrodynamic Actuators. 288
9.3 Piezoelectric Actuators 288
9.3.1 The Piezoelectric Effect 289
9.3.2 Designs and Properties of Piezoelectric Actuators . . . 294
9.3.3 Design of Piezoelectric Actuators for Haptic Systems. 298
9.3.4 Procedure for the Design of Piezoelectric Actuators. 299
9.3.5 Piezoelectric Actuators in Haptic Systems. 304
9.4 Electromagnetic Actuators. 314
9.4.1 Magnetic Energy 314
9.4.2 Design of Magnetic Circuits 317
9.4.3 Examples for Electromagnetic Actuators. 321
9.4.4 Magnetic Actuators in Haptic Devices 324
9.4.5 Conclusion on the Design of Magnetic Actuators. 326
9.5 Electrostatic Actuators 327
9.5.1 Definition of the Electric Field. 327
9.5.2 Designs of Capacitive Actuators with Air-Gap. 329
9.5.3 Dielectric Elastomer Actuators. 335
9.5.4 Designs of Dielectric Elastomer Actuators. 338
9.5.5 Electrorheological Fluids. 342
9.6 Special Designs of Haptic Actuators. 350
9.6.1 Haptic-Kinaesthetic Devices 350
9.6.2 Haptic-Tactile Devices 356
References 364

10 Sensor Design . . . 373
Jacqueline Rausch, Thorsten A. Kern and Christian Hatzfeld
10.1 Force Sensors . . . 373
10.1.1 Constraints . . . 374
10.1.2 Sensing Principles . . . 381
10.1.3 Selection of a Suitable Sensor . . . 413
10.2 Positioning Sensors . . . 419
10.2.1 Basic Principles of Position Measurement . . . 419
10.2.2 Requirements in the Context of Haptics . . . 421
10.2.3 Optical Sensors . . . 422
10.2.4 Magnetic Sensors . . . 425
10.2.5 Other Displacement Sensors . . . 427
10.2.6 Electronics for Absolute Positions Sensors . . . 428
10.2.7 Acceleration and Velocity Measurement . . . 429
10.2.8 Conclusion on Position Measurement . . . 432
10.3 Touch Sensors . . . 433
10.3.1 Resistive Touch Sensors . . . 433
10.3.2 Capacitive Touch Sensors . . . 434
10.3.3 Other Principles . . . 435
10.4 Imaging Sensors . . . 436
10.5 Conclusion . . . 436
References . . . 437

11 Interface Design . . . 443
Thorsten A. Kern
11.1 Border Frequency of the Transmission Chain . . . 444
11.1.1 Bandwidth in a Telemanipulation System . . . 444
11.1.2 Bandwidth in a Simulator System . . . 445
11.1.3 Data Rates and Latencies . . . 446
11.2 Concepts for Bandwidth Reduction . . . 447
11.2.1 Analysis of the Required Dynamics . . . 447
11.2.2 Local Haptic Model in the Controller . . . 447
11.2.3 Event-Based Haptics . . . 448
11.2.4 Movement Extrapolation . . . 450
11.2.5 Compensation of Extreme Dead Times . . . 450
11.2.6 Compression . . . 450
11.3 Technical Standard Interfaces . . . 451
11.3.1 Serial Port . . . 451
11.3.2 Parallel Port . . . 452
11.3.3 USB . . . 453
11.3.4 FireWire: IEEE 1394 . . . 454
11.3.5 Ethernet . . . 454
11.3.6 Measurement Equipment and Multifunctional Interface Cards . . . 455

11.3.7 HIL Systems 455
11.4 Final Remarks on Interface Technology 455
References 456

12 Software Design for Virtual Reality Applications 457
Alexander Rettig
12.1 Overview About the Subject "Virtual Reality" 458
12.1.1 Immersion 458
12.1.2 Natural Interaction 458
12.1.3 Natural Object Behavior 459
12.2 Design and Architecture of VR Systems 461
12.2.1 Hardware Components 461
12.2.2 Device Integration and Device Abstraction 462
12.2.3 Software Components 464
12.2.4 Simulation 466
12.2.5 Subsystems for Rendering 469
12.2.6 Decoupling of the Haptic Renderer from Other Sense Modalities 471
12.2.7 Haptic Interaction Metaphors 473
12.3 Algorithms 474
12.3.1 Virtual Wall 476
12.3.2 "Penalty" Methods 479
12.3.3 Constraint-Based Methods 481
12.3.4 6 DoF Interaction: Voxmap-PointShell Algorithm 484
12.3.5 Collision Detection 490
12.4 Software Packages for Haptic Applications 497
12.5 Perception-Based Concepts for VR software 499
12.5.1 Event-Based Haptics 499
12.5.2 Pseudo-haptic Feedback 499
12.6 Conclusion 500
References 500

13 Evaluation of Haptic Systems 503
Carsten Neupert and Christian Hatzfeld
13.1 System-Centered Evaluation Methods 504
13.1.1 Workspace 505
13.1.2 Output Force-Depending Values 505
13.1.3 Output Motion-Depending Values 508
13.1.4 Mechanical Properties 508
13.1.5 Impedance Measurements 509
13.1.6 Special Properties 511
13.1.7 Measurement of Psychophysical Parameters 511
13.2 Task-Centered Evaluation Methods 512
13.2.1 Task Performance Tests 512

13.2.2 Identification of Haptic Properties and Signals 514
13.2.3 Information Input Capacity (Fitts' Law) 516
13.3 User-Centered Evaluation Methods 518
13.3.1 Workload 518
13.3.2 Subjective Evaluation 520
13.3.3 Learning Effects 521
13.3.4 Effects on Performance in Other Domains 521
13.4 Conclusion 522
References 522

14 Examples of Haptic System Development 525
Limin Zeng, Gerhard Weber, Ingo Zoller, Peter Lotz, Thorsten A. Kern, Jörg Reisinger, Thorsten Meiss, Thomas Opitz, Tim Rossner and Nataliya Stefanova
14.1 Tactile You-Are-Here Maps 526
14.1.1 Introduction 526
14.1.2 The TacYAH Map Prototype 527
14.1.3 Evaluation 532
14.1.4 Conclusion and Outlook 532
14.2 Automotive Interface with Tactile Feedback 532
14.2.1 Context 533
14.2.2 The Floating TouchPad of Mercedes Benz 534
14.2.3 Actuator Design 536
14.2.4 Evaluation 541
14.2.5 Discussion and Outlook 544
14.3 HapCath: Haptic Catheter 546
14.3.1 Introduction 546
14.3.2 Deriving Requirements 547
14.3.3 Design and Development 548
14.3.4 Verification and Validation 551
14.3.5 Conclusion and Outlook 552
References 552

15 Conclusion 555

Appendix A: Impedance Values of Grasps 557

Appendix B: URLs 559

Glossary 565

Index 567

Symbols

a	Sensory background noise (Weber's Law) (–)
a	Acceleration $\left(\frac{\mathrm{m}}{\mathrm{s}^2}\right)$
$\mathbf{a}$	Vector, summarizing actuator displacement and angles a_i (–)
A	Area, cross section (m^2)
$A(j\omega)$	Amplitude response (Chap. 7) (dB)
$\mathbf{A}$	Matrix of a linear system of equations (–)
α	Positive number (–)
α	Angle, Euler rotation (around the x-axis) (degree, radian)
α_{VK}	Coefficient of thermal expansion (K^{-1})
b	Wave impedance
B, B_0	Magnetic flux density (T)
$\mathbf{B}_r$	Remanence flux density (T)
$\mathbf{B}$	Matrix of a linear system of equations (–)
β	Angle, Euler rotation (around the y-axis) (degree, radian)
c_{index}	Arbitrary constant, further defined by index (–)
c	Spring constant (–)
c_θ	Threshold parameter of the psychometric function (–)
c_σ	Sensitivity parameter of the psychometric function (–)
c_λ	Decision criterion (Signal Detection Theory) (–)
C_{ijlm}	Elastic constants $\left(\frac{\mathrm{m}^2}{\mathrm{N}}\right)$
C, C_Q	Capacity $\left(\mathrm{F} = \frac{\mathrm{A \cdot s}}{\mathrm{V}}\right)$
C_b	Coupling capacity (at mechanical full-stop) (F)
C	Transmission elements, controller (Chap. 7) (–)
$\mathbf{C}$	Matrix of a linear system of equations (–)
$\frac{\Delta C}{C_0}$	Capacity change (–)
$\mathbb{C}$	Complex numbers (–)

d	Damping/friction $\left(\frac{\text{N}}{\text{m}\cdot\text{s}}\right)$
d	Distance, deflection, diameter (m)
$d_{ij,k}$, d_{im}	Piezoelectric charge constant $\left(\frac{\text{V}}{\text{m}}\right)$
$\frac{d_t}{d'}$	Detectability (Signal Detection Theory) (–)
D	Density
D	Dielectric flux density (A s m^{-2})
$\mathbf{D}$	Dielectric displacement/electrical displacement density $\left(\frac{\text{C}}{\text{m}^2}\right)$
$\mathbf{D}$	(transmission-) Matrix of a linear system of equations (–)
ΔD	Position-discrete resolution (–)
δ	Phase difference (Sect. 10.1) (–)
e	Piezoelectric voltage coefficient $\left(\frac{\text{A}\cdot\text{s}}{\text{m}^2}\right)$
$\mathbf{e}_i$	Directional unit vector (–)
E	E-modulus, modulus of elasticity $\left(\frac{\text{N}}{\text{m}^2}\right)$
E	Electrical field strength $\left(\frac{\text{V}}{\text{m}}\right)$
$\underline{e}_T$	Absolute transparency error (Sect. 7.4.2) (–)
$\underline{e}_T'$	Relative transparency error (Sect. 7.4.2) (–)
E_{ref}	Reference field strength, with C_s of an ERF being given $\frac{\text{V}}{\text{m}}$
$\mathbf{E}$	Electrical field $\left(\frac{\text{V}}{\text{m}}\right)$
ε	Permittivity $(\varepsilon = \varepsilon_0 \cdot \varepsilon_r)\left(\frac{\text{A}\cdot\text{s}}{\text{V}\cdot\text{m}}\right)$
ε	Relative dielectric constant of piezoelectric material (at constant mechanical tension) $\left(\frac{\text{A}\cdot\text{s}}{\text{V}\cdot\text{m}}\right)$
ε	Remaining error (Chap. 7) (–)
ε_0	Electrical field constant $\left(\varepsilon_0 = 8,854 \cdot 10^{-12}\frac{\text{C}}{\text{V}\cdot\text{m}}\right)$
ε_r	Relative permittivity $\left(\varepsilon_r = \frac{E_0}{E}\right)$ (–)
f	Frequency (Hz)
f_0, f_R	Resonance-frequency (Hz)
f_b, f_g	Border-frequency (Hz)
f_{tot}	Sum of all joint degress-of-freedom of a mechanism (–)
$f_{i,\ldots,g}$	Degree-of-freedom of the *i*th joint in a mechanism (–)
f_{id}	Sum of all identical links in a mechanism (–)
f_{ink}	Dynamics of the detection of all increments for positioning measurement (Hz)
$f(\cdot)$	Static non-linearity (–)
F	Bearing-/movement-DOF of a mechanism (–)
F	Force (–)
ΔF	Force-resolution (N)

Φ	Magnetic flux $(\text{Wb} = \text{V} \cdot \text{s})$
$\phi(j\omega)$	Phase plot (degree)
ϕ	Roll angle, rotation (around z-axis) (degree, radian)
φ	Angle (degree)
φ_R	Phase margin (degree)
Φ	Stimulus (–)
Ψ	Subjective percept (–)
g	Number of joints in a mechanism (Chap. 8) (–)
g	Piezoelectric constant $\left(\frac{\text{V}\cdot\text{m}}{\text{N}}\right)$
$g(x, u, t)$	Transfer function (time domain)
$G(s)$, $\underline{G}$	Transfer function in LAPLACE domain (–)
γ	Angle, Euler rotation (around the z-axis) (degrees, radians)
$\dot{\gamma}$	Shear-rate (s^{-1})
h	Height (m)
h	Viscous damping/friction (network theory, see Table 4.1) (–)
$h(t)$	Transfer function (–)
$\underline{h}$	Mobility $\underline{h} = \frac{1}{\underline{Z}}\left(\frac{\text{m}}{\text{N}\cdot\text{s}}\right)$
h	Element of the complex hybrid matrix **H** (–)
$\mathbf{H}$	Complex hybrid matrix (Chap. 7) (–)
H_c	Coercitive field strength $\left(\frac{\text{A}}{\text{m}}\right)$
$\mathbb{H}$	Hamilton numbers (–)
i, $\underline{i}$	(AC) Current (A)
I	(DC) Current (–)
$\mathbf{I}$, $\mathbf{I}'$	Interaction path *intention* (Sect. 2.3) (–)
I_D	Index of difficulty (Sect. 13.2) (–)
I_p	Index of performance (Sect. 13.2) (–)
I	Moment of inertia (m^4)
j, i	Imaginary unit, $i = \sqrt{-1} \in \mathbb{C}$ (–)
J	Current density $\left(\frac{\text{A}}{\text{m}^2}\right)$
$\mathbf{J} = \frac{\partial \mathbf{x}}{\partial \mathbf{q}}$	JACOBIAN matrix defined by the relation of actuator and TCP speeds (–)
k	Spring constant, mechanical stiffness, elasticity $(\text{N}\,\text{m}^{-1})$
k	Geometrical design dependent constant of ERFs $(\text{m} \cdot \text{s})$
k	Fill-factor of a coil (≥ 1) (–)
k	Coupling-factor or k-factor (Sect. 10.1) (–)
k	Number of chains in a mechanism (–)
k_M	Motor constant (–)
K_{krit}	Critical amplification
K_R	Amplification of a proportional controller
κ	Conditioning number of a mechanism (–)

l	Length (m)
L	Inductivity $\left(\mathrm{H} = \frac{\mathrm{V \cdot s}}{\mathrm{A}}\right)$
λ	Pole of a transfer function (–)
λ	Wavelength (m)
λ	Eigenvalue of a matrix (–)
m	Mass (kg)
M	Torque (Nm)
μ	Movability of a charge-carrier $\left(\frac{\mathrm{m^2}}{\mathrm{V \cdot s}}\right)$
μ	Frictional coefficient (–)
μ	Mean value (–)
μ	Magnetic permeability $(\mu = \mu_0 \cdot \mu_r)\left(\frac{\mathrm{V \cdot s}}{\mathrm{A \cdot m}}\right)$
μ_0	Magnetic field constant $\mu_0 = 4\pi \cdot 10^{-7}\frac{\mathrm{V \cdot s}}{\mathrm{A \cdot m}}$
μ_r	Relative permeability (–)
n, N	Number $\in \mathbb{N}$ (–)
$n = \frac{1}{k}$	Compliance (m N^{-1})
n_0, n_i	Refraction index (–)
$\mathbb{N}$	Natural numbers
ν	Global conditioning index (–)
$\omega = 2\pi f$	Angular frequency (rad s^{-1})
$\underline{\omega}$, $\underline{\Omega}$	Angular velocity $\left(\frac{\mathrm{rad}}{\mathrm{s}}\right)$
p	Pressure $\left(\frac{\mathrm{N}}{\mathrm{m^2}}\right)$
p	Probability (–)
p_L	Lapse rate of the psychometric function (–)
p_G	Guess rate of the psychometric function (–)
p_ψ	Psychometric function (–)
P	Dielectric polarization $\left(\frac{\mathrm{C}}{\mathrm{m^2}}\right)$
P	Power (–)
P_g	Degree of parallelism (–)
$\mathbf{P}$, $\mathbf{P}'$	Interaction path *Perception* (Sect. 2.3) (–)
π	Piezoresistive coefficient $\left(\frac{\mathrm{m^2}}{\mathrm{N}}\right)$
π_l	Piezoresistive coefficient in longitudinal direction $\left(\frac{\mathrm{m^2}}{\mathrm{N}}\right)$
π_q	Piezoresistive coefficient in transversal direction $\left(\frac{\mathrm{m^2}}{\mathrm{N}}\right)$
ψ	Yaw angle, rotation around x-axis (degree, radian)
Ψ	Subjective percept (–)
q, Q	Electrical charge ($\mathrm{C} = \mathrm{A} \cdot \mathrm{s}$)
q_i, $i \in \mathbb{N}$	Driven joint i
q	Fluidic volume flow (–)

$\mathbf{q}$	Vector of actor coordinates (–)
r	Distance, radius (m)
$r_i,\ i \in \mathbb{N}$	Active resistors $\left(\Omega = \frac{\mathrm{V}}{\mathrm{A}}\right)$
R	Electrical resistance (Ω)
R_m	Magnetic resistance/reluctance $\left(\frac{\mathrm{A}}{\mathrm{V \cdot s}}\right)$
$\mathbb{R}$	Real numbers (–)
$\Re$	Real part (–)
$\frac{dR}{R_0}$	Relative resistance change (–)
ΔR_{inch}	Position resolution given in dots-per-inch (dpi)
ΔR_{mm}	Position resolution given in millimeter (mm)
ρ	Density $\left(\frac{\mathrm{kg}}{\mathrm{m}^3}\right)$
ρ	Small number ≥ 0 (–)
ρ	Specific resistance/conductivity ($\Omega \cdot \mathrm{m}$)
$s(t)$, $\underline{S}$	Arbitrary signal in time and frequency domain (–)
s	Elasticity coefficient at a constant field strength $\left(\frac{\mathrm{m}^2}{\mathrm{N}}\right)$
s	LAPLACE operator, $s = \sigma + j\omega$ (–)
S	Mechanical stress (m m^{-1})
S	Number of constraints in a mechanism (–)
σ	Conductivity, $\sigma = \frac{1}{\rho}\left(\frac{\mathrm{S}}{\mathrm{m}} = \frac{\mathrm{A}}{\mathrm{V \cdot m}}\right)$
σ	Singular value of a matrix (–)
t	Time/point in time (s)
tr	Transmission ratio of a gear (–)
T	Mechanical tension $\left(\frac{\mathrm{N}}{\mathrm{m}^2}\right)$
T	Time constant, time delay (s)
τ	Shear force (Chap. 9) (N)
τ	Time constant of the step response of an electrical transmission system $\left(\tau = \frac{L}{R}, \tau = \frac{1}{RC}\right)$ (s)
τ	Torque (Chap. 8) (N m)
θ	Pitch angle, rotation about the y-axis (degree, radians)
Θ	Magnetomotive force (A)
ϑ	Temperature (K)
$u(t)$	(AC) Voltage (V)
U	(DC) Voltage (V)
$\mathbf{u}$	Multidimensional input value of a linear system (–)
v	Velocity (m s^{-1})
V	Magnetic tension, magnetic voltage (A)
V	Volume (m^3)
V_x	LYAPUNOV function (Chap. 7) (–)
$V(\mathbf{x})$	Scalar nonlinear positive definite storage function of system states $\mathbf{x}$
$\dot{V}$	Volume flow $\left(\frac{\mathrm{m}^3}{\mathrm{s}}\right)$

ΔV	Volume-element (m^3)
w	General value for in- and output values (–)
$\mathbf{w}$	Unity vector (–)
W	Work, energy $\left(J = \frac{kg \cdot m^2}{s^2}\right)$
x	Distance, displacement, translation, amplitude, elongation, position (m)
$\mathbf{x} = (x, y, z)$	Cartesian coordinates (–)
$\mathbf{x}$	Inner states of a linear system (–)
$\mathbf{x}$	Vector of TCP coordinates (position and orientation) (–)
Δx	Position resolution (m)
X	Transformation constant (–)
ξ	Displacement (m)
y	Control value (–)
y	Output (–)
$\mathbf{y}$	Multidimensional output value of a linear system (–)
Y	Gyratoric transformation constant (–)
$\underline{Y}$	Mechanical admittance $\left(\frac{m}{N \cdot s}\right)$
z	Disturbance variable (–)
$\underline{Z}$	Mechanical impedance $\left(\frac{N \cdot s}{m}\right)$
$\underline{Z}$	Electrical impedance ($V\ A^{-1}$)

Indices and Distinctions

The usage of the most relevant indices and distinctions used throughout the book is shown using the replacement character ■

$■_0$	Base or reference value
$■_E$	Referring to the real or VR environment
$■_H$	Referring to the master side of a teleoperator (probably derived from "handle")
$■_M$	Referring to the master device of a haptic system
$■_{max}$	Maximum value
$■_{min}$	Minimum value
$■_{rot}$	Referring to a rotational value
$■_S$	Referring to the slave device of a haptic system
$■_T$	Referring to the master side of a teleoperator
$■^T$	Transformed vector or matrix
$■_{trans}$	Referring to a translational value
$■_{\mathrm{user}}$	Referring to the user of a haptic system
$\delta■$	Small change, differential
$\Delta■$	Discretized element
$■_\theta$	Referring to a psychophysical threshold

X	Vector or matrix
■(t)	Time-depending value
$\underline{■}$	Complex value with amplitude/phase or real/imaginary part
$\dot{■}$	Derivative with respect to time

Abbreviations

AAL	Ambient Assisted Living, term for technical systems used to support needy people in daily life
ALARP	As Low as Reasonably Practicable, general decision principle used in risk analysis
API	Application Programming Interface
COTS	Commercial off-the-shelf Products or Devices
DoF	Degrees of freedom, the number of independent motions that can be carried out by a body or mechanism
EMG	Electromyography, non-invasive method to record muscular activity based on the electric potential generated by muscle cells
ERF	Electro-Rheological Fluid, fluid changing its rheological properties when exposed to an eletric field
FDA	United States Food and Drug Administration
FEM	Finite Element Method or Finite Element Model
FTA	Fault Tree Analysis, method to identify risks during the development of a system
GUM	Guide to the Expression of Uncertainty in Measurement, ISO/IEC Guide 98-3
HCI	Human–Computer-Interaction
IEC	International Electrotechnical Commission, international organization issuing standards and conformity assessment for electrical, electronical and related technologies, www.iec.ch
IEEE	Institute of Electrical and Electronics Engineers, professional organization that arranges conferences, publishes journals (like the IEEE Transactions on Haptics) and technology standards. www.ieee.org
ISO	International Organization for Standardization, organization for the development of technical standards with head office in Geneva, CH. www.iso.org
IT	Information Transfer, measure of the ability of a haptic display to convey information from system to user

JND	Just Noticeable Difference, psychometric parameter describing the smallest detectable difference $\Delta\Phi$ from a base stimulus Φ_0 that can be detected by a person
LCT	Lane Change Test, test to simulate in vehicle secondary task demands, standardized in ISO 26022
MRF	Magneto-Rheological Fluid, fluid changing its rheological properties when exposed to an magnetic field
MRI	Magnetic Resonance Imaging, non-invasive imaging technique based on the magnetic spin properties of—mostly—hydrogen atoms
PDE	Partial Differential Equations
PWM	Pulse-Width-Modulation
QS	Quasi-Static, i.e., with a small low border frequency limited by the duration of the measurement
SDT	Signal Detection Theory, approach to describe decision and perception processes mainly based on statistical modeling
SIL	Safety Integrity Level, assessment of the reliability of safety functions according to IEC 61508
SISO	Single Input, Single Output (model)
TCP	Tool Center Point, reference point for kinematic and dynamic measures
TPM	Translational Parallel Machines, mechanisms, whose TCP can only move in three Cartesian coordinates (x, y, z)
TPTA	Telepresence and Teleaction
VR	Virtual Reality

Contributors

Christian Hatzfeld joined the Institute of Electromechanical Design of Technische Universität Darmstadt as a research and teaching assistant in 2008. He received his doctoral degree in 2013 for a work on the perception of vibrotactile forces. Since then, he is the leader of the "Haptic Systems" group. His research interests include development and design methods for task-specific haptic systems and the utilization of human perception properties to alleviate the technical design.

Henry Haus graduated in Electrical Engineering from the Technische Universität Darmstadt in 2010. There he is working as a teaching and research assistant at the Institute of Electromechanical Design. His research focuses on the fabrication of stacked dielectric elastomer transducers and their integration in tactile user interfaces. Currently he is group leader of the electroactive polymers group.

Sebastian Kassner received his doctoral degree (Dr.-Ing.) from Technische Universität Darmstadt in 2013. He joined the university's Institute of Electromechanical Design in 2007 as a teaching and research assistant. His research was focused on haptic human–machine interfaces for robotic surgical systems in the field of minimally invasive surgery. His special interest is the application of the electromechanical network theory on the design process of haptic devices. He served as an expert in ISO's committee "Tactile and Haptic Interactions" (TC159/SC4/WG9). In 2012 he joined SIEMENS AG as a product manager.

Thorsten A. Kern is employed at Continental Automotive GmbH as the director of a department for head-up display development. Earlier, he was as a group leader responsible for the development and production of actuators for visual and haptic applications, especially tactile devices. He was also working in the area of medical simulation technology and VR-applications. His research focused on engineering methods to quantify the haptic impression of actuators and devices. He received his doctoral degree in Electrical Engineering in 2006 from the Technische Universität Darmstadt for the design of a haptic assistive system for cardio-catheterizations.

Marc Matysek graduated in Electrical Engineering from the Technische Universität Darmstadt in 2003. He was awarded his doctorate degree for his research focusing on the technology of dielectric elastomer actuators and their use in applications as tactile displays. He is currently working as an actuation expert at Continental Automotive GmbH.

Oliver Meckel received his degree in Mechanical Engineering in 2002 from Technische Universität Darmstadt. He is working as group leader of the technical department at the Wittenstein motion control GmbH. He is responsible for the development, simulation, testing, and technical product support of electrical gear motors and actuator systems. Earlier, he was working at the Institute for Flight Systems and Control at Technische Universität Darmstadt. His research was focused on adaptive control systems and the development of Unmanned Autonomous Vehicles (UAV).

Carsten Neupert received his diploma in Electrical Engineering and Information Technology from the Technische Universität Darmstadt in 2011. He is currently working as a research associate at the Institute of Electromechanical Design. His main research topic is the development methodology of task-specific haptic user interfaces for telerobotic surgery systems.

Thomas Opitz graduated in Business Administration and Electrical Engineering from the Technische Universität Darmstadt in 2009. There, he is working as a teaching and research assistant at the Institute of Electromechanical Design since 2009. His research focuses on development and control of haptic assistive systems for medical applications.

Jacqueline Rausch is currently employed at Roche Diagnostics GmbH as a Test Design Engineer. Earlier, she worked in the field of novel strain sensing technologies for adaptronic systems. She received her doctoral degree (Dr.-Ing.) from Technische Universität Darmstadt in 2012. She joined the university's Institute of Electromechanical Design in 2006 as a teaching and research assistant. Her research was focused on design and application of miniaturized piezoresistive strain sensing elements—among others, for robotic surgical systems in the field of minimally invasive surgery. Her special interest is the design of multi-axis force sensors for haptic devices.

Alexander Rettig received his diploma in Mathematics at the Technical Universität Darmstadt in 1998. He is working for ask—Innovative Visualisierungslösungen GmbH as a visualization specialist. Earlier, he was working at PolyDimensions GmbH in the field of surgical simulators and as a research associate at the Fraunhofer Institute for Computer Graphics Research (IGD) Darmstadt, where he focused on the integration of haptics into virtual reality systems and development of virtual reality applications.

Stephanie Sindlinger received her doctorate degree from the Technische Universität Darmstadt in 2011 for her work on actuation principles for miniaturized haptic interfaces for medical use. She is currently working in the drive development department of Roche Diagnostics GmbH.

Further Contributions

Further contributions to this book were made by

Dr.-Ing. Peter Lotz
Continental Automotive GmbH, Babenhausen, Germany

Dr.-Ing. Thorsten Meiß
EvoSense Research & Development GmbH, Darmstadt, Germany

Dr.-Ing. Jörg Reisinger
Daimler AG, Sindelfingen, Germany

Tim Rossner
Institute of Electromechanical Design, Technische Universität Darmstadt, Germany

Nataliya Stefanova
Institute of Electromechanical Design, Technische Universität Darmstadt, Germany

Prof. Dr. rer. nat. Gerhard Weber
Human-Computer Interaction Research Group, Technische Universität Dresden, Germany

Dr.-Ing. Limin Zeng
Human-Computer Interaction Research Group, Technische Universität Dresden, Germany

Ingo Zoller, Ph.D.
Continental Automotive GmbH, Babenhausen, Germany

Part I
Basics

Chapter 1
Motivation and Application of Haptic Systems

Christian Hatzfeld and Thorsten A. Kern

Abstract This chapter introduces the philosophical and social aspects of the human haptic sense as a basis for systems addressing this human sensory channel. Several definitions of haptics as a perception and interaction modality are reviewed to serve as a common basis in the course of the book. Typical application areas such as telepresence, training, and interaction with virtual environments and communications are presented, and typical haptic systems from these are reviewed. The use of haptics in technical systems is the topic of this book. But what is *haptics* in the first place? A common and general definition is given as

> **Definition** *Haptics* Haptics describes the sense of touch and movement and the (mechanical) interactions involving these.

but this will probably not suffice for the purpose of this book. This chapter gives a detailed insight into the definition of haptics (Sect. 1.2) and introduces four general classes of applications for haptic systems (Sect. 1.3) as the motivation for the design of haptic systems and—ultimately—for this book. Before that, we give a short summary of the philosophical and social aspects of this human sense (Sect. 1.1). These topics are not addressed any further in this book, but should be kept in mind by every engineer working on haptics.

C. Hatzfeld (✉)
Institute of Electromechanical Design, Technische Universität Darmstadt, Merckstr. 25, 64283 Darmstadt, Germany
e-mail: c.hatzfeld@hapticdevices.eu

T.A. Kern
Continental Automotive GmbH, VDO-Straße 1, 64832 Babenhausen, Germany
e-mail: t.kern@hapticdevices.eu

C. Hatzfeld and T.A. Kern (eds.), *Engineering Haptic Devices*,
Springer Series on Touch and Haptic Systems, DOI 10.1007/978-1-4471-6518-7_1

1.1 Philosophical and Social Aspects

An engineer tends to describe haptics primarily in terms of forces, elongations, frequencies, mechanical tensions, and shear forces. This of course makes sense and is important for the technical design process. However, haptics starts before that. Haptic perception ranges from minor interactions in everyday life, e.g., drinking from a glass or writing this text, to a means of social communication, e.g., shaking hands or giving someone a pat on the shoulder, and very personal and private interpersonal experiences. This section deals with the spectrum and influence haptics has on humans beyond the technological descriptions. It is also a hint for the development engineer, to be responsible and conscious when considering the capabilities to fool the haptic sense.

1.1.1 Haptics as a Physical Being's Boundary

Haptics is derived from the Greek term "haptios" and describes "something which can be touched." In fact, the consciousness about and understanding of the haptic sense has changed many times in the history of humanity. ARISTOTELES puts the sense of touch in the last place when naming the five senses:

1. Sight
2. Hearing
3. Smell
4. Taste
5. Touch

Nevertheless, he attests this sense a high importance concerning its indispensability as early as 350 BC [2]:

> Some classes of animals have all the senses, some only certain of them, others only one, the most indispensable, touch.

The social estimation of the sense of touch experienced all imaginable phases. Frequently, it was afflicted with the blemish of squalor, as lust is transmitted by it [91]:

> Sight differs from touch by its virginity, such as hearing differs from smell and taste: and in the same way their lust-sensation differs.

It was also called the sense of excess [33]. In a general subdivision between lower and higher senses, touch was almost constantly ranged within the lower class. In Western civilization, the church once stigmatized this sense as forbidden due to the pleasure which can be gained by it. However, in the eighteenth century, the public opinion changed and KANT [49] is cited with the following statement:

> This sense is the only one with an immediate exterior perception; due to this it is the most important and the most teaching one, but also the roughest. Without this sensing organ we would be able to grasp our physical shape, whose perception the other two first class senses (sight and hearing) have to be referred to, to generate some knowledge from experience.

KANT thus emphasizes the central function of the sense of touch. It is capable of teaching the spatial perception of our environment. Only touch enables us to feel and classify impressions collected with the help of other senses, put them into context, and understand spatial concepts. Although stereoscopic vision and hearing develop early, the first-time interpretation of what we see and hear requires connection between both impressions perceived independently and information about distances between objects. This can only be provided by a sense, which can bridge the space between a being and an object. Such a sense is the sense of touch. The skin, being a part of this sense, covers a human's complete surface and defines his or her physical boundary, the physical being.

1.1.2 Formation of the Sense of Touch

As mentioned in the previous section, the sense of touch has numerous functions. The knowledge of these functions enables the engineer to formulate demands on the technical system. It is helpful to consider the whole range of purposes the haptic sense serves. However, at this point, we do not yet choose an approach by measuring its characteristics, but observe the properties of objects discriminated by it.

The sense of touch is not only specialized in the perception of the physical boundaries of the body, as said before, but also in the analysis of the immediate surroundings, including the objects present and their properties. Humans and their predecessors had to be able to discriminate, e.g., the structure of fruits and leaves by touch, in order to identify their ripeness or whether they were edible or not, e.g., a furry berry among smooth ones. The haptic sense enables us to identify a potentially harming structure, e.g., a spiny seed, and to be careful when touching it, in order to obtain its content despite its dangerous needles.

For this reason, the sense of touch has been optimized for the perception and discrimination of surface properties, e.g., roughness. Surface properties may range from smooth ceramic-like or lacquered surfaces with structural widths in the range of some micrometers, to somewhat structured surfaces such as coated tables and rough surfaces as in coarsely woven cord textiles with mesh apertures in the range of several millimeters. Humans have developed a typical way to interact with these surfaces enabling them to draw conclusions based on the underlying perception mechanism. A human moves his or her finger along a surface (see Fig. 1.1), allowing shear forces

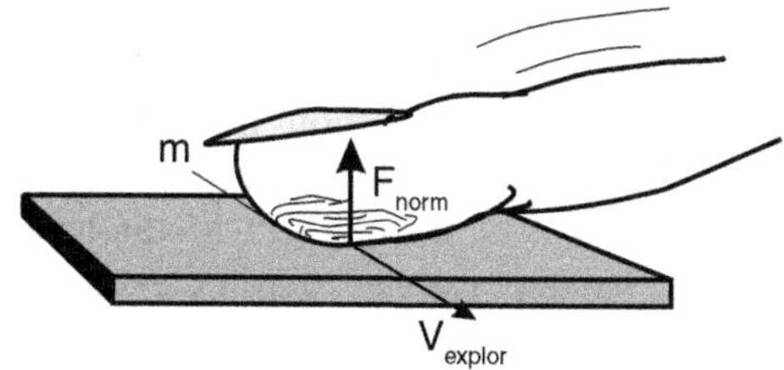

Fig. 1.1 Illustration for the interaction of movements, normal forces on the finger pad, and frictional coupling

to be coupled to the skin. The level of the shear forces is dependent on the quality of the frictional coupling between the object's surface and the skin. It is a summary of the tangential elasticity of the skin depending on the normal preload resulting from the touch F_{norm} and the velocity v_{explor} of the movement and quality of the coupling factor μ.

Anyone who has ever designed a technical frictional coupling mechanism knows that without additional structures or adhesive materials, viscous friction between two surfaces can hardly reach a factor of $\mu_r \geq 0.1$. Nevertheless, nature, in order to be able to couple shear force more efficiently into the skin, has "invented" a special structure in the most important body part for touching and exploration: the fingerprint. The epidermal ridges couple shearing forces efficiently to the skin, as by the bars, a bending moment is transmitted into its upper layers. Additionally, these bars allow to form closures within structural widths of similar size, which means nothing else but canting between the object handled and the hand's skin. At first glance, this is a surprising function of this structure. A second look reminds one of the fact that nature does not introduce any structure without a deeper purpose.

Two practical facts result from this knowledge: First of all, the understanding of shear forces' coupling to the skin has come into focus of current research [29] and has resulted in improvement in the design process of tactile devices. Secondly, this knowledge can be applied to improve the measuring accuracy of commercial force sensors by building ridge-like structures [87].

Another aspect of the haptic sense and probably a evolutionary advantage is the ability to use tools. Certain mechanoreceptors in the skin (see Sect. 2.1 for details) detect high-frequency vibrations that occur when handling a (stiff) tool. Detection of this high-frequency vibrations allows to identify different surface properties and to detect contact situations and collisions [26].

1.1.3 Touchable Art and Haptic Aesthetics

Especially in the twentieth century, art deals with the sense of touch and plays with its meaning. Drastically, the furry cup (see Fig. 1.2) makes you aware of the significance of haptic texture for the perception of surfaces and surface structures. Whereas the general form of the cup remains visible and recognizable, the originally plane ceramic surface is covered with fur.

In 1968, the "Pad- and Touch-Cinema" (see Fig. 1.3) allowed visitors to touch VALIE EXPORT's naked skin for 12 s through a box being covered by a curtain all the time. According to the artist, this was the only valid approach to experience sexuality without the aspect of voyeurism [30]. These are just a few examples of how art and artists played with the various aspects of haptic perception.

As with virtual worlds and surroundings, also haptic interaction has characteristics of art. In 2004, ISHII from MIT Media Laboratory and IWATA from the University of Tsukuba demonstrated startling exhibits of "tangible user interfaces" as shown

Fig. 1.2 MERET OPPENHEIM: furry cup, 1936 [30, 60]

Fig. 1.3 VALIE EXPORT: Pad- and touch-cinema, 1968 [30]

in Fig. 1.4. These interfaces couple visual displays with haptically reconfigurable objects to provide intuitive human–machine interfaces.

Despite the artistic aspect of such installations, recent research evaluates new interaction possibilities for ↪human–computer interaction (HCI)[1] based on such concepts:

- In [90], picture frames are used as tangible objects to initiate a video call to relatives and friends, when placed on a defined space on a special table cloth.
- With TOUCHÉ, *Disney Research* presents a capacitive sensing principle to use almost every object as a touch input device [74]. It is intended to push the development of immersive computers that disappear in objects.

[1] Please note that entries in the glossary and abbreviations are denoted by a ↪ throughout the book.

Fig. 1.4 Example for *Tangible Bits*, with different data streams accessible by opening bottles. In this case, single instrumental voices are combined to a trio [43]

In technical applications, the personal feeling of haptic aesthetics becomes more and more important. Car manufacturers investigate the perceived quality of interfaces [5, 67] to create a touchable brand identity, and there are whole companies claiming to "*make percepts measurable*" [8] and designers provide toolkits to evaluate characteristics of knobs and switches [45]. However, the underlying mechanisms of the assessment of haptic aesthetics are not fully understood. While the general approach to all studies is basically the same, using multidimensional scaling and multidimensional regression algorithms to combine subjective assessments and objective measurements [70], details remain ambiguous, especially regarding dynamics, objective measurement parameters, and measuring conditions as investigated by the authors of this chapter [39, 51].

Recently, Carbon and Jakesch [16] published a comprehensive approach based on object properties and the assessment of familiarities. This topic will probably remain a fascinating field of research for interdisciplinary teams from engineering and psychology.

1.2 Technical Definitions of Haptics

To use the haptic sense in a technical manner, some agreements on terms and concepts have to be made. This section deals with some general definitions and classifications of haptic interactions and haptic perception and is the basis for the following Chap. 2, which will delve deeper into topics of perception and interaction.

1.2.1 Definitions of Haptic Interactions

The haptic system empowers humans to interact with real or virtual environments by means of mechanical, sensory, motor, and cognitive abilities [46]. An interaction consists of one or more operations, that can be generally classified into *motion control* and *perception* [54]. The operations in these classes are called *primitives*, since they cannot be divided and further classified.

The perception class includes the primitives *detection*, *discrimination*, *identification*, and *scaling* of haptic information [27]. The analysis of these primitives is conducted by the scientific discipline called ↪ psychophysics. To further describe the primitives of the description class, the term ↪ stimulus has to be defined:

Definition *Stimulus (pl. stimuli)* Excitation or signal that is used in a psychophysical procedure. It is normally denoted with the symbol Φ. The term is also used in other contexts, when a (haptic) signal without further specification is presented to a user.

Typical stimuli in haptics are forces, vibrations, stiffnesses, or objects with specific properties. This definition allows us a closer look at the perception primitives, since each single primitive can only be applied to certain haptic stimuli, as explained below.

Detection The detection primitive describes how the presence of a stimulus is detected by a human, respectively, a user. Depending on the interaction conditions, stimuli can be detected or not detected. This depends not only on the sensory organs involved (see Sect. 2.1) but also on the neural processing. Only if a stimulus is detected, the other perception primitives can be applied.

Discrimination If more than one stimulus is present *and* detected, the primitive discrimination describes how information is perceived, which are included in different properties of the signal (like frequency or amplitude of a vibration) or object (like hardness, texture, mass).

Identification As well as the discrimination primitive, also the identification primitive is based on more than one present and detected stimuli. These stimuli are however not compared to each other, but with practical or abstract knowledge to allow a classification of the information contained in the stimuli. An example for such a task is the identification of geometric properties of objects like size and global form.

Scaling Scaling is the fourth primitive of perception as generally described by psychophysicists. This primitive describes the behavior of scales when properties of stimuli and objects are rated [81]. While scaling is only of secondary meaning for the description of interactions, it can provide useful information about signal magnitudes in the design process.

The motor control class can be divided into different operations as well. In this class, the primitives *travel*, *selection* and *modification* exist [11]. They can be better explained, if they are linked to general interaction tasks [11, 44].

Travel The movement or travel of limbs, the whole body, or virtual substitutes (avatar) is used to search for or reach a destination or an object, to explore (unknown) environments or to change one's position. Changing of a movement already in progress is included in this primitive.

Selection Especially in virtual environments, marking and/or selection of an object or a function is a vital primitive. It allows for direct interaction in this environment in the first place.

Modification The modification primitive is based on a selection of a function or an object. It describes a change in orientation, position or other properties of an object as well as the combination of more than one object to a single one.

When using motor control primitives, not only the operation itself but the aim of the operation have to be considered for accurate description of an interaction. If, for example, a computer is operated with a mouse as an input device and an icon on the screen is selected, this interaction could be described as a travel primitive or as a selection primitive. A closer look will reveal that the travel primitive is used to reach an object on the screen. This object is selected in a following step. If this interaction should be executed with a new kind of haptic device, the travel primitive is probably considered subordinate to the selection primitive.

Based on these two classes of interaction primitives, SAMUR [72] introduces a ↪taxonomy of haptic interaction. It is given in Fig. 1.5 and allows the classification of haptic interaction. A classification of a haptic interaction is useful for the design of new haptic systems: Requirements can be derived easily (see Chap. 5), analogies can be identified and used in the design of system components, and evaluation is alleviated (see Chap. 13).

Next to the analysis of haptic interaction based on interaction primitives, other psychophysically motivated approaches exist:

- LEDERMAN AND KLATZKY [56] proposed a classification of haptic interaction primitives in two operation classes: identification (The *What-System*) and localization (The *Where-System*).
- HOLLINS [41] proposes a distinction of primitives based on the spatial and temporal resolution of perception (and the combinations thereof) on one side and a class of "haptic" interactions on the other side. The latter correspond roughly to the above-mentioned motion control primitives.

The application of the taxonomy of haptic interactions as given in Fig. 1.5 to the development of task-specific haptic systems seems to be more straightforward as the application of the approaches by LEDERMAN AND KLATZKY and HOLLINS as stated in the above listing. Therefore, these are not pursued any further in this book.

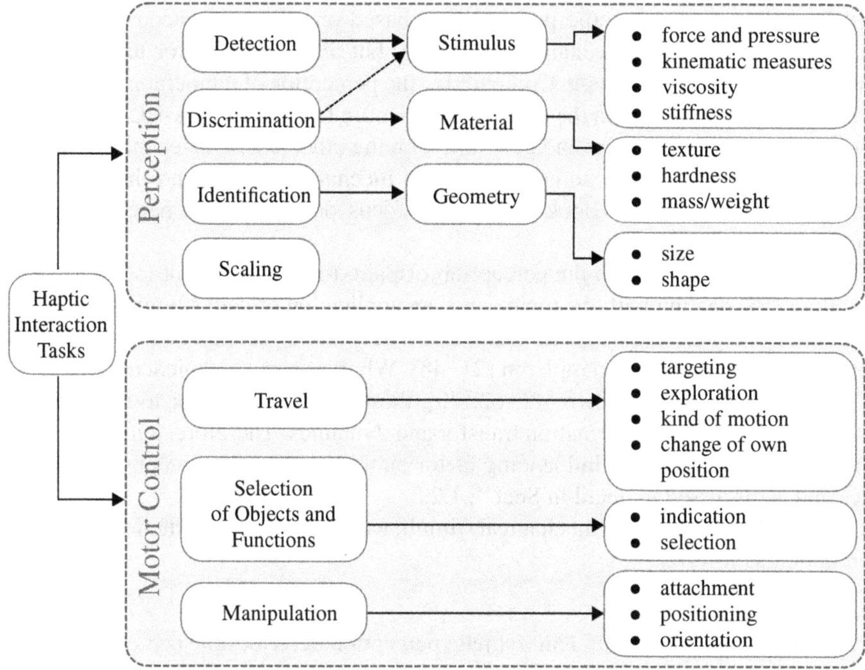

Fig. 1.5 Taxonomy of haptic interaction. Figure based on [27, 72]

1.2.2 Taxonomy of Haptic Perception

Up till now, one of the main taxonomies in the haptic literature has not been addressed: The classification based on ↪ kinaesthetic and ↪ tactile perception properties. It is physiologically based and defines perception solely on the location of the sensory receptors. It is defined in the standard ISO 9241-910 [44] and is given in Fig. 1.6.

Fig. 1.6 Taxonomy of haptic perception as defined in [44]

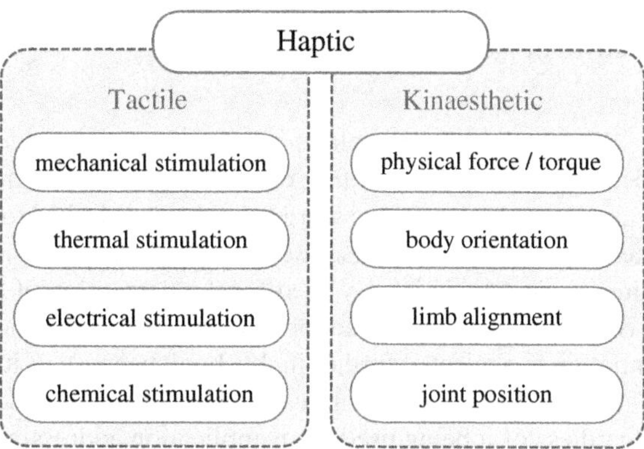

With this definition, tactile perception is based on all ↪ cutaneous receptors. These include not only mechanical receptors, but also receptors for temperature, chemicals (i.e., taste), and pain. Compared to the perception of temperature and pain, mechanical interaction is on the one side much more feasible for task-specific haptic systems in terms of usability and generality; on the other side, it is technically more demanding because of the complexity of the mechanoreceptors and the inherited dynamics. Therefore, this book will lay its focus on mechanical perception and interaction.

For processes leading to the perception of pain, the authors point to the special literature [55] dealing with the topic, since an application of pain stimuli in a haptic system for everyday use seems unlikely. The perception of temperature and possible applications is given, for example, in [21, 48]. Whereas some technical applications of thermal displays are known [42, 62, 73], these seem to be minor to mechanical interaction in terms of information transfer and dynamics. Therefore, temperature is primarily considered as an influencing factor on the mechanical perception capabilities and is discussed in detail in Sect. 2.1.2.

With the confinement on mechanical stimuli, we can define kinaesthetic and tactile perception as follows:

Definition *Kinaesthetic* Kinaesthetic perception describes the perception of the operational state of the human locomotor system, particularly joint positions, limb alignment, body orientation, and muscle tension. For kinaesthetic perception, there are dedicated sensory receptors in muscles, tendons, and joints as detailed in Sect. 2.1. Regarding the taxonomy of haptic interactions, kinaesthetic sensing primarily involves the motion control primitives, since signals from kinaesthetic receptors are needed in the biological control loop for positioning of limbs.

Definition *Tactile* Tactile perception describes the perception based on sensory receptors located in the human skin. Compared to kinaesthetic receptors, they exhibit larger dynamics and are primarily involved in the perception primitives of haptic interaction.

While originally the terms *tactile* and *kinaesthetic* were strictly defined by the location and functions of the sensory receptors, they are now used in a more general way. While the root of the word *kinesthesia* is linked to the description of movement, the term *kinaesthetic* is also used to describe static conditions nowadays [18]. Sometimes, kinaesthetic is only used for the perception of properties of limbs, while the term *proprioception* is used for properties regarding the whole body [58]. This differentiation is neglected further in this book because of its minor technical importance. The term *tactile* often describes any kind of sensor or actuator with spatial resolution, regardless of it being used in an application addressing tactile perception as defined above. While these examples are only of minor importance for the design of haptic systems, the following usage of the terms is an important adaption of the definitions. Primarily based on the dynamic properties of tactile and kinaesthetic perception, the definition of these terms is extended to haptic interactions in general nowadays. The reader may note that the following description is not accurate in terms of temporal sequence of the cited works, but focuses on works with relevant contributions to the present use of the terms *kinaesthetic* and *tactile*.

Based on the works of SHIMOGA, the dynamics of kinaesthetic perception are set equal to the motion capabilities of the locomotor system [32]. The dynamics of tactile perception are bordered at about 1–2 kHz for practical reasons. Higher frequencies can be perceived [28, 78], but it is questioned whether they have significant contribution to perception [86, p. 3]. As further explained in Sect. 2.4.3, this limitation is technically reasonable and necessary for the design of the electromechanical parts of haptic systems. Figure 1.7 shows this dynamic consideration of haptic interaction based on characteristic values from [13, 77, 78].

To extend this dynamic model of perception to a more general definition of interactions, DANIEL AND MCAREE [20] proposed a bidirectional, asymmetric model with a low-frequency (<30 Hz) channel for the exchange of energy and a high-frequency channel for the exchange of information with general implications on the design of haptic interfaces. The mapping based on dynamic properties is meaningful to a greater extent, since users can be considered as mechanical passive systems for frequencies above the dynamics of the active movement capabilities of the locomotion system [40]. This is explained in detail in Chap. 3. Altogether, these aspects (dynamics of perception and movement capabilities, exchange paths of energy and information, and the modeling of the user as active and passive load to a system) lead to the, nowadays, widely accepted model for the partition of haptic interaction in low-frequency kinaesthetic interaction and high-frequency tactile perception.

Both taxonomies of haptic interaction as seen in Fig. 1.5 and haptic perception as seen in Fig. 1.6 and extended in Fig. 1.7 are relevant sources for standard vocabulary in haptic system design. This is needed in the design of haptic systems, since it will simplify and standardize descriptions of haptic interactions. These are necessary to describe the intended functions of a task-specific haptic system and will be described in detail in Sect. 5.2. Further definitions and concepts of haptic interaction and perception are given in Chap. 2 in detail. In the next part of this chapter, possible applications for haptic systems that will become part of the human haptic interaction with systems and environments are presented.

1.3 Application Areas of Haptic Systems

Haptic systems can be found in a multitude of applications. In this section, four general application areas are identified. The benefits and technical challenges of haptic systems in these areas are given. In the latter Sect. 2.3, these application

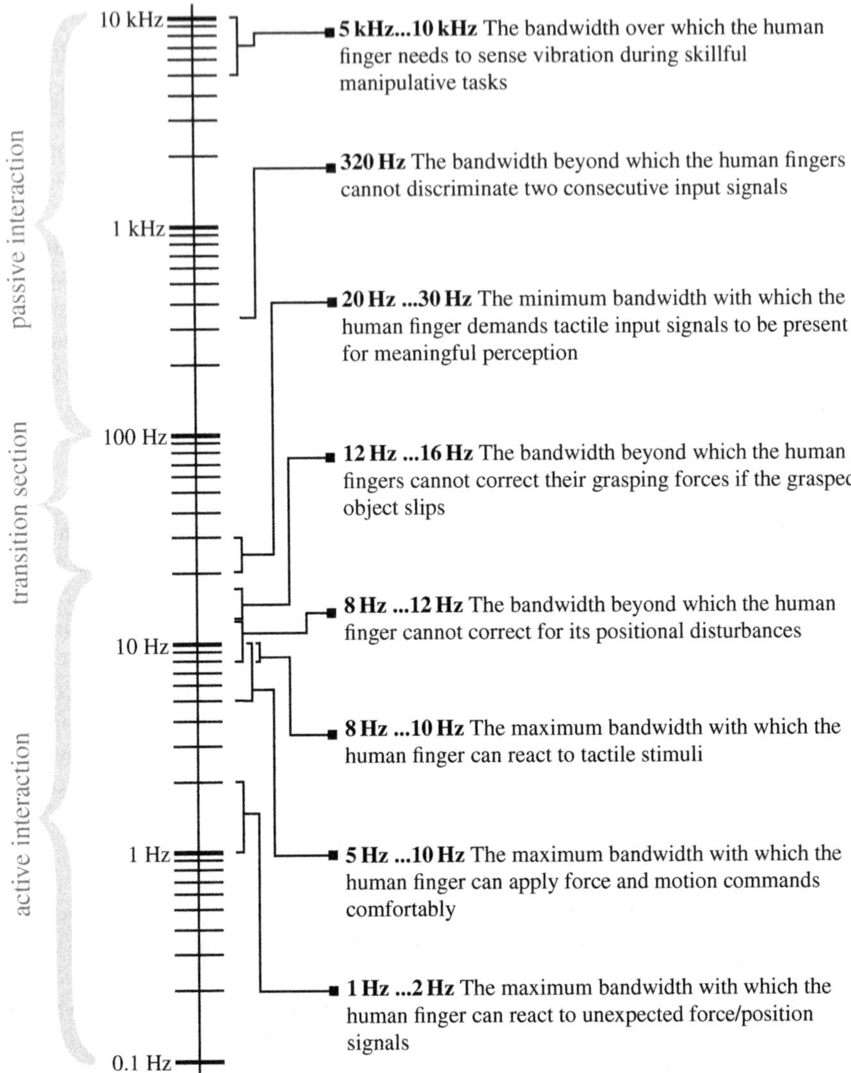

Fig. 1.7 kinaesthetic and tactile haptic interaction. Figure is based on data from [13, 77, 78]

areas are combined with a general model of human–system–environment interaction, leading to an interaction-based definition of basic system structures.

1.3.1 Telepresence, Teleaction, and Assistive Systems

Did you ever think about touching a lion in a cage?

With a ↪ telepresence and teleaction (TPTA) system, you could do just that without exposing yourself to risks, since they provide the possibility to interact mechanically with remote environments (We neglect the case of the lion feeling disturbed by the fondling...).

In a strict definition of TPTA systems, there is no direct mechanical coupling between operator and manipulated environment, but only via the TPTA system. Thus, transmission of haptic signals is possible in the first place, since mechanical interaction is converted to other domains (mainly electrical) and can be transmitted easily. They are often equipped with additional multimodal features, mainly a one-directional visual channel displaying the environment to the operator of the TPTA system.

Examples include systems for underwater assembly, when visual cues are useless because of dispersed particles in the water [22], scaled support of micro- and nano-positioning [1, 25], and surgical applications [23, 89]. The use of TPTA systems reduces task completion time and minimizes errors and handling forces compared to systems without a haptic feedback [61]. In surgical applications, new combinations of insofar incompatible techniques are possible, for example, palpation in minimal invasive surgery. Studies also show a safety increase for patients [88].

Most known TPTA systems are used for research applications. Figure 1.8 shows an approach by *Quanser*, supplying a haptic interface and a robot manipulator arm both supported by the QUARC control suite (see Sect. 12.4 for further information). Based on this combination, versatile bilateral teleoperation scenarios can be designed, as, for example, NEUROARM, a teleoperation system for neurological interventions

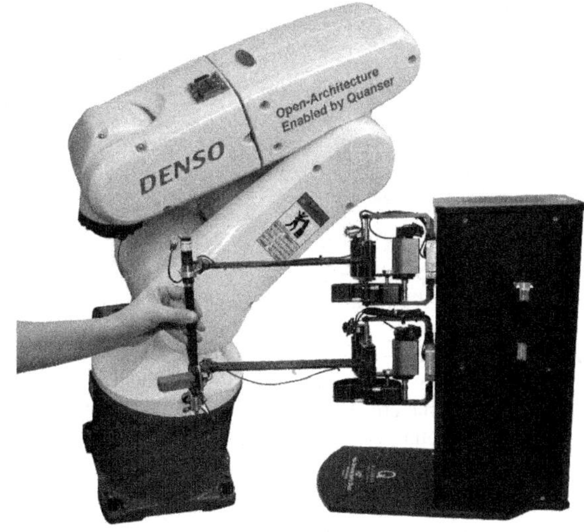

Fig. 1.8 Versatile teleoperation by *Quanser*: HD^2 haptic interface with 7 DoF of haptic feedback and DENSO OPEN ARCHITECTURE robot with 6 DoF. Image courtesy of *Quanser*, Markham, Ontario, CA

[83]. Example interventions include the removal of brain tumors, which require high position accuracy and real-time integration of ↪ magnetic resonance imaging (MRI) images.

The development of TPTA systems is technically most challenging. This is caused by the unknown properties of the environment, having influence on system stability, the required high accuracy of sensors and actuators to present artifact-free haptic impressions, and data transmission over long distances with additional aspects of packeted transmission, (packet) losses, and latency.

A special type of TPTA systems are so-called ↪ comanipulators, which are mainly used in medical applications [23]. Despite the mechanical interaction over the TPTA system, additional environment manipulation (and feedback) can be exerted by parts of the system (a detailed definition based on the description of the interaction can be found in Sect. 2.3). Examples for such comanipulators are INKOMAN and HapCath, which are developed at the *Institute for Electromechanical Design*.[2]

The HapCath system that adds haptic feedback to cardiovascular interventions is presented in detail as an example in Sect. 14.3. Figure 1.9 displays the INKOMAN instrument, which is the result of the joint research project SOMIT-FUSION funded by the German Ministry of Education and Research. It is an extension of a laparoscopic instrument with a parallel kinematic structure [69], which provides additional ↪ degrees of freedom (DoF) of a universal tool platform [75]. This allows minimal invasive interventions at previously unreachable regions of the liver. By integrating a multi-component force sensor in the tool platform [66], interaction forces between instrument and liver can be displayed to the user [50]. This allows techniques like palpation to identify vessels or cancerous tissue. With the general form of a laparoscopic instrument, additional interaction forces can be exerted by

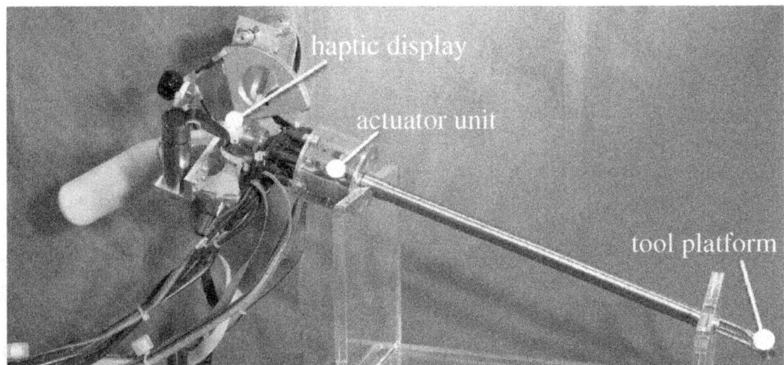

Fig. 1.9 INKOMAN—intracorporal manipulator for minimal invasive abdomen interventions with increased flexibility. The figure shows the handheld instrument with a haptic display based on a delta kinematic structure. The parallel kinematic structure used to move the tool platform is driven by ultrasonic traveling wave motors. Figure adapted from [50]

[2] Affiliation of the majority of the authors of this book.

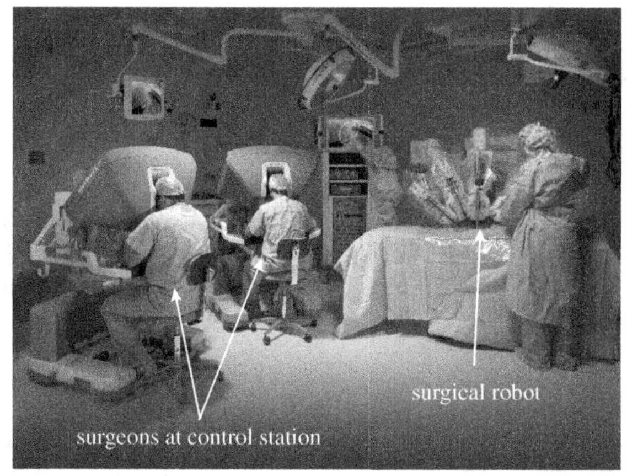

Fig. 1.10 DAVINCI SURGICAL SYSTEM, picture courtesy of Intuitive Surgical Inc., Sunnyvale, CA, USA

the surgeon by moving the complete instrument, and it is therefore classified as a comanipulation system.

TPTA systems are the main focus of research activities, probably since there are only small markets with a high potential for such systems. An exception is medical applications, where non-directly coupled instruments promise higher safety and efficient usage, for example, by avoiding collisions between different instruments or lowering contact and grip forces [53, 88]. Also, automated procedures like knot tying can be accelerated and conducted more reliably [9]. However, the distinction between a haptic TPTA system and a robotic system for medical use is a thin line: The aforementioned functions do not require haptic feedback. This explains the large number of existing medical robotic systems in research and industry [63, 64], dominated by the well-known DAVINCI SURGICAL SYSTEM by *Intuitive Surgical Inc.* as shown in Fig. 1.10. This system was developed for urological and gynecological interventions and incorporates a handling console with three-dimensional view of the operation area and a considerable number of instruments, which are directed by the surgeon on the console and actuated with cable drives [36]. There is no haptic feedback for this system available, although there are promising approaches as discussed in Sect. 2.4.4.

For consumer application, *Holland haptics* just recently announced FREBBLE [59], a device intended to convey the feeling of holding someone's hand over the Internet. This is also an interesting hardware concept as a low-cost teleoperation device.

1.3.2 Virtual Environments

The second main application area for haptic systems is interaction with virtual environments. Since this is a large field of applications, we will have a closer look at different areas, where interaction with generated situations is used to a wider extent.

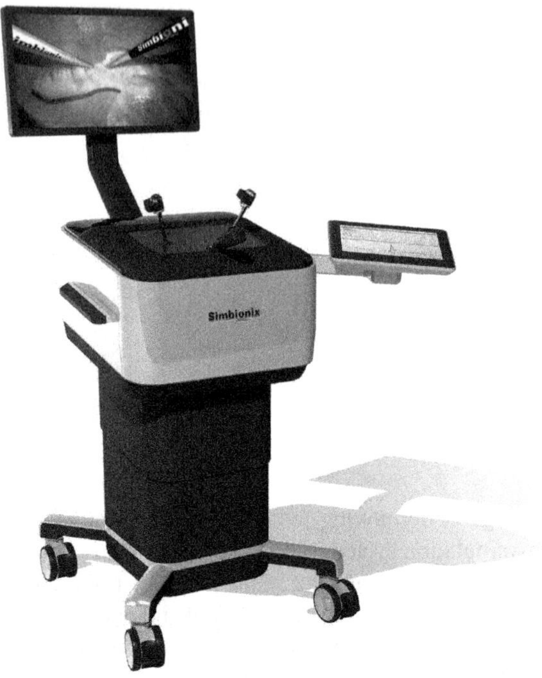

Fig. 1.11 Laparoscopic simulator LAP MENTOR III. The system was designed to simulate interventions in the abdomen. Picture courtesy of *Simbionix* USA, Cleveland, OH, USA

Medical Training A large number of systems is designed to provide medical training without jeopardizing a real patient [19]. In addition to haptic feedback, this systems generally provide also visual and acoustic feedback to generate a realistic impression of the simulated procedure. You can find systems to train the diagnosis of joint lesions [68] and simulators for endoscopic, laparoscopic, and intravascular interventions [72]. Figure 1.11 shows an example of such a surgical simulator. Surgeons trained on simulators show a better task performance in several studies [3, 7]. In addition, simulators can be used very early in medical training, since they do not put patients at risk and have a higher availability.

Industrial Design In industrial design applications, virtual environments are used to simulate assembly operations and subjective evaluation of prototypes. Although there are much less applications than in medical training, this area pushes technology development: Some requirements can only be met with new concepts such as admittance systems and form displays. One of these is the HAPTIC STRIP, which consists of a bendable and twistable surface that can be additionally positioned in 6 DoF in space [10]. It is shown in Fig. 1.12 and can be used to display large-scale forms of new designs without having to manufacture a prototype.

Multimodal Information Displays Since the haptic sense was developed to analyze objects and the environment, similar application with a high demand of intuitive access to information can be found in literature. Haptic systems are used

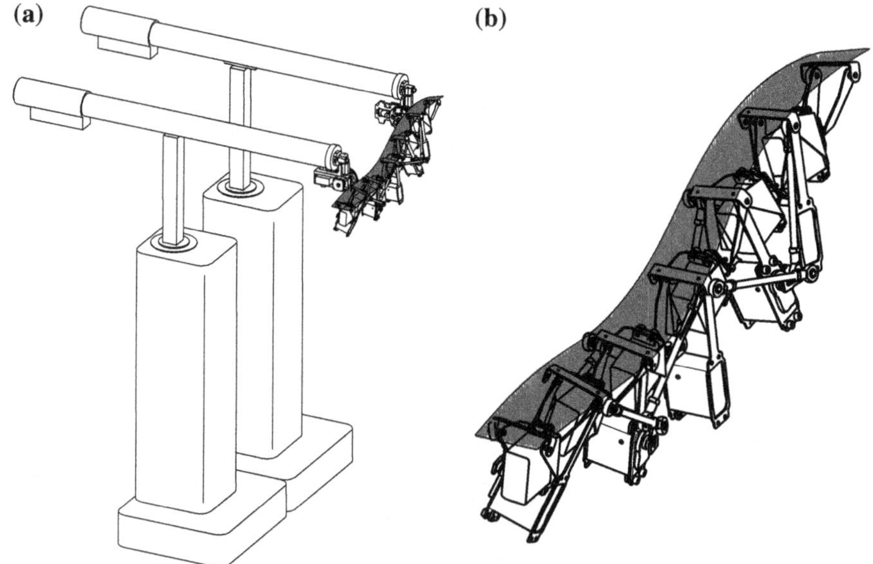

Fig. 1.12 The haptic strip system. The strip is mounted on two HapticMaster admittance type interfaces. Capacitive sensors on the strip surface sense the user's touch. Figure is based on [10]

to display large amount of information in biology and chemistry [15, Chap. 9] and are also used as means for the synthesis of complex molecules [12]. For this application, the human ability to detect patterns (in visual representations) is used for a coarse positioning of synthesis partners, whereas micro-positioning is supported by haptic representation of the intermolecular forces.

Another example for multimodal display of information was recently presented by *Microsoft Research* [79]. The TOUCHMOVER is an actuated screen with haptic feedback that can be used to display object and material properties or to intuitively access volumetric data like, for example, ↪MRI scans. Figure 1.13 shows this application of the system. Annotations are marked visually and haptically with a detent, allowing for intuitive access and collaboration.

Consumer Electronics For the integration of haptic feedback in computer games, *Novint Technologies, Inc.* presented the FALCON haptic interface in 2006. It is based on a delta parallel kinematic structure and distinguishes itself through a competitive price tag at around 300$. This device is also used in several research projects like, for example, [76], because of the low price and the support in several ↪ application programming interface (API) (see Sect. 12.4 for further information).

To provide a more intense music experience, haptic systems conveying low-frequency acoustic signals to musicians are known, for example the BUTT KICKER by *The Guitammer Company* as shown in Fig. 1.14. The system delivers low-frequency signals specific to a single person, thus lowering the overall sound

Fig. 1.13 TOUCHMOVER with user exploring MRI data. Picture courtesy of *Microsoft Research*, Redmond, WA, USA

Fig. 1.14 Electrodynamic actuator BUTT KICKER for generating low-frequency oscillations on a drum seat

pressure on a musical stage. To allow for the touch of fabric over the internet, the HAPTEX [37] project developed rendering algorithms as well as interface hardware.

Compared to the design of TPTA systems, the development of haptic interfaces for interactions with virtual environments seems to be slightly less complex, since more knowledge about the interaction environment is present in the design process. However, new aspects such as derivation and allocation of the environment data arise with these applications. Because of the wider spread of such systems, cost efficiency has to be taken into account.

1.3.3 Noninvasive Medical Applications

Based on specific values of haptic perception, diagnosis of certain illnesses and dysfunctions can be made. Certain types of eating disorders [34, 35] and diabetic neuropathy [62] are accompanied with diminished haptic perception capabilities. They can therefore be diagnosed with a measurement of the perception or motor exertion parameters and comparison with the population mean. Next to diagnosis, haptic perception parameters can also be used as a progress indicator in stroke [4] and limb [92] rehabilitation.

For these purposes, cost-efficient systems with robust and efficient measurement protocols are needed. Because feedback from the user can be received with any means, development is easier than the development of TPTA or VR systems. These systems are the foci of several research groups; up till now, there is no system for comprehensive use in the market.

1.3.4 Communication

The fourth and by numbers the largest application area of haptic systems is basic communications. The most prominent example is probably on your desk or in your pocket—the vibration function of your phone. Compared to communication based on visual and acoustic signals, haptics give the opportunity to convey information in a discrete way and offer the possibility of spatial resolution. Communication via the haptic sense tends to be intuitive, since feedback arises at the point the user is interacting with. A simple example is a switch, which will give a haptic feedback when pressed.

Therefore, haptics are an attractive communication channel in demanding environments, for example, when driving a car. Several studies show that haptic communication tends to distract users less from critical operations than the use of other channels such as vision or audition [71, 80]. Applications include assistive systems for navigation in military applications [31]; a practical example for an adaptive haptic user interface for automotive use is given in Sect. 14.2. With the increasing number of steer by wire applications and the vision of autonomous driving vehicles, the haptic channel is identified as a possibility to raise awareness of the driver in possibly dangerous situations as investigated in [57].

More recently, the increasing use of consumer electronics with touch screens triggers a demand for technologies to add haptic feedback. It is intended to facilitate the use without recurring visual status inspection. Solutions for such applications include the usage of a number of different actuation principles, which will be the focus of Chap. 9.

Another application area is tactile interfaces for the blind and visually impaired [47, 82]. Despite displaying Braille characters, tactile interfaces offer navigation support (see, for example, the HAPTIMAP [38] project providing toolkits for standard mobile terminals or the tactile You-Are-Here-Maps presented in Sect. 14.1) or interactions with graphic interfaces [6, 65]. Recent studies imply positive results for the support of elder people in navigation tasks via tactile cues compared to just visual or auditory cues [52]. Figure 1.15 gives examples of haptic systems used for communication applications.

Despite the analysis of energy-efficient actuation principles for mobile usage, scientific research in this area addresses the design of haptic icons for information

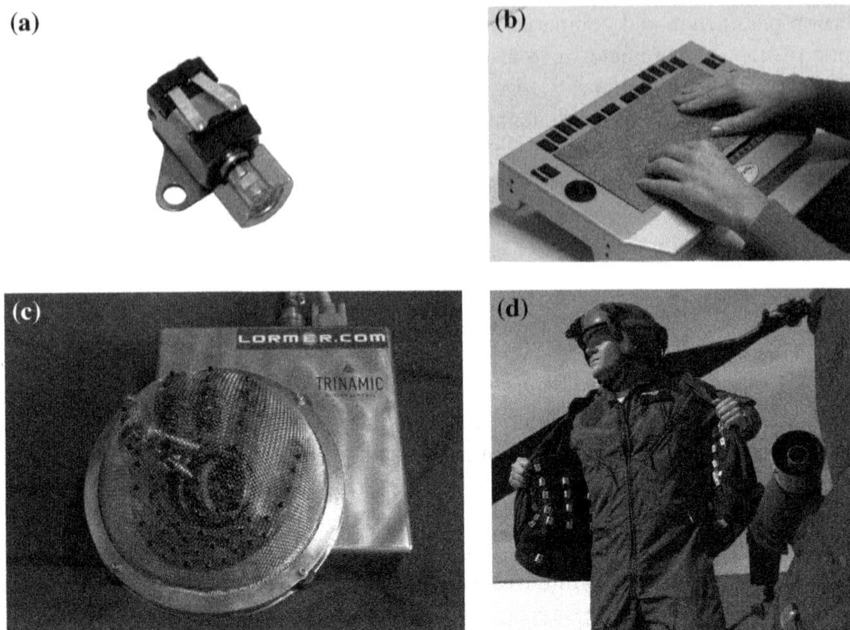

Fig. 1.15 Components and systems for communication via the haptic sense. **a** Vibration motor for mobile devices. **b** HYPERBRAILLE system for displaying graphic information for visually impaired users, image courtesy of *metec AG*, Stuttgart, Germany. **c** LORMER system as machine–human interface conveying text information using the lorm alphabet on the palm and hand of the user, image courtesy of Thomas Rupp, http://www.lormer.de. **d** TACTILE TORSO DISPLAY, vest for displaying flight information on pilots' torso, image courtesy of *TNO*, Soesterberg, The Netherlands

transfer. Sometimes also called *tactons*, *hapticons*, or *tactile icons*, the influence of rhythm, signal form, frequency, and localization is investigated [14, 24]. Up till now, information transfer rates of 2 … 12 bit per second were reported [17, 85], although the latter require a special haptic interface called the TACTUATOR designed for communication applications [84].

For completeness, also passive systems such as a computer keyboard, trackballs, and mice are part of this application area, since they convey information given in the form of a motion control operation to a (computer) system. Although there exists some kind of haptic feedback, it is not dependent on the interaction, but solely on the physical characteristics of the haptic system such as inertia, damping, or friction.

1.3.5 Why Use a Haptic System?

The reasons one might want to use a haptic system are numerous: Perhaps you want to improve the task performance or lower the error rate in a manipulation scenario, address a previously unused sensory channel to convey additional information, or gain advantages over a competitor in an innovation-driven market. This book does not answer the question whether haptics is able to fulfill the wishes and intentions connected to these reasons, but focuses on the design of a specific haptic system for the intended application.

Although there are many guidelines on how to implement haptic and multimodal feedback for optimal task performance (they will be addressed in Sect. 5.1.2), there are only limited sources on how to decide whether a haptic feedback is usable for an application. As an exception, ACKER [1] provides some criteria for the application of telepresence technologies in industrial applications.

1.4 Conclusions

Technical systems addressing the haptic sense cover a wide range of applications. Since this book focuses on the design process of task-specific haptic interfaces, the following chapters will first focus on in-depth analysis of haptic interaction in Chap. 2 and the role of the user in a haptic system in Chap. 3, before a detailed analysis of the development and structure of haptic systems is presented in Chaps. 4 and 6. This provides the basis for the second part of the book, which will deal with the actual design of a task-specific haptic system.

Recommended Background Reading

[16] Carbon, C.-C. & Jakesch, M.: **A model for haptic aesthetic processing and its implications for design**. Proceedings of the IEEE, 2013.
General model about the development of haptic aesthetics and the implications for the design of products.

[33] Grunwald, M.: **Human Haptic Perception: Basics and Applications**. Birkhäuser, Basel, CH, 2008.
General collection about the haptic sense with chapters about theory and history of haptics, neuro-physiological basics and psychological aspects of haptics.

References

1. Acker A (2011) Anwendungspotential von Telepräsenz-und Teleaktionssystemen für die Präzisionsmontage. Dissertation, Technische Universität München, München. http://www.mediatum.ub.tum.de/doc/1007163/1007163.pdf
2. Adams RJ (1999) Stable haptic interaction with virtual environments. PhD thesis, University of Washington, Washington

3. Ahlberg G et al (2007) Proficiency-based virtual reality training significantly reduces the error rate for residents during their first 10 laparoscopic cholecystectomies. Am J Surg 193(6):797–804. doi:10.1016/j.amjsurg.2006.06.050
4. Allin S, Matsuoka Y, Klatzky R (2002) Measuring just noticeable differences for haptic force feedback: implications for rehabilitation. In: Proceedings of the 10th symposium on haptic interfaces for virtual environments and teleoperator systems, Orlando, FL, USA. doi:10.1109/HAPTIC.2002.998972
5. Anuguelov N (2009) Haptische und akustische Kenngrößen zur Objektivierung und Optimierung der Wertanmutung von Schaltern und Bedienfeldern für den KFZ-Innenraum. Dissertation, Technische Universität Dresden. http://www.d-nb.info/1007551739
6. Asque C, Day A, Laycock S (2012) Cursor navigation using haptics for motion-impaired computer users. In: Haptics: perception, devices, mobility, and communication. Springer, Heidelberg, pp 13–24. doi:10.1007/978-3-642-31401-8_2
7. Bajka M et al (2009) Evaluation of a new virtual-reality training simulator for hysteroscopy. Surg Endosc 23(9):2026–2033. doi:10.1007/s00464-008-9927-7
8. Battenberg Robotic GmbH & Co. KG (2007) Wahrnehmungen messbar machen. Company information: http://www.messrobotic.de/fileadmin/pdf/Produkte-MessroboticStandard-Wahrnehmungen.pdf
9. Bauernschmitt R et al (2005) Towards robotic heart surgery: introduction of autonomous procedures into an experimental surgical telemanipulator system. Int J Med Robot Comput Assist Surg 1(3):74–79. doi:10.1002/rcs.3010.1002/rcs.30
10. Bordegoni M et al (2010) A force and touch sensitive self-deformable haptic strip for exploration and deformation of digital surfaces. In: Kappers AML, Bergmann-Tiest WM, van der Helm CT (eds) Haptics: generating and perceiving tangible sensations. LNCS 6192. Proceed ings of the eurohaptics conference, Amsterdam, NL. Springer, Heidelberg, pp 65–72. doi:10.1007/978-3-642-14075-4_10
11. Bowman D, Hodges L (1999) Formalizing the design, evaluation, and application of interaction techniques for immersive virtual environments. J Vis Lang Comput 10(1):37–53. doi:10.1006/jvlc.1998.0111
12. Brooks FP et al (1990) Project GROPE—haptic displays for scientific visualization. ACM SIGGRAPH Comput Graph 24:177–185. doi:10.1145/97880.97899
13. Brooks TL (1990) Telerobotic response requirements. In: IEEE international conference on systems, man and cybernetics, Los Angeles, CA, USA. doi:10.1109/ICSMC.1990.142071
14. Brown L, Brewster S, Purchase H (2006) Multidimensional tactons for non-visual information presentation in mobile devices. In: Conference on human-computer interaction with mobile devices and services, pp 231–238. doi:10.1145/1152215.1152265
15. Burdea GC, Coiffet P (2003) Virtual reality technology. In: Burdea GC, Coiffet P (eds) Wiley-Interscience, New York, NY, USA. ISBN: 978-0471360896
16. Carbon C-C, Jakesch M (2013) A model for haptic aesthetic processing and its implications for design. Proc IEEE 101(9):2123–2133. doi:10.1109/JPROC.2012.2219831
17. Cholewiak SA, Tan HZ, Ebert DS (2008) Haptic identification of stiffness and force magnitude. In: Symposium on haptic interfaces for virtual environments and teleoperator systems. Reno, NE, USA. doi:10.1109/HAPTICS.2008.4479918
18. Clark F, Horch K (1986) Kinesthesia. In: Boff KR, Kaufman L, Thomas JP Handbook of perception and human performance. Wiley-Interscience, New York, NY, USA, pp 13.1–13.61. ISBN: 978-0471829577
19. Coles TR, Meglan D, John NW (2010) The role of haptics in medical training simulators: a survey of the state-of-the-art. IEEE Trans Haptics 4:51–66. doi:10.1109/TOH.2010.19
20. Daniel R, McAree P (1998) Fundamental limits of performance for force reflecting teleoperation. Int J Rob Res 17(8):811–830. doi:10.1177/027836499801700801
21. Darian-Smith I, Johnson K (1977) Thermal sensibility and thermoreceptors. J Invest Dermatol 69(1):146–153. doi:10.1111/1523-1747.ep12497936
22. Dennerlein J, Millman P, Howe R (1997) Vibrotactile feedback for industrial telemanipulators. In: 6th Annual symposium on haptic Interfaces for virtual environment and teleoperator systems. http://www.biorobotics.harvard.edu/pubs/1997/vibrotactile.pdf

23. Eb Vander Poorten E, Demeester E, Lammertse P (2012) Haptic feedback for medical applications, a survey. In: Actuator conference, Bremen. https://www.radhar.eu/publications/e.-vander-poorten-actuator12-haptic-feedback-formedical-applications-a-survey
24. Enriquez M, MacLean K, Chita C (2006) Haptic phonemes: basic building blocks of haptic communication. In: Proceedings of the 8th international conference on multimodal interfaces (ICMI). Banff, Alberta. ACM, Canada, pp 302–309. doi:10.1145/1180995.1181053
25. Estevez P et al (2010) A haptic tele-operated system for microassembly. In: Ratchev S (ed) Precision assembly technologies and systems. Springer, Heidelberg, pp 13–20. doi:10.1007/978-3-642-11598-1_2
26. Fiene J, Kuchenbecker KJ, Niemeyer G (2006) Event-based haptic tapping with grip force compensation. In: IEEE symposium on haptic interfaces for virtual environment and teleoperator systems. doi:10.1109/HAPTIC.2006.1627063
27. Gall S, Beins B, Feldman J (2001) Psychophysics. The Gale encyclopedia of psychology, Gale. ISBN: 978-0787603724
28. Gault RH (1927) On the upper limit of vibrational frequency that can be recognized by touch. Sci New Ser 65(1686):403–404. doi:10.1126/science.65.1686.403
29. Gerling G, Thomas G (2005) The effect of fingertip microstructures on tactile edge perception. In: Haptic interfaces for virtual environment and teleoperator systems, First joint eurohaptics conference and symposium on WHC, pp 63–72. doi:10.1109/WHC.2005.129
30. Getzinger G (2006) Haptik—rekonstruktion eines verlustes. Naturerkenntnis, Hautsinn. Profil, Wien, p 146. ISBN: 3-89019-521-0
31. Gilson R, Redden E, Elliott L (2007) Remote tactile displays for future soldiers. Technical report, Aberdeen Proving Ground, MD. U.S. Army Research Laboratory, USA http://www.dtic.mil/cgi-bin/GetTRDoc?Location=U2&doc=GetTRDoc.pdf&AD=ADA468473
32. Goertz RC (1964) Manipulator systems development at ANL. In: Proceedings of the 12th conference on remote systems technology, ANS, pp 117–136
33. Grunwald M (2008) Human haptic perception: basics and applications. Birkhäuser, Switzerland. ISBN: 3764376112
34. Grunwald M (2008) Haptic perception in anorexia nervosa. In: Grünwald M (ed) Human haptic perception: basics and applications. Birkhäuser, Basel. ISBN: 3764376112
35. Grunwald M et al (2001) Haptic perception in anorexia nervosa before and after weight gain. J Clin Exp Neuropsychol 23(4):520–529. doi:10.1076/jcen.23.4.520.1229
36. Guthart G, Salisbury J Jr (2000) The intuitive telesurgery system: overview and application. In: IEEE international conference on robotics and automation (ICRA), San Francisco, CA, USA, vol 1, pp 618–621. doi:10.1109/ROBOT.2000.844121
37. Haptex (2007) Grant No. IST-6549, last visited 07.03.2012. European Union. http://www.haptex.miralab.unige.ch
38. HaptiMap—Haptic, audio and visual interfaces for maps and location based services (2013) European Union Grant No. FP7-ICT-224675. www.haptimap.org
39. Hatzfeld C, Kern TA, Werthschüzky R (2010) Improving the prediction of haptic impression user ratings using perception-based weighting methods: experimental evaluation. In: Kappers AML et al (ed) Haptics: generating and perceiving tangible sensations, LNCS 6191. Proceedings of the eurohaptics conference, Amsterdam, NL. Springer, Heidelberg, pp 93–98. doi:10.1007/978-3-642-14064-8_14
40. Hogan N (1989) Controlling impedance at the man/machine interface. In: IEEE international conference on robotics and automation (ICRA), Scottsdale, AZ, USA, pp 1626–1631. doi:10.1109/ROBOT.1989.100210
41. Hollins M (2002) Touch and haptics. In: Pashler H (ed) Steven's handbook of experimental psychology. Wiley, New York, pp 585–618. ISBN: 978–0471377771. doi:10.1002/0471214426.pas0114
42. Ino S et al (1993) A tactile display for presenting quality of materials by changing the temperature of skin surface. In: IEEE international workshop on robot and human communication. doi:10.1109/ROMAN.1993.367718

43. Ishii H (2004) Bottles: a transparent interface as a tribute to Mark Weiser. In: IEICE: transactions on information and systems, pp E87–D.6. doi:10.1145/346152.346159
44. ISO 9241 (2011) Ergonomics of human system interaction—part 910: framework for tactile and haptic interaction. ISO
45. Jagodzinski R, Wintergerst G (2009) A toolkit for developing haptic interfaces—control dial based on electromagnetic brake. In: UIST, Victoria, BC, CA. http://www.acm.org/uist/archive/adjunct/2009/pdf/demos/paper184.pdf
46. Jandura L, Srinivasan M (1994) Experiments on human performance in torque discrimination and control. In: Dynamic systems and control, ASME, DSC-55 1, pp 369–375. http://www.rle.mit.edu/touchlab/publications/1994_002.pdf
47. Johansson G (2008) Haptics as a substitute for vision. In: Hersh MA, Johnson MA (eds) Assistive technology for visually impaired and blind people. Springer, Heidelberg, pp 135–166. ISBN: 978-1846288661
48. Jones L, Berris M (2002) The psychophysics of temperature perception and thermal interface design. In: 10th symposium on haptic interfaces for virtual environments and teleoperator systems, Orlando, FL, USA. doi:10.1109/HAPTIC.2002.998951
49. Kant I (1983) Anthropologie in pragmatischer Hinsicht. Philosophische anthropologie, Charakter. Reclam, Stuttgart, p 389. ISBN: 3-15-007541-6
50. Kassner S (2013) Haptische mensch-maschine-schnittstelle für ein laparoskopisches Chirurgiesystem. Dissertation, Technische Universität Darmstadt. http://www.tubiblio.ulb.tudarmstadt.de/63334
51. Kern T (2010) Requirements and design considerations for the measurement of haptic object properties. In: Linz A (ed) Procedia engineering. Proceedings of the eurosensors conference, vol 5, pp 592–596. doi:10.1016/j.proeng.2010.09.179
52. Kim S et al (2012) Route guidance modality for elder driver navigation. In: Pervasive computing, LNCS 7319. Springer, Heidelberg, pp 179–196. doi:10.1007/978-3-642-31205-2_12
53. King C-H et al (2009) Tactile feedback induces reduced grasping force in robot-assisted surgery. IEEE Trans Haptics 2(2):103–110. doi:10.1109/TOH.2009.4
54. Kirkpatrick A, Douglas S (2002) Application-based evaluation of haptic interfaces. In: 10th symposium on haptic interfaces for virtual environment and teleoperator systems, Orlando, FL, USA, pp 32–39. doi:10.1109/HAPTIC.2002.998938
55. Kruger L, Friedman MP, Carterette EC (eds) (1996) Pain and touch. Academic Press, Maryland Heights
56. Lederman SJ, Klatzky RL (2009) Haptic perception: a tutorial. Attention Percept Psychophysics 71(7):1439. doi:10.3758/APP.71.7.1439
57. Liedecke C, Baumann G, Reuss H-C (2014) Potential of the foot as a haptic interface for future communication and vehicle controlling. In: 10th ITS European congress, Helsinki, FIN
58. Loomis JM, Lederman SJ (1986) Tactual perception. Handb Percept Human Perform 2: 2
59. My Frebble (2014) Last visited March 2014. http://www.myfrebble.com
60. Néret G (1998) Erotik in der Kunst des 20. Jahrhunderts. ErotikMotiv; Kunst; Geschichte 1900–199213.1c. Taschen, Tokyo, p 200. ISBN: 3-8228-7853-7
61. Nitsch V, Färber B (2012) A meta-analysis of the effects of haptic interfaces on task performance with teleoperation systems. IEEE Trans Haptics 6:387–398. doi:10.1109/ToH.2012.62
62. Norrsell U et al (2001) Tactile directional sensibility and diabetic neuropathy. Muscle Nerve 24(11):1496–1502. doi:10.1002/mus.1174
63. Pott P, Scharf H, Schwarz M (2005) Today's state of the art in surgical robotics. Comput Aided Surg 10(2):101–132. doi:10.1080/10929080500228753
64. Pott P, Schwarz M (2010) State of the art of medical robotics—areas of application. In: Jahrestagung der DGBMT 44
65. Prescher D, Weber G, Spindler M (2010) A tactile windowing system for blind users. In: Proceedings of the 12th international ACM SIGACCESS conference on computers and accessibility. ACM, Orlando, pp 91–98. doi:10.1145/1878803.1878821
66. Rausch J (2012) Entwicklung und Anwendung miniaturisierter piezoresistiver Dehnungsmesselemente. Dissertation, Technische Universität Darmstadt. http://www.tuprints.ulb.tudarmstadt.de/3003/1/Dissertation-Rausch-online.pdf

67. Reisinger J (2009) Parametrisierung der Haptik von handbetätigten Stellteilen. Dissertation, Technische UniversitätMünchen. https://www.mediatum.ub.tum.de/doc/654165/654165.pdf
68. Riener R et al (2002) Orthopädischer trainingssimulator mit haptischem feedback. Automatisierungstechnik 50:296–301. doi:10.1524/auto.2002.50.6.296
69. Röse A (2011) Parallelkinematische Mechanismen zum intrakoporalen Einsatz in der laparoskopischen Chirurgie. Dissertation, Technische Universität Darmstadt. http://www.tuprints.ulb.tu-darmstadt.de/2493/1/Dissertation_Roese_Druckversion.pdf
70. Rösler F, Battenberg G, Schüttler F (2009) Subjektive Empfindungen und objektive Charakteristika von Bedienelementen. Automobiltechnische Zeitschrift 4:292–297. doi:10.1007/BF03222068
71. Ryu J et al (2010) Vibrotactile feedback for information delivery in the vehicle. IEEE Trans Haptics 3(2):138–149. doi:10.1109/TOH.2010.1
72. Samur E (2012) Performance metrics for haptic interfaces. Springer, Heidelberg. ISBN 978–1447142249. doi:10.1007/978-1-4471-4225-6
73. Sato K, Maeno T (2012) Presentation of sudden temperature change using spatially divided warm and cool stimuli. In: Haptics: perception, devices, mobility, and communication. Springer, Heidelberg, pp 457–468. doi:10.1007/978-3-642-31401-8_41
74. Sato M, Poupyrev I, Harrison C (2012) Touché: enhancing touch interaction on humans, screens, liquids, and everyday objects. In: Proceedings of the 2012 ACMannual conference on human factors in computing systems, pp 483–492. doi:10.1145/2207676.2207743
75. Schlaak HF et al (2008) A novel laparoscopic instrument with multiple degrees of freedom and intuitive control. In: 4th European conference of the international federation for medical and biological engineering, Antwerpen. doi:10.1007/978-3-540-89208-3_394
76. Shah A et al (2010) How to build an inexpensive 5-DOF haptic device using two novint falcons. In: Kappers AML, Bergmann-Tiest WM, van der Helm FC (eds) Haptics: generating and perceiving tangible sensations, LNCS 6191. Proceedings of the eurohaptics conference, Amsterdam, NL. Springer, Heidelberg, pp 136–143. doi:10.1007/978-3-642-14064-8_21
77. Shimoga K (1993) A survey of perceptual feedback issues in dexterous telemanipulation part I. Finger force feedback. In: Proceedings of the IEEE virtual reality annual international symposium, Seattle, WA, USA, pp 263–270. doi:10.1109/VRAIS.1993.380770
78. Shimoga K (1993) A survey of perceptual feedback issues in dexterous telemanipulation part II. Finger touch feedback. In: Proceedings of the IEEE virtual reality annual international symposium, Seattle, WA, USA, pp 271–279. doi:10.1109/VRAIS.1993.380769
79. Sinclair M, Pahud M, Benko H (2013) Touch mover: actuated 3D touchscreen with haptic feedback. In: Proceedings of the 2013 ACM international conference on interactive tabletops and surfaces. ACM, pp 287–296. doi:10.1145/2512349.2512805
80. Spence C, Ho C (2008) Tactile and multisensory spatial warning signals for drivers. IEEE Trans Haptics 1:121–129. doi:10.1109/TOH.2008.14
81. Stevens SS (1975) Psychophysics. In: Stevens G (ed) Piscataway. Transaction Books, USA. ISBN: 978-0887386435
82. Summers I (ed) (1996) Tactile aids for the hearing impaired. Whurr, London, GB. ISBN 978-1870332170
83. Sutherland GR et al (2013) The evolution of neuroArm. Neurosurgery 72:A27–A32. doi:10.1227/NEU.0b013e318270da19
84. Tan H, Rabinowitz W (1996) A new multi-finger tactual display. J Acoust Soc Am 99(4):2477–2500. doi:10.1121/1.415560
85. Tan H et al (2003) Temporal masking of multidimensional tactual stimuli. J Acoust Soc Am, Part 1116(9):3295–3308. doi:10.1121/1.1623788
86. Verrillo RT, Gescheider GA (1992) Perception via the sense of touch. In: Summers IR (ed) Tactile aids for the hearing impaired. Whurr, London, pp 1–36. ISBN: 978-1-870332-17-0
87. Vasarhelyi G et al (2006) Effects of the elastic cover on tactile sensor arrays. Sens Actuators A Phys 132(1):245–251. doi:10.1016/j.sna.2006.01.009
88. Wagner C, Stylopoulos N, Howe R (2002) The role of force feedback in surgery: analysis of blunt dissection. In: 10th symposium on haptic interfaces for virtual environment and teleoperator systems. doi:10.1109/HAPTIC.2002.998943

89. Westebring-van der Putten EP et al (2008) Haptics in minimally invasive surgery—a review. Minim Invasive Ther Allied Technol 17(1):3–16. doi:10.1080/13645700701820242
90. Wilde E et al (2013) Technisch vermitteltes soziales Handeln im Alter-ein Gestaltungsprozess. In: Lebensqualität im Wandel von Demografie und Technik. doi:10.1007/s10278-011-9416-8
91. Wolf U (2007) Aristoteles' nikomachische ethik. Wiss. Buchges., Darmstadt
92. Yang X et al (2007) Hand tele-rehabilitation in haptic virtual environment. In: IEEE international conference on robotics and biomimetics (ROBIO). Sanya C, pp 145–149. doi:10.1109/ROBIO.2007.4522150

Chapter 2
Haptics as an Interaction Modality

Christian Hatzfeld

Abstract This chapter focuses on the biological and behavioral basics of the haptic modality. On one side, several concepts for describing interaction are presented in Sect. 2.2, and on the other side, the physiological and psychophysical bases of haptic perception are discussed in Sect. 2.1. The goal of this chapter is to provide a common basis to describe interactions and to convey a basic understanding of perception, and the description via psychophysical parameters. Both aspects are relevant to the formal description of the purpose of a haptic system and the derivation of requirements, which are further explained in Chap. 5. Several conclusions arising from the description of perception and interaction are given in Sect. 2.4.

2.1 Haptic Perception

This section give a brief summary of relevant topics from the scientific disciplines dealing with haptic perception. It is intended to reflect the current state of the art as a necessary extension for an engineer designing a haptic system. Physiologists and psychophysicists are therefore asked to forgive simplifications and impreciseness. For all engineers, Fig. 2.1 gives a general block diagram of haptic perception that forms a conscious ↪ percept from a ↪ stimulus.

Analyzing each block of this diagram, the mechanical properties of the skin as stimuli transmitting apparatus are dealt with in Chap. 3. Section 2.1.1 deals with the characteristics of mechanoreceptors in the skin and locomotion system, while Sect. 2.1.2 introduces the psychophysical methods used to evaluate these characteristics. In Sects. 2.1.3 and 2.1.4, threshold and super-threshold parameters of human haptic perception are presented.

C. Hatzfeld (✉)
Institute of Electromechanical Design, Technische Universität Darmstadt,
Merckstr. 25, 64283 Darmstadt, Germany
e-mail: c.hatzfeld@hapticdevices.eu

C. Hatzfeld and T.A. Kern (eds.), *Engineering Haptic Devices*,
Springer Series on Touch and Haptic Systems, DOI 10.1007/978-1-4471-6518-7_2

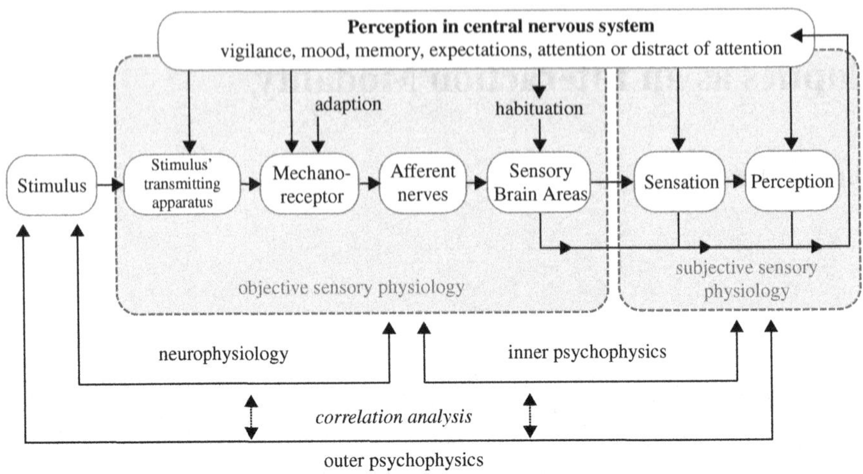

Fig. 2.1 Block diagram of haptic perception and the corresponding scientific areas investigating the relationships between single parts as defined in [45, 140]

2.1.1 Physiological Basis

This section deals with the physiological properties of the tactile and kinaesthetic receptors as defined in the previous chapter (Sect. 1.2.2). We do not cover neural activity in detail, but only look at a general model that is useful for a closer look at multimodal systems.

2.1.1.1 Tactile Receptors and Their functions

From a histological view, there are four different sensory cell types in glabrous skin and two additional sensory cell types in hairy skin. They are located in the top 2 mm of the skin as shown in Fig. 2.2. The sensory cells in glabrous skin are named after their discoverers, while the additional cells in hairy skin have functional names [31]. Because of the complex mechanical transmission properties of the skin and other body parts such as vessels and bone, compression and shear forces in the skin and—for high-frequency stimuli—surface waves are expected in the skin as a reaction to external mechanical stimuli. These lead to various pressure and tension distributions in the skin that are detected differently by individual sensory cells. In general, sensory cells near the skin surface will react only to adjacently applied stimuli, while cells localized more deeply like the Ruffini endings and Pacinian corpuscles will react also to stimuli applied farther away. These differences are presented in the following for the well-researched receptors in glabrous skin. For hairy skin, less information is available. While tactile disks are assumed to exhibit similar properties as Merkel cells because of the same histology as the receptor, hair follicle receptors are attributed to detect movements on the skin surface. The following sections concentrate on tactile receptor cells in glabrous skin because of the higher technical relevance of these areas.

To investigate the behavior of an individual sensory cell, a single nerve fiber is contacted with an electrode and electrical impulses in the fiber are recorded as shown in Fig. 2.3 [81]. The results in the following paragraphs are based on measurements that are complicating to conduct on a living organism or on a human test person.

From Sensory Cells to Mechanoreceptors and Channels

When reviewing the literature on the physiology of haptic perception, several terms are used seemingly interchangeable with each other. Since microneurography often does not allow a distinct mapping of a sensory cell to the contacted nerve fiber, a formal separation between a single sensory cell as given in Fig. 2.2 and the term mechanoreceptor has been established [211]. A ↪ mechanoreceptor is defined as

> **Definition** *Mechanoreceptor* An entity consisting of one or more sensory cells, the corresponding nerve fibers, and the connection to the central nervous system.

The classification of a mechanoreceptor is based on the size of the ↪ receptive fields and the adaptation behavior of the receptor when a constant pressure stimulus is applied. The receptive field denotes the area on the skin, on which an external

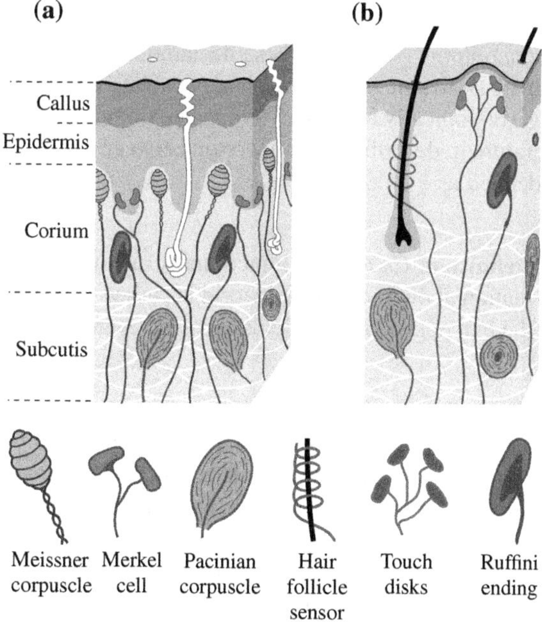

Fig. 2.2 Histology of tactile receptors in **a** glabrous and **b** hairy skin. Figure adapted from [211]

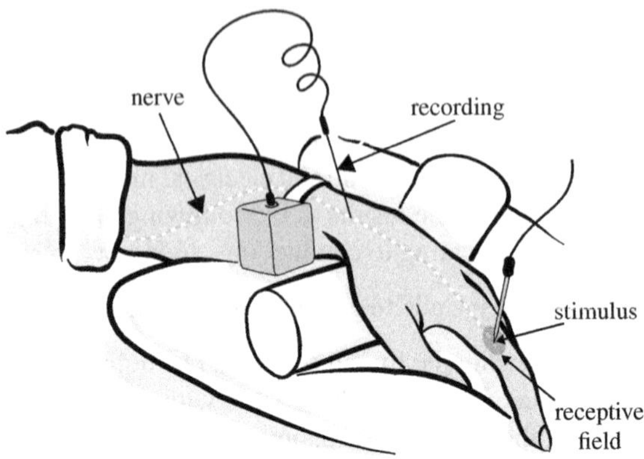

Fig. 2.3 Recording of electrical impulses of a single mechanoreceptor with microneurography. Figure adapted from [81]

Table 2.1 Receptive fields of the tactile mechanoreceptors in glabrous skin

Receptor type	Sensory cell	Size (mm^2)	Boundary
SA-I	Merkel disk	7–12	Distinct
SA-II	Ruffini endings	50–75	Diffuse
RA-I	Meissner corpuscle	7–19	Distinct
RA-II (PC)	Pacinian corpuscle	100–300	Diffuse

Table gives the size and form of the receptive fields based on data from [23, 38, 103, 153, 211]

mechanical stimulus will evoke a nervous impulse on a single nerve fiber. The size of the receptive field depends on the number of sensory cells that are connected to the investigated nerve fiber. Tactile mechanoreceptors exhibit either small (normally indicated with *I*) or large receptive fields (indicated with *II*). The adaptation behavior is classified as *slowly adapting* (*SA*) or *rapidly adapting* [*RA, sometimes also called fast adapting* (*FA*)]. With these declarations, four mechanoreceptors can be defined that are shown in Table 2.1.This nomenclature is based on a biological view. Next to these biologically motivated terms, one finds the term *channel* in the psychophysical literature to describe the connection between sensory cells and the brain. A channel is defined as

Definition *Channel* "Functional/structural pathway in which specific information about the external world is passed to some location in the brain where the perception of a particular sensory event occurs" (*Quote from* [15, p. 49]).

The difference in the definition of mechanoreceptors is the integration of functional processes such as masking in the channel model. In general, the terms for channels and mechanoreceptors are used synonymously. The channels are named NP-I (RA-I receptor, NP stands for non-Pacinian), NP-II (SA-II receptor), NP-III (SA-I receptor), and PC (RA-II receptor) [17]. There is experimental evidence for the presence and involvement of four channels in haptic perception in glabrous skin, but only three channels in hairy skin [18]. When using the channel model to describe

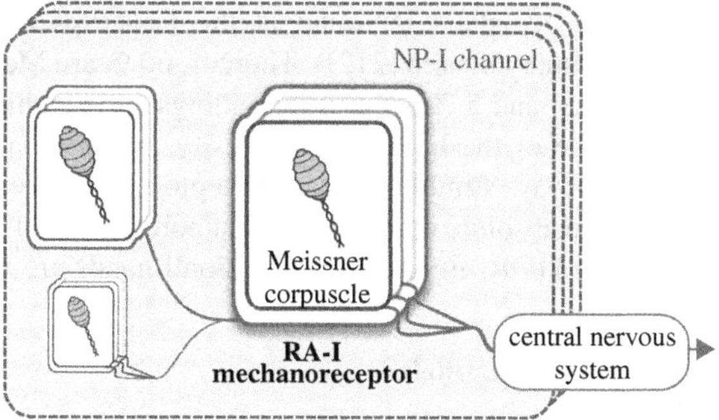

Fig. 2.4 Relation of the terms sensory cell, mechanoreceptor, and channel using the NP-I channel as example. The NP-I channel consists of RA-I mechanoreceptors that are based on Meissner corpuscles as sensory cells. NP-I and NP-III channels process signals of multiple sensory cells, while NP-II and PC channels are based on signals of single sensory cells [17]

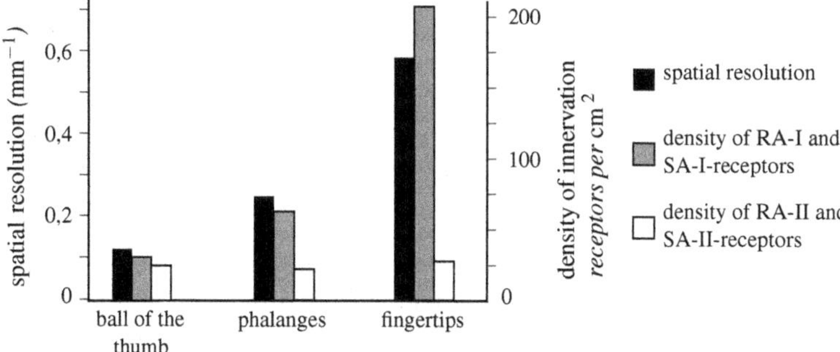

Fig. 2.5 Innervation density of mechanoreceptors near and far from the skin surface. The greater innervation density of Merkel and Meissner receptors leads to higher spatial solution of quasi-static stimuli. Figure based on [211]

haptic interaction, one has to be aware that certain aspects of interaction, such as surface properties and reactions, to static stimuli cannot be explained fully by it [122, Chap. 4]. In this book, this discrepancy is not discussed in detail in favor of a primitive-based description of interactions that involve perception and motion control. An overview of the different terms for the description of tactile perception is given in Fig. 2.4.

Spatial Distribution of Mechanoreceptors

The spatial distribution of the different mechanoreceptors depends on the skin region considered. For the skin of the hand, there is a varying distribution depending on the depth of the mechanoreceptors in the skin: Near-surface receptors (RA-I, SA-I) show higher density in the fingertips than in the palm, and deeper localized receptors show only light dependency on the skin region. This is shown in Fig. 2.5.

The highest density of receptors is found at the fingertips and adds up to 250 receptors/cm^2 [13] (primary source [212]). Thereof, 60 % are Meissner corpuscles, 60 % are Merkel disks, and 5 % are Ruffini endings and Pacinian corpuscles, respectively [193]. Because of the high spatial density, it can be assumed that a mechanical stimulus will always stimulate several receptors of different types. However, not the density, but the absolute number of mechanoreceptors of different users is approximately the same [179]. Because of this, small hands are more sensitive than large hands.

Functions of Receptors and Channels

Next to the physiological and histological differences in the mechanoreceptors described above, channels differ in additional and functional properties [15, 17]:

Frequency Dependency Channels are sensitive in different frequency ranges. While NP-II and NP-III channels are sensitive to (quasi-)static and low-frequency stimuli (up to about 20–50 Hz), the PC channel becomes sensitive at about 50 Hz and detects stimuli up to a frequency of 10 kHz. The main reason for the frequency dependency lies in the biomechanical structure of the receptors [67, 81]. To investigate frequency dependence of channels, psychophysical measurements using masking schemes are used. This leads to different results for the sensitivity and frequency selectivity of the several channels, depending on the measurement procedures.

Thresholds Each channel exhibits thresholds that are independent of the other channels. An aggregation of the information from different channels takes place in the central nervous system [15]. There is also no evidence of cross talk between channels [112]. Recent studies find evidence for a linear behavior of the channels and the aggregation process (see Sect. 2.1.4).

Summation Properties Summation describes the property of a channel, to consider more than one temporal or spatial contiguous stimuli as a single stimulus. The reasons for summation are given by the neural activities to conduct impulses through the nerve fibers [42]. There are also assumptions of summation mechanisms in the central nervous system to compensate for sensitivity differences in the different receptors in a channel [66, 68]. Studies show no spatial summation for (quasi-)static stimuli [136], but only for dynamic ones [126].

Temperature Dependency Thresholds and frequency dependency are influenced by temperature. A stronger dependency on temperature is attributed to NP-II and PC channels compared to NP-I and NP-III [17]. Other studies assess temperature-induced threshold changes at the glabrous skin of the hand starting at a frequency of 125 Hz, but find no effect on the hairy skin of the forearm [217]. Next to this, also the mechanical properties of the skin exhibit temperature dependency of the mechanical properties [78].

Table 2.2 gives an overview of the discussed properties for each channel. It also includes information about the coding of channels referring to kinematic measures such as deflection, velocity (change of deflection), and acceleration (change of velocity) [23, 81].

Based on these properties, functions of the different channels in perception and interaction can be identified [115–117, 178].

NP-I (RA-I, Meissner corpuscle) Most sensory cells in human skin belong to the NP-I channel. They exhibit a lower spatial resolution than SA-I receptors, but have a higher sensibility and a slightly larger bandwidth. The corresponding sensory cells are called Meissner corpuscles and exhibit a biomechanical structure that makes them insensitive to quasi-static stimuli.
The RA-I receptors are sensitive to stimuli acting tangential to the skin surface. They are important for the detection of slip of handheld objects and the associated sensomotoric control of grip forces. Together with the PC channel, they are relevant for the detection of frequencies of vibrations [147, 165].
The NP-I channel can detect bumps of height of just $2\,\mu$m on an otherwise flat surface, if there is a relative movement between surface and skin. This movement leads to a deformation of the papillae on the skin by the object. Reaction forces and deformations are located in a frequency bandwidth that will activate the RA-I receptors [150]. Similarly, filter properties of surface structures are used for the design of force sensors [192].

NP-II (SA-II, Ruffini ending) The SA-II receptors in this channel are sensitive to lateral elongation of the skin. They detect the direction of an external force, for example, while holding a tool. The NP-II channel is more sensitive than the NP-III channel, but has a much lower spatial resolution.
This channel also transmits information about the position of limbs, when joint flexion induces skin elongation. The SA-II receptors are therefore also relevant for kinaesthetic perception. With specific stimulation of the NP-II channel, an illusion of the position of limbs can be generated [44, 99].

NP-III (SA-I, Merkel disk) The NP-III channel with Merkel disks right under the skin surface is sensitive to strains in normal and shear directions. Because of the slow adaptation of the channel, the high density of sensory cells and the high spatial resolution (less than 0.5 mm although the receptive field is larger than that), it is used to detect elevations, corners, and curvatures. It is therefore the basis for the detection of object properties such as form and texture.
Because of the coding of intensity and intensity changes, this channel is responsible (together with the RA-I channel) for reading ↪ Braille. Studies also show an effect of the channel when wrist and elbow forces are to be controlled [123].

PC (RA-II, Pacinian corpuscle) The PC channel with rapidly adapting RA-II receptors exhibits the largest receptive fields and the largest sensitivity bandwidth. It is mainly used to detect vibrations arising in the usage of tools. These vibrations originate in the contact of the tool with the environment and are transmitted by the tool itself. They allow the identification of surface properties with a stiff tool,

Table 2.2 Selection of relevant parameters of tactile sensory channels

Channel	Sensory cell	Receptor type	Sensitive bandwidth (Hz)	Spatial resolution (mm)	Coding	Summation properties	Temperature dependency
Sources	[17]	[17]	[15, 17, 23, 103, 168]	[80]	[23, 81]	[17, 122]	[17]
NP-I	Meissner corpuscle	RA-I	10–50	3–5	Velocity	Spatial	Sightly
NP-II	Ruffini ending	SA-II	QS–50	10	Deflection	Temporal	Existent
NP-III	Merkel disk	SA-I	QS–20	0.5	Deflection, velocity	*Not known*	Slightly
PC	Pacinian corpuscle	RA-II	40–1,000	>20	Acceleration	Spatial and temporal	Existent

Averaged values particularly common ranges are given, when more than one source is considered. For sizes of receptive fields, see Table 2.1
Used abbreviation ↪ quasi-static (QS)

for example [50]. Because of the very high sensitivity (vibrational amplitudes of just a few nanometers can be detected by the PC channel [17]), also sensory cells located further away from the application of stimuli contribute to perception by reacting to surface wave propagation [31, 162]. To suppress the influence of dynamic forces arising in the movement of limbs on the perception of the PC channel, the Pacinian corpuscles exhibit a strong high-pass characteristic with slopes up to 60 dB per decade. This is realized by the biomechanical structure of the sensory cells.

In interactions, RA-II receptors signal that something is happening, but do not necessarily contribute to the actual interaction. In addition, contributions of the PC channel to the detection of surface roughness and texture are assumed [38].

2.1.1.2 Kinaesthetic Receptors and Their functions

For kinaesthetic perception, there are two known receptor groups [34, 80, 119]. The so-called *neuromuscular spindles* consist of muscle fibers with wound around nerve fibers that are placed parallel to the skeletal muscles. Because of this placement, strain of the skeletal muscles can be detected. Histologically, they consist of two systems, the nuclear bag fibers and the nuclear chain fibers, that react to intensity change and intensity [107].

The second group of receptors are *Golgi tendon organs*. These are located mechanically in series to the skeletal muscles and detect mechanical tension. They are used to control the force the muscle exerts as well as the maximum muscle tension. Special forms of the Golgi organs exist in joints, where extreme joint position and ligament tension are detected [80]. They react mostly on intensity. Figure 2.6 shows these three types of receptors.

The dynamic requirements on kinaesthetic sensors are lesser compared to tactile sensors, since the extremities exhibit a low-pass behavior. The requirements with regard to relative resolution are comparable to the tactile system. Proximal joints exhibit higher absolute resolution than distal joints. The hip joint can detect angle changes as low as 0.22°, while finger joint resolution increases to 4.4° [34]. This is because of the greater influence of proximal joints on the position error of an extremity. The position accuracy increases with increasing movement velocity [119].

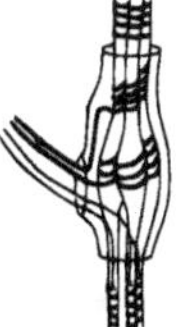

Fig. 2.6 Kinaesthetic sensors for muscular tension (Golgi tendon organ) and strain (bag and chain fibers). Figure adapted from [196]

Kinaesthetic perception is supported by information from the NP-II channel, from the vestibular organ responsible for body balance, and from visual control by the eye. Different from the tactile system, the kinaesthetic system does not code intensities or their changes, but exhibits some sort of sense for the effort needed to perform a movement [201, 210]. For applications such as rotary knobs, this means that a description based on movement energy with regard to rotary angle does correlate better with user ratings than the widespread description based on torques with regard to rotary angle.

2.1.1.3 Other Sensory Receptors

The skin also includes sensory receptors for thermal energy [39] as well as pain receptors. The latter are attributed a protection function, signaling pain when tissue is mechanically damaged [144]. Both aspects are not discussed in detail in this book because of the minor technical importance.

2.1.1.4 Neural Processing

Haptic information detected by the tactile and kinaesthetic mechanoreceptors is coded and transmitted by action potentials on the axons of the involved neurons to the central nervous system. The coding of the information resembles the properties given in Table 2.2 and is illustrated in Fig. 2.7.

The biochemical processes taking place in the cells are responsible for the temporal summation (when several action potentials reach a dendrite of a neuron within a short interval) and the spatial summation (when action potentials of more than one receptor arrive at the same neuron) of mechanoreceptor signals. Each individual action potential would not be strong enough to evoke a relay of the signal through the neuron [41]. In the rest of this book, further neurophysiological considerations are of minor importance, but can be assessed in a standard physiology or neurophysiology textbook.

A more interesting question in the design of haptic systems is the synthesis of different information from haptic, visual, and acoustic senses to an unconscious or conscious ↪ percept, and a resulting action. While the neural processes are investigated in-depth [114], there is no comprehensively confirmed theory about the processing in the central nervous system (spinal cord and brain).

Current research favors a Bayesian framework that also incorporates former experiences when assessing information [48, 184]. Figure 2.8 gives a schematic description of this process that is confirmed by current studies [70, 172].

2.1.2 Psychophysical Description of Perception

The investigation of perception processes, that is, the link between an objectively measurable ↪ stimulus and a subjective ↪ percept, is the task of psychophysics,

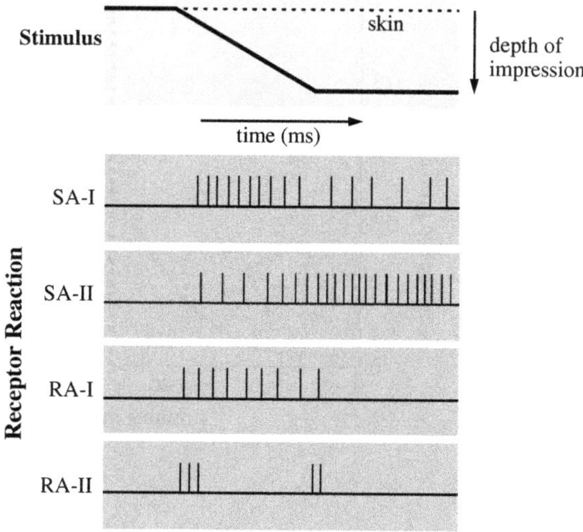

Fig. 2.7 Action potentials of sensory cells when a single stimulus is exerted on the skin. Figure adapted from [81]

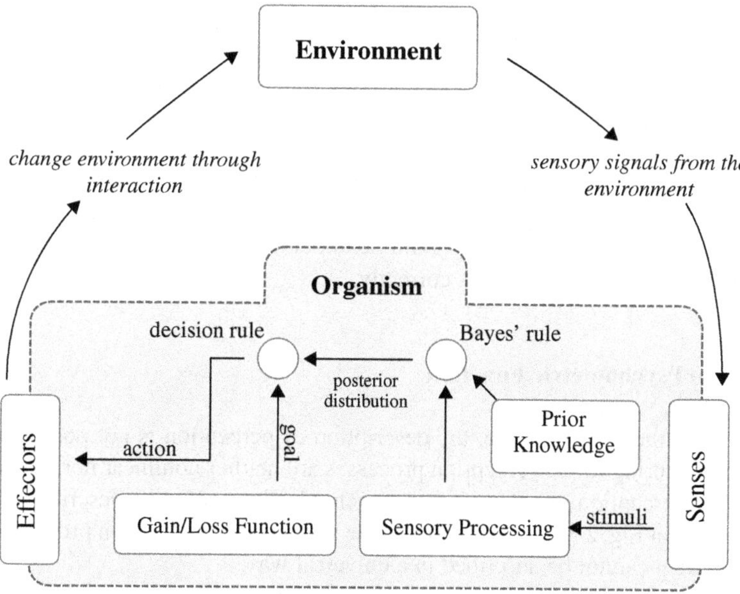

Fig. 2.8 Sensory integration according to HELBIG [95]. Prior knowledge is combined with current sensory impressions for a percept of the situation. Based on a gain/loss analysis, a decision is made and an interaction using the effectors (limbs, speech) is initiated

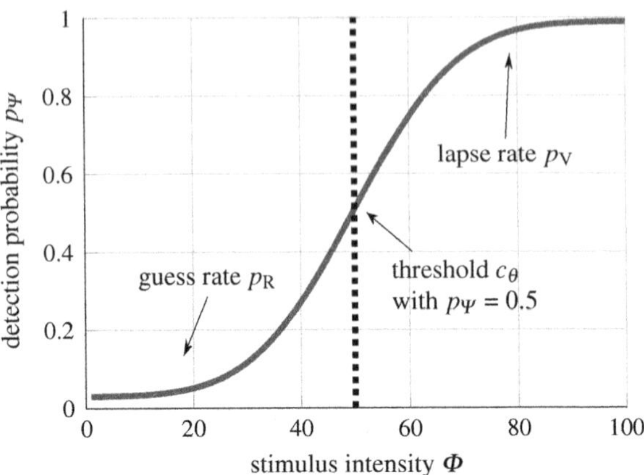

Fig. 2.9 Psychometric function based on a normal distribution with $c_\theta = 50$ stimulus units, guess rate p_G, and lapse rate p_V

a section of experimental psychology. It was established by FECHNER in the late mid-nineteenth century [49]. As shown in Fig. 2.1, there are several components in psychophysical studies. *Inner psychophysics* deals with the connection of neural activity and the formation of percepts, while *outer psychophysics* investigates the reactions to an outer stimulus. These components were established by FECHNER [45]. Nowadays, modern technologies also allow the investigation of neurophysiological problems linking outer stimuli with neural activity and the analysis of correlations between neurophysiology, inner psychophysics, and outer psychophysics.

For the design of haptic systems, we concentrate on outer psychophysics, since only physical properties of stimuli and the corresponding subjective percepts allow the derivation of design parameters and design goals. Therefore, the remainder of this chapter will only deal with procedures and parameters from outer psychophysics. It describes the main principles that should be understood by every systems engineer to interpret psychophysical studies correctly.

2.1.2.1 The Psychometric Function

Regardless of the kind of sense, the description of perception is not possible with general engineering tools. Perception processes are neither nonlinear nor stationary, because the perception process of inner psychophysics cannot be described in that way. Looking at Fig. 2.8, this is obvious since weighting and decision processes and risk assessment cannot be described in a universal way.

Because of this, perception processes in outer psychophysics are not described by specific values but by probability functions. From these functions, specific values can be extracted. Figure 2.9 gives an example of such a $\hookrightarrow$ psychometric function. On the x-axis, the intensity of an arbitrary stimulus Φ is plotted, while the y-axis gives the probability p for a test person detecting the stimulus with that intensity.

According to [223], the psychometric function p_Ψ has a general mathematical description according to Eq. (2.1).

$$p_\Psi(\Phi, c_\theta, c_\sigma, p_G, p_L) = p_G + (1 - p_G - p_L) \cdot f(\Phi, c_\theta, c_\sigma) \tag{2.1}$$

with a stimulus Φ and the following parameters:

Base Function f The base function determines the form of the psychometric function. In literature, you can find different approaches for the base function. Often a cumulative normal distribution (see Eq. 2.2), sigmoid functions (see Eq. 2.3), and Weibull distributions (see Eq. 2.4) are used:

$$f_{\mathrm{cdf}}(c_\theta, c_\sigma, \Phi) = \frac{1}{c_\sigma\sqrt{2\pi}} \int_{-\infty}^{\Phi} e^{\frac{-(t-c_\theta)^2}{2c_\sigma^2}}\, \mathrm{d}t \tag{2.2}$$

$$f_{\mathrm{sig}}(c_\theta, c_\sigma, \Phi) = \frac{1}{1 + e^{-\frac{c_\sigma}{c_\theta}(\Phi - c_\theta)}} \tag{2.3}$$

$$f_{\mathrm{wei}}(c_\theta, c_\sigma, \Phi) = 1 - e^{-(\frac{\Phi}{c_\theta})^{c_\sigma}} \tag{2.4}$$

Nowadays, there is no computational limit for the calculation of functions and extracting values; therefore, the choice of a base function depends on prior experiences of the experimenter. When investigating the visual sense, a Weibull distribution will better fit data [84], when working with ↪ signal detection theory (SDT), a normal distribution is assumed [224]. Sigmoid functions were often used in early simulation studies because of the low computational effort needed to calculate psychometric functions. The current state of the art in mathematics allows for non-model-based description of the psychometric function. In psychophysics, these approaches have not been seen very often till now, whereas first studies show a comparable performance of these techniques compared to a model-based description [51].

Base Function Parameters $(c_\theta,\ c_\sigma)$ These two parameters can be treated demonstratively as the perception threshold and the sensitivity of the test person. c_θ gives the threshold, denoting the stimulus Φ with a detection probability of 0.5. The sensitivity parameter c_σ is given as the slope of the psychometric function at the threshold level. A high sensitivity will yield a large slope, and a low sensitivity will yield a flatter curve, resulting in many false detections.

Guess Rate p_G The guess rate gives the portion of false-positive answers for very low stimulus intensities that cannot be detected by the test person normally. Such false-positive answers can arise in guessing, erroneous answers, or abstraction of the test person. In simulating psychometric functions, the guess rate is used to model force-choice answering paradigms.

Lapse Rate p_L The lapse rate models false-negative answers when a large stimulus intensity is not detected by a test person. The main reason for lapses is inattentiveness of the test person during a psychophysical procedure.

To find a psychometric function, psychophysics knows a bundle of procedures that are addressed in the following section. It has to be kept in mind that all of the above and the following are valid not only for a stimulus with a changing intensity, but also for any other kind of changing signal parameters such as frequency, energy, and proportions between different signals.

2.1.2.2 Psychometric Procedures

The general goal of a psychometric procedure is the determination of a psychometric function p_Ψ as a whole or a single point $(\Phi \,|\, p_\Psi(\Phi))$ defined by a given probability p_Ψ. In general, each run of a psychometric procedure consists of several trials, in which a stimulus is presented to a test person and a reaction is recorded in a predefined way. Figure 2.10 gives a general taxonomy to classify psychometric procedures.

Each procedure consists of a measuring method and an answering paradigm. The method determines the course of a psychometric experiment, particularly the start intensity of the stimulus and the changes during each run, the conditions to stop a trial, and the calculation rule to obtain a psychometric function from the measured data. The answering paradigm defines the way a test person is presented the stimulus and the available answering options. The choice of a suitable procedure is not the topic of this book, but an interesting topic nevertheless. Further information about the simulation of procedures and the definition of suitable quality criteria can be found in [90, 173].

Methods

The first, nowadays called "classical," methods were developed back in the nineteenth century. The most familiar are called *Method of Constant Stimuli*, *Method of Limits*, and *Method of Adjustment*. They have barely practical relevance in today's

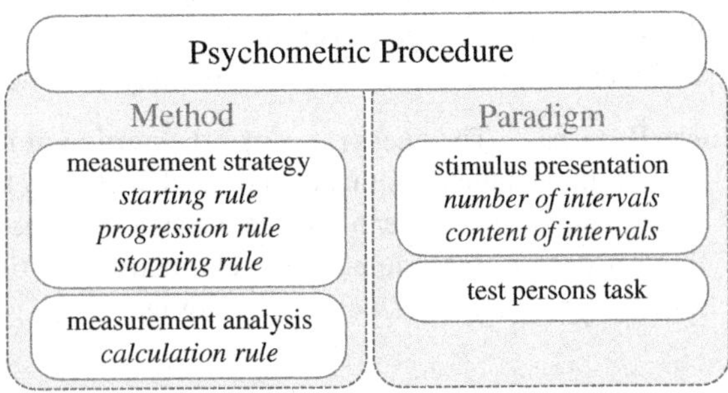

Fig. 2.10 Taxonomy of psychometric procedures. The figure is based on classifications in [24]

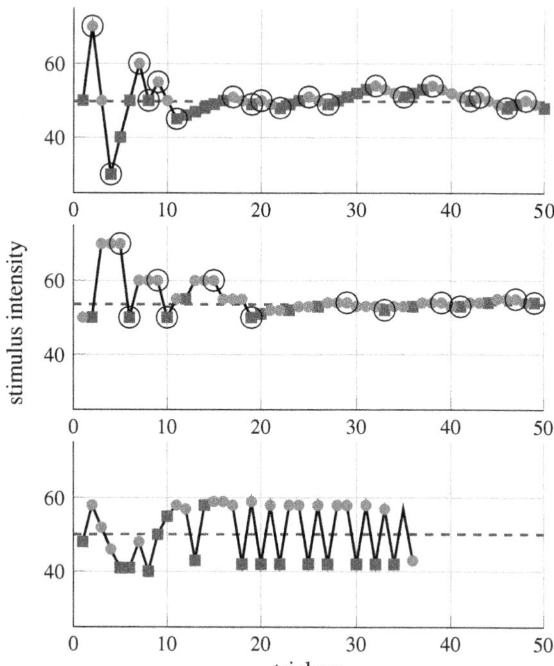

Fig. 2.11 Simulated runs of common psychometric methods. The *upper graph* shows a *simple up-down staircase*, theoretically converging at $c_\theta = 50$, and the *middle graph* shows a *transformed up-down staircase* with a *1up-3down* progression rule and a theoretical convergence level of $c_\theta = 57.47$. The *lower graph* shows a run from a Ψ*-method*, also converging at $c_\theta = 50$. Simulated answers of the subject are shown in *green circles* (correct answer) and *red squares* (incorrect answer), and staircase reversals are *circled*. The *dotted line* indicates the calculated threshold

experiments, so they are not detailed here any further. Please refer to [182] for a detailed explanation. Modern methods are derived from these classical methods, but are generally adaptive in such a way that the progression rule depends on the answers of the test person in the course of the experiment [158]. They can be classified into heuristic- and model-based methods.

Heuristic-Based Methods These methods are based on predetermined rules that are applied to the answers of the test person to change the stimulus intensity in the course of the psychophysical experiment (change in the progression rule). Stopping and starting rules are normally fixed (by the total number of trials for example) for each experiment beforehand. The most widely spread heuristic method is the so-called *staircase method*. It is based on the classic *Method of Limits* and tries to nest the investigated threshold with the intensities of the test stimulus. Figure 2.11 gives two examples of a staircase method with different progression rules. It becomes clear that the name of this method originates in this kind of display of the test stimulus intensities over the test trials.

The definition of progression rules is based on the number of correct answers of the test person leading to a lower test stimulus and the number of false answers leading to a higher test stimulus. The original staircase method, also called *Simple Up-Down Staircase* (see upper part of Fig. 2.11), will change stimulus intensities after every trail, such converging at a threshold with a detection probability of

Table 2.3 Probability of convergence of adaptive staircase methods with different progression rules

Progression rule	Number of false answers for raising stimulus intensity	Number of correct answers for lowering stimulus intensity	Number of convergence probability
1 up–1 down	1	1	$p(X) = 0.500$
1 up–2 down	1	2	$p(X) = 0.707$
2 up–1 down	2	1	$p(X) = 0.293$
1 up–3 down	1	3	$p(X) = 0.794$
1 up–4 down	1	4	$p(X) = 0.841$
4 up–1 down	4	1	$p(X) = 0.159$

Table based on [160]

0.5 [160]. The request for other detection probabilities led to another form of the staircase method, the so-called *Transformed Up-Down Staircase Method*. For these methods, the progression rule is changed and needs, for example, more than one correct answer for a downward change in the test stimulus. Figure 2.11 gives an example of a 1up-3 down progression rule, lowering the test stimulus after three correct answers and raising it for every false answer.

In [160], LEVITT calculates the convergence probability for several progression rules. Table 2.3 gives the convergence probabilities for common progression rules. To interpret studies on haptic perception incorporating experiments with staircase methods, the convergence probability has to be taken into account. However, newer studies cast some doubts on this interpretation of the progression rule [53], arguing that the amount of intensity change is much more relevant for the convergence of a staircase than the progression rule. For system design, one therefore has to resort to larger assessment factors in the interpretation of these kinds of data.

The calculation of a threshold is normally carried out as a mean of the last stimulus intensities leading to a reversal in the staircase direction. Typical values are, for example, 12 reversals for the calculation of the threshold and 16 reversals as a stopping criterion for the whole experiment run.

Another important heuristic method is the so-called *parameter estimation by sequential testing* (*PEST*) *method* [208]. The heuristic keeps a given stimulus intensity until some assessment of the reliability of the answers can be made. The method was designed to yield high accuracy with a small number of trials. One of the main disadvantages of this method is the calculation rule that only considers the very last stimulus. However, several modern adaptations such as *ZEST* and *QUEST* try to overcome some of these disadvantages [137].

Model-Based Methods A model of the psychometric function is the basis of these methods that measure or estimate the parameters of the function as given in Eq. (2.1). Most methods incorporate some kind of prior knowledge of the psychometric function from experience or previous experiments (Bayes approach) and use different kinds of estimators (maximum likelihood, probit estimation). Examples for these methods are *ML test* [84] that use a maximum-likelihood estimator for determination of the function parameter. The end of each experiment run is determined by the confidence interval of the estimated parameters. If the interval is smaller than a given value, the experiment run is stopped.

In 1999, KONTSEVICH AND TYLER introduced the *Ψ-Method* [143], combining promising elements from several other methods. This method is not very prominent in haptics research, but is considered the most sophisticated method in psychophysics in general [139, 182]. It is able to estimate a threshold in as little as about 30 trials and sensitivity in about 300 trials.

One of the general advantages of a model-based method is the calculation of a whole psychometric function, and not only of a single threshold. Therefore, more than one psychometric parameter can be calculated from a single experiment. Adversely, one should have confidence in the model used in the method for the investigated sense. As said above, data show a slight advantage for Weibull-based models for the visual sense [84], while studies of the author in this chapter yield better results for a logistic function for the assessment of force perception thresholds [87].

Paradigms

Answering paradigms describe the way a test person will give the answer to a stimulus in such a way that the procedure can react according to its inherent rules. The theoretical basis for answering paradigms is given in the ↪ SDT, a statistical approach to describe arbitrary decision processes that are often applied to sensory processes. It is based on the assumption that not only stimuli, but also noise will contribute to perception. In the perception continuum, this is represented by a noise distribution (mainly Gaussian). If there is no stimulus present, the noise distribution is present in neural activity and processing, and if a stimulus is present, the noise distribution is added to the stimulus. Figure 2.12 shows this theoretical basis of ↪ SDT.

Near the absolute detection threshold, both distributions will overlap. In this area on the perception continuum, it is indistinguishable whether a neural activity comes from a stimulus or just from innate noise. To decide whether a stimulus is present or not, the test person constructs a decision criterion c_λ. If an input signal is greater than this criterion, a stimulus is identified, and smaller inputs are neglected. Unfortunately, this decision criterion varies with time and other external conditions. Therefore, one aim of ↪ SDT is to investigate the behavior of a test person regarding this criterion. The detectability d' arising from the SDT can be used to calculate comparable sensitivity parameters for different test persons and to compare studies with different psychometric procedures.

With these implementations, one can differentiate liberal (low decision criterion) from conservative (high decision criterion) test persons. For example, studies show that the consumption of alcohol will not change the sensitivity of a person, but will

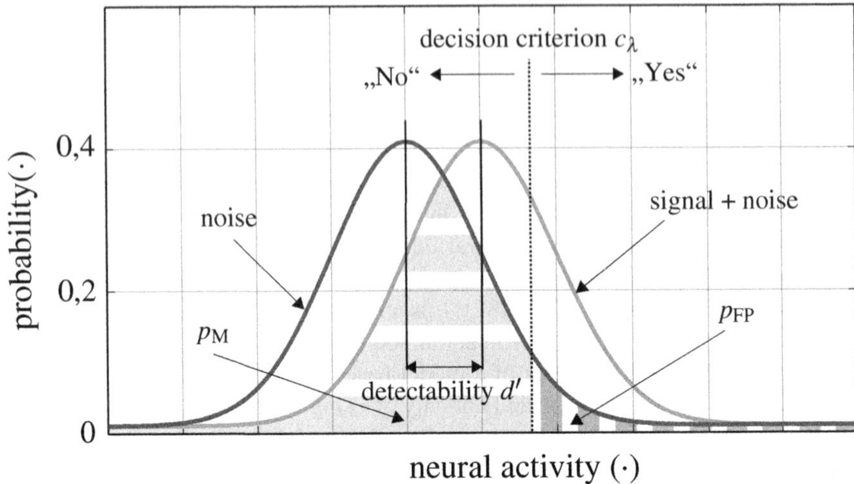

Fig. 2.12 Baseline model of the signal detection theory (*SDT*). The model consists of two neural activity distributions for noise and signal + noise. The noise distribution is always present and can be interpreted as sensory background noise. If an additional stimulus is presented, the whole distribution shifts upward. The subject decides based on a decision criterion c_λ, whether a neural activity is based on a stimulus or not. The subject shown here exhibits a badly placed conservative decision criterion: Many signals are missed (*horizontal striped area* p_M), but only a few false-positive answers are recorded (*vertical striped area* p_{FP}). The detectability d', defined as the span between the midpoints of both distributions, is independent of the decision criterion of the subject and can be used as an objective measure of the difficulty of a detection experiment

influence the decision criterion to become more liberal. This leads to better detection of smaller stimuli, but will also produce more false-positive answers. Based on this stochastic approach of decision theory, answering paradigms can be defined, which are used to minimize the influence of varying decision criteria. The most common paradigms are described in the following.

Yes/No Paradigm The easiest paradigm is the simple *Yes/No paradigm*. A test person will, for example, answer “yes” or “no” to the question of whether a stimulus was present or not. Obviously, a varying decision criterion will affect this answer. One has to trade this disadvantage for the shorter time this paradigm needs to present a stimulus compared to other paradigms.

Forced-Choice Paradigm These paradigms arise directly from ↪ SDT to find an objective measure of a subjective assessment. To achieve this, each trial includes more than one alternative with a test stimulus and the test person is compelled to give a positive answer in every trial, for example, which interval contained a stimulus or which stimulus had the largest intensity. This paradigm can be combined with most of the methods mentioned above. In general, forced-choice paradigms are denoted by *xAFC* (*x alternative forced choice*) or *xIFC* (*x interval forced choice*). The first abbreviation is used, when several alternatives are

presented to a test person, while the term interval is used for a temporal sequence of the alternatives. x denotes the number of alternatives, respectively, intervals. Naturally, forced-choice paradigms increase the guessing rate in the psychometric function with an additional probability of $\frac{1}{x}$. This has to be considered in the mapping of experimental results back to a psychometric function.
An example experiment setup would include five reference stimuli and one unknown test stimulus given to the test person. The test person would have do decide which reference stimulus corresponds to the test stimulus. If stimuli were miniature golf balls with different compliance, this experiment can be classified as a 5AFC paradigm.

Unforced-Choice Paradigm In [129], KÄRNBACH describes an adaptive procedure that does not require a forced choice, but also allows an "I don't know" answer. This procedure leads to more reliable results from test persons without extensive experience in simulations. Particularly in experiments incorporating a comparison of different stimuli, this answering paradigm could provide a more intuitive approach to the experiment for the test person and could therefore lead to more motivation and better results. Based on Fig. 2.10, this unforced-choice option belongs to the paradigm definition, but has to be incorporated in the method rules as well. Therefore, this paradigm is only found in a limited number of studies, but finds also application in recent studies of haptic perception [185].

2.1.2.3 Psychometric Parameters

In most cases, not the whole psychometric function, but characteristic values are sufficient for usage in the design of haptic systems. The most important parameters are described in this section.

Absolute Thresholds

These parameters describe the human ability to detect a stimulus at all. They are defined as the stimulus intensity Φ with a detection probability of $p_{\Psi} = 0.5$ [164, Chap. 5]. However, since many psychometric procedures do not converge at this probability, most studies call their results threshold, regardless of the convergence probability.

For the design of haptic systems, absolute thresholds give absolute margins for sensors and actuators for noise and, otherwise, induced errors: A vibration that is "detected" by a sensor because of inherent noise in the sensor signal processing or displayed by an actuator is acceptable as long as the user of a haptic system does not feel it. Therefore, reliable assessment of these thresholds is important to define suitable requirements. On the other hand, absolute thresholds define a lower limit in communication applications: Each coded information has to be at least as intense as the absolute threshold to be detectable, even if one probably will choose some considerably higher intensity level to ensure detection even in distracting environments.

Differential Thresholds

Differential thresholds describe the human ability to differentiate between two stimuli that differ in only one property. The first differential thresholds were recorded by WEBER at the beginning of the last century [221]. He investigated the differential threshold of weight perception by placing a mass (reference stimulus Φ_0) of a test person's hand and adding additional mass $\Delta\Phi$ until the test person reported a higher weight. The additional mass needed to evoke this perception of a higher weight is called the *difference limen* (*DL*).

Further studies showed that the quotient of $\Delta\Phi$ and Φ_0 would be constant in a wide range of reference stimulus intensities. This behavior is called *Weber's law*, and the *Weber fraction* given in Eq. (2.5) is also called ↪ just noticeable difference (JND).

$$\text{JND} := \frac{\Delta\Phi}{\Phi_0 + a} \tag{2.5}$$

The ↪ JND is generally given in percent (%) or decibel (dB) with respect to the reference stimulus Φ_0. Since further studies of Weber's law showed an increase in JNDs for low reference stimuli near the absolute threshold, the additional parameter a was introduced. It is generally interpreted as sensory background noise in the perception process [56, Chap. 1], which is a similarity to the basic assumption of the ↪ SDT. The resulting change of the JND near the absolute thresholds is so large that a consideration in the design of technical systems is advisable.

It is generally agreed that the JND denotes the amount of stimulus change that is detected as greater, half the time. In the literature, one can find two different approaches to measure a JND in a psychophysical experiment. It has to be noted that these approaches do not necessarily measure the 50 % point of the psychometric function:

- Direct comparison of a reference stimulus Φ_0 with a test stimulus $\Phi_0 + \Delta\Phi$. The stimulus controlled by the psychometric procedure is necessarily $\Delta\Phi$, and test persons have to assess if the test stimulus is greater than the test stimulus. The JND is calculated according to the procedures calculation rule; the convergence probability has to be taken into account when interpreting and using the JND.
- According to [164, Chap. 5], the JND can also be determined using two points of a psychometric function as given in Eq. (2.6):

$$\text{JND} := \Phi(p_\Psi = 0.75) - \Phi(p_\Psi = 0.25) \tag{2.6}$$

This definition is useful if one cannot control the stimulus intensity freely during the test and has to measure a complete psychometric function with fixed stimuli (e.g., real objects with defined texture, curvature, or roughness) or with long adaptation times of the test person.

It may be noted that both approaches do not necessarily lead to the same numerical value of a JND. For certain classes of experiments, special terms for the differential thresholds have been coined. They are briefly described in the following:

Point of Subjective Equality (PSE) In experiments with a fundamental difference between test and reference stimulus in addition to stimulus change, the differential threshold with $p_{\Psi} = 0.5$ is also called *point of subjective equality* (*PSE*). At this stimulus configuration, the test person cannot discriminate the two stimuli. An example could be the assessment of the intensity of two vibrations that are coupled into the skin normally and laterally. The fundamental difference is the coupling direction and the intensities of both stimuli are adjusted in a way that the intensity is perceived as equal by the test person.

Successiveness Limen (SL) If two stimuli are presented to a test person, they are only perceived as two different stimuli if there is a certain time period between them. This time period is called *successiveness limen* (*SL*). For mechanical pulses, SL can be determined to about 5 ms, while direct stimulation of nerve fibers will exhibit an SL of about 50 ms [122, Chap. 4].

Two-point threshold The two-point threshold describes the distance between the application points of two stimuli that are needed to make this stimuli distinguishable from another. The smallest two-point thresholds can be found at the tongue and lips (< 1 mm for static stimuli). At the fingertip, thresholds of 1–2 mm can be found. Other body areas exhibit two-point thresholds of several centimeters as shown in Fig. 2.13 [81]. The *spatial resolution* is the reciprocal value of the two-point threshold.

Just Tolerable Difference (JTD) The just tolerable difference denotes the difference between two stimuli, which is differentiable but still tolerable for the test subject. It is also termed the *quality JND* and depends more on individual appraisal and judgment of the subjects than on the abilities of the sensory system. This measure can be used to determine system properties that are acceptable to a large number of users, as it is done in various other sensory modalities such as taste [35] or vision [108].

The knowledge of differential thresholds has a major meaning. JNDs give the range of signals and stimuli that cannot be distinguished by the user, i.e., a limit for the reproducibility of a system. Proper consideration of JNDs in product design will yield systems with good user ratings and minimized technical requirements as shown in [54].

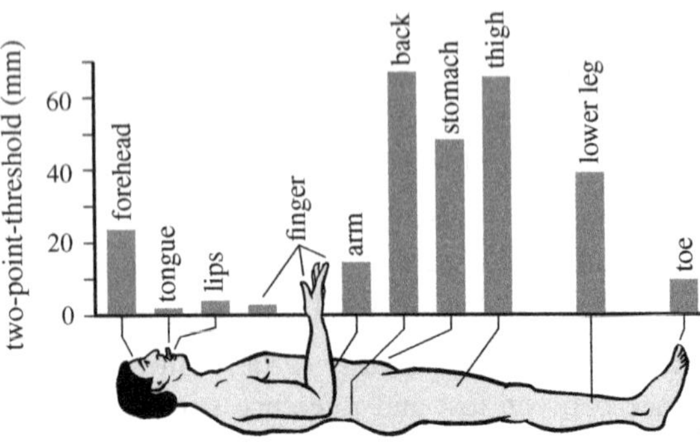

Fig. 2.13 Two-point threshold at various locations. Figure adapted from [81]

Description of Scaling Behavior

In the mid-nineteenth century, FECHNER formulated a relation between objectively measurable stimulus Φ and the subjective percept Ψ based on *Weber's law* in Eq. 2.5. He set the JND equal to a non-measurable increment of the subjective percept and integrated over several stimulus intensities defined by increments of the JND.[1] This leads to *Fechner's law* as given in Eq. 2.7:

$$\Psi = c \log \Phi \tag{2.7}$$

In Eq. (2.7), c is a constant depending on the investigated sensory system. However, *Fechner's law* is based on two assumptions that are rendered invalid in further studies: As shown above, *Weber's law* is not universally applicable for all stimulus intensities. Secondly, an increment as high as the current JND will not evoke an increment in perception [56, Chap. 1]. In the mid-twentieth century, S.S. STEVENS proposed the *power law*, a new formulation of the relation between objective stimuli and subjective percepts, based on experimental data that could not be explained by *Fechner's law*:

$$\Psi = c\Phi^a \tag{2.8}$$

In Eq. (2.8), c is a scaling parameter as well, which is often neglected in further analysis. The parameter a denotes the coupling between subjective perception and objective measurable stimuli and depends on the individual experiment. By logarithmization of Eq. (2.8), it can be calculated as the slope of the resulting straight line $\log \Psi = \log c + a \log \Phi$ [56, Chap. 13]. The *power law* can be summarized briefly as *"a constant percentage change in the stimulus produces a constant percentage change in the sensed effect"* [199, p. 16]. To analyze these changes, particular psychometric procedures can be used, as for example, found in [199]. Typical values in haptics include $a = 3.5$ for the intensity of electrical stimulation at the fingertip

[1] An elaborate derivation can be found in [56, Chap. 1].

(a 20 % increase will double the perceived intensity) and $a = 0.95$ for a vibration with a frequency of 60 Hz (i.e., a declining relation).

2.1.2.4 Factors Influencing Haptic Perception

From different studies of haptic perception, external influencing factors that affect absolute and differential thresholds are known. They originate in the properties of the different blocks given in Fig. 2.1. For the design of haptic systems, they have to be considered as disturbance variables or can be used to purposefully manipulate the usage conditions. An example may be the design of a grip or the control of a minimum or maximum grip force at the end-effector of a haptic system. The following list will give the technically relevant influencing factors.

Temperature Temperature will influence the mechanical properties of the skin [78]. Furthermore, perception channels exhibit a temperature dependance as given in Table 2.2. The absolute perception threshold is affected by temperature and the lowest thresholds are observed at about body temperature [16, 69]. This effect increases for higher frequencies [76, 217], denoting a higher temperature dependence of mechanoreceptors with greater receptive fields.

Age With increasing age, the perception capabilities of high-frequency vibration decrease. This is observed in different studies for finger and palm [52, 60, 64, 72, 215, 216]. The change in form and spatial distribution of mechanoreceptors, especially of Pacinian corpuscles, is deemed the cause for this effect.

Contact Area Because of the different receptive fields of mechanoreceptors, the contact area is an important influencing parameter on haptic perception. With small contact areas, the thresholds for high-frequency vibrations increase (higher intensities are needed for detection), while the lowest thresholds can only be measured with large contact areas about the size of a fingertip or greater.
Furthermore, not the density of mechanoreceptors but the absolute number is approximately constant among test persons. Therefore, the size of the hand is relevant for perception capabilities, and smaller hands will be more sensitive [179]. Since there is a (slight) correlation between sex and hand size, this is the reason for some contradictory studies on the dependency of haptic perception on the sex of the test person [64, 73, 214, 215, 222].

Other Factors Several other factors with influences on haptic perception thresholds such as the menstrual cycle [60, 69, 96], diseases such as *bulimia* or *anorexia nervosa* [79], skin moisture [218], and the influence of drinks and tobacco can be identified. In the design of haptic systems, these factors cannot be incorporated in a meaningful way, since they can neither be controlled nor be influenced in system design or usage.

2.1.2.5 What Do We Feel?

To investigate perception, an exact physical representation of the stimulus must be known. In auditory perception, this is sound pressure that will affect the eardrum and will be conducted via the middle ear to the nerve fibers in the cochlea and the organ of Corti. Visual perception is based on the detection of photons of a particular wavelength in the cones and rods in the retina.

In haptics, one will find different physical representations of stimuli, namely forces F and kinematic measures such as acceleration a, velocity v, or deflection d. The usage of a certain representation mainly depends on the purpose of the study or the system: Forces are sometimes easier to describe and measure because of their characteristics as a flux coordinate defined at a single point. Kinematic measures exhibit characteristics of a differential coordinate, i.e., they can only be measured in relation to a prior-defined reference. Many studies (especially of dynamic stimuli) are based on kinematic measures, since their definitions do not depend on the mechanical properties of the test person. GREENSPAN showed psychophysical measurements with less variation when stimuli were defined by kinematic measures compared to force [77].

However, there is evidence that humans do not only feel forces or kinematic measures. Perception is most likely based on the distribution of mechanical energy in the skin where the mechanoreceptors are located. This distribution cannot be described with reasonable effort in detail (although there are some attempts to FE modeling of the human skin [2, 37, 219, 220]); furthermore, it cannot be produced as a controlled stimuli for psychophysical experiments.

A common approach is to consider the human skin as a mechanical system whose properties are not changed by haptic interaction. This is supported by studies conducted by HOGAN who showed that a human limb can be modeled as a passive mechanical impedance for frequencies higher than the maximum frequency of human motion capabilities. In that case, forces and kinematic measures coupled with the skin are related via the mechanical impedance $\underline{z}_{\text{user}}$ according to Eq. (2.9)

$$\frac{\underline{F}}{\underline{v}} = \underline{z}_{\text{user}} \tag{2.9}$$

with $v = \frac{\mathrm{d}d}{\mathrm{d}t} = \int a\,\mathrm{d}t$. Applied to perception, this means that each force perception threshold could be calculated from other thresholds defined by kinematic measures via the mechanical impedance of the test person. This relation is used in a couple of studies [110, 111] to calculate force perception thresholds from deflection-based measurements. Own studies of the author used force-based measurements to experimentally prove the relation given in Eq. (2.9) [89].

One can therefore conclude that perception is based on the complex distribution of mechanical energy in the skin. For the design of haptic systems, a simplified consideration of the user as a source of mechanical energy with own mechanical parameters as given by mechanical impedance $\underline{z}_{\text{user}}$ is applicable. Furthermore, this model is also valid for the description of perception, linking perception parameters by mechanical impedance as well. Some important psychometric parameters are given in the following section of this chapter, and a detailed view on the modeling of the user is given in Chap. 3.

2.1.3 Characteristic Values of Haptic Perception

There are numerous studies investigating haptic perception. For the design process of haptic systems, the focus does not lie on single biological receptors but rather on the human's combined perception resulting from the sum of all tactile and kinaesthetic mechanoreceptors. As outlined in the following chapters, a dedicated investigation of perception thresholds is advisable for the selected grip configuration of a haptic system. This section gives some results of the most important ones, but is not complete. It is ordered according to the type of psychometric parameter. To interpret the results of different studies correctly, Fig. 2.14 gives some explanation of the anatomical terms for skin location and skeleton parts.

2.1.3.1 Absolute Thresholds

One of the most advanced studies of haptic perception was carried out by the group of GESCHEIDER ET AL. The most popular curve likely is the absolute perception

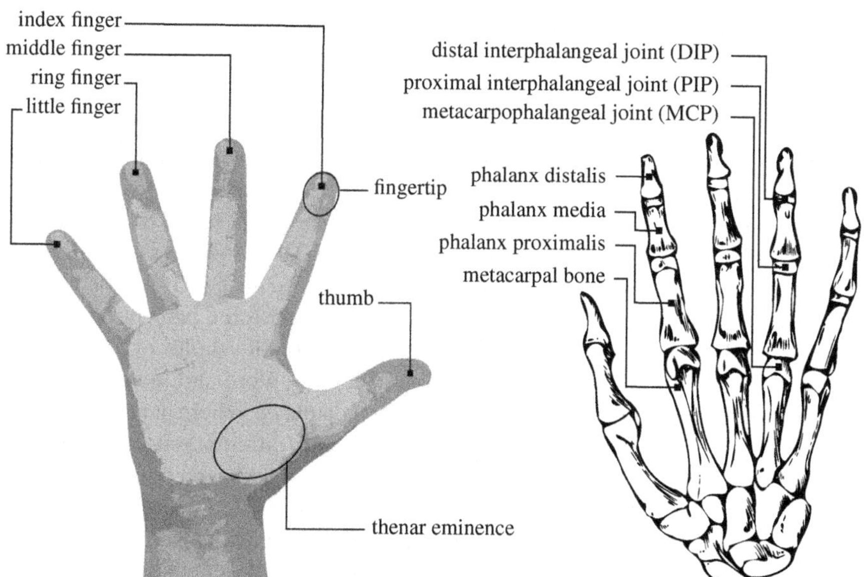

Fig. 2.14 Anatomical terms for skin areas and skeleton parts of the human hand

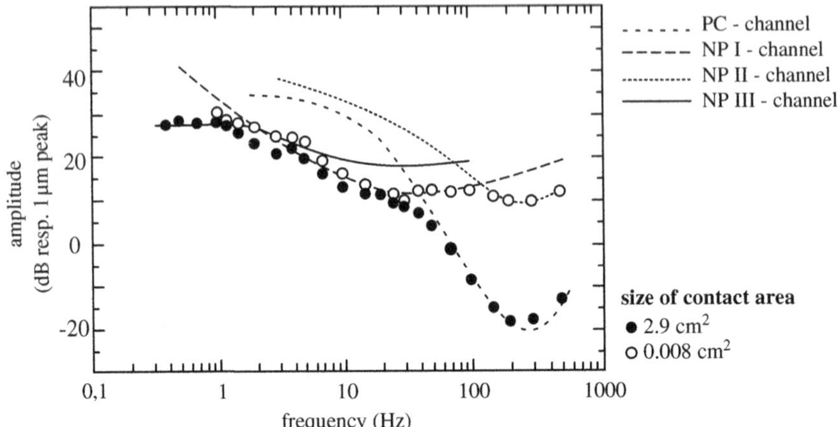

Fig. 2.15 Absolute threshold of tactile perception channels at the thenar eminence with respect to contact size. Measurements were taken with closed-loop velocity control of the stimuli. To address individual channels, combinations of frequencies, intensities, contact areas, and masking effects are employed. The psychometric procedure used converges at a detection probability of $p = 0.75$. The figure is adapted from [57]

threshold of vibrotactile stimuli defined by deflections of the skin at the thenar eminence as given in Fig. 2.15 [57]. Since the channel model arises in the work of this group, a lot of their studies deal with these channels and their properties. In Fig. 2.15, some properties of this model can be seen: The thresholds are influenced by receptive fields, and the highly sensitive RA-II receptors are only exited with large contact areas; in addition, the most sensitive channel is responsible for the detection of a stimulus. Other work include the investigation of the perception properties of the fingertip [67] and intensive studies of masking and summation properties [60].

Other relevant studies were conducted by ISRAR ET AL., investigating vibrotactile deflection thresholds of hands holding a stylus [110] and a sphere [111], some common grip configurations of haptic interfaces. They investigate seven frequencies in the range of 10–500 Hz with an adaptive staircase (1up-3 down progression rule) and a 3IFC paradigm and find absolute thresholds of 0.2–0.3 μm at 160 Hz. The studies include the calculation of the mechanical impedance and force perception thresholds as well. BRISBEN ET AL. investigated the perception thresholds of vibrotactile deflections tangential to the skin, a condition becoming more and more important when dealing with tactile feedback on touchscreen displays. Whole hand grasps and single digits were investigated with an adaptive staircase (different progression rules) and 2IFC and 3IFC paradigms. They additionally investigate perception thresholds for 40 and 300 Hz stimuli at different locations on the hand and with different contact areas. Newer studies by GLEESON ET AL. investigate the properties of several stimuli parameters such as velocity, acceleration, and total deflection [71] on the perception of shear stimuli. They found accuracy of direction perception depending on both speed and total displacement of the stimulus, with accuracy rates greater

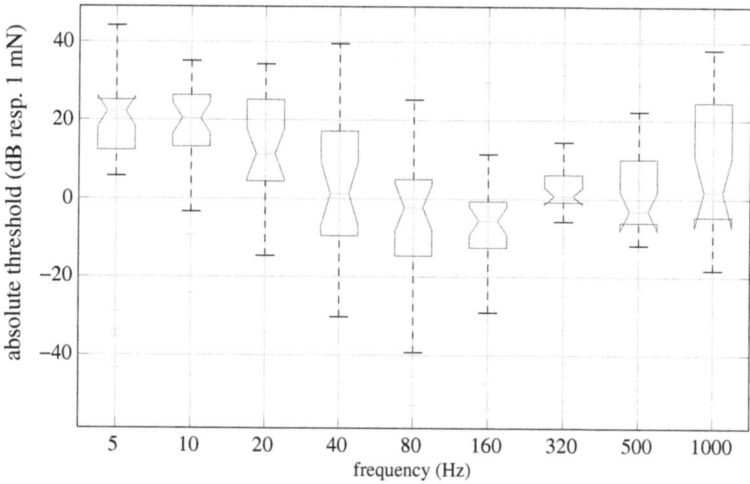

Fig. 2.16 Absolute force perception threshold based on experiments with 27 test persons, measured with a quasi-static preload of 1 N. Thresholds are obtained with an adaptive staircase procedure converging at a detection probability of 0.707 with a 3IFC paradigm. Data are given as boxplot, since not all data for each frequency are normal distributed. The *boxplot* denotes the median (*horizontal line*), the interquartile range (IQR, *closed box* defined by the 0.25 and 0.75 quantile), data range (*dotted line*), and outliers (data points with more than 1.5 IQRs from the 0.25 to 0.75 quantile). The *indentation* denotes the confidence interval of the median ($\alpha = 0.05$). Data taken from [87, 89]

than 95 % occurring at tangential displacement of 0.2 mm and a displacement speed of 1 mms^{-1}. The study further includes analysis of priming and learning effects and the application to skin stretch-based communication devices.

One of the most important effects on haptic perception originates in the size of the contact area. All of the above-mentioned studies show lower perception thresholds for frequencies around 200 Hz with larger contact areas. However, this effect seems to be limited by the minimum area required to arouse mechanoreceptors in the PC channel, which is probably about 3 cm^2, corresponding to a contactor diameter of about 20 mm. When more than one finger is involved in the interaction, the authors of [136] did not find a summation effect of thresholds.

Regarding the perception of forces, the corresponding absolute thresholds can be calculated according to Eq. (2.9). There are few studies dedicated to the absolute perception of forces. THORNBURY AND MISTRETTA investigate the sensitivity to tactile forces applied by a modified version of von Frey filaments. They find a significant influence of age on the absolute threshold that is most likely related to the decrease in mechanoreceptor density. Young subjects (mean age 31 years) exhibit absolute thresholds of 140 μN, while older subjects have higher thresholds of about 660 μN, measured with a staircase method, constant stimuli intensities, and a 2IFC paradigm. Since the stimuli were applied manually by the experimenter, application dynamics cannot be determined from the study but probably contribute to the very low reported thresholds. ABBINK AND VAN DER HELM investigated absolute force

Table 2.4 Selected absolute thresholds of the human hand

Base item	Threshold	Body part	Value	Source
Static stimuli	Skin deformation	Fingertip	10 μm[a]	[127]
	Two-point threshold	Fingertip	2–3 mm[b]	[23, 127]
		Palm	10–11 mm	[127, 194]
	Force	Fingertip	0.8 mN	[23]
		Palm	1.5 mN	[23]
	Pressure	Fingertip	$0.2\,N/cm^2$	[194]
Dynamic stimuli	Frequency, upper limit	Finger (tactile)	5–10 kHz	[21, 23]
		Whole body (kinaesthetic)	20–30 Hz	[21]
	Maximum sensitivity	Fingertip, palm	At 200–300 Hz[c]	[60]
	Amplitude	Fingertip, palm	0.1–0.2 μm (normal stimulation) at 200–300 Hz[d]	[60]
		Whole hand, grasping	0.2–0.3 μm at 150–200 Hz (tangential stimulation)[e]	[20]
		Sphere, stylus	0.2–0.3 μm at 160 Hz[f]	[110, 111]
	Two-point threshold	Fingertip	0.8 mm[g]	[149]

[a] If movement is permitted, isolated surface structures of 0.85 μm height can be perceived [142, 150]. If surface roughness is to be detected, stimuli as low as 0.06 μm are perceived [150]
[b] The two-point threshold decreases, if the two stimuli are presented shortly after another. A position change of a stimulus can be resolved spatially up to ten times better than the static two-point threshold [127]
[c] The perception threshold is strongly dependent on the vibration frequency, the location of the stimulus, and the size of the contact area [60]
[d] Amplitudes larger than 0.1 mm are perceived as annoying at the fingertips [105]. A stimulation with constant frequency and amplitude results in desensitization, increasing up to a numb feeling, which may last several minutes after the end of the stimulation [27, 128]
[e] Whole hand grasping a cylinder with a diameter of 32 mm. Vibrations were created along the cylinder axis
[f] Sphere with a diameter of 2 inches was grasped with the *phalanx distalis* of all fingers, and the stylus is taken from a PHANToM haptic interface and held with three fingers
[g] A correct detection probability of at least 0.75 was measured for 12 frequencies, ranging from 1 to 560 Hz in 22 subjects

perceptions on the foot with different footwear (socks, sneaker, bowling shoe) for low-frequency stimuli (<1 Hz) and a static preload of 25 N. They find the lowest perception thresholds of 8 N in the sock condition, whereas the perception threshold is defined with a detection probability greater than 0.98. Also motivated by the small number of studies, the author of this chapter measured perception thresholds for vibrotactile forces up to 1,000 Hz as shown in Fig. 2.16.

In summary, one can find numerous studies determining absolute thresholds for the perception of stimuli defined by deflections. Fewer studies are conducted on

Table 2.5 Relevant parameters and results of studies of dynamic force JNDs

Source	Reference stimulus (N)	Stimulus frequency	Interaction condition[a]	JND[b] (%)
[4]	2.25	Not given	Active	10
[40]	2.5; 3.5	Not given	Active	12
[98]	Not applicable[c]	Up to 200 Hz (estimated[c])	Active	10
[102]	0.3; 0.5; 1; 2.5	Quasi-static	Passive	43–15
[181]	1.5	Quasi-static	Passive	10
[180]	1; 2	100–500 Hz (discrete frequencies)	Passive	23–13

Table based on [160]

[a] In active conditions, test subjects were required to apply movement on their own, while in passive conditions, only the measurement setup exerts forces on the subject

[b] Ordering according to reference force ordering

[c] JNDs are based on an experiment, where subjects could interact freely with a custom haptic interface described in [98]

the absolute perception of forces. Table 2.4 summarizes some values of absolute perception thresholds for the human hand.

2.1.3.2 Differential Thresholds

For haptics, several studies furnish evidence of the applicability of Weber's law as stated in Eq. (2.5). GESCHEIDER ET AL. [62] as well as VERRILLO ET AL. [213] measure ↪ JNDs of 1–3 dB for deflection-defined stimuli with reference stimuli of 5–40 dB above absolute threshold for frequency ranges exceeding 250 Hz. The measurements of GESCHEIDER ET AL. are based on broadband and single-frequency stimulus excitation. They show an independence of channels for the JND, whereas no fully constant JND was determined for high reference levels. This is addressed as *"a near miss to Weber's law"* by the authors [62], but this observation should not have a significant impact on the design of haptic systems.

Regarding the JND of forces, several studies were conducted with an active exertion of forces by the test person. JONES measures JNDs of about 7 % from matching force experiments of the elbow flexor muscles [124], a value that is confirmed by PANG ET AL. [176]. However, one cannot determine the measurement dynamics from the experimental setup, based on Fig. 1.7; a maximum bandwidth of 10–15 Hz seems to be likely. From other studies evaluating the perception of direction and perception-inspired compression algorithms (see Sect. 2.4.4), estimations of the JND for forces can be made. This is summarized in Table 2.5. All studies show JNDs over 10 % for reference stimuli well above the absolute threshold and increasing JNDs for reference stimuli near the absolute threshold.

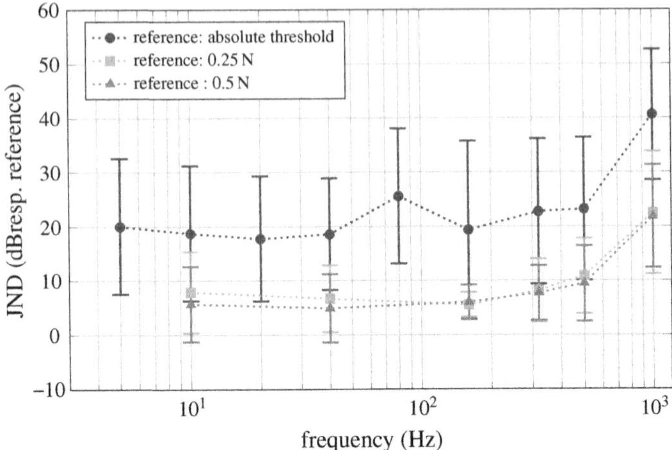

Fig. 2.17 Just noticeable differences of dynamic force. JNDs were calculated with an adaptive staircase procedure converging at a detection probability of 0.707 and a 3IFC paradigm from studies conducted with 29 test persons (absolute threshold reference) and 36 test persons (0.25 and 0.5 N reference conditions), respectively. The test setup is described in [86], and a static preload of 1 N was used. Data taken from [87, 88]

Own studies of the author of this chapter evaluated the JND for dynamic forces in the range from 5 to 1,000 Hz. As reference stimuli, the individual perception threshold and fixed values of 0.25 and 0.5 N were used. The results are given in Fig. 2.17. They show no channel dependence (despite a significantly higher value for the JND at 1,000 Hz) and affirm the increasing JND for reference stimuli near the absolute threshold. However, with about 4–8 dB for frequencies less than 1,000 Hz, the JND in the 0.25 and 0.5 N condition is higher than the previously reported values.

Jones and Hunter investigated the perception of stiffness and viscosity and found JNDs of 23 % for stiffness [120] and 34 % for viscosity [121] with a matching procedure using both forearms with stimuli generated by linear motors. The JND for stiffness is similar to that given in other studies as reported in [32, 177]. Further differential thresholds for the perception of haptic measures by the human hand are given in Table 2.6.

2.1.3.3 Object Properties

The properties of arbitrary objects are closely related to the interaction primitives. Typical exploration techniques to detect object properties are dealt with in the following section. Despite the basic perception of form, size, texture, hardness, and weight of an object, there are a couple of other properties relevant to the design of haptic systems. Bergmann Tiest reviews a large number of studies regarding the material properties of roughness, compliance, temperature, and slipperiness. The results are relevant for the design of devices to display such properties, and the rep-

Table 2.6 Selected differential thresholds of the human hand

Base item	Threshold	Body part	Value	Source
Static stimuli	Force	Finger span	5–10 %	[177]
	Deflection	Fingertip	10–25 %	[14]
	Length	Finger span	3–10 %	[177]
	Compliance	Finger span	5–15 %	[177]
	Pressure	Wrist	4–19 % [a]	[202]
	Torque	Thumb, index finger	13 %	[32, 113]
	Position resolution (kinaesthetic) [b]	finger joint	2.5°	[202]
		Wrist, elbow	2°	[202]
		Shoulder	0.8°	[202]
	Force direction	Pen-hold posture[c]	25–35°	[6, 204]
Dynamic stimuli	Vibration amplitude at 160 Hz	fingertip	16 %	[36]
	Frequency resolution	Fingertip(tactile)	8–10 % [d]	[14]
	successiveness limen	mechanoreceptor property	5 ms [e]	[23]

[a] Experiment was made with a reference pressure of 1.8 N/cm^2 at the dorsal side of the wrist. JND increased strongly with reduced contact area: 4.4 % at 5.06 cm^2 and 18.8 % at 1.27 cm^2 [202]
[b] Test subject's limbs were positioned by the experimenter with no active movement involved
[c] A PHANToM haptic interface was used in both studies [d] The capability to differ stimuli is reduced after 320 Hz [21]
[e] If one has to decide which of two stimuli was applied first, a minimum time of 20 ms has to be between the onset of the two stimuli [23]

resentation of compliance is especially relevant for interaction with virtual realities. Key points of the analysis are outlined in the following based on [8], whereas primary sources and some other references are cited as well. KLATZKY ET AL. also review the perception of object properties and algorithms to render these properties to engineering applications [138]. The work of SAMUR, summarizing several studies on the perception of object properties, could be of further interest [191].

Roughness Roughness is one of the most studied object properties in haptic perception. The perception of roughness is based on uneven pressure distribution on the skin surface for static touch conditions and the vibrations arising when stroking a surface or object dynamically. It was shown that finer textures with particle sizes smaller than 100 μm can only be detected in dynamic touch conditions, while coarser textures can be detected in static conditions, too. Active and passive touch conditions have no effect on the perceived roughness. This is called the *duplex theory of roughness perception* [100]. However, not only sensitive bandwidth and touch condition have influence on the human ability to perceive roughness. Other studies found influences of the contact force, other stimuli in the tested set, and the friction between surface and skin. Regarding differential thresholds, KNOWLES ET AL. found JNDs of 10–20 % for friction in rotary knobs

[141], and PROVANCHER ET AL. recorded JNDs of 19–28 % for sliding virtual blocks against each other [183].
Scaling experiments showed that roughness can be identified as the opposite to smoothness. In similar experiments, no effect of visual cues was found and a power function exponent (see Eq. 2.8) of 1.5 was measured. In a nutshell, the perception of roughness appears to be a complex ravel of not only material properties but also interaction conditions such as friction and contact force. This makes the modeling of roughness challenging; on the other hand, there are a vast number of possibilities to display roughness properties in technical systems [28].

Compliance This property describes the mechanical reaction of a material to an external force. It can be described by Young's modulus or—technically more relevant—the stiffness of an object that combines material and geometric properties of an object as shown further in Eq. (2.10). When evaluating physical stiffness with the perceived compliance, a power function exponent of 0.8 was calculated and softness and hardness were identified as opposites. For the perception of softness, cutaneous and kinaesthetic cues are used, while cutaneous information is both necessary and sufficient. Studies by BERGMANN TIEST AND KAPPERS determined that soft materials were mostly judged by the stiffness information, i.e., the relationship of material deformation with exerted force, while harder stimuli are judged by the surface deformation [9].
Several other studies show that the perception of hardness, i.e., the converse of compliance of an object, is better modeled by the relation of the temporal change of forces compared to the penetration velocity than by the normally employed relation of force with velocity [151]. This has to be considered in the rendering of such objects and is therefore dealt with in Chap. 12. To render a haptic contact perceived as undeformable, necessary stiffnesses from 2.45 Nm^{-1} [202] to 0.2Nm^{-1} [148] are reported.

Slipperiness Slipperiness is not researched very deeply until now. It is physically strongly related to friction and roughness. The detection of slipperiness is important for the adjustment of grip forces when interacting with objects. While an accurate perception of slipperiness requires some relative movement, microslip movements of a grasped object that are sensed with cutaneous receptors are made responsible for the adjustment of grip forces [25]. Studies show forces just 10 % higher than the minimum force needed to prevent slip. The adjustment occurs with a reaction time of 80–100 ms, which is faster than a deliberate adjustment [104].

Viscosity Not necessarily an object property, the ratio of shear stress to shear rate is relevant for virtual representation of fluids and viscoelastic materials. Based on real viscous fluids stirred with the finger and a wooden spatula, Weber fractions of about 0.3 were determined for high viscosities with increasing values for low viscosities [10]. Regarding scaling parameters, power function exponents for stirring silicone fluids of 0.42 are reported [199, Chap. 1].

Curvature While curvature itself is not necessarily relevant for the design of haptic systems, the detection capabilities of humans are quite astonishing. In [75], subjects were able to report a curvature with a base-to-peak height of just 90 μm on a strip of length 20 mm. However, researchers suggest a measure of base-to-peak height in relation to half the strip length to generate a robust measure for curvature perception that can be interpreted as the perceivable gradient. This measure leads to a unitless parameter of value 0.09. Differential thresholds are reported to be about 10 % for convex curvatures with radii range starting from about 3 mm [74]. In the same study, convex curvatures of radius 204 mm could discriminate from flat surfaces, and for concave curvatures, a threshold of 185 mm could be assessed with a detection probability of 0.75.

Temperature However, not only an object property, but also basic properties of temperature perception are summarized here, partly based on [122]. Humans can detect intensity differences in warming pulses as low as 0.1 % (base temperature of 34 °C, warming pulse with base intensity of 6 °C) [118]. Changes of 0.16 °C for warmth and 0.12 °C for cold from a base temperature of 33 °C can be detected at the fingertip and are still lower for the thenar eminence. When skin temperature changes slowly with less than 0.1 $°\mathrm{C\,s^{-1}}$, changes of up to 6 °C in a zone of 30–36 °C cannot be detected. More rapid changes will make small changes perceivable. The technical use of temperature perception is limited by a temperature of 44 °C, where damage is done to the human skin [83].
Perceptually more relevant is the thermal energy transfer from skin into the object. Humans are able to discriminate a change in heat transfer of about 35–45 %. Because of different thermal conductivity and specific heat capacity, different materials can be identified by static touch alone. On this heat transfer mechanism, modeling, rendering, and displaying thermal information is discussed in a number of studies cited in [8]. For technical applications, JONES AND HO discuss known models and the implications of the design of thermal displays [125].

Despite these general properties, there are a vast number of more complex object properties that arise largely in the interpretation of the user. It is difficult to find clear technological terms for these interpretations. In the literature, one can find an approach to describe this interpretation: Users are asked to rate objects on different scales called semantic differentials. Based on these ratings, a multidimensional scaling procedure will identify similar scales [19]. Regarding surface properties, HOLLINS showed three general dimensions perceived by a user: rough ↔ smooth, hard ↔ soft, and a third, not exactly definable dimension (probably elastic compressibility) [101]. This approach is also successfully used in the evaluation of passive haptic interfaces [189]. The accurate display of surface properties is still a relevant topic in haptic system design. Readers interested in this topic are referred to the work of WIERTLEWSKI [225] and the results of the HAPTEX PROJECT [82].

2.1.3.4 Scaling Parameters

Another important psychophysical measure is the interpretation of the intensity of different stimuli by the user, normally termed scaling. Particularly for tactile applications, the perception of the intensity of normally and laterally applied stimuli is of importance. One of the first comparisons of the perception of tangential and normal stimuli was carried out by BIGGS AND SRINIVASAN . They found a 1.7–3.3 times higher sensitivity for tangential displacements compared to normal stimulation at both the forearm and fingerpads. They conclude that tangential displacement is the better choice for peak-displacement-limited actuators, while normal displacement should be chosen for actuators limited in peak forces. One notes that this is caused not by higher sensitivity but by differences in the mechanical impedance for normal and tangential stimuli [11].

Classical psychophysical evaluation of scaling behavior was reported by HUGONY in the last century. Figure 2.18 shows the result as curves of equally perceived intensity, denoting stimulus amplitudes for different frequencies perceived as equally intense by the user. Such curves can be applied to generate targeted intensity changes of complex stimuli: A slight amplitude increase for low-frequency components will evoke the same perceived intensity than a much larger amplitude change of mid- and high-frequency components. This behavior can be optimized with regard to the energy consumption of the actuators in a haptic system.

The results further imply perception dynamics as high as 50 dB (defined as difference between absolute threshold and nuisance level) that are confirmed by newer studies like [213], stating a dynamic of 55 dB. Other results from the study imply an amplitude JND of 10–25 % and a JND of 8–10 % for frequency. This goes along well with the above-reported results.

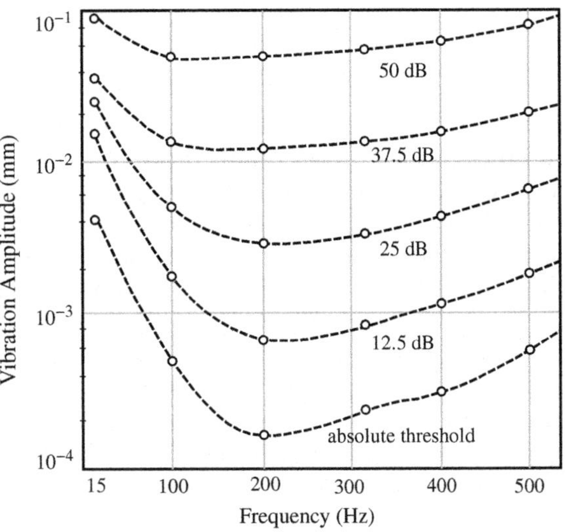

Fig. 2.18 Curves of equal perceived intensity at the fingertip. The *lowest curve* denotes the measured absolute threshold, while the *upper curve* is perceived as nuisance. Figure is adapted from [105]

Table 2.7 Possible variables in haptic perception experiments

Type	Examples
Dependent variables	Psychophysical properties
Independent variables	Contact area, contact force, masking stimuli, other treatments
Controllable variables	Skin moisture, skin temperature, test person's age, systematic errors of the setup
Confounding variables	Fatigue, test person's experience, multimodal interaction, unacquainted factors

2.1.3.5 Some Words on the Quality of Studies of Haptic Perception

Studies of haptic perception are conducted by scientists from various backgrounds. Depending on the formal training and customs in different disciplines, the author experienced a large variety of qualities of haptic perception studies. Based on his own training in measurement and instrumentation and his own studies dealing with haptic perception, the following hints are given on how to assess the quality of a perception study for further use in the design of haptic systems.

Measurement Goal and Hypothesis A hypothesis should be stated for each experiment. Hypotheses formulated in terms of well-established psychophysical properties like those described above (see Sect. 2.1) are preferable for later comparisons of the study results. Further, external influencing variables should be considered in hypothesis formulation. In general, one can differentiate dependent, independent, controllable, and confounding variables as shown in Table 2.7 for investigations of haptic perception.

Independent variables are addressed in the formulation of the hypotheses and are varied during the experiment. Depending on the hypothesis, known influencing variables can be considered as independent or controllable. Controllable variables have a known effect on the result of the experiment and should therefore be measured or closely watched. Possible means are keeping the test setup at constant temperature and a preselection of test subjects based on age, body length, weight, etc. Confounding variables contribute to the measurement error and cannot be completely taken care of.

Measurement Setup and Errors The measurement setup of a haptic perception study should be well fit for the investigated perception parameter or the intended result. This means, for example, that all parts of the measurement setup should exhibit adequate frequency response, rated ranges, and sampling rates for the expected values in the experiment.

The design and construction of the setup should be neat to prevent unwanted effects and errors such as electromagnetic disturbance by other equipment in the laboratory. Setups should favorably be fully automated to prevent errors induced by the experimenter.

The setup should be documented including all procedures and measurements of systematic and random errors. Based on a model of the measurement setup and its components, an analysis of systematic error propagation as well as a documented calibration of the setup and its components with known input signals and a null signal should be conducted. Longtime stability, reproducibility, external influences, and random errors should be analyzed and documented. Application of standardized methods such as the ↪ Guide to the Expression of Uncertainty in Measurement (GUM) [109] is preferable. If possible, systematic errors should be corrected.

Measurement Procedure There should be a considerable number of test persons in a study. A dedicated statistical analysis with less than 15 subjects seems to be questionable and should at least be explained in the study, explicitly addressing the type II error of the experiment design [132]. Larger numbers of 30 and more subjects are advisable.
Regarding psychophysical procedures, a previously reported and favorably adaptive procedure should be used. Newer studies should only use non-adaptive procedures in case of non-changeable stimuli (like gratings on real objects). The report of pretests and the impact on the design of the final study should be discussed in the documentation. Interactions with other sensual modalities such as vision and audition should be kept in mind and eventually controlled, for example, by earplugs and masking noise.

Analysis Datasets not included in the analysis should be addressed, and the criteria for this decision must be reported. All results should be analyzed statistically and the location parameters of the results should be given (i.e., mean and standard error for normal distributed results, and median and IQR for not normal distributed results). If external parameters are included in the study, an analysis of variance (ANOVA) as well as post hoc tests for significance of treatment group averages should be conducted and reported. If other analysis tools, such as confusion matrix, are used effort should be put into a statistical analysis of the significance of the result. Errors in the measurement setup should be addressed in the analysis.
If possible, results should be compared to other studies with similar setups and intention. When large differences occur, a detailed discussion of these differences and suggestions for further studies is advisable. To enable further studies based on the experiment results, test results for all effects (not only the significant ones) should be reported, as they can be used to determine effect sizes (useful for sample size calculations, see [132]) and to conduct metastudies [170].

If all of the above hints are considered, most conference proceedings would not report results but only measurement setups and their characterization. However, keeping the criteria for good measurement setups in mind will improve the quality of results and broaden the usage possibilities of the study results.

2.1.4 Further Aspects of Haptic Perception

Despite the classic psychophysical questions *(detection, discrimination, identification, scaling)*, there are a couple of other aspects relevant to the design of haptic systems. Some of these are discussed briefly in the following.

2.1.4.1 Effects of Multiple Stimulation

When more than one stimulus is applied in close temporal or spatial proximity to the first stimulus, several effects of multiple stimulation are known. The following list is based on VERRILLO [213]:

Masking This effect describes the decline of the detection ability of a stimulus when an additional, disturbing stimulus, the so-called masker, is present at the same location in temporal proximity. Masking can occur when masker stimuli are present before, while, and after the actual stimulus is presented. The masking properties depend on frequency and amplitude of the masker as well as on the age of the test person and the receptor channel involved. When the masker is presented right before the test stimulus (<1 s), the amount of masking depends on the time offset between masker and stimulus [166, 206], which is specific for each receptor channel. If a dedicated masker is used, specific receptor channels can be addressed. This is an important procedure to investigate properties of individual channels [59]. Masking finds application in the perception-based compression of data streams, and for acoustics, this is one of the main elements of the MP3 format [169].

Enhancement This effect occurs when a conditioning stimulus causes a stimulus in temporal succession to appear to be of greater intensity.

Summation When two or more stimuli are presented closely in time, the combination of the sensation magnitude is described as summation.

Suppression This effect is basically a masking effect, when both stimuli are presented at different locations.

For haptics, especially masking effects were investigated mainly by the group of GESCHEIDER ET AL. [58, 59, 61, 63, 65, 110, 206]. Studies of other effects are not known to the author. At the moment these multiple stimulation effects have to be considered as side effects in haptic interaction. Except for the analysis of receptor channels, there is no direct use of one known to the author.

2.1.4.2 Linearity of Haptic Perception

Recent studies imply that channels of haptic perception not only have independent thresholds [7], but also resemble a linear system. CHOLEWIAK ET AL. investigated

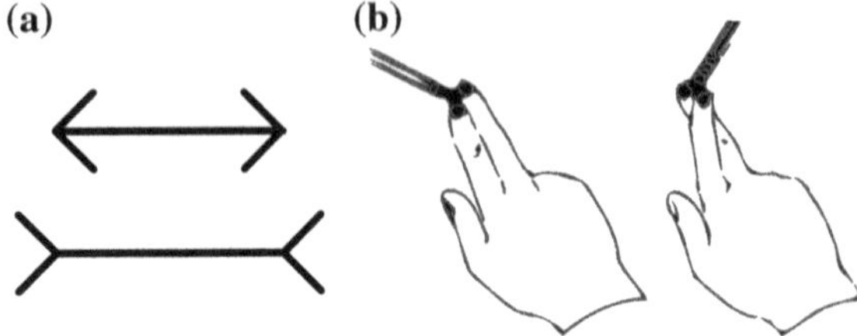

Fig. 2.19 Examples of haptic illusions **a** MÜLLER–LYER *illusion*, **b** ARISTOTELES *illusion*

spatially displayed gratings and found the necessity for each spatial frequency harmonic to be higher than the perception threshold at that frequency to be perceived by the user [33]. These results allow to consider error margins and detection thresholds independently for each frequency in the design of haptic systems [190].

A first application of this property of haptic perception was presented by ALLERKAMP ET AL. in the design of a haptic system to describe surface properties of textiles: Analogous to the spectral decomposition of an arbitrary color into red, green, and blue, textures were analyzed to be represented by two dedicated vibration frequencies for single receptor types [3]. This approach minimizes the hardware and data storage effort to present complex surface properties.

2.1.4.3 Anisotropy of Haptic Perception

Despite the above-mentioned differences in the scaling of lateral and tangential stimuli on the skin, there is also an anisotropy of kinaesthetic perception and interaction capabilities [1, 40, 130]. The perception and control of proximal movements (toward the body) are worse than movements in distal direction (away from the body). This property can be of meaning in the ergonomic design of workplaces with haptic interfaces and in tests and evaluations based on Cartesian coordinates.

2.1.4.4 Fooling the Sense of Touch

In addition to acoustics and vision, there are a couple of haptic illusions. They are generated by anatomical properties, neural processing, or misinterpretation of percepts such as a conflict of visual and haptic perception [6]. Since many visual illusions can be found in haptics too, and because of the similar neural processing and interpretation mechanisms, an explanation analogous to the visual system is anticipated [55]. As HAYWARD [91] puts it, "Perceptual illusions have furnished considerable material for study and amusement": Two examples of basic haptic illusions are given in Fig. 2.19. The MÜLLER–LYER illusion on the left side is borrowed from visual perception, but can also be proved for haptic stimuli. Both lines are perceived as different lengths because of the arrow heads, even if they have the same length. The

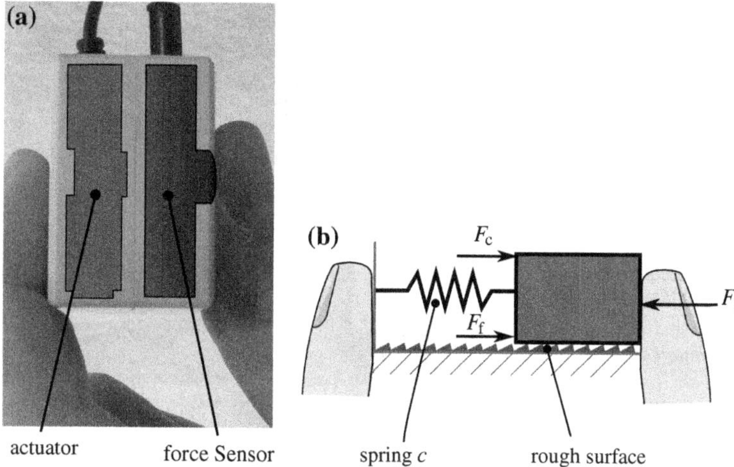

Fig. 2.20 KOOBOH **a** outer form and internal components and **b** internal system model. Picture courtesy by J. Kildal, *Nokia Research Center*, Espoo, FIN

ARISTOTELES illusion can be reproduced easily by the reader: Touching an object like a pencil with crossed fingers will evoke the illusion of two objects. If a wall is touched instead of an object, a straight wall will be perceived as a corner and vice versa. Further illusions can be found in the works of HAYWARD and LEDERMAN [91, 155].

Example: Kooboh

An application of haptic illusions in the design of haptic systems was presented by KILDAL in 2012 [134]. KOOBOH consists of a solid, non-deformable box with an integrated force sensor and a vibration actuator as shown in Fig. 2.20. The control software simulates an internal system model containing a spring connected with a (massless) object sliding on a rough surface.

The user applies a force F_a to the system, normally resulting in a deflection $d = \frac{F_a}{c}$ of the spring c. When the object is moved because of the applied force, a frictional force F_f would be generated, depending on the texture of the rough surface and the position of the object. Since the box is non-deformable, the reaction of the (virtual) spring cannot be felt. But because the applied force is measured by the force sensor, the theoretical deflection of the object and the resulting friction force F_f can be calculated. Depending on the structure of the rough surface, F_f will exhibit periodical, high-frequency contents that can be displayed by the integrated actuator. The user interprets these two contradictory percepts as a fully functional model as shown in Fig. 2.20, efficiently neglecting that the system does not move.

Pseudo-Haptic Feedback

A important technical application of another kind of haptic illusion is the use of disagreeing information about the visual and haptic channel. Termed "pseudo-

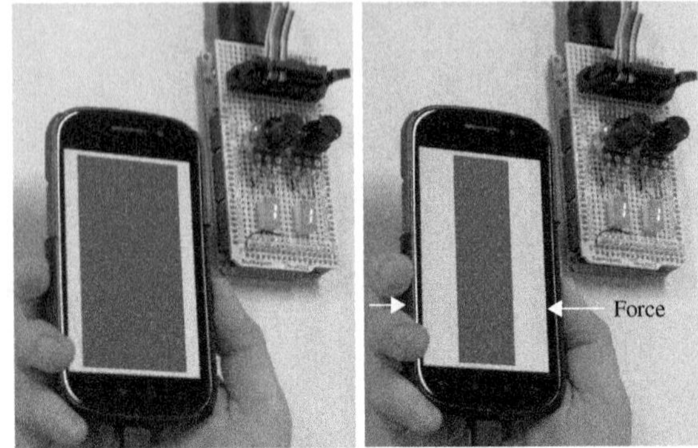

Fig. 2.21 Pseudo-haptic feedback in a mobile application [135]. Force exerted on the mobile device by the user is measured with pressure sensors. Based on this force, the deformation on the screen is calculated based on a virtual stiffness, leading to the impression of a compliant device. Pictures courtesy by T. Nojima, *University of Electro-Communication*, Tokyo, JP

haptic feedback," it is used in virtual environments to simulate properties such as stiffness, texture, or mass with limited or distorted haptic feedback and accurate visual feedback [152, 184]. A simple example is given by KIMURA ET AL. in [135] as shown in Fig. 2.21. A visual representation of a spring is displayed on a mobile phone equipped with a force sensor. The deformation of the visual representation depends on the force applied and the virtual stiffness of the displayed spring. Changing the virtual stiffness leads to a different visual representation and a feeling of different springs, although the user will always press the unchanged mobile phone case.

2.1.4.5 Haptic Icons and Categorized Information

All of the above is based on continuous stimuli and their perception. Another important aspect is the perception of categorized information that comes to use mainly in communication applications. Probably the most prominent example is the vibration alarm on a smart phone that can be configured with different patterns for signaling a message or a call. Several groups have investigated the basic properties of such haptic icons (sometimes also called tactons or hapticons) [22, 47, 163]. They found different combinations out of waveform, frequency, pattern, and spatial location suitable to create a set of distinguishable haptic icons based on multidimensional scaling analysis.

The use of categorized information in haptic systems introduces another measure of human perception, i.e., information transfer (IT) [203]. This measure describes how much distinguishable information can be displayed with haptic signals defined by combinations of the above-mentioned signal properties. However, it is no pure measure of perception, but also depends on the haptic system used. Because of this, it qualifies as an evaluation measure for haptic communication systems as detailed in Chap. 13. Reported information transfer ranges from 1.4 to 1.5 bits for the differentiation of force magnitude and stiffness [32] up to 12 bits for multi-axis systems, especially designed for haptic communications of deaf–blind people [205, 206].

2.2 Concepts of Interaction

In daily life, only the least haptic interactions of man with the environment can be classified as solely passive, which are purely passive perception procedures. Most interactions are a combination of motion and perception to implement a prior-defined intention. For the design of haptic systems, generally agreed on terms are needed to describe the intended functions of a system. In this section, some common approaches for this purpose are described. The section ends with a list of motion capabilities of the human locomotor system.

The taxonomy of haptic interaction by SAMUR, as given in Sect. 1.2.2, is one possibility. It was developed for evaluation of systems interacting with virtual environments and is therefore most suitable for their description. Other interactions can be described by combinations of the taxonomy elements as well, but lack intuition when describing everyday interactions. Stepping slightly away from the technical basis of SAMUR'S taxonomy and turning toward the functional meaning of haptic interaction for man and his environment, one will find the exploration theory of LEDERMAN and KLATZKY outlined in the following section. Further concepts such as active touch and passive touch as well as gestures are described, which are commonly used as input modality on touchscreens and other hardware with similar functionality.

2.2.1 Haptic Exploration of Objects

One of the most important tasks of haptic interaction is the exploration of unknown objects to assess their properties and usefulness. Not only tactile information, but also kinaesthetic perception contributes to these assessments. One of the most relevant sources for the evaluation of surfaces is the relative movement between the skin and the object.

In [157], LEDERMAN AND KLATZKY identify different exploratory procedures that are used to investigate unknown objects. Figure 2.22 shows the six most important procedures [154]. Table 2.8 gives an insight into costs and benefits when assessing certain object properties.

2.2.2 Active and Passive Touch

The above-described combination of movement action and perception is of such fundamental meaning that two terms have been established to describe such interaction.

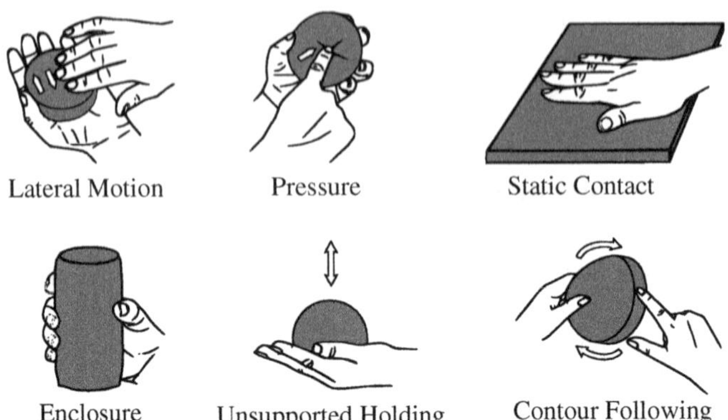

Fig. 2.22 Important exploratory procedures. Figure adapted from [154]

Table 2.8 Correlation of exploratory procedures to ascertainable object properties according to [122]

	Texture	Hardness	Temp.	Weight	Vol.	mF	eF	Duration (s)	Active DoF
Lateral motion	■	□	□					3	2
Pressure	□	■	□					2	1
Static contact	□		■		□	□		<1	0
Unsupported holding				■	□	□		2	2
Enclosure	□			□	■	■		2	1
Contour following	□			□	□	□	■	11	3

■ Denotes properties that can be asserted optimally by the exploration technique, □ Denotes properties that are asserted in a sufficient way
Used abbreviations: temp. temperature, *vol.* volume, *mF* macroscopic form, *eF* exact form

Definition *Active Touch* Active touch describes the interaction with an object where a relative movement between user and object is controlled by the user.

Definition *Passive Touch* Passive touch describes the interaction with an object when relative movement is induced by external means, for example by the experimental setup.

Both conditions can be summarized as *dynamic touch*, while the touch of objects without a relative movement is defined as *static touch* [100]. This differentiation is indeed seldom used. Active touch is generally considered superior to passive touch in its performance. LEDERMAN and KLATZKY attribute this to the different focus of the observer [154, p. 1439]:

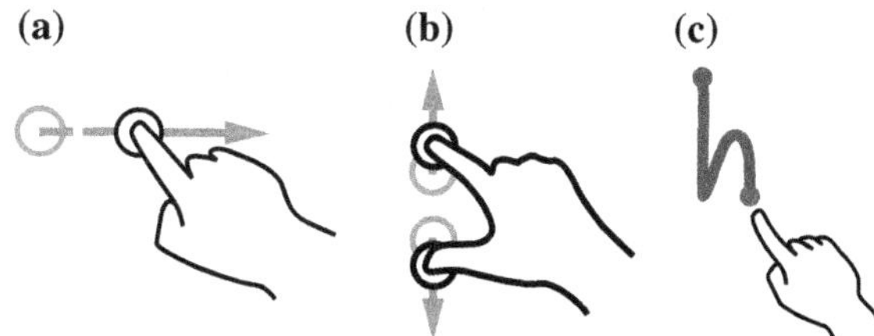

Fig. 2.23 Gesture examples for touch input devices. **a** Horizontal flicker movement, **b** two-finger scaling, and **c** input gesture for the letter *h*. Pictures by *Gestureworks* (www.gestureworks.com), used with permission

> Being passively touched tends to focus the observer's attention on his or her subjective bodily sensations, whereas contact resulting from active exploration tends to guide the observer's attention to properties of the external environment.

Studies show independence of the assessment of material and system properties from the exploration type (active or passive touch condition) [156, 161]. Active touch delivers better performance for the exploration of geometric properties [154, 200]. From a technical viewpoint, the implementation of active exploration techniques is a challenge, since transmitted signals have to be synchronized with the relative movement.

2.2.3 Gestures

Gestures are a form of nonverbal communication that are studied in a large number of scientific disciplines such as social sciences, history, communication, and rhetoric and—quite lately—human–computer interaction. Concentrating on the latter, one can find gestures when using pointing devices such as mice, joysticks, and trackballs. Recently, gestures for touch-based devices have become more prominent. Some examples are given in Fig. 2.23, and for further information, see [131] for a taxonomy of gestures in human–computer interaction. An informative list on all kinds of gestures can be found in the Wikipedia under the reference term "List of Gestures."

Gestures can be used as robust input means in complex environments, as for example, in the car as shown with touch-based gestures in Sect. 14.2 or based on a camera image [174]. For use in haptic interaction, gestures have further meaning when interacting with virtual environments, as discrete input options in mobile applications and in connection with specialized haptic interfaces such as AIREAL [197], which combine a 3D camera with haptic feedback through an air vortex, or the ULTRAHAPTICS project, which generates haptic feedback in free air by superposing the signals from a matrix of ultrasound emitters [29]. At the moment of writing of this book, ISO/TC 159/SC 4/WG 9 is working on a new standard on the usage of gestures in tactile and haptic interaction (ISO 9241-960).

2.2.4 Human Movement Capabilities

Since users will interact with haptic systems, the capabilities of their movement have to be taken into account.

2.2.4.1 Dynamic Properties of the Locomotor System

While anatomy answers questions regarding the possible movement ranges (see [209] for example), there are a few studies dealing with the dynamic abilities of humans. TAN ET AL. [202] conducted a study to investigate the maximum controllable force and the average force control resolution. They found maximum controllable forces that could be maintained for at least 5 s in the range of 16–51 N for the joints of the hand and forces in the range of 35–102 N for wrist, elbow, and shoulder joints. Forces about half as large as the maximum force could be controlled with an accuracy of 0.7–3.4 %. This study is based on just three test persons, but other studies find similar values, for example, when grasping a cylindrical grip with forces ranging from 7 N (proximal phalanx of the little finger) to 99 N (tip of the thumb) [85]. AN ET AL. find female's hand strengths in the range of 60–80 % of male's hand strengths [5].[2]

Regarding velocities, HASSER derives velocities of 60–105 $\mathrm{cms^{-1}}$ for the tip of the extended index finger [85] and about 17 $\mathrm{rad\ s^{-1}}$ for the MCP and PIP joints. BROOKS [21] reports maximum velocities of 1.1 $\mathrm{ms^{-1}}$ and maximum accelerations of 12.2 $\mathrm{ms^{-2}}$ from a survey of 12 experts of telerobotic systems.

2.2.4.2 Properties of Interaction with Objects

When touching a surface, users show exploration velocities of about 2 $\mathrm{cm\,s^{-2}}$ (with a range of 1 to 25 $\mathrm{cm\,s^{-1}}$) and contact forces ranging from 0.3 to 4.5N [26]. Other studies confirm this range for tapping with a stylus [175] and when evaluating the roughness of objects [207]. SMITH ET AL. found average normal forces of 0.49 to 0.64 N for exploring raised and recessed tactile targets on surfaces with the index finger. Recessed targets were explored with slightly larger forces and lower exploring speed (7.67 $\mathrm{cm\,s^{-1}}$ compared to 8.6 $\mathrm{cm\,s^{-1}}$ for raised targets). Increased friction between finger and explored object lead to higher tangential forces. While the average tangential force in normal condition was 0.42 N, the tangential force was raised to 0.65 N in the increased friction condition (realized by a sucrose coating of the fingertip) [195].

For minimal-invasive surgery procedures with a tool-mediated contact, radial (with respect to the endoscopic tool axis) forces up to 6 N and axial forces up to 16.5 N were measured by RAUSCH ET AL. High forces of about 4 N on average were recorded for tasks involving holding, pressing, and pulling of tissue, and low forces were used for tasks such as laser cutting and coagulation, all measured with a

[2] Unfortunately, the number of test subjects involved in the studies is not reported.

force-measuring endoscope operated by medical professionals as reported in [186, 187]. Tasks were carried out with movement frequency components of up to 9.5 Hz, which is in line with the above-reported values (see Fig. 1.7).

HANNAFORD provides a database with measurements of force and torque for activities of daily living like writing and dialing with a cell phone, among others [188]. The database can be found at http://brl.ee.washington.edu/HapticsArchive/index.html and provides raw data and data-handling scripts for own research and is open to external contributions.

2.3 Interaction Using Haptic Systems

In this section, interactions using haptic systems are discussed and the nomenclature for haptic systems is derived from these interactions. The definitions are derived from the general usage in the haptic community and a number of publications by different authors [43, 92–94, 133] as well as logically extended based on the interaction model shown in Fig. 2.24.

While used in a general way so far, the term *haptic systems* is defined as follows:

> **Definition** *Haptic Systems* Systems interacting with a human user using the means of haptic perception and interaction. Although modalities such as temperature and pain belong to the haptic sense, too, *haptic systems* refer only to pure mechanical interaction in this book. In many cases, the term *haptic device* is synonymously used for haptic systems.

In this sense, haptic systems cover not only the fundamental haptic inputs and outputs, but also the system control instances needed to drive actuators, read out sensors, and take care of data processing. This is in accordance with known definitions of mechatronic systems like that of CELLIER [30]:

> A system is characterized by the fact that we can say what belongs to it and what does not, and by the fact that we can specify how it interacts with its environment. System definitions can furthermore be hierarchical. We can take the piece from before, cut out a yet smaller part of it and we have a new system.

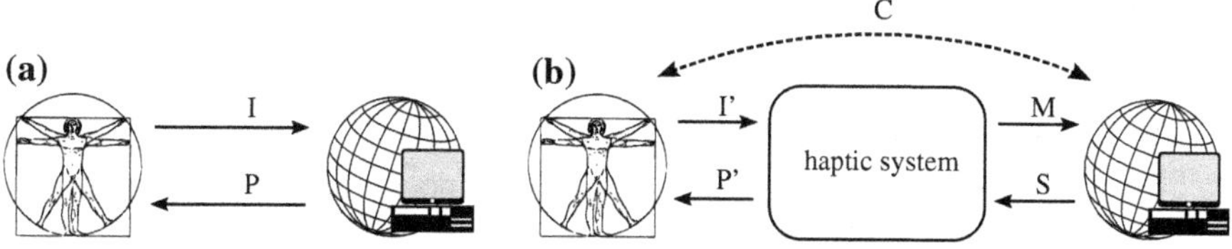

Fig. 2.24 Haptic interaction between humans and environment. **a** Direct haptic interaction and **b** utilization of haptic systems. The interaction paths are denoted as follows: *I* intention, *P* perception, *M* manipulation, *S* sensing, *C* comanipulation/other senses

The terms *system*, *device*, and *component* are not defined clearly on an interdisciplinary basis. Dependent on one's point of view, the same object can be a "device" for a hardware designer, a "system" for a software engineer, or just another "component" of another hardware engineer. These terms are therefore also used in different contexts in this book.

Compared to other perception modalities, haptics offers the only bidirectional communication means between the human user and the environment [12, p. 94]. A *user* is defined as

Definition *User* A person interacting (haptically) with a (haptic) system. The user can convey intentions to the system and receive (haptic) information depending on the *application* of the system. In this sense, a *test person* or *subject* in a psychophysical experiment is a user as well, but not all users can be considered as subjects.

In this book, a haptic system is always considered to have a specific *application* as for example, those outlined in Sect. 1.3. We therefore also define this term as follows:

Definition *Application* Intended utilization of a haptic system.

One has to keep in mind that this definition includes ↪ commercial off-the-shelf (COTS) haptic interfaces coupled to a computer with a software program to visualize biochemical components as well as the use of a specially designed haptic display as a physical interface. Particularly in this section, the term *application* has therefore to be considered context sensitive.

Figure 2.24 gives a schematic integration of an arbitrary haptic system in the interaction between a human user and a (virtual) environment. Based on this, one can identify typical classes of haptic systems.

2.3.1 Haptic Displays and General Input Devices

The probably most basic haptic system, shown in Fig. 2.25, is a

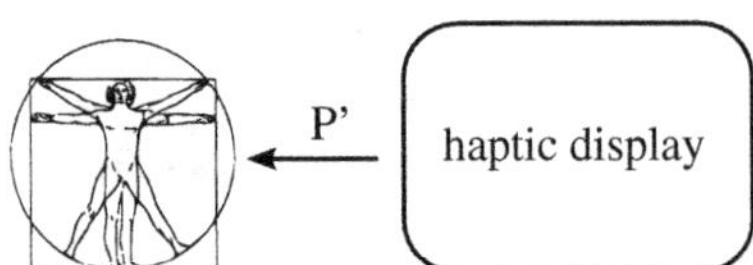

Fig. 2.25 Interaction scheme of a haptic display

Definition *Haptic Display* A haptic display solely addresses the interaction path **P** with actuating functions. Mechanical reactions of the human user have no direct influence on the information displayed by the haptic display, since user actions are not recorded and cannot be provided to the application.

Haptic displays are used to convey information originating in status information about the system incorporating the display. Typical applications are ↪ Braille row displays and—of course—the vibration alarm in mobile devices. Since the overlap to the next class of systems is somehow fuzzy, for the rest of this book, a haptic display is defined as a device that only incorporates actuating functions but no sensory functions (except the internal ones needed for the correct functionality of the actuating part). These types of devices are mainly used in communication applications as those shown in Fig. 1.15a, c and d. Often, a haptic display can be seen as a mechatronic component of a haptic system with additional functionality, for example, an *assistive system* described in the next section.

For completeness, also systems addressing only the interaction path **I** can be identified. These are basically general input devices such as buttons, keyboards, switches, touchscreens and mouse that record intentions of the user mechanically and convey them to an application. Being mechanical components themselves, they naturally exhibit mechanical reactions felt as haptic feedback by the user, but are normally independent of the application. For example, the haptic feedback from a computer keyboard is the same for the *F1* key or for the *Return* key, while the effects of these intentions are quite different. Therefore, general input devices are defined as devices with a predominant input functionality that can be used in different applications and a subordinated haptic feedback—independent from the application and resulting unintended from the real mechanical design of the input device. With the focus on generality, specialized input devices such as emergency stop buttons are excluded, since they exhibit a defined haptic feedback to convey the current state of the input device.

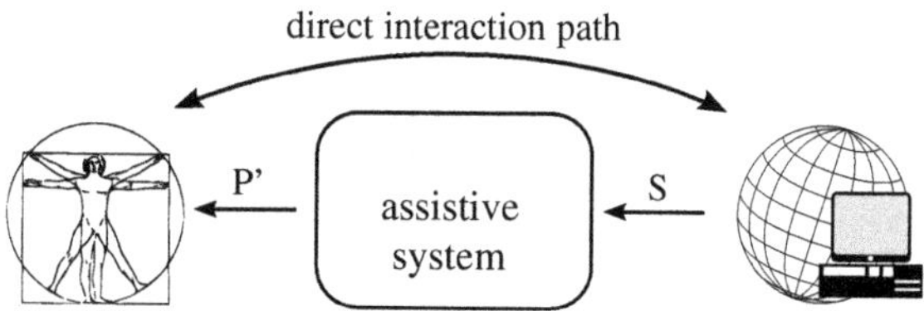

Fig. 2.26 Interaction scheme of an assistive system that adds haptic feedback to an existing direct interaction

2.3.2 Assistive Systems

This class of haptic systems shown in Fig. 2.26 is based on haptic displays, but also include an application-dependent sensory function.

Definition *Haptic-Assistive System* System that adds haptic information to a natural, non-technical-mediated haptic interaction on path **P** based on sensory input on path **S**.

Assistive systems are a main application area for haptic displays. The sensory input of assistive systems is not necessarily of a mechanical kind. However, compared to a haptic display as described above, an assistive system will add to existing, natural haptic interaction (i.e., without any technical means).

2.3.3 Haptic Interfaces

If an intention recording function and an *intended* haptic feedback functionality is combined, another class of haptic systems can be defined as shown in Fig. 2.27:

Definition *Haptic Interface* Haptic interfaces address the interaction path **P** with actuating functions, but also record the user's intentions along the interaction path **I** with dedicated sensory functions. These data are fed to the application and evoke commands to the system or visualization under control. Depending on the application, a mechanical user input can result in direct haptic feedback.

Haptic interfaces are mostly used as universal operating devices to convey interactions with different artificial or real environments. Typical applications with task-specific interfaces include stall-warning sidesticks in aircraft and force feedback joysticks in consumer applications. Another application is the interaction with virtual environments that is normally achieved with a large number of ↪ COTS haptic interfaces. These can also be used in a variety of other interaction tasks, and some

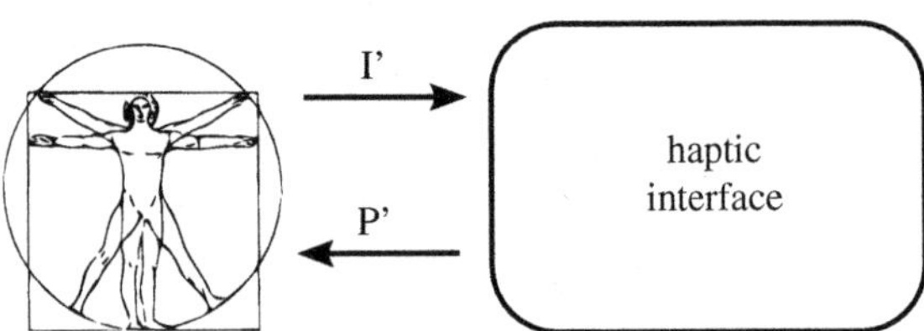

Fig. 2.27 Relevant interaction paths of a haptic interface

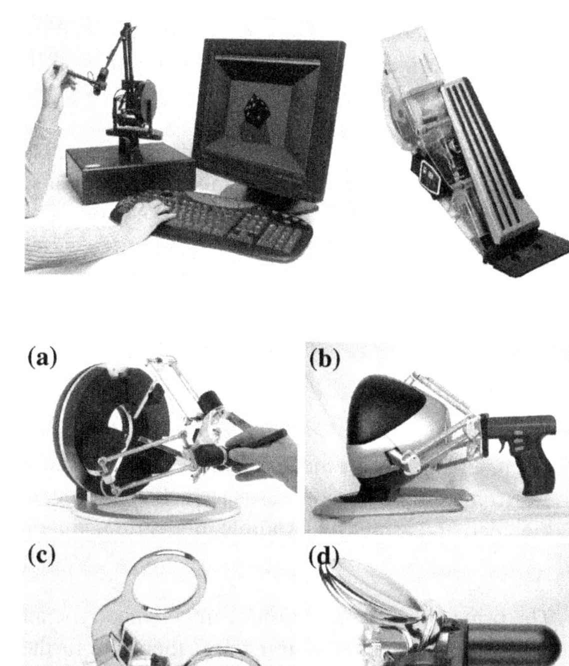

Fig. 2.28 Two haptic interfaces, PHANToM PREMIUM 1.5 (*3D Systems Geomagic Solutions*, Rock Hill, SC, USA) and ACCELERATOR FORCE FEEDBACK PEDAL (AFFP, *Continental Automotive*, Hannover, Germany). Images used with permission

Fig. 2.29 Realizations of tool-mediated contact in commercial haptic interfaces. **a** OMEGA.6 with a stylus interface (*Force Dimension*, Nyon, Switzerland), **b** FALCON with a pistol-like grip for gaming applications (*Novint*, Rockville Centre, NY, USA), and **c** and **d** pinch and scissor grip interfaces for the PHANToM PREMIUM (*3D Systems Geomagic Solutions*, Rock Hill, SC, USA). Images used with permission

applications were outlined in Sect. 1.3. Some ↪ COTS haptic interfaces as well as an example for a task-specific haptic interface for driving assistance is shown in Fig. 2.28. Other task-specific haptic interfaces are developed for use in medical training systems.

In general, ↪ COTS devices support input and output at only a single point in the workspace. The position of this ↪ tool center point (TCP) in the workspace of the device is sent to the application, and all haptic feedback are generated with respect to this point. Since the interaction with a single point is somewhat not intuitive, most devices supply contact tools such as styluses or pinch grips, which mediate the feedback to the user. This grip configuration is a relevant design parameter and is further addressed in Sect. 3.1.3. Figure 2.29 shows some typical grip situations of ↪ COTS devices with such *tool-mediated contact.*

2.3.3.1 System Structures

To fulfill the request for independent channels for input (user intention) and output (haptic feedback) of the haptic interface and the physical constraint of energy conservation, one can define exact physical representations of the input and output of haptic interfaces. This leads to two fundamental types of haptic systems that are defined by their mechanical inputs and outputs as follows:

Definition *Impedance-Type System* Impedance-type systems (or just *impedance systems*) exhibit a mechanical input in the form of a kinematic measure and a mechanical output in the form of a force or torque. In case of a haptic interface, the mechanical input (in most cases, the position of the ↪ TCP) is conveyed as an electronic output to be used in other parts of the application.

Definition *Admittance-Type System* Admittance-type systems exhibit a mechanical input in the form of a force or a torque that is conveyed as an electronic output in most cases as well. The mechanical output is defined by a kinematic measure, for example, deflection, velocity, or acceleration.

The principal differentiation of impedance type and admittance type of systems is fundamental to haptic systems. It is therefore further detailed in Chap. 6.

2.3.3.2 Force Feedback Devices

The term *force feedback* is often used for the description of haptic interfaces, especially in advertising force feedback joysticks, steering wheels, and other consumer products. A detailed analysis of these systems yields the following characteristics for the majority of such systems:

System Structure Because of the output of forces, they resemble impedance-type systems as defined above.

Dynamics Force feedback systems address the whole dynamic range of haptic interaction, but do not convey spatially distributed (tactile) information.

Contact Situation Most force feedback systems do not allow for natural exploration, but convey information via a tool.

These characteristics show a quite deep level of detail. The only comparable other term with similar detail depth is perhaps *tactile feedback*, mostly defining spatially distributed feedback in the dynamic range of passive interaction (see Fig. 1.7). However, these terms are used so widely in technical and non-technical applications with different and not agreed on definitions that they are used in this book in favor of other, clearly defined terms. In that case, force feedback devices would be better described as *impedance-type interfaces with tool-mediated haptic feedback*. Since this is a scientific book, the longer term is preferred to an unclear definition.

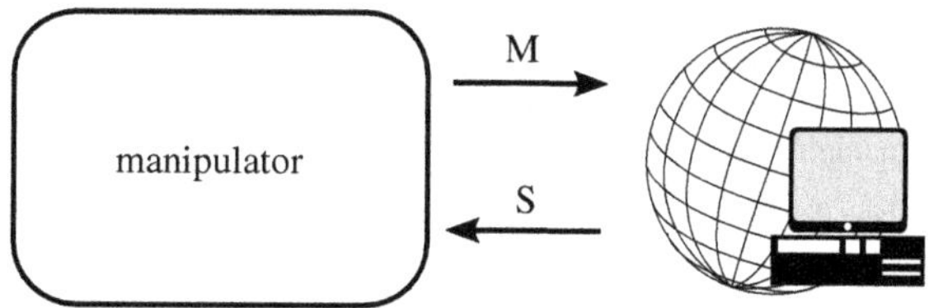

Fig. 2.30 Interaction scheme of a manipulator

2.3.4 Manipulators

There are only a limited number of systems from outside the haptic community that can be classified as impedance systems. For admittance systems, one can find haptic interfaces (for example, the haptic master interface shown in Fig. 1.12 is an admittance-type interface) as well as mechanical manipulators from other fields. For example, industrial robots are normally designed as admittance systems that can be commanded to a certain position and can measure reaction forces if equipped properly. In the here-presented nomenclature of haptic system design, such robots can be defined as manipulators:

Definition *Manipulator* A technical system that uses interaction path **M** to manipulate or interact with an object or (remote) environment. Sensing capabilities (interaction path **S**) are used for the internal system control of the manipulator and/or for generating haptic feedback to a user.

Figure 2.30 shows the corresponding interaction scheme.

2.3.5 Teleoperators

The combination of a haptic interface and a manipulator yields the class of teleoperation systems with the interaction scheme shown in Fig. 2.31.

Definition *Teleoperation Systems* A combined system recording the user intentions on path $\mathbf{I}'$ via the manipulation path **M** to a real environment, measuring interactions on the sensing path **S** and providing haptic feedback to the user via the perception path $\mathbf{P}'$.

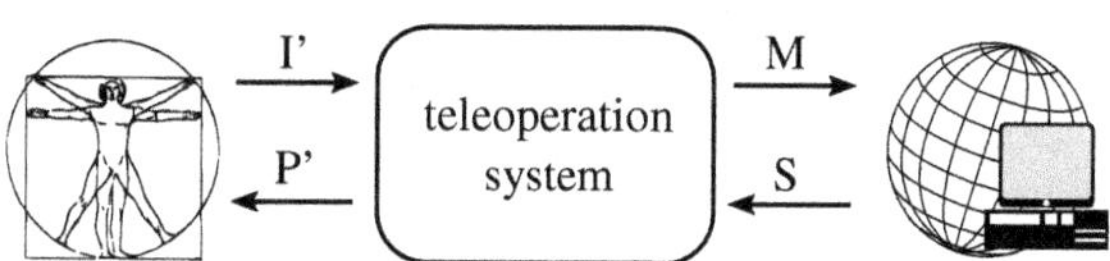

Fig. 2.31 Teleoperation interaction scheme

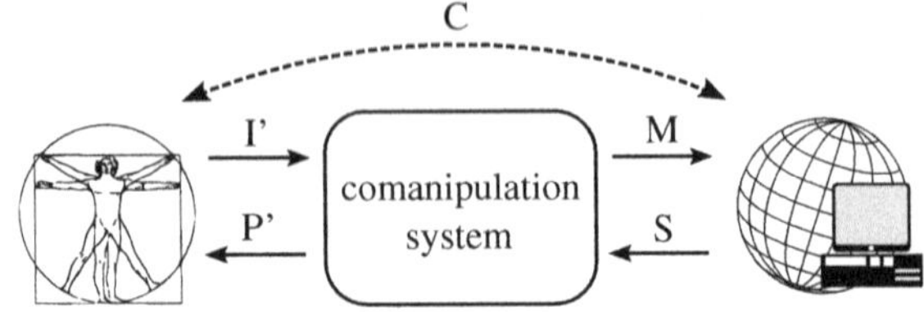

Fig. 2.32 Interaction scheme for comanipulators with additional direct feedback via comanipulation path C

An extension of teleoperators is the class of ↪ telepresence and teleaction (TPTA) systems that include additional feedback from other senses such as vision and/or audition. Both terms are used synonymously sometimes. Teleoperation systems allow a spatially separated interaction of the user with a remote physical environment. The simplest system is archived by coupling an impedance-type haptic interface with an admittance-type manipulator, since inputs and outputs correspond correctly. Often, couplings of impedance–impedance systems are used because of the availability of components, which generate higher demands on the system controller.

2.3.6 Comanipulators

If additional mechanical interaction paths are present, telepresence systems turn into a class of systems called comanipulators [43]:

Definition *Comanipulation System* Telepresence system with an additional direct mechanical link between user and the environment or object interacted with.

Comanipulator systems are often used in medical applications, since they minimize the technical effort compared to a pure teleoperator because of less moving mass, fewer active ↪ DoF, and minimized workspaces, but also induce new challenges for the control and stability of a system. In an application, the user moves the reference frame of the haptic system.

Compared to the above-mentioned assistive systems, comanipulators exhibit a full teleoperational interaction scheme with additional direct feedback, while assistive systems add additional haptic feedback to a non-technical-mediated interaction between user and application. This is shown in Fig. 2.32.

2.3.7 Haptic System Control

To make the above-described systems usable in an application, another definition of more technical nature has to be introduced:

> **Definition** *Haptic System Control* The haptic system control is that part of a real system that not only controls the single mechanical and electrical components to ensure proper sensing, manipulating, and displaying haptic information, but also takes care of the connection to other parts of the haptic system. This may be, for example, the connection between a haptic interface and a manipulator or the interface to some virtual reality software.

While the pure control aspects are addressed in Chap. 7, one has to consider also other design tools and information structures such as event-based haptics (see Sects. 11.2.3), pseudo-haptics (see Sects. 2.1.4.4 and 12.5), and the general connection to rendering software (see Chap. 12) using a real interface (see Chap. 11). In this book and in other sources, one will also find the term *haptic controller* used synonymously for the whole complex of the here-described haptic system control.

2.4 Engineering Conclusions

Based on the above, one can conclude a general structure for interaction with haptic systems and assign certain attributes to the different input and output channels of a haptic system. This is shown in Fig. 2.33 that extends Fig. 2.24. In the figure, the output channel of the haptic system toward the user is separated in mainly tactile and kinaesthetic sensing channels. This is done with respect to the explanations given in Sect. 1.2.1 and with the knowledge that there are many haptic interfaces that will fit into this classification, which will also be used further on in this book occasionally. The parameters given in Fig. 2.33 give an informative basis for interaction with haptic systems.

In the remainder of this section, several conclusions for the design of task-specific haptic systems are given based on the properties of haptic interaction.

2.4.1 A Frequency-Dependent Model of Haptic Properties

Haptics, and especially tactile feedback, is a dynamic impression. There are little to no static components. Without exploring scientific findings, a simple impression of the dynamic range covered by haptic and tactile feedback can be estimated by taking a look at different daily interactions (Fig. 2.34).

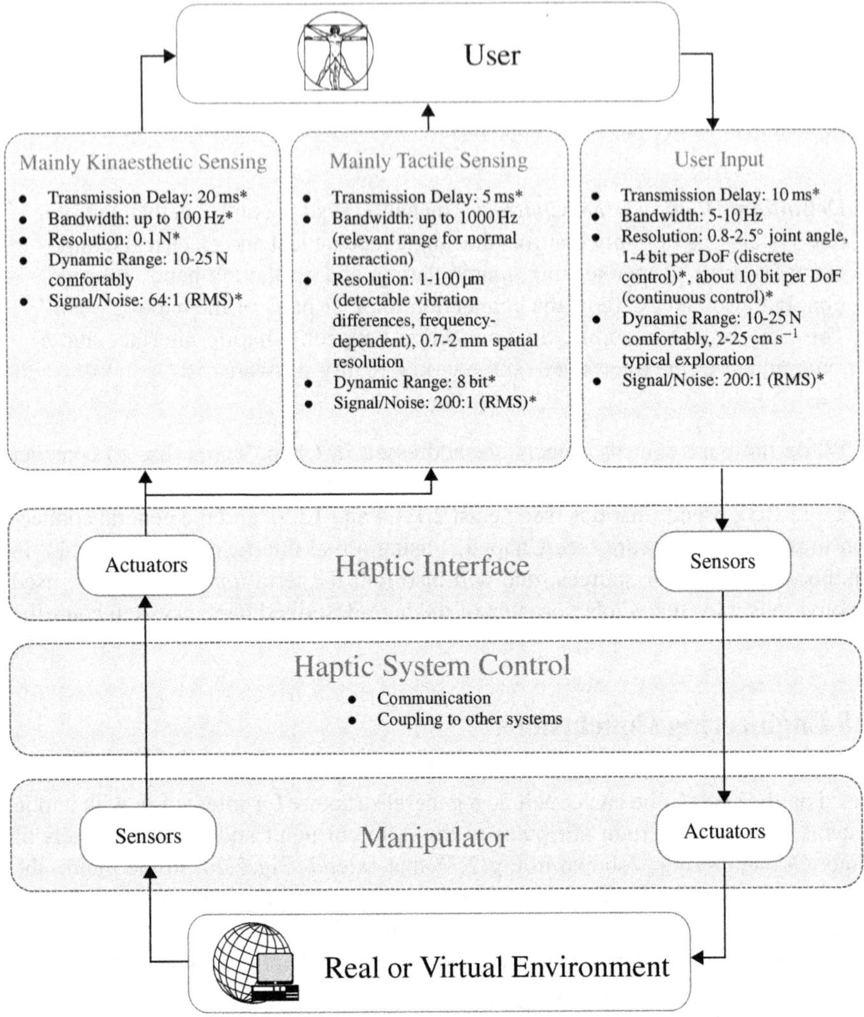

Fig. 2.33 General input and output ports for a haptic system in interaction with the human hand. Figure is based on [23, 46], values from [46] are based on surveys among experts and are labeled with an *asterisk* (*); other values are taken from the different sources stated above.

When handling an object, the first impression to be explored is its *weight*. There is probably none who has not been caught by surprise at least once lifting an object which in the end was lighter than expected. The impression is usually of comparably low frequency and typically directly linked to the active touch and movement applied to the object.

Exploring an object with the finger to determine its fundamental *shape* is the next interaction type in terms of its dynamics. When touching such objects, a global

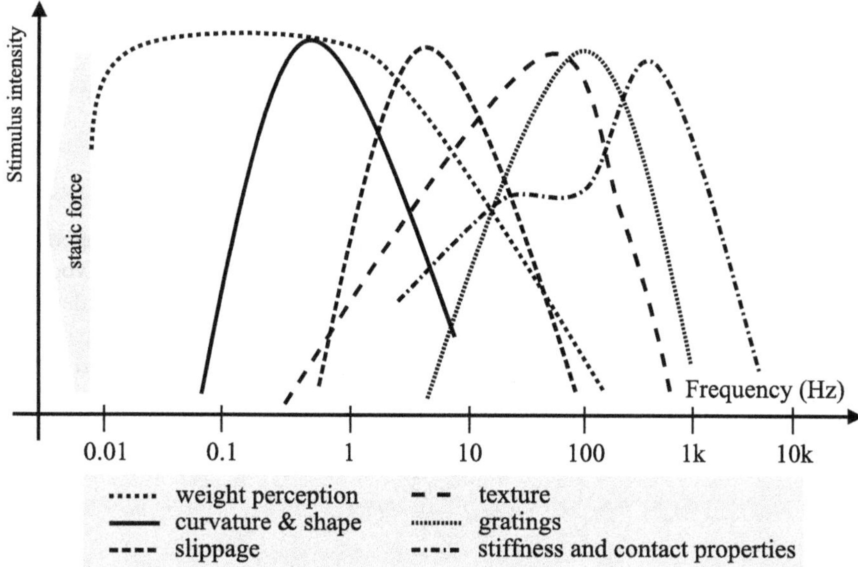

Fig. 2.34 Concept of modalities and frequency dependency

deformation of the finger and a tangential load to its surface are relevant to create such an impression. There has been research performed by HAYWARD showing that the pure inclination of the finger's surface already creates an impression of shape. However, still being quite global, the dynamic information coded in this property is not very rich.

Dynamics increases when it becomes urgent to react. One of the most critical situations our biology of touch is well prepared for is the detection of *slippage*. Constant control of normal forces to the object prevents it from slipping out of our grasp. Being highly sensitive to shear and stick-slip, this capability enables us to gently interact with our surrounding.

When it comes to *slippage-textured* surfaces, their dynamics must also be mentioned. Their frequency of course depends on geometrical properties, however, their exploration during active touch typically happens in the range above 100 Hz. Within this sensitive area, discrimination capabilities of *textures* are naturally most sensitive, as the vibrotactile sensitivity of the human finger is climbing to its highest level.

Whether *gratings* do differ from *textures* is something that can be discussed endlessly. The principal excitation of the tactile sensory orchestra may be identical, however, *gratings* are more like a dirac pulse, whereas *textures* are more comparable to a continuous signal.

Last but not least, hard contacts and the properties they reveal about an object reflect the most dynamic signal processing a haptic interaction may have. And surprisingly, a strong impact on an object reveals more about its volume and structural properties as any gentle interaction can ever show. Therefore, *stiffness* is worth an own set of thoughts in the following section.

2.4.2 Stiffnesses

Already the initial touch of a material gives us information about its haptic properties. A human is able to immediately discriminate, whether he or she is touching a wooden table, a piece of rubber, or a concrete wall with his or her fingertip. Besides the acoustic and thermal properties, especially the tactile and kinaesthetic feedback play a significant role. Based on the simplified assumption of a double-sided fixed plate, its stiffness k can be identified by use of Young's modulus E according to Eq. (2.10) [159].

$$k = 2\,\frac{b\,h^3}{l^3} \cdot E \tag{2.10}$$

Figure 2.35a shows the calculation of stiffnesses for a plate of an edge length of 1 m and a thickness of 40 mm of different materials. In comparison, the stiffnesses of commercially available haptic systems are given in (Fig. 2.35b). It is obvious that these stiffnesses of haptic devices are factors of 10 lower than the stiffnesses of concrete, everyday objects such as tables and walls. However, stiffness is just <u>one</u> criterion for the design of a good haptic system and should not be overestimated. The wide range of stiffnesses reported to be needed for rendering of undeformable surfaces as shown in Sect. 2.1.3 is strong evidence of the interdependency of several different parameters. The comparison above makes us aware of the fact that a pure reproduction of solid objects can hardly be realized with a single technical system. It rather takes a combination of stiff and dynamic hardware, as especially the dynamic interaction in high-frequency areas dominates the quality of haptics, which has extensively been discussed in the previous section.

2.4.3 One Kilohertz: Significance for the Mechanical Design

As stated above, haptic perception ranges up to a frequency of 10 kHz, whereby the area of the highest sensitivity lies between 100 and 1 kHz. This wide range of haptic perception enables us to perceive microstructures on surfaces with the same accuracy as enabling us to identify the point of impact when drumming with our fingers on a table.

For a rough calculation, a model based on Fig. 2.36 is considered to be a parallel circuit between a mass m and a spring k. Assuming an identical "virtual" volume V of material and taking the individual density ϱ for qualitative comparison, the border frequency f_b for a step response can be calculated according to Eq. (2.11).

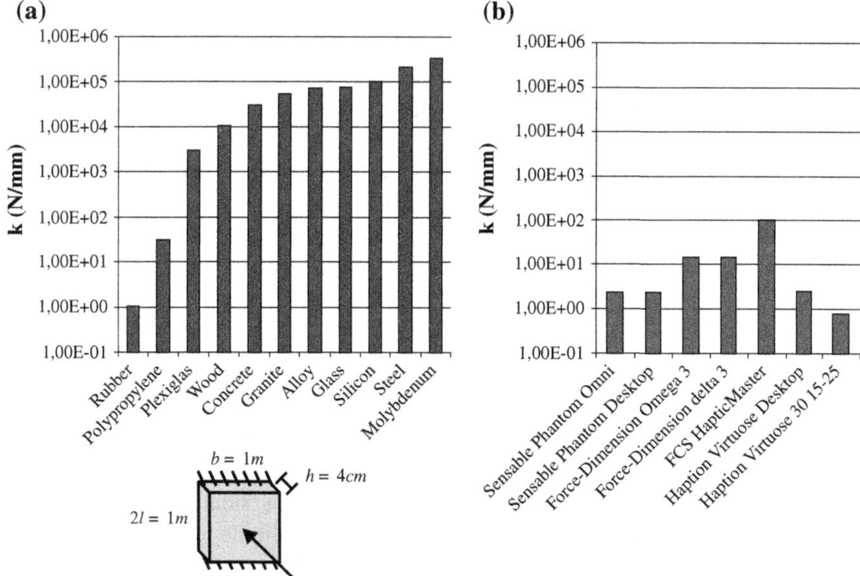

Fig. 2.35 **a** Comparison between stiffnesses of a $1 \times 1 \times 0.04\,\text{m}^3$ plate of different materials and **b** realizable stiffnesses by commercial haptic systems

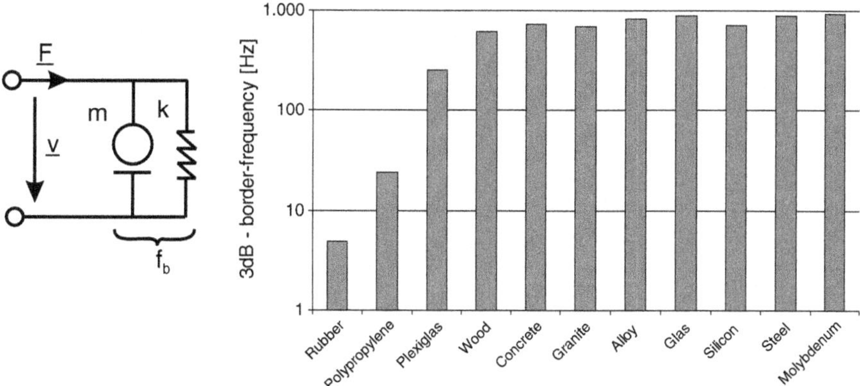

Fig. 2.36 3 dB border frequency f_b of an excitation of a simple mechanical model parametrized as different materials

$$f_b = \frac{1}{2\pi}\sqrt{\frac{k}{m}} = \frac{1}{2\pi}\sqrt{\frac{k}{V\varrho}} \tag{2.11}$$

Figure 2.36 shows the border frequencies of a selection of materials. Only in case of rubber and soft plastics, border frequencies of below 100 Hz appear. Harder plastic material (Plexiglas) and all other materials show border frequencies above 700 Hz.

One obvious interpretation would state that any qualitatively good simulation of such a collision demands at least such bandwidth of dynamics within the signal-conditioning elements and the mechanical system.

As a consequence, a frequent recommendation for the design of haptic systems is the transmission of a full bandwidth of 1 kHz (and in some sources even up to 10 kHz). This requirement is valid with respect to software and communications engineering, as sampling systems and algorithmic can achieve such frequencies easily today. Considering the mechanical part of the design, we see that dynamics of 1 kHz is enormous, maybe even utopian. Figure 2.37 gives another rough calculation of oscillating force amplitude according to Eq. (2.12).

$$F_0 = \left| \underline{x} \cdot (2\pi f)^2 \cdot m \right| \tag{2.12}$$

The basis of the analysis is a force source generating an output force $\underline{F}_0$. The load of this system is a mass (e.g., a knob) of 10 g (!!). The system does not have any additional load, i.e., it does not have to generate any haptically active force to a user. A periodic oscillation of a frequency f and an amplitude $\underline{x}$ is assumed. With expected amplitudes for the oscillation of 1 mm at 10 Hz, a force of approximately 10 mN is necessary. At a frequency of 100 Hz, there is already a force of 2–3 N needed. At a frequency of 700 Hz, the force already increases to 100 N and this is what happens when moving a mass of 10 grams. Of course, in combination with a user impedance as load, the amplitude of the oscillation will decrease in areas of below 100 μm, proportionally decreasing the necessary force. But this calculation should make aware of the simple fact that the energetic design and power management of electromechanical systems with application in the area of haptics needs to be done very carefully.

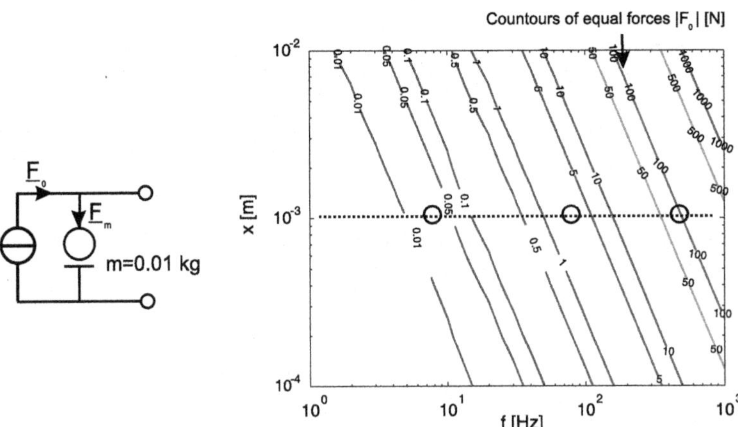

Fig. 2.37 Equipotential line of necessary forces in dependency of amplitude and frequency of the acceleration of a mass with 10 g

The design of a technical haptic system is always a compromise between bandwidth, stiffness, dynamics of signal conditioning, and maximum force amplitudes. Even with simple systems, the design process leads the engineer to the borders of what is physically possible. Therefore, it is necessary to have a good model for the user according to his being a load to the mechanical system and according to his or her haptic perception. This model enables the engineer to carry out an optimized design of the technical system, and its generation is the focus point of Chap. 3.

2.4.4 *Perception-Inspired Concepts for Haptic System Design*

At the end of this chapter, two examples illustrate the technical importance of an understanding of perception and interaction concepts. The examples present two technical applications that purposefully use unique properties of the haptic sensory channel to design innovative and better haptic systems.

2.4.4.1 *Example:* Event-Based Haptics

Based on the bidirectional view of haptic interactions (see Sect. 1.2.2) with a low-frequency kinaesthetic interaction channel and a high-frequency tactile perception channel, KONTARINIS and HOWE published a new combination of kinaesthetic haptic interfaces with additional sensors and actuators for higher frequencies. Tests included their use in the virtual representation and exploration of objects [171] as well as in teleoperation systems.

Based on this work, NIEMEYER ET AL. proposed *event-based haptics* as a concept for increasing realism in virtual reality applications [106]. In superposing the kinaesthetic reactions of a haptic interface with high-bandwidth transient signals for certain events such as touching a virtual surface, the haptic quality of this contact situation can be improved considerably [145]. The superposed signals are recorded using accelerometers and played back open-loop, if a predefined interaction event takes places.

This concept proved as a valuable tool for rendering of haptic interactions with virtual environments. Rendering quality is increased with a comparatively small hardware effort in the form of additions to (existing) kinaesthetic user interfaces. Technically not an addition to an existing kinaesthetic system, but still based on the event-based haptics approach, the VERROTOUCH System by KUCHENBECKER ET AL. was developed as an addition to the DAVINCI Surgical System. It adds tactile and auditory feedback based on vibrations measured at the end of the minimal-invasive instrument attached to the robot [146]. These vibrations are processed and played back using vibratory motors attached to the DAVINCI controls and additional auditory speakers.

The system shown in Fig. 2.38 conveys the properties of rough surfaces and contact events with manipulated objects. The augmented interaction was evaluated

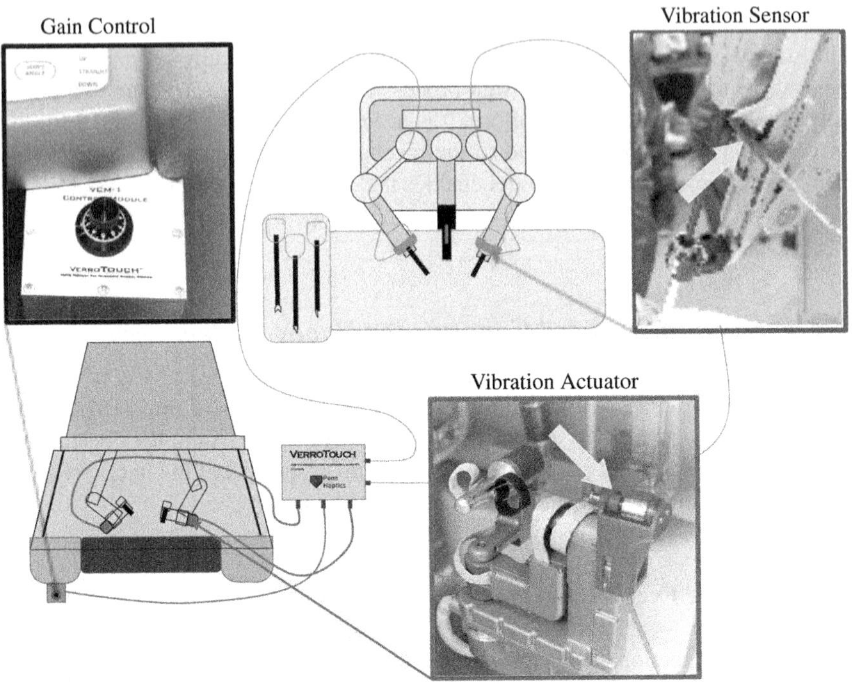

Fig. 2.38 Integration of VERROTOUCH into the DAVINCI SURGICAL SYSTEM. Figure adapted from [146]

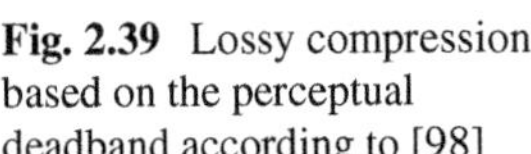

Fig. 2.39 Lossy compression based on the perceptual deadband according to [98]

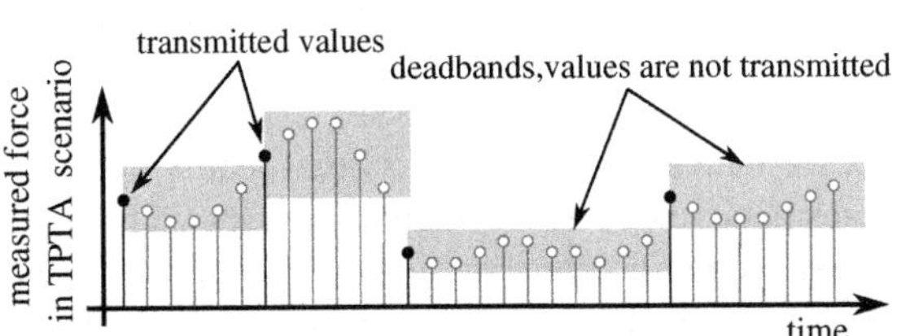

positively in a study with 11 surgeons [167]. Objective task metrics showed neither improvement nor impairment of the tested tasks.

2.4.4.2 *Example:* Perceptual Deadband Coding

Perceptual deadband coding (PD) is a perception-oriented approach to minimize the amount of haptic data that has to be transmitted in real-time applications such as teleoperation [97, 98]. To achieve this data reduction, new data are only transmitted from the slave to the master side, if the change compared to the preceding data point is greater than the ↪ JND. The perceptual deadband coding is illustrated for the one-dimensional case in Fig. 2.39, but can be extended easily to more-dimensional so-called dead zones [198].

Recommended Background Reading

- **Attention, Perception, & Psychophysics**, Volume 63, Issue 8, 2001. *Special edition about the psychometric function and psychophysical procedures.*

[122] Jones, L. & Lederman, S.: **Human Hand Function**. Oxford University Press, 2006.
Extensive analysis about the human hand including perception and interaction topics.

[182] Prins, N. & Kingdom, F. A. A.: **Psychophysics: a Practical Introduction**. Academic Press, Maryland Heights, MO, USA, 2010.
Psychophysics textbook with a good overview about modern psychometric procedures and the underlying statistics.

[224] Wickens, T. D.: **Elementary Signal Detection Theory**. Oxford University Press, Oxford, GB, 2002.
Modern and quite entertaining book about modern signal detection theory.

References

1. Acker A (2011) Anwendungspotential von Telepräsenz-und Teleaktionssystemen für die Präzisionsmontage. Dissertation. Technische Universität München. http://mediatum.ub.tum.de/doc/1007163/1007163.pdf
2. Adams M (2011) Project summary nanobiotact. Technical Report EU FP6
3. Allerkamp D et al (2007) A vibrotactile approach to tactile rendering. Vis Comput 23(2):97–108. doi:10.1007/s00371-006-0031-5
4. Allin S, Matsuoka Y, Klatzky R (2002) Measuring just noticeable differences for haptic force feedback: implications for rehabilitation. In: Proceedings of the 10th symposium on haptic interfaces for virtual environments & teleoperator systems. Orlando, FL, USA. doi:10.1109/HAPTIC.2002.998972
5. An L, Askew, Chao E (1986) Biomechanics and functional assessment of upper extremities. In: Karwowski W (ed) Trends in ergonomics/human factors III. Elsevier, North-Holland, pp 573–580. ISBN: 978-0444700360
6. Barbagli F et al (2006) Haptic Discrimination of force direction and the influence of visual information. ACM Trans Appl Percep (TAP) 3(2):135. doi:10.1145/1141897.1141901
7. Bensmaïa S, Hollins M, Yau J (2005) Vibrotactile intensity and frequency information in the pacinian system: a psychophysical model. Attention Percep Psychophys 67(5):828–841. doi:10.3758/BF03193536
8. Bergmann Tiest WM (2010) Tactual perception of material properties. Vision Res 50(24):2775–2782. doi:10.1016/j.visres.2010.10.005
9. Bergmann Tiest WM, Kappers A (2009) Cues for haptic perception of compliance. IEEE Trans Haptics 2(4):189–199. doi:10.1109/TOH.2009.16

10. Bergmann Tiest WM, Vrijling AC, Kappers AM (2010) Haptic perception of viscosity. In: Haptics: generating and perceiving tangible sensations. Springer, Berlin, pp 29–34. doi:10.1007/978-3-642-14064-8_5
11. Biggs J, Srinivasan MA (2002) Tangential versus normal displacements of skin: relative effectiveness for producing tactile sensations. In: 10th symposium on haptic interfaces for virtual environments and teleoperator systems. doi:10.1109/HAPTIC.2002.998949
12. Biggs S, Srinivasan M (2002) Haptic interfaces. In: Stanney K, Hale KS (eds) Handbook of virtual environments. Lawrence Erlbaum, London, pp 93–116. ISBN: 978-0805832709
13. Birch AS and Srinivasan MA (1999) Experimental determination of the viscoelastic properties of the human fingerpad. Technical Report, Touch Lab, Massachusetts Institute of Technology, Cambridge, MA, USA. http://dspace.mit.edu/bitstream/handle/1721.1/4127/RLE-TR-632-41961944.pdf
14. Blume HJ, Boelcke R (1990) Mechanokutane Sprachvermittlung. Reihe 10 137, vol 137. VDI-Verl, Düsseldorf. ISBN: 3-18-143710-7
15. Bolanowski SJ (1996) Information processing channels in the sense of touch. In: Franzén O, Johansson R, Terenius L (eds) Somesthesis and the neurobiology of the somatosensory cortex. Birkhäuser, Basel, pp 49–58. ISBN: 978–0817653224. doi:10.1007/978-3-0348-9016-8_5
16. Bolanowski SJ, Verrillo RT (1982) Temperature and criterion effects in a somatosensory subsystem: a neurophysiological and psychophysical study. J Neurophysiol 48(3): 836–855. http://www.ncbi.nlm.nih.gov/pubmed/7131055
17. Bolanowski SJ et al (1988) Four channels mediate the mechanical aspects of touch. J Acoust Soc Am 84(5):1680–1694. doi:10.1121/1.397184
18. Bolanowski S, Gescheider G, Verrillo R (1994) Hairy skin: psychophysical channels and their physiological substrates. Somatosens Motor Res 11(3):279–290. doi:10.3109/08990229409051395
19. Borg I, Groenen PJF (2005) Modern multidimensional scaling—theory and applications. Springer, Heidelberg
20. Brisben AJ, Hsiao SS, Johnson KO (1999) Detection of vibration transmitted through an object grasped in the hand. J Neurophysiol 81:1548–1558. http://jn.physiology.org/content/81/4/1548
21. Brooks TL (1990) Telerobotic response requirements. In: IEEE international conference on systems, man and cybernetics. Los Angeles, CA, USA. doi:10.1109/ICSMC.1990.142071
22. Brown L, Brewster S, Purchase H (2006) Multidimensional tactons for non-visual information presentation in mobile devices. In: Conference on human–computer interaction with mobile devices and services, pp 231–238. doi:10.1145/1152215.1152265
23. Burdea GC (1996) Force and touch feedback for virtual reality. Wiley-Interscience, New York
24. Buus S (2002)Psychophysical methods and other factors that affect the outcome of psychoacoustic measurements. In: Genetics and the function of the auditory system: proceedings of the 19th danavox symposium, Copenhagen, DK, pp 183–225
25. Cadoret G, Smith AM (1996) Friction, not texture, dictates grip forces used during object manipulation. J Neurophysiol 75(5):1963–1969. http://jn.physiology.org/content/75/5/1963
26. Caldwell D, Tsagarakis N, Giesler C (1999) An integrated tactile/shear feedback array for stimulation of finger mechanoreceptor. In: IEEE international conference on robotics and automation, vol 1, pp 287–292. doi:10.1109/ROBOT.1999.769991
27. Caldwell DG, Lawther S, Wardle A (1996) Multi-modal cutaneous tactile feedback. In: Proceedings of the IEEE international conference on intelligent robots and systems, pp 465–472. doi:10.1109/IROS.1996.570820
28. Campion G, Hayward V (2008) On the synthesis of haptic textures. IEEE Trans Robot 24(3):527–536. ISSN 1552–3098. doi:10.1109/TRO.2008.924255
29. Carter T et al (2013) UltraHaptics: multi-point mid-air haptic feedback for touch surfaces. In: Proceedings of the 26th annual ACM symposium on user interface software and technology. ACM 2013, pp 505–514. doi:10.1145/2501988.2502018
30. Cellier FE (1991) Continuous system modeling. Springer, Berlin, pp 23–60, ISBN: 9780387975023

31. Cholewiak RW, Collins AA (1991) Sensory and physiological bases of touch. In: Heller MA, SchiffThe W (eds) Psychology of touch. Lawrence Erlbaum, London, pp 23–60. ISBN: 0805807500
32. Cholewiak SA, Tan HZ, Ebert DS (2008) Haptic identification of stiffness and force magnitude. In: Symposium on haptic interfaces for virtual environments and teleoperator systems. Reno, NE, USA. doi:10.1109/HAPTICS.2008.4479918
33. Cholewiak SA et al (2010) A frequency-domain analysis of haptic gratings. IEEE Trans Haptics 3:3–14. doi:10.1109/TOH.2009.36
34. Clark F, Horch K (1986) Kinesthesia. In: Boff KR, Kaufman L, Thomas JP (eds) Handbook of perception and human performance. Wiley-Interscience, New York, pp 13.1–13.61. ISBN: 978-0471829577
35. Conner M et al (1988) Individualized optimization of the salt content of white bread for acceptability. J Food Sci 53(2):549–554. doi:10.1111/j.1365-2621.1988.tb07753.x
36. Craig JC (1972) Difference threshold for intensity of tactile stimuli. Percep Psychophys 11(2):150–152. doi:10.3758/BF03210362
37. Dandekar K, Raju B, Srinivasan M (2003) 3-D finite-element models of human and monkey fingertips to investigate the mechanics of tactile sense. J Biomech Eng 125:682. doi:10.1115/1.1613673
38. Dargahi J, Najarian S (2004) Human tactile perception as a standard for artificial tactile sensing—a review. Int J Med Robot Comput Assis Surg 1(1):23–35. doi:10.1002/rcs.3
39. Darian-Smith I, Johnson K (1977) Thermal sensibility and thermoreceptors. J Invest Dermatol 69(1):146–153. doi:10.1111/1523-1747.ep12497936
40. Dorjgotov E et al (2008) Force amplitude perception in six orthogonal directions. In: Symposium on haptic interfaces for virtual environments and teleoperator systems, Reno, NE, USA. doi:10.1109/HAPTICS.2008.4479927
41. Draguhn A (2009) Membranpotenzial und Signalübertragung in Zellverbänden. In: Klinke R et al (eds) Physiologie, Thieme, pp 60–97. ISBN: 9783137960065
42. Dudel J (2006) Synaptische übertragung. In: Schmidt F, Schaible HG (eds) Neuro- und Sinnesphysiologie. Springer, Heidelberg. doi:10.1007/3-540-29491-0
43. Eb Vander Poorten E, Demeester E, Lammertse P (2012) Haptic feedback for medical applications, a Survey. In: Actuator conference, Bremen. https://www.radhar.eu/publications/e.-vander-poorten-actuator12-haptic-feedback-formedical-applications-a-survey
44. Edin BB, Abbs JH (1991) Finger movement responses of cutaneous mechanoreceptors in the dorsal skin of the human hand. J Neurophysiol 65(3):657. http://www.humanneuro.physiol.umu.se/PDF-SCIENCE/1991_edin_abbs_DIST.pdf
45. Ehrenstein WH, Ehrenstein A (1999) Psychophysical methods. In: Windhorst U, Johansson H (eds) Modern techniques in neuroscience research. Springer, Heidelberg, pp 1211–1241. doi:10.1007/3-540-29491-0
46. Ellis S (1994) What are virtual environments? IEEE Comput Graph Appl 14(1): 17–22. ISSN 0272–1716. doi:10.1109/38.250914
47. Enriquez M, MacLean K, Chita C (2006) Haptic phonemes: basic building blocks of haptic communication. In: Proceedings of the 8th international conference on multimodal interfaces (ICMI), Banff, Alberta, Canada, ACM 2006, pp 302–309. doi:10.1145/1180995.1181053
48. Ernst M, Bülthoff H (2004) Merging the senses into a robust percept. Trends Cogn Sci 8(4):162–169. doi:10.1016/j.tics.2004.02.002
49. Fechner GT (1860) Elemente der Psychophysik. Breitkopf und Härtel, Leipzig
50. Fiene J, Kuchenbecker KJ, Niemeyer G (2006) Event-based haptic tapping with grip force compensation. In: IEEE symposium on haptic interfaces for virtual environment and teleoperator systems. doi:10.1109/HAPTIC.2006.1627063
51. Foster DH, Zychaluk K (2009) Model-free estimation of the psychometric function. Attention Percep Psychophys 71(6):1414–1425. doi:10.3758/APP.71.6.1414
52. Frisina R, Gescheider G (1977) Comparison of child and adult vibrotactile thresholds as a function of frequency and duration. Percep Psychophys 22(1):100–103. doi:10.3758/BF03206086

53. Garcia E.a-Pérez M (1998) Forced-choice staircases with fixed step sizes: asymptotic and small-sample properties. Vision Res 38(12):1861–1881. doi:10.1016/S0042-6989(97)00340-4
54. Garneau CJ, Parkinson MB (2013) Considering just noticeable difference in assessments of physical accommodation for product design. In: Ergonomics ahead-of-print, pp 1–12. http://www.ncbi.nlm.nih.gov/pubmed/24099095
55. Gentaz E, Hatwell Y (2008) Haptic perceptual illusions. In: Grunwald M (ed) Human haptic perception. Birkhäuser, Basel. ISBN 978-3-7643-7611-6. doi:10.1007/978-3-7643-7612-3_17
56. Gescheider GA (1997) Psychophysics—the fundamentals. Lawrence Erlbaum, Mahwah
57. Gescheider GA, Bolanowski SJ, Hardick KR (2001) The frequency selectivity of information—processing channels in the tactile sensory system. Somatosens Motor Res 18(3):191–201. doi:10.1080/01421590120072187
58. Gescheider GA, Migel N (1995) Some temporal parameters in vibrotactile forward masking. J Acoust Soc Am 98(6):3195–3199. doi:10.1121/1.413809
59. Gescheider GA, O'Malley MJ, Verrillo RT (1983) Vibrotactile forward masking: evidence for channel independence. J Acoust Soc Am 74(2):474–485. doi:10.1121/1.389813
60. Gescheider GA, Wright JH, Verillo RT (2009) Information-processing channels in the tactile sensory system. Psychology Press, New York
61. Gescheider GA et al (1985) Vibrotactile forward masking: psychophysical evidence for a triplex theory of cutaneous mechanoreception. J Acoust Soc Am 78(2):534–543. doi:10.1121/1.392475
62. Gescheider GA et al (1990) Vibrotactile intensity discrimination measured by three methods. J Acoust Soc Am 87(1):330–338. doi:10.1121/1.399300
63. Gescheider GA et al (1992) Vibrotactile forward masking as a function of age. J Acoust Soc Am 91(3):1690–1696. doi:10.1121/1.402448
64. Gescheider GA et al (1994) The effects of aging on information-processing channels in the sense of touch: I. Absolute sensitivity. Somatosens Motor Res 11(4):345–357. doi:10.3109/08990229409028878
65. Gescheider GA et al (1995) Vibrotactile forward masking: effects of the amplitude and duration of the masking simulus. J Acoust Soc Am 98(6):3188–3194. doi:10.1121/1.413808
66. Gescheider GA et al (1999) Vibrotactile temporal summation: probability summation or neural integration? Somatos Motor Res 16:229–242. doi:10.1080/08990229970483
67. Gescheider GA et al (2002) A four-channel analysis of the tactile sensitivity of the fingertip: frequency selectivity, spatial summation and temporal summation. Somatosens Motor Res 19:114–124. doi:10.1080/08990220220131505
68. Gescheider GA et al (2005) Spatial summation in the tactile sensory system: probability summation and neural integration. Somatosens Motor Res 22:255–268. doi:10.1080/08990220500420236
69. Gescheider G et al (1984) Effects of the menstrual cycle on vibrotactile sensitivity. Percep Psychophys 36(6):586–592. doi:10.3758/BF03207520
70. Giachritsis C, Wright R, Wing A (2010) The contribution of proprioceptive and cutaneous cues in weight perception: early evidence for maximum-likelihood integration. In: Kappers AML, Bergmann-Tiest WM, van der Helm FC (eds) Haptics: generating and perceiving tangible sensations. LNCS, vol 6191. Proceedings of the eurohaptics conference, Amsterdam, NL. Springer, Heidelberg, pp 11–16. doi:10.1007/978-3-642-14064-8_2
71. Gleeson BT, Horschel SK, Provancher WR (2010) Perception of direction for applied tangential skin displacement: effects of speed, displacement and repetition. IEEE Trans Haptics 3(3):177–188. ISSN 1939-1412. doi:10.1109/TOH.2010.20
72. Goble A, Collins A, Cholewiak R (1996) Vibrotactile threshold in young and old observers: the effects of spatial summation and the presence of a rigid surround. J Acoust Soc Am 99(4):2256–2269. doi:10.1121/1.415413
73. Goff G et al (1965) Vibration perception in normal man and medical patients. J Neurol Neurosurg Psychiatry 28:503–509. doi:10.1136/jnnp.28.6.503

74. Goodwin A, John K, Marceglia A (1991) Tactile discrimination of curvature by humans using only cutaneous information from the fingerpads. Exp Brain Res 86(3):663–672. doi:10.1007/BF00230540
75. Gordon IE, Morison V (1982) The haptic perception of curvature. Percep Psychophys 31(5):446–450. doi:10.3758/BF03204854
76. Green B (1977) The effect of skin temperature on vibrotactile sensitivity. Percep Psychophys 21(3):243–248. doi:10.3758/BF03214234
77. Greenspan J (1984) A comparison of force and depth of skin indentation upon psychophysical functions of tactile intensity. Somatosens Motor Res 2(1):33–48. http://www.ncbi.nlm.nih.gov/pubmed/6505462
78. Greenspan JD, Bolanowski SJ (1996) The psychophysics of tactile perception and its peripheral physiological basis. In: Kruger L, Friedman MP, Carterette EC (1996) Pain and touch. Academic Press, Maryland Heights. ISBN 978-0124269101
79. Grunwald M et al (2001) Haptic perception in anorexia nervosa before and after weight gain. Jo Clin Exp Neuropsychol 23(4):520–529. doi:10.1076/jcen.23.4.520.1229
80. Hale KS, Stanney KM (2004) Deriving haptic design guidelines from human physiological, psychophysical, and neurological foundations. IEEE Comput Graph Appl 24(2):33–39. doi:10.1109/MCG.2004.1274059
81. Handwerker HO (2006) Somatosensorik. In: Schmidt F, Schaible HG (eds) Neuro- und Sinnesphysiologie. Springer, Heidelberg. doi:10.1007/3-540-29491-0
82. Haptex. Grant No. IST-6549, last visited 07.03.2012. European Union. 2007. http://haptex.miralab.unige.ch/
83. Hardy JD, Goodell H, Wolff HG (1951) The influence of skin temperature upon the pain threshold as evoked by thermal radiation. Science. doi:10.1126/science.115.2992.499
84. Harvey LO (1986) Efficient estimation of sensory thresholds. Behav Res Methods 18(6):623–632. doi:10.1163/156856897X00159
85. Hasser CJ (1995) Force-reflecting anthropomorphic hand masters. Technical Report, AL/CF-TR-1995 0110, Armstrong Laboratory. US Air Force. http://oai.dtic.mil/oai/oai?verb=getRecord&metadataPrefix=html&identifier=ADA316017
86. Hatzfeld C, Kern TA, Werthschützky R (2010) Design and evaluation of a measuring system for human force perception parameters. Sens Actuators Phys 162(2):202–209. doi:10.1016/j.sna.2010.01.026
87. Hatzfeld C (2013) Experimentelle Analyse der menschlichen Kraftwahrnehmung als ingenieurtechnische Entwurfsgrundlage für haptische Systeme. Dissertation, Technische Universität Darmstadt. http://tuprints.ulb.tu-darmstadt.de/3392/. Dr. Hut Verlag, München. ISBN 978-3-8439-1033-0
88. Hatzfeld C, Werthschützky R (2012) Just noticeable differences of low-intensity vibrotactile forces at the fingertip. In: Isokoski P, Springare J (eds) Haptics: perception, devices, mobility, and communication. LNCS 7282. Proceedings of the eurohaptics conference, Tampere, FIN. Springer, Heidelberg. doi:10.1007/978-3-642-31404-9_8
89. C. Hatzfeld and R. Werthschützky. Mechanical Impedance as Coupling Parameter of Force and Deflection Perception: Experimental Evaluation. In: Isokoski P, Springare J (eds) Haptics: perception, devices, mobility, and communication. LNCS 7282. Proceedings of the eurohaptics conference, Tampere, FIN. Springer, Heidelberg. doi:10.1007/978-3-642-31401-8_18
90. Hatzfeld C, Werthschützky R (2013) Simulation und Auswahl psychometrischer Verfahren zur Ermittlung von Kennwerten menschlicher Wahrnehmung. In: Knapp W, Gebhardt M (eds) Messtechnisches Symposium des Arbeitskreises der Hochschullehrer für Messtechnik, vol XXVII. Shaker, Aachen, Sept 2013, pp 51–62
91. Hayward V (2008) A brief taxonomy of tactileillusions and demonstrations that can be done in a hardware store. Brain Res Bull 75(6):742–752. doi:10.1016/j.brainresbull.2008.01.008
92. Hayward V, Astley OR (1996) Performance measures for haptic interfaces. Robot Res 1:195–207. doi:10.1007/978-1-4471-0765-1_22
93. Hayward V, MacLean KE (2007) Do it yourself haptics: part I. IEEE Robot Autom Mag 14(4):88–104. doi:10.1109/M-RA.2007.907921

94. Hayward V, MacLean KE (2007) Do it yourself haptics: part II. IEEE Robot Autom Mag 15:104–119. doi:10.1109/M-RA.2007.914919
95. Helbig HB, Ernst MO (2008) Haptic perception in interaction with other senses. In: Grunwald (ed) Human haptic perception—basics and applications. Birkhäuser, pp 235–249. ISBN 978-3764376116
96. Henkin R (1974) Sensory changes during the menstrual cycle. In: Ferin M et al (eds) Biorhythms and human reproduction. Wiley, New York, pp 277–285. ISBN 978-0471257615. http://www.ncbi.nlm.nih.gov/pubmed/11465979
97. Hinterseer P (2009) Compression and transmission of haptic data in telepresence and teleaction systems. Dissertation, Technische Universität München. http://mediatum.ub.tum.de/doc/676484/676484.pdf
98. Hinterseer P et al (2008) Perception-based data reduction and transmission of haptic data in telepresence and teleaction systems. IEEE Trans Sig Process 56(2):588–597. doi:10.1109/TSP.2007.906746
99. Hollins M (2002) Touch and haptics. In: Pashler H (ed) Steven's handbook of experimental psychology. Wiley, New York, pp 585–618. ISBN 978–0471377771. doi:10.1002/0471214426.pas0114
100. Hollins M, Risner SR (2000) Evidence for the duplex theory of tactile texture perception. Percep Psychophys 62(4):695–705. doi:10.3758/BF03206916
101. Hollins M et al (1993) Perceptual dimensions of tactile surface texture: a multidimensional scaling analysis. Attention Percep Psychophys 54(6):697–705. doi:10.3758/BF03211795
102. Höver R et al (2009) Computationally efficient techniques for data-driven haptic rendering. In: Third joint euroHaptics conference and symposium on haptic interfaces for virtual environment and eeleoperator systems (WorldHaptics conference), Salt Lake City, UT, USA, pp 39–44. doi:10.1109/WHC.2009.4810814
103. Howe RD (1994) Tactile sensing and control of robotic manipulation. Adv Robot 8:245–261. doi:10.1163/156855394X00356
104. Howe RD (1992) A force-reflecting teleoperated hand system for the study of tactile sensing in precision manipulation. In: IEEE international conference on robotics and automation. IEEE, pp 1321–1326. doi:10.1109/ROBOT.1992.220166
105. Hugony A (1935) über die Empfindung von Schwingungen mittels des Taststinns. Zeitschrift für Biologie 96:548–553
106. Hwang J, Williams M, Niemeyer G (2004) Toward event-based haptics: rendering contact using open-loop force pulse. In: 12th international symposium on haptic interfaces for virtual environment and teleoperator systems, Chicago, IL, USA. doi:10.1109/HAPTIC.2004.1287174
107. Illert M, Kuhtz-Buschbeck JP (2006) Motorisches system. In: Schmidt F, Schaible FG (eds) Neuro- und Sinnesphysiologie. Springer, Berlin. ISBN 978-3-540-25700-4. doi:10.1007/3-540-29491-0
108. ISO 20462 (2012) Photography—psychophysical experimental methods for estimating image quality—Part 3: quality ruler method, ISO
109. ISO/IEC Guide 98–3 (2008) Uncertainty of measurement ? Part 3: guide to the expression of uncertainty in measurement. Genf, CH: ISO
110. Israr A, Choi S, Tan HZ (2006) Detection threshold and mechanical impedance of the hand in a pen-hold posture. In: Peking C (eds) International conference on intelligent robots and systems (IROS), pp 472–477. doi:10.1109/IROS.2006.282353
111. Israr A, Choi S, Tan HZ (2007) Mechanical impedance of the hand holding a spherical tool at threshold and suprathreshold stimulation levels. In: Second joint euroHaptics conference and symposium on haptic interfaces for virtual environment and teleoperator systems (WorldHaptics conference), Tsukuba. doi:10.1109/WHC.2007.81
112. Israr A, Tan HZ (2006) Frequency and amplitude discrimination along the kinesthetic-cutaneous continuum in the presence of masking stimuli. J Acoust Soc Am 120(5):2789–2800. doi:10.1121/1.2354022

113. Jandura L, Srinivasan M (1994) Experiments on human performance in torque discrimination and control. In: Dynamic systems and control, ASME, DSC-55 1, pp 369–375. http://www.rle.mit.edu/touchlab/publications/1994_002.pdf
114. Johansson RS, Birznieks I (2004) First spikes in ensembles of human tactile afferents code complex spatial fingertip events. Nat Neurosci 7(2):170–177. doi:10.1038/nn1177
115. Johnson KO (2001) The roles and functions of cutaneous mechanoreceptors. Curr Opin Neurobiol 11(4):455–461. doi:10.1016/S0959-4388(00)00234-8
116. Johnson K (2002) Neural basis of haptic perception. In: Pashler H (ed) Steven's handbook of experimental psychology. Wiley, New York, pp 537–583. doi:10.1002/0471214426.pas0113
117. Johnson K, Yoshioka T, Vega-Bermudez F (2000) Tactile functions of mechanoreceptive afferents innervating the hand. J Clin Neurophysiol 17(6):539. doi:10.1097/00004691-200011000-00002
118. Johnson K et al (1979) Coding of incremental changes in skin temperature by a population of warm fibers in the monkey: correlation with intensity discrimination in man. J Neurophysiol 42(5):1332–1353. http://www.ncbi.nlm.nih.gov/pubmed/114610
119. Jones L (2000) Kinesthetic sensing. MIT Press, Cambridge. http://citeseerx.ist.psu.edu/viewdoc/summary?doi=10.1.1.133.5356
120. Jones L, Hunter I (1990) A perceptual analysis of stiffness. Exp Brain Res 79(1):150–156. doi:10.1007/BF00228884
121. Jones L, Hunter I (1993) A perceptual analysis of viscosity. Exp Brain Res 94(2):343–351. doi:10.1007/BF00230304
122. Jones L, Lederman S (2006) Human hand function. Oxford University Press, Oxford
123. Jones L, Piateski E (2006) Contribution of tactile feedback from the hand to the perception of force. Exp Brain Res 168:289–302. doi:10.1007/s00221-005-0259-8
124. Jones LA (1989) Matching forces: constant errors and differential thresholds. Perception 18(5):681–687. doi:10.1068/p180681
125. Jones LA, Ho H-N (2008) Warm or cool, large or small? The challenge of thermal displays. IEEE Trans Haptics 1(1):53–70. doi:10.1109/TOH.2008.2
126. Jung J, Ryu J, Choi S (2007) Physical and perceptual characteristics of vibration rendering in mobile device. ACM Trans Appl Perception
127. Kaczmarek KA, Bach-Y-Rita P (1995) Tactile displays. In: Barfield W, Furness T (eds) Virtual environments and advanced interface design. Oxford University Press, New-York, pp 349–414. ISBN 978-0195075557
128. Kaczmarek K et al (1991) Electrotactile and vibrotactile displays for sensory substitution systems. IEEE Trans Biomed Eng 38(1):1–16. doi:10.1109/10.68204
129. Kaernbach C (2001) Adaptive threshold estimation with unforced-choice tasks. Percep Psychophys 63(8):1377–1388. doi:10.3758/BF03194549
130. Kappers AML, Koenderink JJ (1999) Haptic perception of spatial relations. Perception 28(6):781–795. doi:10.1068/p2930
131. Karam M, Schraefel MC (2005) A taxonomy of gestures in human computer interactions. Techincal Report, University of Southampton. http://eprints.soton.ac.uk/id/eprint/261149
132. Keppel G (1991) Design and analysis: a researcher's handbook. Pearson Education, Old Tappan
133. Kern T (2006) Haptisches Assistenzsystem für diagnostische und therapeutische Katheterisierungen. PhD thesis, Techische Universität Darmstadt, Institut für Elektromechanische Konstruktionen. http://tuprints.ulb.tu-darmstadt.de/761/
134. Kildal J (2012) Kooboh: variable tangible properties in a handheld haptic-illusion box. In: Isokoski P, Springare J (eds) Haptics: perception, devices, mobility, and communication. Proceedings of the eurohaptics conference, Tampere. Springer, Heidelberg, pp 191–194. doi:10.1007/978-3-642-31404-9_33
135. Kimura T, Nojima T (2012) Pseudo-haptic feedback on softness induced by grasping motion. In: Haptics: perception, devices, mobility, and communication. Springer, Berlin, pp 202–205. doi:10.1007/978-3-642-31404-9_36

136. King HH, Donlin R, Hannaford B (2010) Perceptual thresholds for single vs. multi-finger haptic interaction. In: IEEE haptics symposium, Waltham, MA, USA, pp 95–99. doi:10.1109/HAPTIC.2010.5444670
137. King-Smith PE et al (1994) Efficient and unbiased modifications of the QUEST threshold method: theory, simulations, experimental evaluation and practical implementation. Vision Res 34(7):885–912. doi:10.1016/0042-6989(94)90039-6
138. Klatzky RL, Pawluk D, Peer A (2013) Haptic perception of material properties and implications for applications. Proc IEEE 101:2081–2092. doi:10.1109/JPROC.2013.2248691
139. Klein S (2001) Measuring, estimating, and understanding the psychometric function: a commentary. Attention Percep Psychophys 63(8):1421–1455. doi:10.3758/BF03194552
140. Klinke R (2010) Das zentrale Nervensystem—Grundlage bewussten Menschseins. In: Klinke R et al (eds) Physiologie. Thieme, pp 623–642. ISBN 978-3137960065
141. Knowles WB, Sheridan TB (1966) The feel of rotary controls: friction and inertial. human factors. J Human Fact Ergon Soc 8(3):209–215. doi:10.1177/001872086600800303
142. Kontarinis DA, Howe RD (1993) Tactile display of contact shape in dextrous telemanipulation. In: ASME winter annual meeting: advances in robotics, mechatronics and haptic interfaces, New Orleans, DSC-vol 49, pp 81–88
143. Kontsevich L, Tyler C (1999) Bayesian adaptive estimation of psychometric slope and threshold. Vision Res 39(16):2729–2737. doi:10.1016/S0042-6989(98)00285-5
144. Kruger L, Friedman MP, Carterette EC (eds) (1996) Pain and touch. Academic Press, Maryland Heights. ISBN 978-0123992390
145. Kuchenbecker KJ, Fiene J, Niemeyer G (2006) Improving contact realism through event-based haptic feedback. IEEE Trans Visual Comput Graph 12(2):219–230. doi:10.1109/TVCG.2006.32
146. Kuchenbecker K et al (2010) VerroTouch: high-frequency acceleration feedback for telerobotic surgery. In: Kappers AML, Bergmann-Tiest WM, van der Helm FC (eds) Haptics: generating and perceiving tangible sensations. Proceedings of the eurohaptics conference, Amsterdam, NL. Springer, Heidelberg, pp 189–196. doi:10.1007/978-3-642-14064-8_28
147. Kuroki S, Watanabe J, Nishida S (2012) Dissociation of vibrotactile frequency discrimination performances for supra-threshold and near-threshold vibrations. In: Isokoski P, Springare J (eds) Haptics: perception, devices, mobility, and communication. Proceedings of the eurohaptics conference, Tampere. Springer, Heidelberg, pp 79–84. doi:10.1007/978-3-642-31404-9_14
148. Kwon DS et al (2001) Realistic force reflection in a spine biopsy simulator. In: IEEE international conference on robotics and automation, vol 2, pp 1358–1363. doi:10.1109/ROBOT.2001.932799
149. Kyung KH et al (2005) Perceptual and biomechanical frequency response of human dkin: implication for design of tactile displays. In: First joint eurohaptics conference and symposium on haptic interfaces for virtual environment and teleoperator systems (WorldHaptics conference). doi:10.1109/WHC.2005.105
150. LaMotte RH, Srinivasan M (1991) Surface microgeometry: tactile perception and neural encoding. In: Franzen O, Westman J (eds) Information processing in the somatosensory system. MacMillan Press, London, pp 49–58. ISBN 978-0333524930
151. Lawrence DA et al (2000) Rate-hardness: a new performance metric for haptic interfaces. IEEE Trans Robot Autom 16:357–371. doi:10.1109/70.864228
152. Lécuyer A (2009) Simulating haptic feedback using vision: a survey of research and applications of pseudo-haptic feedback. Pres Teleoper Virt Environ 18(1):39–53. doi:10.1162/pres.18.1.39
153. Lederman SJ (1991) Skin and touch. In: Dulbecco R (ed) Encyclopedia of human biology. Academic Press, Maryland Heights, pp 51–63. ISBN 978-0122267475
154. Lederman SJ, Klatzky RL (2009) Haptic perception: a tutorial. Attention Percep Psychophys 71(7):1439. doi:10.3758/APP.71.7.1439
155. Lederman S, Jones L (2011) Tactile and Haptic Illusions. IEEE Transactions on Haptics 4(4):273–294. doi:10.1109/TOH.2011.2

156. Lederman SJ (1981) The perception of surface roughness by active and passive touch. Bull Psychonomic Soc 18(5):253–255. doi:10.3758/BF03333619
157. Lederman SJ, Klatzky RL (1987) Hand movements: a window into haptic object recognition. Cogn psychol 19(3):342–368. doi:10.1016/0010-0285(87)90008-9
158. Leek MR (2001) Adaptive procedures in psychophysical research. Percep Psychophys 63(8):1279–1292. doi:10.3758/BF03194543
159. Lenk A et al (2011) Electromechanical systems in microtechnology and mechatronics: electrical, mechanical and acoustic networks, their interactions and applications. Springer, Heidelberg. ISBN 978-3-642-10806-8
160. Levitt H (1971) Transformed up-down methods in psychoacoustics. J Acous Soc Am 49(2):467–477. doi:10.1121/1.1912375
161. Li Y et al (2008) Passive and active kinesthetic perception just noticeable difference for natural frequency of virtual dynamic systems. In: Symposium on haptic interfaces for virtual environments and teleoperator systems, Reno. doi:10.1109/HAPTICS.2008.4479908
162. Libouton X et al (2012) Tactile roughness discrimination of the finger pad relies primarily on vibration sensitive afferents not necessarily located in the hand. Behav Brain Res 229(1):273–279. doi:10.1016/j.bbr.2012.01.018
163. MacLean K and Enriquez M (2003) Perceptual design of haptic icons. In: Proceedings of EuroHaptics. http://citeseerx.ist.psu.edu/viewdoc/summary?10.1.1.138.6172.
164. Macmillan N, Creelman C (2005) Detection theory: a user's guide. Lawrence Erlbaum, London
165. Mahns D et al (2006) Vibrotactile frequency discrimination in human hairy skin. J Neurophysiol 95(3):1442. doi:10.1152/jn.00483.2005
166. Makous JC, Gescheider GA, Bolanowski SJ (1996) Decay in the effect of vibrotactile masking. J Acoust Soc Am 99(2):1124–1129. doi:10.1121/1.414597
167. McMahan W et al (2011) Tool contact acceleration feedback for telerobotic surgery. IEEE Trans Haptics 4(3):210–220. doi:10.1109/TOH.2011.31
168. Morioka M, Griffin MJ (2005) Thresholds for the perception of hand-transmitted vibration: dependence on contact area and contact location. Somatosens Motor Res 22:281–297. doi:10.1080/08990220500420400
169. Musmann H (2006) Genesis of the MP3 audio coding standard. IEEE Trans Consum Electron 52(3):1043–1049. doi:10.1109/TCE.2006.1706505
170. Nitsch V, Färber B (2012) A meta-analysis of the effects of haptic interfaces on task performance with teleoperation systems. IEEE Trans Haptics 6:387–398. doi:10.1109/ToH.2012.62
171. Okamura AM, Dennerlein JT, RD Howe (1998) Vibration feedback models for virtual environments. In: International conference on robotics & automation. doi:10.1109/ROBOT.1998.677050
172. Okazaki R, Kajimoto H, Hayward V (2012) Vibrotactile stimulation can affect auditory loudness: a pilot study. In: Isokoski P, Springare J (eds) Haptics: perception, devices, mobility, and communication. LNCS 7282. Proceedings of the eurohaptics conference, Tampere. Springer, Heidelberg, pp 103–108. doi:10.1007/978-3-642-31404-9_18
173. Otto S, Weinzierl S (2009) Comparative simulations of adaptive psychometric procedures. In: Jahrestagung der Deutschen Gesellschaft für Akustik Dt Ges für Akustik, Rotterdam, pp 1276–1279. ISBN 9783980865968
174. Paek TS, Bahl P, Foehr OH (2011) Interacting with a mobile device within a vehicle using gestures. 20130155237 A1
175. Pai D, Rizun P (2003) The WHaT: a wireless haptic texture sensor. In: 11th symposium on haptic interfaces for virtual environment and teleoperator systems, pp 3–9. doi:10.1109/HAPTIC.2003.1191210
176. Pang X, Tan H, Durlach N (1991) Manual discrimination of force using active finger motion. Percep Psychophys 49(6):531–540. doi:10.3758/BF03212187
177. Pang X, Tan H, Durlach N (1992) Manual resolution of length, force and compliance. In: ASME DSC Adv Robot 42:13–18. https://engineering.purdue.edu/hongtan/pubs/PDFfiles/C05_Tan_ASME1992.pdf

178. Pare M, Carnahan H, Smith A (2002) Magnitude estimation of tangential force applied to the fingerpad. Exp Brain Res 142(3):342–348. doi:10.1007/s00221-001-0939-y
179. Peters R, Hackeman E, Goldreich D (2009) Diminutive digits discern delicate details: fingertip size and the sex difference in tactile spatial acuity. J Neurosci 29(50):15756. doi:10.1523/?JNEUROSCI.3684-09.2009
180. Pongrac H (2008) Vibrotactile perception: examining the coding of vibrations and the just noticeable difference under various conditions. Multimedia Syst 13(4):297–307. doi:10.1007/s00530-007-0105-x
181. Pongrac H et al (2006) Limitations of human 3D force discrimination. In: Proceedings of human—centered robotics systems. http://citeseerx.ist.psu.edu/viewdoc/download?doi:10.1.1.68.8597%26rep=rep1%26type=pdf.
182. Prins N, Kingdom FAA (2010) Psychophysics: a practical introduction. Academic Press, Maryland Heights
183. Provancher WR, Sylvester ND (2009) Fingerpad skin stretch increases the perception of virtual friction. IEEE Trans Haptics 2(4):212–223. doi:10.1109/TOH.2009.34
184. Pusch A, Lécuyer A (2011) Pseudo-haptics: from the theoretical foundations to practical system design guidelines. In: Proceedings of the 13th international conference on multimodal interfaces, ACM 2011, pp 57–64. doi:10.1145/2070481.2070494
185. Rank M et al (2012) Masking effects for damping JND. In: Isokoski P, Springare J (eds) Haptics: perception, devices, mobility, and communication. LNCS 7282. Proceedings of the eurohaptics conference, Tampere. Springer, Heidelberg, pp 145–150. doi:10.1007/978-3-642-31404-9_25
186. Rausch J et al (2006) INKOMAN-analysis of mechanical behaviour of liver tissue during intracorporal interaction. In: Gemeinsame Jahrestagung der Deutschen, Österreichischen und Schweizerischen Gesellschaften für Biomedizinische Technik 6(9)
187. Rausch J (2005) Analyse der mechanischen Eigenschaften von Lebergewebe bei intrakorporaler Interaktion. Diploma Thesis, Darmstadt: Technische Universität Darmstadt, Institut für ElektromechanischeKonstruktionen. http://tubiblio.ulb.tu-darmstadt.de/53792/
188. Redmond B et al (2010) Haptic characteristics of some activities of daily living. In: Haptics Symposium, IEEE, pp 71–76. doi:10.1109/HAPTIC.2010.5444674
189. Rösler F, Battenberg G, Schüttler F (2009) Subjektive Empfindungen und objektive Charakteristika von Bedienelementen. Automobiltechnische Zeitschrift 4:292–297. doi:10.1007/BF03222068
190. Salisbury C et al (2011) What you can't feel won't hurt you: evaluating haptic hardware using a haptic contrast sensitivity function. IEEE Trans Haptics 4(2):134–146. doi:10.1109/TOH.2011.5
191. Samur E (2010) Systematic evaluation methodology and performance metrics for haptic interfaces. Dissertation, école Polytechnique Fédérale de Lausanne. doi:10.5075/epfl-thesis-4648. http://infoscience.epfl.ch/record/145888
192. Scheibert J et al (2004) A novel biomimetic haptic sensor to study the physics of touch. In: Colloque Mé, nanotransduction, Paris. http://www.lps.ens.fr/scheibert/MT2004.pdf
193. Seow KC (1988) Physiology of touch, grip and gait. In: Webster JG (ed) Tactile sensors for robotics and medicine. Wiley, New York, pp 13–40. ISBN 978-0471606079
194. Shimoga K (1993) A survey of perceptual feedback issues in dexterous telemanipulation part I. Finger force feedback. In: Proceedings of the IEEE virtual reality annual international symposium, Seattle, WA, USA, pp 263–270. doi:10.1109/VRAIS.1993.380770
195. Smith AM, Gosselin G, Houde B (2002) Deployment of fingertip forces in tactile exploration. Exp Brain Res 147(2):209–218. doi:10.1007/s00221-002-1240-4
196. Smith CU (2000) The biology of sensory systems. Wiley, Chichester, p 445. ISBN 0-471-89090-1
197. Sodhi R et al (2013) AIREAL: interactive tactile experiences in free air. ACM Trans Graph (TOG) 32(4):134. doi:10.1145/2461912.2462007
198. Steinbach E et al (2011) Haptic data compression and communication. Sig Process Mag 28(1):87–96. doi:10.1109/MSP.2010.938753

199. Stevens SS (1975) Psychophysics. Transaction Books, Piscataway. ISBN 978-0887386435
200. Symmons M et al (2005) Active versus passive touch in three dimensions. In: First joint eurohaptics conference and symposium on haptic interfaces for virtual environment and teleoperator systems (WorldHaptics Conference), Pisa. doi:10.1109/WHC.2005.20
201. Tan HZ et al (1993) Manual resolution of compliance when work and force cues are minimized. In: Advances in robotics, mechatronics and haptic interfaces, pp 99–104. http://citeseerx.ist.psu.edu/viewdoc/download?10.1.1.50.2758&rep=rep1&type=pdf
202. Tan HZ et al (1994) Human factors for the design of force-reflecting haptic interfaces. In: ASME DSC Dyn Syst Control 55(1):353–359. http://touchlab.mit.edu/publications/1994_004.pdf
203. Tan HZ et al (1999) Information transmission with a multifinger tactual display. Percep Psychophys 61(6):993–1008. doi:10.3758/BF03207608
204. Tan HZ et al (2006) Force direction discrimination is not influenced by reference force direction. In: Haptics-e 1, pp 1–6. http://jks-folks.stanford.edu/papers/Haptic-Discrimination.pdf
205. Tan H, Rabinowitz W (1996) A new multi-finger tactual display. J Acous Soc Am 99(4):2477–2500. doi:10.1121/1.415560
206. Tan H et al (2003) Temporal masking of multidimensional tactual stimuli. J Acous Soc Am 116(9):3295–3308. doi:10.1121/1.1623788
207. Tanaka Y et al (2012) Contact force during active roughness perception. In: Isokoski P, Springare J (eds) Haptics: perception, devices, mobility, and communication, vol 7282. LNCS. Proceedings of the eurohaptics conference, Tampere. Springer, Heidelberg, pp 163–168. doi:10.1007/978-3-642-31404-9_28
208. Taylor M, Creelman C (1967) PEST: efficient estimates on probability functions. J Acous Soc Am 41(4):782–787. doi:10.1121/1.1910407
209. Tillmann BN (2010) Atlas der Anatomie des Menschen. Springer, Berlin. ISBN 978-3-642-02679-9
210. Toffin D et al (2003) Perception and reproduction of force direction in the horizontal plane. J Neurophys 90:3040–3053. doi:10.1152/jn.00271.2003
211. Treede RD (2007) Das somatosensorische system. In: Schmidt RF, Lang F (eds) Physiologie des Menschen. Springer, Heidelberg, pp 296–323. doi:10.1007/978-3-540-32910-7_14
212. Å. Vallbo A, Johansson R et al (1984) Properties of cutaneous mechanoreceptors in the human hand related to touch sensation. Human Neurobiol 3(1):3–14. http://www.ncbi.nlm.nih.gov/pubmed/6330008
213. Verrillo RT, Gescheider GA (1992) Perception via the sense of touch. In: Summers IR (ed) Tactile aids for the hearing impaired. Whurr, London, pp 1–36. ISBN 978-1-870332- 17-0
214. Verrillo R (1979) Comparison of vibrotactile threshold and suprathreshold responses in men and women. Percep Psychophys 26(1):20–24. doi:10.3758/BF03199857
215. Verrillo R (1980) Age related changes in the sensitivity to vibration. J Gerontol 35(2):185–193. doi:10.1093/geronj/35.2.185
216. Verrillo R (1982) Effects of aging on the suprathreshold responses to vibration. Percep Psychophys 32(1):61–68. doi:10.3758/BF03204869
217. Verrillo R, Bolanowski S (1986) The effects of skin temperature on the psychophysical responses to vibration on glabrous and hairy skin. J Acous Soc Am 80:528. doi:10.1121/1.394047
218. Verrillo R et al (1998) Effects of hydration on tactile sensation. Somatos Motor Res 15(2):93–108. doi:10.1080/08990229870826
219. Wagner M, Gerling G, Scanlon J (2008) Validation of a 3-D finite element human fingerpad model composed of anatomically accurate tissue layers. In: Symposium on haptic interfaces for virtual environment and teleoperator systems, IEEE 2008, pp 101–105. doi:10.1109/HAPTICS.2008.4479922
220. Wang Z et al (2012) A 3-D nonhomogeneous FE model of human fingertip based on MRI measurements. IEEE Transactions on Instrument Measure 61(12):3147–3157. doi:10.1109/TIM.2012.2205102
221. Weber EH (1905) Tastsinn und Gemeingefühl. Engelmann. ISBN 9783836402491 (Reprint)

222. Weinstein S (1968) Intensive and extensive aspects of tactile sensitivity as a function of body part, sex, and laterality. In: First international symposium on the skin senses
223. Wichmann F, Hill N (2001) The psychometric function: I. Fitting, sampling, and goodness of fit. Percep Psychophys 63(8):1293. doi:10.3758/BF03194544
224. Wickens TD (2002) Elementary signal detection theory. Oxford University Press, Oxford
225. Wiertlewski M (2013) Reproduction of tactual textures: transducers, mechanics, and signal encoding. Springer, Berlin. ISBN 978-1-4471-4840-1

Chapter 3
The User's Role in Haptic System Design

Thorsten A. Kern and Christian Hatzfeld

Abstract A good mechanical design has to consider the user in his or her mechanical properties. The first part of this chapter deals with the discussion of the user as a mechanical *load* on the haptic device. The corresponding model is split into two independent elements depending on the frequency range of the oscillation. Methods and measurement setups for the derivation of mechanical impedance of the user are reviewed, and a thorough analysis of impedance for different grip configurations is presented. In the second part of the chapter, the user is considered as the ultimate measure of quality for a haptic system. The relation of psychophysical parameters like the absolute threshold or the JND to engineering quality measures like resolution, errors, and reproducibility is described and application depending quality measures like haptic transparency are introduced.

3.1 The User as Mechanical Load

Thorsten A. Kern

3.1.1 Mapping of Frequency Ranges onto the User's Mechanical Model

The area of active haptic interaction movements—made in a conscious and controlled way by the user—is of limited range. Sources concerning the dynamics of human movements differ as outlined in the preceding chapters. The fastest conscious move-

---

T.A. Kern (✉)
Continental Automotive GmbH, VDO-Straße 1, 64832 Babenhausen, Germany
e-mail: t.kern@hapticdevices.eu

C. Hatzfeld
Institute of Electromechanical Design, Technische Universität Darmstadt,
Merckstr. 25, 64283 Darmstadt, Germany
e-mail: c.hatzfeld@hapticdevices.eu

C. Hatzfeld and T.A. Kern (eds.), *Engineering Haptic Devices*,
Springer Series on Touch and Haptic Systems, DOI 10.1007/978-1-4471-6518-7_3

ment performed by humans is done with the fingers. Movements for typing of up to 8 Hz can be observed.[1] As these values refer to a ten-finger interaction, they have to be modified slightly. However, as the border frequency of a movement lies above the pure number of a repetitive event, an assumption of the upper border frequency of 10 Hz for active, controlled movement covers most cases.

The major part of the spectrum of haptic perception is passive (*passive haptic interaction*, see Fig. 1.7). The user does not have any active influence or feedback within this passive frequency range. In fact, the user is able to modify his properties as a mechanical load by altering the force when holding a knob. But although this change influences the higher frequency range, the change itself happens with lower dynamics within the dynamic range of active haptic interaction. A look at haptic systems addressing tactile and kinaesthetic interaction channels shows that the above modeling has slightly different impacts:

- The output values of kinaesthetic systems F_{out} (Fig. 3.1a) result in two reactions by the user. First, a spontaneous, not directly controllable movement reaction v_{spo} happens as a result of the mechanical properties of the fingertip (depending on the type of grasp, this can be also the complete interior hand and its skin elasticity). Second, an additional perception of forces takes place. This perception K[2] is weighted according to the actual situation and results in a conscious reaction of the motor parts of the body. These induced reactions v_{ind} summed up with the spontaneous reactions result in the combined output value v_{out} of the user.
- The movements of tactile devices v_{out} (Fig. 3.1b) and the consciously performed movement of the user v_{ind} result in a combined movement and velocity. This elongation acts on the skin, generating the output value F_{out} as a result of its mechanical properties. This conscious movement v_{ind} sums up to v_{out} in the opposite direction of the original movement, as with opposite movement directions, the skin's elongation increases and results in a larger force between user and technical system. Analogously, it subtracts with movements in the same direction, as in this case, the device (or the user, depending on the point of view) evades the acting force trying to keep deformation low and to perceive just a small haptic feedback. According to this model, only the output value F_{out} of the combined movement is perceived and contributes to a willingly induced movement.

If you transfer the model of Fig. 3.1 into an abstract notation, all blocks correspond to the transfer function G_{Hn}. Additionally, it has to be considered that the user's reaction K' is a combined reaction of complex habits and the perception K; therefore, a necessity to simplify this branch of the model becomes eminent. For the purpose of device design and requirement specification, the conscious reaction is modeled by a disturbing variable only limited in bandwidth, resulting in a block diagram

---

[1] 8 Hz corresponds to a typing speed of 480 keystrokes per minute. Four hundred keystrokes are regarded as very good for a professional typist, 300–200 keystrokes are good, and 100 keystrokes can be achieved by most laymen.

[2] K, a variable chosen completely arbitrarily, is a helpful construct for understanding block diagrams rather than having a real neurological analogy.

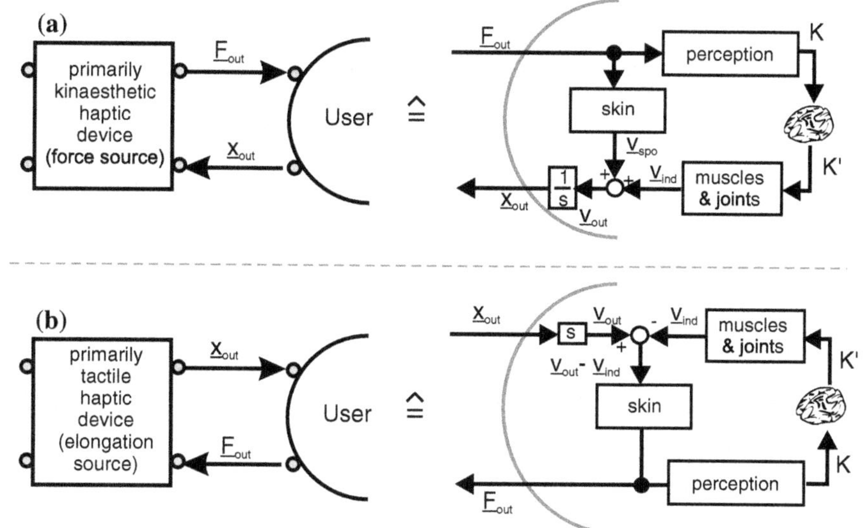

Fig. 3.1 User models as a block structure from kinaesthetic (**a**) and tactile (**b**) systems

according to Fig. 3.2c for kinaesthetic and according to Fig. 3.2d for tactile devices. The transfer function G_{H3} corresponds to the mechanical admittance of the grasp above the border frequency of user interaction f_g.

With regard to the application of the presented models, there are two remarks to be considered:

- The notation in Figs. 3.1 and 3.2 for elongations x and forces F being input and output values of users is just one approach to the description. In fact, an *impedance coupling* exists between user and haptic system, making it impossible to distinguish between input and output parameters. However, the decoupled haptic device is designed for being a position or force source. This in fact is the major motivation to define input and/or output parameters of the user. But there are certain actuators (e.g., ultrasonic devices) which can hardly be defined as being part of either one of these classes. As a consequence, when describing either system, the choice of the leading sign and the direction of arrows should carefully be done!
- The major motivation for this model is the description of a mechanical load for the optimized dimensioning of a haptic system. To guarantee the closed-loop control engineering stability of a simulation or a telemanipulation system, further care has to be taken of the frequency range of active haptic interaction below 10 Hz. Stability analysis in this area can either be achieved by more detailed models or by an observation of input and output values according to their *control engineering passivity*. Further information on this topic can be found in Chap. 7.

The following sections on user impedance give a practical model for the transfer function G_{H3} used in Fig. 3.2.

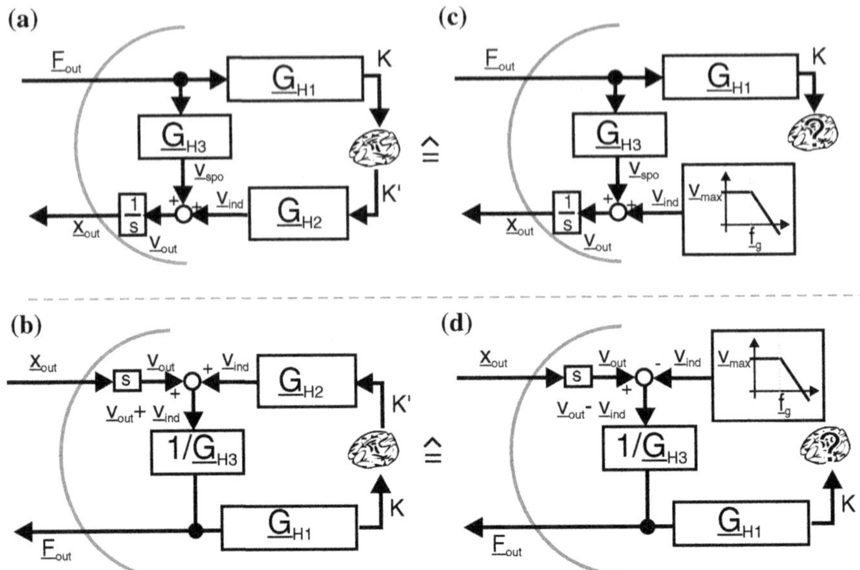

Fig. 3.2 Transformation of the user models' block structures in transfer functions including simplifications of the model for the area of active haptic interaction for kinaesthetic (**a** + **c**) and tactile (**b** + **d**) systems

3.1.2 Modeling the Mechanical Impedance

The user's reaction as part of any haptic interaction combines a conscious, bandwidth-limited portion—the area of active haptic interaction—and a passive portion, mainly resulting from the mechanical properties of fingers, skin, and bones. The influence of this second part stretches across the whole frequency range, but emphasizes the upper area for high frequencies. This section describes the passive part of haptic interaction. The transfer function G_{H3} as in Fig. 3.2 is a component of the impedance coupling with force-input and velocity-output and is therefore a mechanical admittance of the human Y_H respectively the mechanical impedance Z_H.

$$G_{\text{H3}} = \frac{v_{\text{spo}}}{F_{\text{out}}} = \frac{v_{\text{out}} - v_{\text{ind}}}{F_{\text{out}}} = Y_H = \frac{1}{Z_H} \tag{3.1}$$

In the following, this mechanical impedance of the user is specified. The parameter impedance combines all mechanical parameters of an object or system that can be expressed in a linear, time-invariant description, i.e., mass m, compliance k, and damping d. High impedance therefore means that an object has at least one of the three properties:

1. hard and stiff in the sense of spring stiffness,
2. large mass in the sense of inertial force,
3. sticky and tight in the sense of high friction.

In any case, a small movement (velocity v) results in a high force reaction F with high impedances. Low impedance means that the object, the mechanics, is accordingly soft and light. Even high velocities result in small counter forces in this case. The human mechanical impedance is dependent on a number of influence parameters:

- type of grasp being directly influenced by the construction of the handle,
- physiological condition,
- grasping force being directly influenced by the will of the user,
- skin surface properties, for example, skin moisture.

The quantification of human mechanical impedance requires taking as many aspects into account as possible. The type of grasp is defined by the mechanical design of the device. Nevertheless, a selection of typical grasping situations will give a good overview of typical impedances appearing during human–machine interaction. User–individual parameters like physiological condition and skin structure can be covered best by the analysis of a large number of people of different conditions. By choosing this approach, a span of percentiles can be acquired covering the mechanical impedances typically appearing with human users. The "free will" itself, however, is—similar to the area of active haptic interaction—hard if not impossible to be modeled. The time-dependent and unpredictable user impedance dependency on the will can only be compensated if the system is designed to cover all possible impedance couplings of actively influenced touch. Another approach would be to indirectly measure the will to adapt the impedance model of the user within the control loop. Such an indirect measure is, in many typical grasping situations, the force applied between two fingers or even the whole hand holding an object or a handle. In the simplest design, the acquisition of such a force can be done by a so-called *dead-man-switch*, which in 1988 was proposed by HANNAFORD for use in haptic systems [8]. A dead-man-switch is pressed as long as the user holds the control handle in his or her hand. It detects the release of the handle resulting in a change in impedance from Z_H to 0.

3.1.3 Grips and Grasps

There is a nomenclature for different types of grasps shown in Fig. 3.3. The hand is an extremity with 27 bones and 33 muscles. It combines 13 (fingers), respectively, 15 (incl. the wrist) degrees of freedom.[3] Accordingly, the capabilities of man to grasp are extremely versatile.

[3] Thumb: 4 DoF, index finger: 3 DoF, middle finger: 2 DoF (sometimes 3 DoF), ring finger: 2 DoF, small finger: 2 DoF, wrist: 2 DoF. The rotation of the whole hand happens in the forearm and therefore does not count among the degrees of freedom of the hand itself.

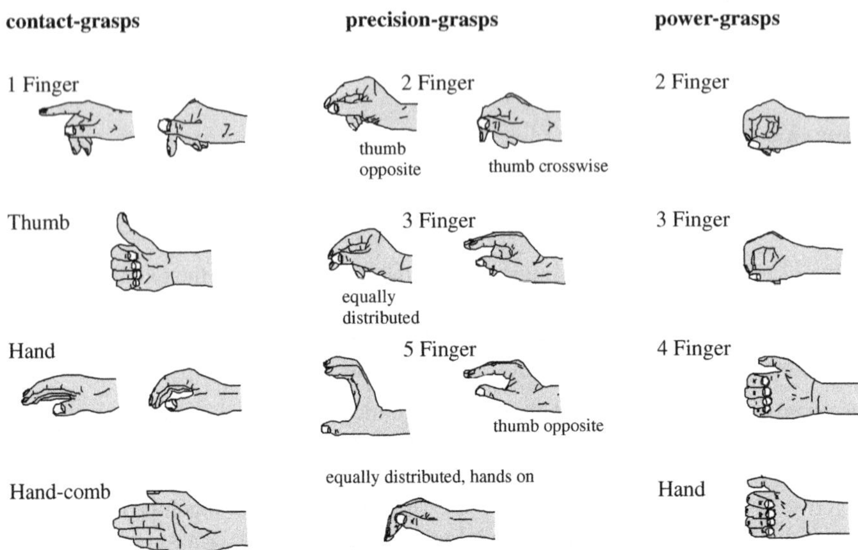

Fig. 3.3 Grip configurations, figure based on [2]

There are three classes of grasps to be distinguished:

- The **contact grasp** describes the touch of an object using the whole hand or major parts of it. Keys and buttons are typically actuated by contact grasps. Even the fingers resting on a keyboard or a piano are called contact grasps. A contact grasp always blocks one direction of movement for an object (which is one half of a degree of freedom). Contact grasps can be regarded as linear only in case of a preload high enough. With light touches, the point of release and the according liftoff of the object are always nonlinear.
- The **precision grasp** describes grasping with several fingers. Typically, a precision grasp locks at least one degree of freedom of the grasped object by form closure with one finger and a counter bearing—often another finger. Additional degrees of freedom are hindered by friction. Precision grasps vary in stiffness of coupling between man and machine. At the same time, they are the most frequent type of grasping.
- The **power grasp** describes an object with at least one finger and a counter bearing, which may be another finger, but is frequently the whole hand. The power grasp aims at locking the grasped object in all degrees of freedom by a combination of form and force closures. Power grasps are—as the name already implies—the stiffest coupling between humans and machines.

Further discrimination of grasps is made by FEIX ET AL. [3] and documented online with the purpose of reducing the mechanical complexity of anthropomorphic hands [4]. The reported taxonomy could be useful for very specialized task-specific

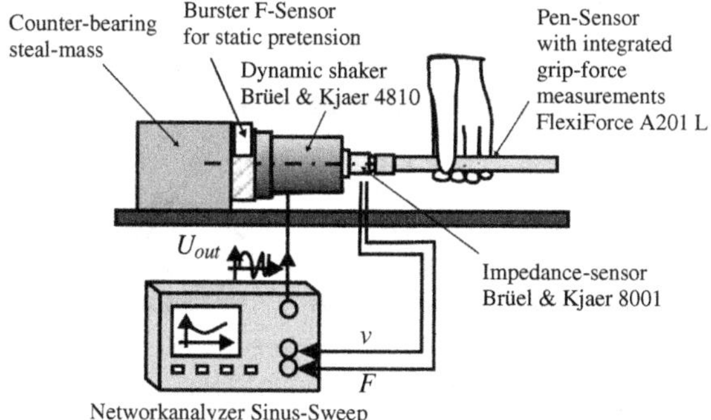

Fig. 3.4 Measurement setup for the acquisition of user impedances according to [16]

systems. For all classes of grasps, measurements of the human's impedance can be performed. According to the approach presented by KERN [16], the measurement method and the models of user impedance are presented including the corresponding model parameters in the following sections.

3.1.4 Measurement Setup and Equipment

The acquisition of mechanical impedances is a well-known problem in measurement technology. The principle of measurement is based on an excitation of the system to be measured by an actuator, simultaneously measuring force and velocity responses of the system. For this purpose, combined force and acceleration sensors (e.g., the impedance sensor 8001 from *Brüel & Kjær*, Nærum, DK) exist, whereby the charge amplifier of the acceleration sensor includes an integrator to generate velocity signals.

In [22], WIERTLEWSKI and HAYWARD argue that measurements with impedance heads are prone to measurement errors because of the mechanical construction of the sensor based on BROWNJOHN ET AL. [1]. However, errors induced by the construction of the measurement head appear at frequencies larger than 2,000 Hz, values that are only seldom used in the design of haptic interfaces. Furthermore, interpersonal variations and calibration of the measurement setup based on a concentrated network parameter approach are used to minimize the errors even for high frequencies in the following.

In general, the impedance of organic systems is *nonlinear* and *time variant*. This nonlinearity is a result of a general viscoelastic behavior of tissue resulting from a combined response of relaxation, conditioning, stretching, and creeping [6]. These effects can be reproduced by mechanical models with concentrated elements. However, they are dependent on the time history of excitation to the measured object. It can be expected that measurements based on step excitation are different from those acquired with a sinusoid sweep. Additionally, the absolute time for measurement has some influence on the measures by conditioning. Both effects are systematic mea-

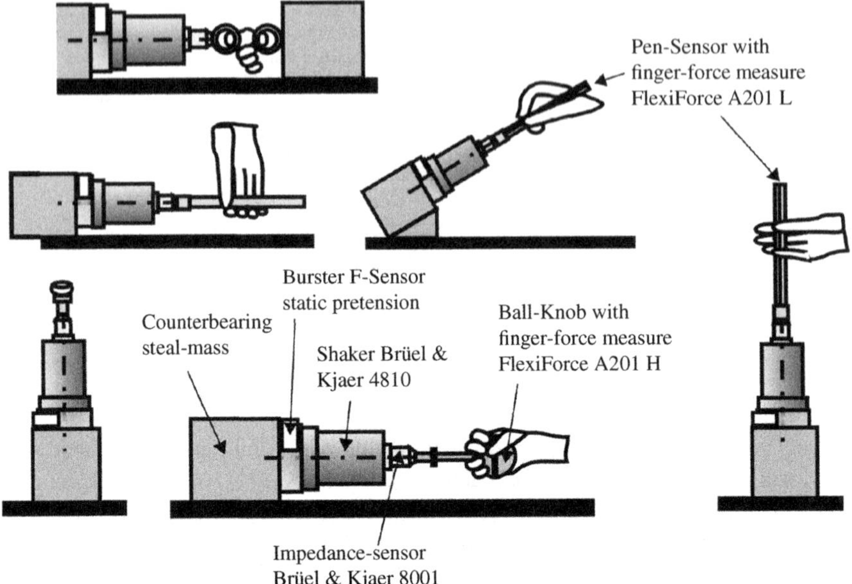

Fig. 3.5 Impedance measurement settings for different grasps

surement errors. Consequently, the models resulting from such measurements are an indication of the technical design process and should always be interpreted with awareness of their variance and errors.

All impedance measures presented here are based on a sinusoid sweep from upper to lower frequencies. The excitation has been made with a defined force of 2 N amplitude at the sensor. The mechanical impedance of the handle has been measured by calibration measurements and was subtracted from the measured values. The impedance sensors are limited concerning their dynamic and amplitude resolution, of course. As a consequence, the maximum frequency up to which a model is valid depends on the type of grasp and its handle used during measure. This limitation is a direct result of the amplitude resolution of the sensors and the necessity at high frequencies to have a significant difference between the user's impedance and the handle's impedance for the model to be built on. The presented model parameters are limited to the acquired frequency range and cannot be applied to lower or higher frequencies. The measurement setup is given in Fig. 3.5.

3.1.5 Models

In order to approximate the human impedance, a number of different approaches were taken in the past (Fig. 3.6). For its description, mechanical models based on concentrated linear elements were chosen. They range from models including active user reactions represented by force sources (Fig. 3.6a), to models with just three elements (Fig. 3.6c) and combined models of different designs. The advantage of a mechanical model compared to a defined transfer function with a certain degree in enumerator and denominator results from the possibility of interpreting the elements of the model as being a picture of physical reality. Elasticities and dampers connected in circuit with the exciting force can be interpreted as coupling to the skin. Additionally, the mechanical model creates very high rankings by its interconnected elements which allow much better fit to measurements than free transfer functions.

KERN [16] defined an eight-element model based on the models in Fig. 3.6 for the interpolation of the performed impedance measures. The model can be characterized by three impedance groups typical for many grasping situations:

Z_3 (Eq. 3.4) models the elasticity and damping of the skin being in direct contact with the handle. Z_1 (Eq. 3.2) is the central element of the model and describes the mechanical properties of the dominating body parts—frequently fingers. Z_2 (Eq. 3.3) gives an insight into the mechanical properties of the limbs, frequently hands, and allows to make assumptions about the preloads in the joints in a certain grasping situation.

$$Z_1 = \frac{s^2 m_2 + k_1 + d_1 s}{s} \tag{3.2}$$

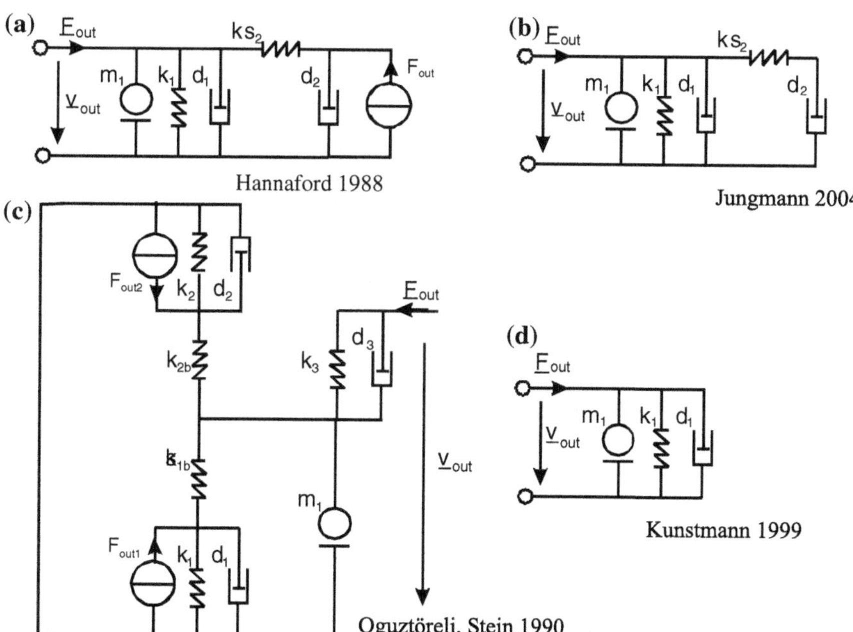

Fig. 3.6 Modeling the user with concentrated elements, **a** [8], **b** [15] **c** [19], **d** [17]

$$Z_2 = \left(\frac{s}{d_2\, s + k_2} + \frac{1}{s m_1} \right)^{-1} \tag{3.3}$$

$$Z_3 = \frac{d_3\, s + k_3}{s} \tag{3.4}$$

$$Z_B = Z_1 + Z_2 \tag{3.5}$$

Combined, the model's transformation is given as

$$Z_\mathrm{H} = Z_3 \| Z_B \tag{3.6}$$

$$Z_\mathrm{H} = \left(\frac{s}{d_3\, s + k_3} + \left(\frac{s^2 m_2 + k_1 + d_1\, s}{s} + \left(\frac{s}{d_2\, s + k_2} + \frac{1}{s m_1} \right)^{-1} \right)^{-1} \right)^{-1} \tag{3.7}$$

3.1.6 Modeling Parameters

For the above model (Eq. 3.7), the mechanical parameters can be identified by measurement and approximations with real values. For the values presented here, approximately 48–194 measurements were made. The automated algorithm combines an evolutionary approximation procedure followed by a curve fit with optimization based on Newton curve fitting, to achieve a final adjustment of the evolutionarily found starting parameters according to the measurement data. The measurements vary according to the mechanical preload—the grasping force—to hold and move the control handles. This mechanical preload was measured by force sensors integrated into the handles. For each measurement, this preload could be regarded as being static and was kept by the subjects with a 5 % range of the nominal value. As a result, the model's parameters could be quantified not only depending on the grasping situation but also depending on the grasping force. The results are given in the following section. The display of the mechanical impedance is given in decibels, whereby 6 dB equals a doubling of impedance. The list of model values for each grasping situation is given in Appendix A.

3.1.6.1 Power Grasps

Within the class of power grasps, three grasping types were analyzed. Impedance between 35 and 45 dB is measured for the grasp of a cylinder (Fig. 3.8) and of a sphere

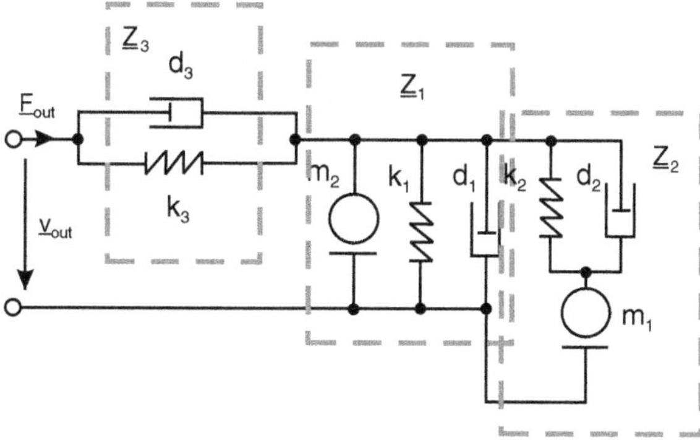

Fig. 3.7 Eight-element model of the user's impedance [16], modeling the passive mechanics for frequencies >20 Hz

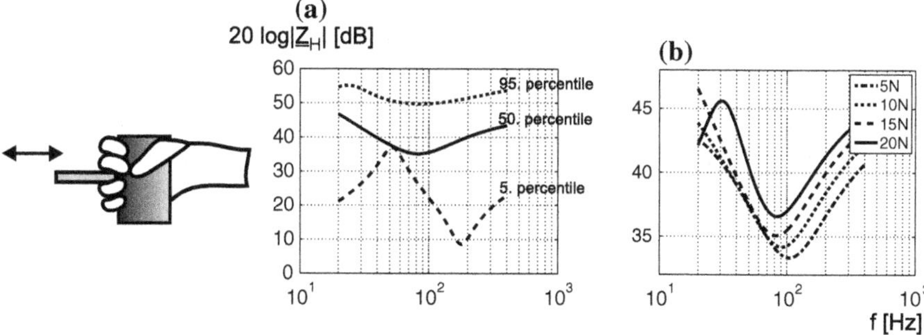

Fig. 3.8 Impedance with percentiles (**a**) and at different force levels (**b**) for power grasps of a cylinder (Ø25 mm, defined for 20–400 Hz)

of similar dimensions (Fig. 3.9) with the whole hand. It shows antiresonance in the area of 80 Hz, which moves for a grasp of the cylinder slightly to higher frequencies with increased grasping force. The percentiles, especially the 5th, reveal that the model is based on a large variance and uncertainty. It is likely that the influence of the subjects' variability in their physical parameters like the size of hands and fingers influences these measurements a great deal.

In case of grasping two rings with thumb and index finger (Fig. 3.10), the results are much more accurate. Impedance ranges between 15 and 35 dB. The antiresonance in the frequency range of 70–100 Hz shows a clear dependency on grasping forces. If we look at the parameters, this change is a result of a variance within the elasticity coefficients k_1 and m_2, building the central parallel resonance of the model. The mechanical system "hand" becomes stiffer (k_1 increases), but the mass m_2 part of the antiresonance diminishes. An easy interpretation of this effect is not obvious. At a value of 10 kg, the mass m_1 builds an almost stiff counter bearing.

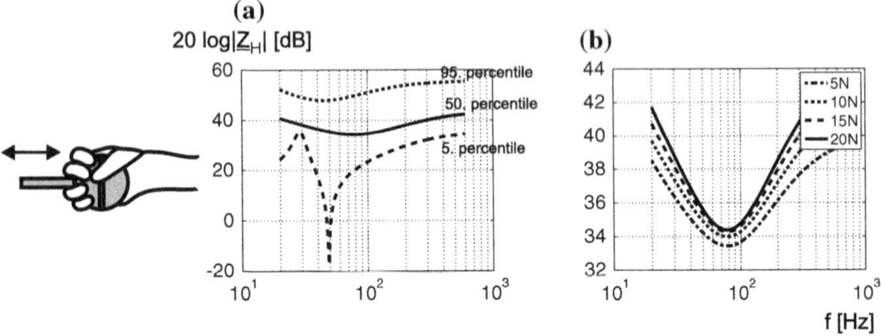

Fig. 3.9 Impedance with percentiles (**a**) and at different force levels (**b**) for a power grasp of a sphere (Ø40 mm, defined for 20–600 Hz)

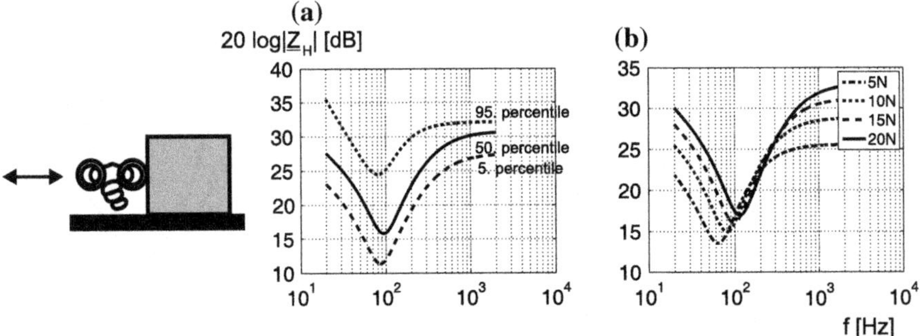

Fig. 3.10 Impedance with percentiles (**a**) and at different force levels (**b**) for a two-fingered power grasp of rings (Ø25 mm of the *inner ring*, defined for 20–1 kHz)

3.1.6.2 Precision Grasps

Within the area of precision grasps, three types of grasps were analyzed. Holding a measurement cylinder similar to a normal pen at an angle of 30° (Fig. 3.11), we find a weak antiresonance in the area of around 150–300 Hz. This antiresonance is dependent on the grasping force and moves from weak forces and high frequencies to large forces and lower frequencies. The general dependency makes sense, as the overall system becomes stiffer (the impedance increases) and the coupling between skin and cylinder becomes more efficient resulting in more masses being moved at higher grasping forces.

The general impedance does not change significantly if the cylinder is held in a position similar to a máobi Chinese pen (Fig. 3.12). However, the dependency on the antiresonance slightly diminishes compared to the above pen hold posture.

This is completely different from the variant of a pen in a horizontal position held by a three-finger grasp (Fig. 3.13). A clear antiresonance with frequencies between

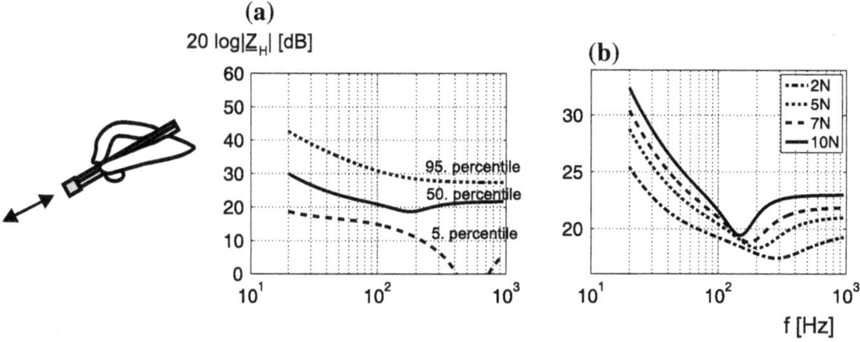

Fig. 3.11 Impedance with percentiles (**a**) and at different force levels (**b**) for a two-fingered precision grasp of a pen-like object held like a pen (Ø10 mm, defined for 20–950 Hz)

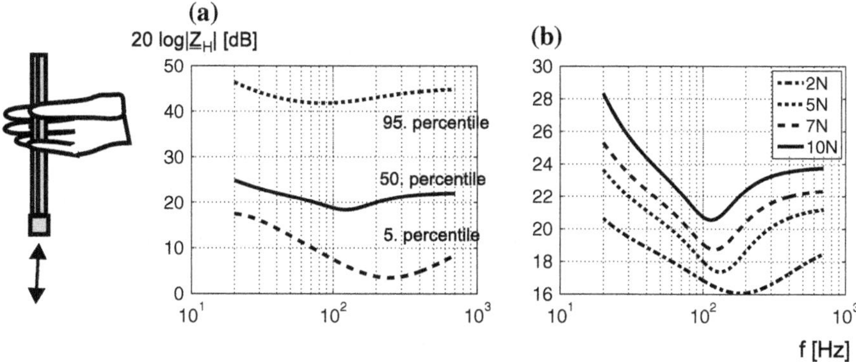

Fig. 3.12 Impedance with percentiles (**a**) and at different force levels (**b**) for a two-fingered precision grasp of a pen-like object held like an "máobi" Chinese pen (Ø10 mm, defined for 20–700 Hz)

80 and 150 Hz appears largely dependent in shape and position on the grasping force. All observable effects in precision grasps can hardly be traced back to the change of a single parameter but are always a combination of many parameters' changes.

3.1.6.3 One-Finger Contact Grasp

All measurements were done on the index finger. Direction of touch, size of touched object, and touch force normal to the skin were varied within this analysis. Figure 3.14a shows the overview of the results for a touch being analyzed in normal direction. The mean impedance varies between 10 and 20 dB with a resonance in the range of 100 Hz. Throughout all measured diameters of contactor size and forces, no significant dependency of the position of the antiresonance on touch forces were noted. However, a global increase in impedance is clearly visible. Observing the

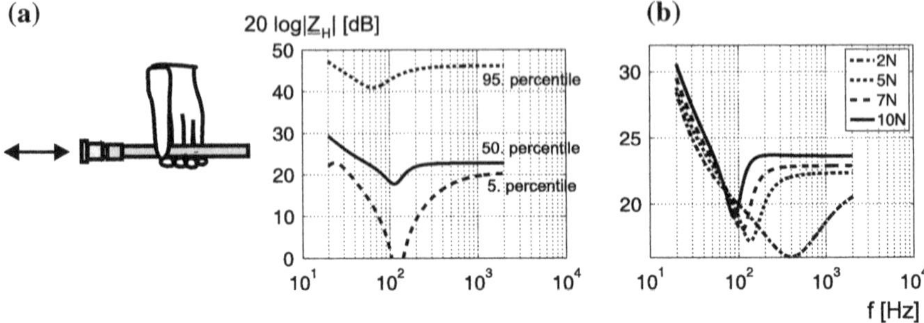

Fig. 3.13 Impedance with percentiles (**a**) and at different force levels (**b**) for a five-fingered precision grasp of a pen-like object in horizontal position (Ø10 mm, defined for 20–2 kHz)

impedance dependent on contactor size, we can recognize an increase in the antiresonance frequency. Additionally, it is fascinating to see that the stiffness decreases with an increase of contact area. The increase in resonance is probably a result of less material and therefore less inertia participating in generating the impedance. The increase in stiffness may be a result of smaller pins deforming the skin more deeply and therefore getting nearer to the bone as a stiff mechanical counter bearing.

In comparison, with measurements performed with a single pin of only 2 mm in diameter (Fig. 3.14b), the general characteristic of the force dependency can be reproduced. Looking at the largest contact element of 15 mm, in diameter, we are aware of a movement of the resonance frequency from 150 Hz to lower values down to 80 Hz for an increase in contact force.

In orthogonal direction, the skin results differ slightly. Figure 3.15a shows a lateral excitation of the finger pad with an obvious increase of impedance at increased force of touch. This rise is mainly the result of an increase of damping parameters and masses. The position of the antiresonance in frequency domain remains constant at around 150 Hz. The picture changes significantly for the impedance in distal direction (Fig. 3.15b). The impedance still increases, but the resonance moves from high frequencies of around 300 Hz to lower frequencies. Damping increases too, resulting in the antiresonance being diminished until nonexistence. At 45° (Fig. 3.15c), a combination of both effects appears. Antiresonance moves to a higher frequency and loses its sharpness compared to the pure lateral excitation. A first trend of change within the position of the antiresonance in frequency domain with higher forces can be identified additionally.

3.1.6.4 Superordinate Comparison of Grasps

It is interesting to compare the impedances among different types of touch and grasps with each other:

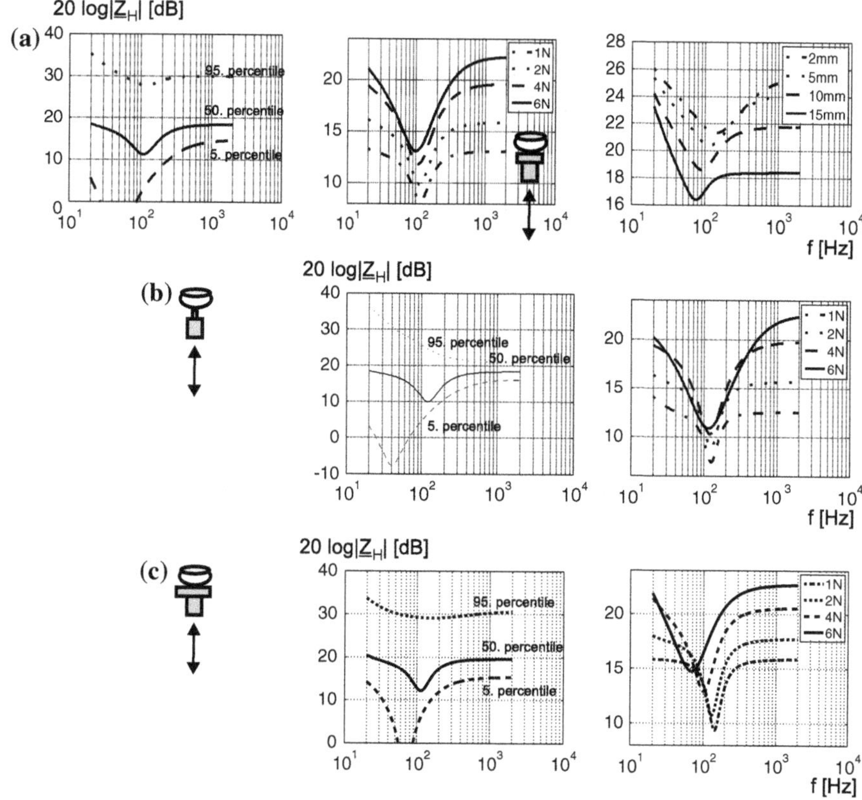

Fig. 3.14 Impedance of finger touch via a *cylindrical plate* for different contact forces (1–6 N) and in dependency from diameter (**a**), for the *smallest plate* (Ø 2 mm) (**b**) and the *largest plate* (Ø 15 mm) (**c**) (defined for 20–2 kHz)

- Almost all raw data and the interpolated models show a decrease of impedance within the lower frequency range of 20 Hz to the maximum of the first antiresonance. As for precision grasps (Figs. 3.11, 3.12, 3.13), normal fingertip excitation (Fig. 3.14), and the power grasp of rings (Fig. 3.10), the gradient equals 20 dB/decade resembling a dominating pure elongation proportional effect of force response— elasticity—within a low frequency range. Within this low-bandwidth area, nonlinear effects of tissue including damping seem to be not very relevant. Looking at such interactions, we can assume that any interaction including joint rotation of a finger is almost purely elastic in a low frequency range.
- Many models show clear antiresonance. Its position varies between 70 Hz at power grasps (Figs. 3.8, 3.9, 3.10) and 200 Hz or even 300 Hz at finger touch analyzed in orthogonal direction (Fig. 3.15). The resonance is a natural effect of any system including a mass and elasticity. Therefore, it is not its existence which is relevant for interpretation, but its shape and the position within the frequency range. As to

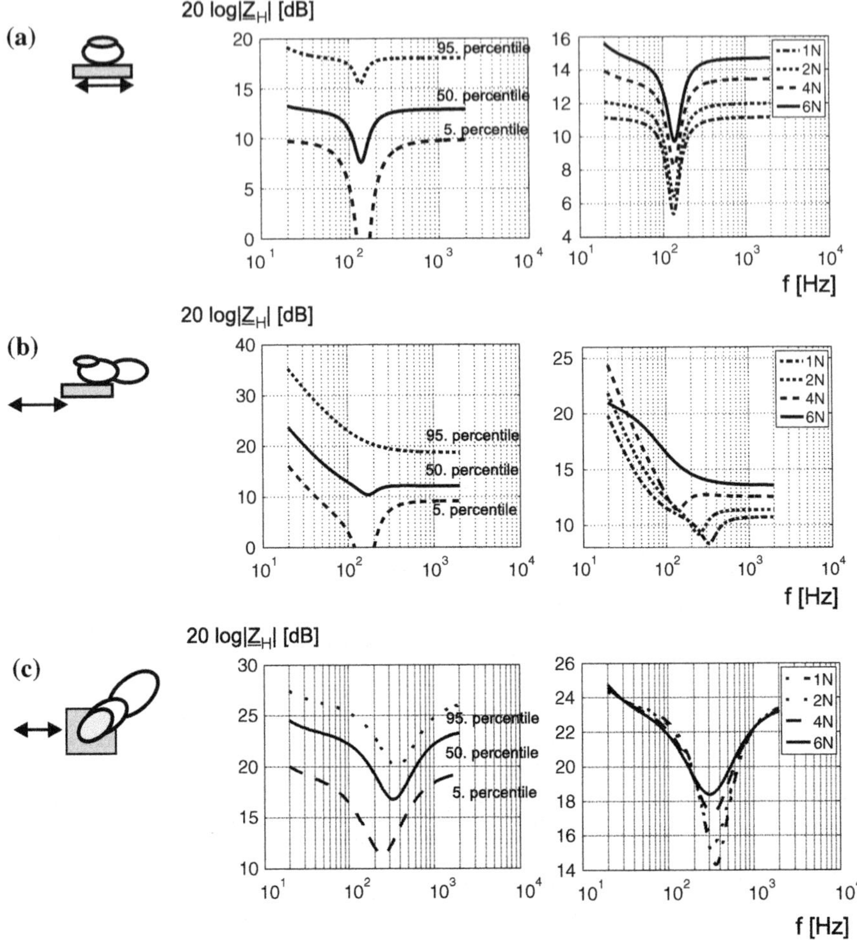

Fig. 3.15 Impedance for finger touch of a plate moving in orthogonal direction to the skin at different force levels (1–6 N) (defined for 20–2 kHz). Movement in lateral direction (**a**), distal direction (**b**), and at 45° (**c**)

positions, especially the power grasp of two rings (Fig. 3.10), the precision grasps of a cylinder in a pen-like position (Fig. 3.11) and in horizontal position (Fig. 3.13) and the touch of an orthogonal moving plate in distal direction (Fig. 3.15c) and a large plate in normal direction (Fig. 3.15c) have a clear dependence on grasping force. The interpretation is not as obvious as in case one. We assume that the normal touch of the plate shows similarities to the contact situation when touching the rings. Additionally, the normal touch is part of the precision grasps mentioned above. In the case of many subjects grasping the horizontal cylinder, it could be observed that the thumb was positioned less orthogonally but more axially to the cylinder, which could excite it primarily in distal direction, thus also contributing to this effect.

- The shape of the antiresonance is another interesting factor. It can be noted that especially in the analysis of finger grasps and thereat orthogonal excitation (Fig. 3.15a), the antiresonance is very narrow. An interpretation is hard to be formulated. It seems that with grasps and especially touches involving less material the antiresonance becomes narrower in shape.
- For all measurements, at high frequencies above the antiresonance, the frequency characteristic becomes linear and constant, which resembles a pure damping effect. This becomes obvious at the pen-hold posture among the precision grasps (Fig. 3.11) and with the lateral displacement in orthogonal direction, (Fig. 3.15a), but is part of any curve and model. Alternatively, inertia could be assumed to dominate the high frequencies, being represented by a linear increase of mechanical impedance. Mainly, power grasps show a tendency to this increase. This measured effect is especially relevant, as it confirms common assumptions that for high-frequency haptic playback with kinaesthetic devices, the user can be assumed as a damping load.
- A last glance should be taken at the absolute height level and the variance of height of the impedance due to preloads. For all grasps, it varies in a range (regarding the median curves only) of 20 dB as a maximum. Impedance is higher for power grasps, slightly lower for precision grasps, and much lower for touches, which is immediately obvious. The change in the preload for one grasp typically displaces the absolute impedance to higher levels. This displacement varies between 4 and 10 dB.

If speculations should be made on still unknown, not yet analyzed types of touches according to the given data, it should be reasonable to assume the following:

A. **Power grasp** The median impedance should be around 36 dB. Model the impedance with a dominating elasticity effect until an antiresonance frequency of 80 Hz, not varying much neither in height nor in position of the antiresonance. Afterward, allow inertia to dominate the model's behavior.

B. **Precision grasp** The median impedance should be around 25 dB. Model the impedance with a dominating elasticity effect until an antiresonance frequency of around 200 Hz. The position of the antiresonance diminishes in an area of 100 Hz due to change in preload. Above that antiresonance, let the impedance become dominated by a damping effect. The height of impedance changes in a range of 5 dB by the force of the grasp.

C. **Finger touch** The median impedance should be around 12 dB. Model the impedance with a well-balanced elasticity and damping effect until an antiresonance frequency of around 150 Hz. The position of the anti-resonance is constant, with the exception of large contact areas moving in normal and distal directions. Above that antiresonance, let the impedance become strongly dominated by a damping effect. The absolute height of impedance changes in an area of up to 10 dB depending on the force during touch.

3.1.7 Comparison with Existing Models

For further insight into and qualification of the results, a comparison with published mechanical properties of grasps and touches is presented in this section. There are two independent trends of impedance analysis in the scientific focus: the measurement of mechanical impedance as a side product of psychophysical studies at threshold level and measurements at higher impedance levels for general haptic interaction. The frequency plots of models and measurements are shown in Fig. 3.16.

In [11], the force detection thresholds for grasping a pen in normal orientation have been analyzed. Figure 3.16a shows an extract of the results compared to the pen-like grasp of a cylinder of the model in Fig. 3.11a. Whereas the general level of impedance does fit, the dynamic range covered by our model is not as big as described in the literature. Analyzing the data as published, we can state that the minimum force measured by ISRAR is ≈60 μN at the point of lowest impedance. A force sensor reliably measuring at this extreme level of sensitivity exceeds the measurement error of our setup and may be the explanation for the difference in the dynamic range

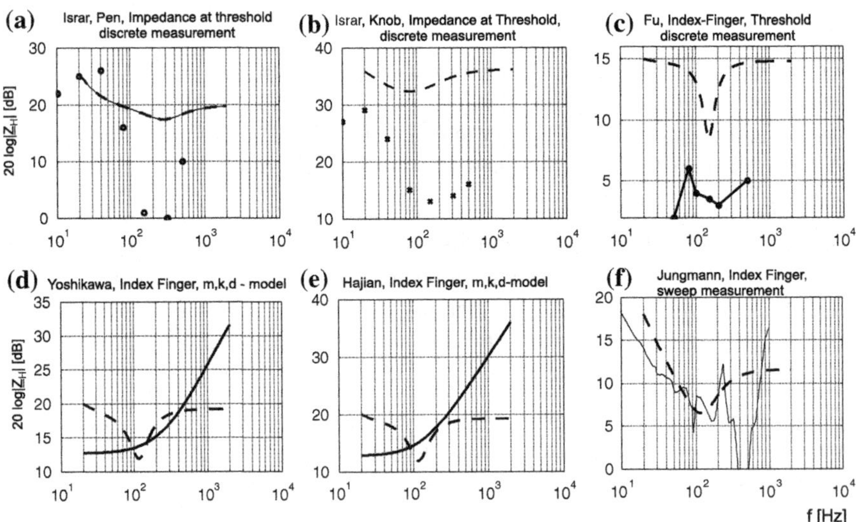

Fig. 3.16 Comparison of the model from Fig. 3.7 with data from similar touches and grasps as published by ISRAR [11, 12], FU [5], YOSHIKAWA [23], HAJIAN [7], JUNGMANN [14]

covered. In another study [12], the force detection threshold of grasping a sphere with the fingertips was analyzed. The absolute force level of interaction during these measurements was in the range of mN. A comparison (Fig. 3.16b) between our model of touching a sphere and these data show a difference in the range of 10–20 dB. However, such small contact forces resemble a large extrapolation of our model data to low forces. The difference can therefore be easily explained by the error resulting from this extrapolation.

FU [5] measured the impedance of the fingertip at a low force of 0.5 N. He advanced an approach published by HAJIAN [7]. A comparison between our model and their data concerning the shape is hardly possible due to the small number of discrete frequencies of this measurement. However, the impedance is again 10 dB lower than that of our touch model of a 5 mm cylinder at normal oscillations similar to Fig. 3.14. Once more, the literature data describes a level of touch force not covered by our measurements , and therefore, the diagram in Fig. 3.16c is an extrapolation of the model of these low forces.

As a conclusion of this comparison, the model presented here cannot necessarily be applied to measurements done at lower force levels. Publications dealing with touch and grasp at reasonable interaction forces reach nearer to the model parameter estimated by our research. YOSHIKAWA [23] published a study of a three-element mechanical model regarding the index finger. The study was based on a time-domain analysis of a mechanical impact generated by a kinaesthetic haptic device. The measured parameters result in a frequency plot (Fig. 3.16d) which is comparable to our model of low frequencies, but shows neither the complexity nor the variability of our model in a high frequency range of above 100 Hz. A similar study in time domain was performed by HAJIAN [7] with slightly different results. Measurements available as raw data from JUNGMANN [14] taken in 2002 come close to our results, although obtained with different equipment.

Besides these frequency plots, the model's parameters allow a comparison with absolute values published in the literature: SERINA [21] made a study on the hysteresis of the fingertips' elongation versus force curve during tapping experiments. This study identified a value for k for pulp stiffness ranging from 2 N/mm at a maximum tapping force of 1 N –7 N/mm at a tapping force of 4 N. This value is about 3–8 times larger than the dominating k_2 in our eight-element model. The results of FU [5] make us assume that there was a systematic error concerning the measurements of SERINA, as the elongation measured at the fingernail does not exclusively correspond to the deformation of the pulp. Therefore, the difference in the values of k between our model and their measurements can become reasonable. Last but not least, MILNER [18] carried out several studies on the mechanical properties of the fingertip in different loading directions. In the relevant loading situation, a value of k ranging from 200 to 500 N/m was identified by him. This is almost perfect within the range of our model's stiffness.

3.1.8 Final Remarks on Impedances

The impedance model as presented here will help in modeling of haptic perception in high frequency ranges above 20 Hz. However, it completely ignores any mechanical properties below this frequency range. This is a direct consequence of the general approach to human–machine interaction presented in Chap. 2 and has to be considered when using this model.

Another aspect to consider is that the above measurements show a large inter-subject variance of impedances. In extreme cases, they span 20 dB meaning nothing else but a factor of 10 between the 5th and the 95th percentiles. Further research on the impedance models will minimize this variance and allow a more precise picture of impedances. But already, this database, although not yet completed, allows to identify helpful trends for human load and haptic devices.

3.2 The User as a Measure of Quality

Christian Hatzfeld

SALISBURY ET AL. postulated a valuable hypothesis for the design of task-specific haptic systems: Their 2011 paper title reads *What You Can't Feel Won't Hurt You: Evaluating Haptic Hardware Using a Haptic Contrast Sensitivity Function* [20]. In this work, they use haptic contrast sensitivity functions (the inverse of the sinusoidal grating detection threshold) to evaluate ↪ COTS devices. With a more general view, the first part of this paper title summarizes the second role of the user and his or her properties in the design of haptic systems: as the instance that determines whether the presented haptic feedback is good enough or not. In this section, this approach is detailed on three aspects of the system design, i.e., resolutions, errors, and the quality of the haptic interaction.

3.2.1 Resolution of Haptic Systems

Resolution is mainly an issue in the selection and design of sensors and actuators, while the latter is also influenced by the kinematic structure used in interfaces and manipulators. In general, sensors on the manipulation side have so sense at least as good as the human user is able to perceive after the information is haptically displayed by the haptic interface. On the interface side, sensors have to be at least as accurate as the reproducibility of the human motor capability, to convey the users' intention correctly. For the actuating part, the attribution is vice versa: Actuators on the manipulating side have to be as accurate as the human motor capability, while the haptic interface has to be as accurate as human perception can resolve.

Unfortunately, this is the worst case for technical development: Sensors (on the manipulating side) and actuators (on the interfacing side) have to be as accurate as human perception. Therefore, accurate readings of *absolute thresholds* are indispensable to determine the necessary resolutions for sensors and actuators, if one wants to build a high-fidelity haptic system. On the other hand, systematic provisions to alter the perception thresholds favorably by changing the contact situation (contact area, contact forces) at the primary interface are possible. This is further detailed in Sect. 5.2.

For applications not involving teleoperation, the requirements are basically the same, but extend to other parts of the system: For interaction with virtual realities, the software has to supply sufficient discretization of the virtual data (a nontrivial problem, especially if small movements and hard contacts are to be simulated), systems for communication have to supply enough mechanical energy that the perception threshold is surpassed to ensure clear transmission of information. Last but definitely not the least, all errors resulting from digital quantization and other system-inherent noise have to be lower than the absolute perception thresholds of the human user.

3.2.2 Errors and Reproducibility

While resolutions are quite a challenge for the design of haptic systems because of the high sensitivity of human haptic perception, the handling of errors is somewhat easier. The basic assumption about the perception of haptic signals with regard to errors and reproducibility is the following: There is no error if there is no difference detectable by the user. This property is expressed by the ↪ JND. *Weber's Law* as stated in Eq. (2.5) facilitates this further: For low references, the acceptable error increases due to increasing differential thresholds. This accommodates the fact that the errors of technical systems and components mainly increase when the reference values decrease.

For large reference values, this relative resolution of human perception is much smaller than the absolute resolution of technical systems that are uniformly distributed along the whole nominal range. This has to be taken into account if information is to be conveyed haptically.

3.2.3 Quality of Haptic Interaction

While resolution and errors are pretty much linked directly to perception parameters, the assessment of haptic quality is somewhat more difficult. It is also based on the assumption that the quality of a haptic interaction is good enough if all intended information is transmitted correctly to the user and no additional information or errors are perceived. The second part can basically be achieved by considering the above-mentioned points regarding errors and resolution. The assessment, whether all information are transmitted correctly, is more difficult since the user and the perceived information have to be taken into account. In general, this is only possible if suitable evaluation methods are used; Chap. 13 gives an overview of such methods with respect to the intended application.

Another example for the evaluation of haptic quality is the concept of haptic transparency for teleoperation systems. This property describes the ability of a haptic system to convey only the intended information (normally defined as the mechanical impedance of the environment at the manipulator side $\underline{Z}_e$) to the user (in terms of the displayed impedance of the haptic interface $\underline{Z}_t$) without displaying the inherent properties of the haptic system. This definition is further detailed in Sect. 7.4.2. Despite the above said, this property can be tested without a user test, but with considerable effort regarding the mechanical measurement setup.

When further considering haptic perception properties, especially ↪ just noticeable differences, the common binary definition of transparency can be transformed into a nominal value with a lot less requirements on the technical system. This concept was developed by HATZFELD, KASSNER, and NEUPERT [9, 10] and is further explained in Sect. 7.4.2.

One should keep in mind that all of the above-mentioned thresholds are generally dependent on frequency and the contact situation in the best case. In the worst case, they are also dependent on the experimental methodology used to obtain them, which will necessarily require a retest of the perception property needed.

Recommended Background Reading

[4] Feix, T.; Pawlik, R.; Schmiedmayer, H.; Romero, J. & Kragic, D.: **A Comprehensive Grasp Taxonomy**. In: Robotics, Science and Systems Conference: Workshop on Understanding the Human Hand for Advancing Robotic Manipulation, 2009.
Thorough Analysis of human grasps, also available online at http://grasp.xief.net/.

[13] Jones, L. & Lederman, S.: **Human Hand Function**. Oxford University Press, 2006.
Extensive analysis about the human hand including perception and interaction topics.

References

1. Brownjohn J et al (1980) Errors in mechanical impedance data obtained with impedance heads. J Sound Vibr 73(3):461–468. doi:10.1016/0022-460X(80)90527-1
2. Bullinger H-J (1978) Einflußfaktoren und Vorgehensweise bei der ergonomischen Arbeitsmittelgestaltung. Universität tuttgart, Habilitation
3. Feix T (2012) Human grasping database. Letzter Abruf. 20.3.2012. http://grasp.xief.net
4. Feix T et al (2009) A comprehensive grasp taxonomy. In: Robotics, science and systems conference: workshop on understanding the human hand for advancing robotic manipulation. http://grasp.xief.net/documents/abstract.pdf

5. Fu C-Y, Oliver M (2005) Direct measurement of index finger mechanical impedance at low force. In: Eurohaptics conference, and symposium on haptic interfaces for virtual environment and teleoperator systems. World haptics 2005. First joint. Embedded systems and physical science center, Motorola Labs., USA, pp 657–659. doi:10.1109/WHC.2005.40
6. Fung Y-C (1993) Biomechanics: mechanical properties of living tissues, 2nd edn. Springer, New York [u.a.], p XVIII, 568. ISBN: 0-387-97947-6, 3-540-97947-6
7. Hajian AZ, Howe RD (1997) Identification of the mechanical impedance at the human finger tip. J Biomech Eng 119:109–114. doi:10.1115/1.2796052
8. Hannaford B, Anderson R (1988) Experimental and simulation studies of hard contact in force reflecting teleoperation. In: IEEE international conference on robotics and automation, vol 1. Jet Propulsion Lab., Caltech, Pasadena, CA, USA, pp 584–589. doi:10.1109/ROBOT.1988.12114
9. Hatzfeld C (2013) Experimentelle analyse der menschlichen Kraftwahrnehmung als ingenieurtechnische Entwurfsgrundlage für haptische Systeme. Dissertation, Technische Universität Darmstadt. http://tuprints.ulb.tu-darmstadt.de/3392/. München: Dr. Hut Verlag. ISBN: 978-3-8439-1033-0
10. Hatzfeld C, Neupert C, Werthschützky R (2013) Systematic consideration of haptic perception in the design of task-specific haptic systems. Biomed Tech 58. doi:10.1515/bmt-2013-4227
11. Israr A, Choi S, Tan HZ (2006) Detection threshold and mechanical impedance of the hand in a pen-hold posture. In: International conference on intelligent robots and systems (IROS), Peking, pp 472–477. doi:10.1109/IROS.2006.282353
12. Israr A, Choi S, Tan HZ (2007) Mechanical impedance of the hand holding a spherical tool at threshold and suprathreshold stimulation levels. In: Second joint EuroHaptics conference and symposium on Haptic interfaces for virtual environment and teleoperator systems (WorldHaptics conference), Tsukaba. doi:10.1109/WHC.2007.81
13. Jones L, Lederman S (2006) Human hand function. Oxford University Press, Oxford. ISBN: 0195173155
14. Jungmann M, Schlaak HF (2002) Taktiles Display mit elektrostatischen Polymeraktoren. In: Konferenzband des 47. Internationalen Wissenschaftlichen Kolloquiums, Technische Universität Ilmenau. http://tubiblio.ulb.tu-darmstadt.de/17485/
15. Jungmann M (2004) Entwicklung elektrostatischer Festkörperaktoren mit elastischen Dielektrika für den Einsatz in taktilen Anzeigefeldern. Dissertation, Technische Universität Darmstadt, p 138. http://tuprints.ulb.tu-darmstadt.de/500/
16. Kern T et al (2006) Study of the influence of varying diameter and grasp-forces on mechanical impedance for the grasp of cylindrical objects. In: Proceedings of the Eurohaptics conference, Paris. doi:10.1007/978-3-540-69057-3_21
17. Kunstmann C (1999) Handhabungssystem mit optimierter Mensch-Maschine-Schnittstelle für die Mikromontage. VDI-Verlag, Düsseldorf. ISBN: 978-3-642-57024-7
18. Milner TE, Franklin DW (1998) Characterization of multijoint finger stiffness: dependence on finger posture and force direction. IEEE Trans Biomed Eng 43(11):1363–1375. doi:10.1109/10.725333
19. Oguztoreli MN, Stein RB (1990) Optimal task performance of antagonistic muscles. Biol Cybern 64(2):87–94. doi:10.1007/BF02331337
20. Salisbury C et al (2011) What you can't feel won't hurt you: evaluating haptic hardware using a haptic contrast sensitivity function. IEEE Trans Haptics 4(2):134–146. doi:10.1109/TOH.2011.5
21. Serina Elaine R, Mote C, Rempel D (1997) Force response of the fingertip pulp to repeated compression—effects of loading rate, loading angle and anthropometry. J Biomech 30(10):1035–1040. doi:10.1016/S0021-9290(97)00065-1
22. Wiertlewski M, Hayward V (2012) Mechanical behavior of the fingertip in the range of frequencies and displacements relevant to touch. J Biomech 45(11):1869–1874. doi:10.1016/j.jbiomech.2012.05.045
23. Yoshikawa T, Ichinoo Y (2003) Impedance identification of human fingers using virtual task environment. In: IEEE/RSJ international conference on intelligent robots and systems (IROS 2003), vol 3, pp 3094–3099. doi:10.1109/IROS.2003.1249632

Chapter 4
Development of Haptic Systems

Christian Hatzfeld and Thorsten A. Kern

Abstract This chapter deals with the general design processes for the development of task-specific haptic systems. Based on known mechatronic development processes such as the V-model, a specialized variant for haptic systems is presented that incorporates a strong focus on the intended interaction and the resulting impacts on the development process. Based on this model, a recommended order of technical decisions in the design process is derived. General design goals of haptic systems are introduced in this chapter as well. These include stability, haptic quality, and usability that have to be incorporated in several stages of the design process. A brief introduction to different forms of technical descriptions for electromechanical systems, control structures, and kinematics is also included in this chapter to provide a common basis for the second part of the book.

4.1 Application of Mechatronic Design Principles to Haptic Systems

Obviously, haptic systems are mechatronic systems incorporating powerful actuators, sophisticated kinematic structures, specialized sensors, and demanding control structures as well as complex software. The development of these parts is normally the focus of specialized areas of specialists, i.e., mechanical engineers, robotic specialists, sensor and instrumentation professionals, control and automation engineers, and software developers. A haptic system engineer should be at least able to understand the basic tasks and procedures of all of these professions, in addition to the required basic knowledge of psychophysics and neurobiology outlined in the previous chapters.

C. Hatzfeld (✉)
Institute of Electromechanical Design, Technische Universität Darmstadt,
Merckstr. 25, 64283 Darmstadt, Germany
e-mail: c.hatzfeld@hapticdevices.eu

T.A. Kern
Continental Automotive GmbH, VDO-Straße 1, 64832 Babenhausen, Germany
e-mail: t.kern@hapticdevices.eu

C. Hatzfeld and T.A. Kern (eds.), *Engineering Haptic Devices*,
Springer Series on Touch and Haptic Systems, DOI 10.1007/978-1-4471-6518-7_4

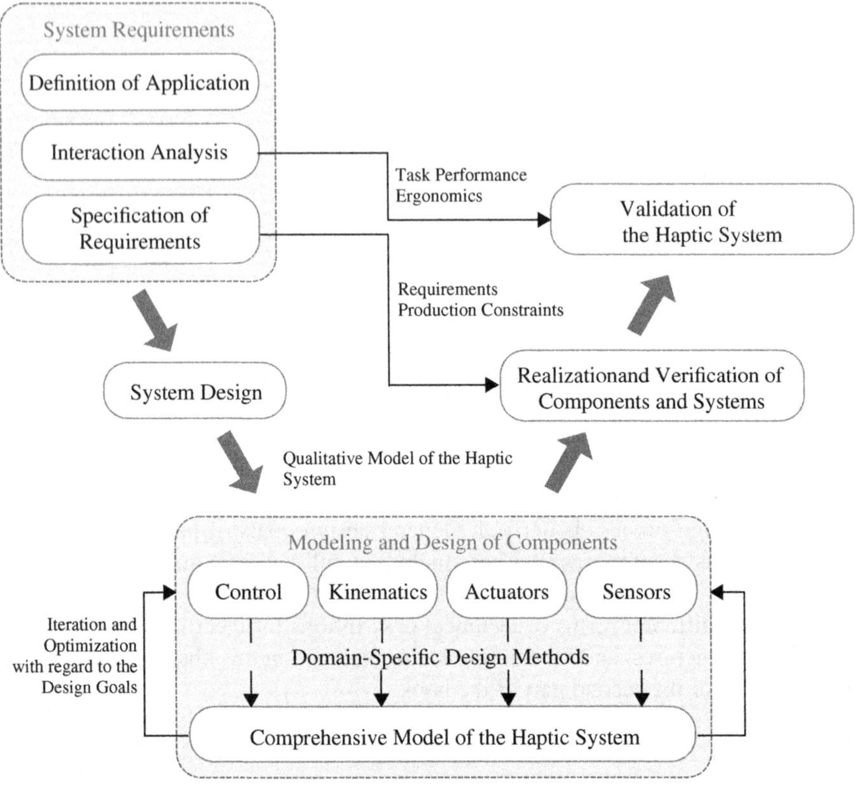

Fig. 4.1 Adaption of the V-model for the design of haptic systems

All of the above-mentioned professions use different methods, but generally agree on the same concepts in developing their parts of a haptic system. These can be integrated in some common known development design methods as for example, the V-MODEL for the development of mechatronic systems [16]. The model was originally developed for pure software design by the Federal Republic of Germany, but adapted to other domains as well. For the design of task-specific haptic systems, the authors detailed and extended some phases of the derivation of technical requirements based on [3] (Interaction Analysis) and [4] (Detailed Modeling of Mechatronic Systems). This adapted model is shown in Fig. 4.1. Based on this, five general stages are derived for the design of haptic systems. These stages are the basis for the further structure of this book and therefore detailed in the following sections.

The V-MODEL knows different variations, depending on the actual usage and scale of the developed systems. In this case, the above-mentioned variation was chosen over existing model variations, to be able to include additional steps in each stage of the V-MODEL. The resulting model is probably nearest to the W-MODEL for the design of adaptronic systems introduced by NATTERMANN AND ANDERL [8], because this model also includes an iteration in the modeling and design stage. It is further based on a comprehensive data management system that includes not only information about interfaces and dependencies of individual components, but also a simulation model of each part. Since there is no comparable data basis for the design of haptic systems (that probably make use of a wider range of physical actuation and sensing principles than adaptronic systems till date), the W-MODEL approach is not directly transferable and more iterations in the modeling and design stage have to be accepted.

4.1.1 Stage 1: System Requirements

The first stage is used for the derivation of system requirements. For the design of task-specific haptic systems, a breakdown in three phases seems useful.

Definition of Application As described in Sect. 2.3, each haptic system should be assigned a well-defined application. This definition is the starting point of haptic system design and comes as a probably vague idea from the client ordering a task-specific haptic system and has to be detailed by the development engineer.

Interaction Analysis Based on the detailed application definition, the intended interaction of the user with the haptic system should be analyzed. For this step, the different concepts of interaction shown in Sect. 2.2 will provide useful vocabulary for the description of interactions. Based on this interaction, the intended grip configuration should be chosen and perceptual parameters for this configuration should be acquired, either from the known literature or by own psychophysical studies. At least, absolute thresholds and the ↪JND should be known for the next steps, along with a model of the mechanical impedance of the intended grip configuration.
Another result of this phase is detailed and quantified interaction goal for application in terms of task performance and ergonomics. Possible categories of these goals are given in Chap. 13. If, for example, a communication system is designed, possible goals could be a certain amount of information transfer (IT) [5] and a decrease in cognitive load in an exemplary application scenario measured by the NASA Task Load Index [2].

Specification of Requirements Based on the predefined steps, a detailed analysis of technical requirements on the task-specific haptic system can be made. This should include all technical relevant parameters for the whole system and each component (i.e., actuators, sensors, kinematic structures, interfaces, control structure, and software design). Chapter 5 provides some helpful clusters depending on different interactions for the derivation of precise requirement definitions.

The result of this stage is at least a detailed requirement list. The necessary steps are detailed in Chap. 5. Further tools for the requirement of engineering can be used as well, but are not detailed further in this book.

4.1.2 Stage 2: System Design

In this stage, the general form and principles used in the system and its components have to be decided on. In general, one can find a vast number of different principles for components of haptic systems. During the technical development of haptic systems, the decisions on single components influence each other intensively. However, this influence is not identical between all components. For the engineer, it is necessary to proceed in the solution identification for each component, after having gained the knowledge of the requirements for the haptic system. It is obvious that according to a systematic development process, each solution has to be compared to the specifications concerning its advantages and disadvantages. The recommended procedure to deal with the components is the basis for the chapter structure of this section of the book and is summarized once again for completeness:

1. Decision about the control engineering structure of the haptic system based on the evaluation of the application (tactile or kinaesthetic), the impedance in idle state (masses >20 g and friction acceptable), and the maximum impedance (stiffness >300 N/m or smaller). This decision is based on the general structures described in Chap. 6 and the control structure of the haptic system described in Chap. 7.
2. Decision about the kinematics based on calculations of the workspace and the expected stiffness as detailed in Chap. 8.
3. Based on the now-known mechanical structure, the actuator design can be made. Chapter 9 deals with this topic, starting with an approximate decision about working principles based on performance criteria and detailed information about the common actuation principles for haptic systems.
4. Dependent on the chosen control engineering structure, the force sensor design can be performed parallel to the actuator design as detailed in Chap. 10.
5. Relatively uncritical for the design is the choice of the kinematic sensors (see Chap. 10).
6. The electronic interfaces are subordinate to all the decisions made before (see Chap. 11).
7. The software design of the haptic rendering itself, in many aspects, is so independent of prior technical decisions that it can be decoupled in almost all aspects from the rest of the design, when all specifications are made. Chapter 12 summarizes some typical topics for the design of this system component.

Nevertheless, it is vital to note that, e.g., the kinematics design cannot be realized completely decoupled from the available space for the device and the forces and torques—or rather the actuator. Additionally, kinematics directly influences any measurement technology as even displacement sensors have limitations on resolution and dynamics. The order suggested above for taking decisions has to be understood as a recommendation for processing the tasks; it does not free the design engineer from the responsibility to keep an overview of the subcomponents and their reciprocal influence.

A good possibility to keep track of this influence is the definition of clear interfaces between single components. This definition should include details about the form of energy and data exchanged between the components and be further detailed in the course of the development process to include a clear definition of, for example, voltage levels, mechanical connections, standard interfaces, and connectors used.

4.1.3 Stage 3: Modeling and Design of Components

4.1.3.1 Modeling of Components

Based on the decisions from the preceding stage, the individual components can be modeled and designed. For this, general domain-specific methods and description forms are normally used, which are further described in Sect. 4.3. This step will first result in a model of the component that will include all relevant design parameters that influence the performance and design of the component. Some of these parameters can be chosen almost completely freely (i.e., control and filter parameters), while others are limited by purchased parts in the system component (one will, for example, only find sensors with different but fixed ranges as well as actuators with fixed supply voltages, etc.).

4.1.3.2 Comprehensive Model of the Haptic System

In a second step, a more general model of the component should be developed, which exhibits similar interfaces to adjacent components like those defined in the preceding Sect. 4.1.2. Furthermore, this model should only include the most relevant design parameters to avoid excessive parameter sets.

When the interfaces of adjacent components match, the models of all components can be combined to a comprehensive model of the haptic system with general haptic input and output definitions (see Fig. 2.33) and relevant design parameters for each individual component. Normally, a large number of components are involved in these comprehensive models. For a teleoperation system, one can roughly calculate two actuators, two kinematic structures, two positioning sensors for actuator control, one force sensor, and the corresponding power and signal processing electronics for *each* ↪DoF with the resulting modeling and simulation effort.

Even if they are very large, such models are advisable to optimize the haptic system with respect to the below-mentioned design goals such as stability and haptic quality. Only with a comprehensive model can one evaluate the intercomponent influences on these design goals. Based on the descriptions of the system structure given in Chap. 7, the optimization of the comprehensive model will lead to additional requirements on the individual components or modifications of the prior-defined interfaces between components. These should also be documented in the requirement list.

One has to keep in mind that all parameters are prone to errors, especially variances with regard to the nominal value and differences between the real part and the (somewhat) simplified model. During optimization of the comprehensive model, robustness of the results with regard to these errors has to be kept in mind.

4.1.3.3 Optimization of Components

Based on the results of the optimized comprehensive model, the individual components of a haptic system can be further optimized. This step is not only needed when there is a change of interface definitions and requirements of single components, but also when it is normally necessary to ensure certain requirements of the system that are not dependent on a single component only. Examples are the overall stiffness of the kinematic structure, the mass of the moving parts of the system, and, of course, the tuning parameters of control loops.

For the optimization of components, typical mechatronic approaches and techniques can be used (see for example [9, 18] and Sect. 4.3). Further aspects such as standard conformity, security, recycling, wearout, and suitability for production have to be taken into account in this stage, too.

In practice, the three parts of *Stage 3: Modeling and Design of Components* are not used sequentially, but with several iterations and branches. Experience and intuition of the developer will guide several aspects influencing the success and duration of this stage, especially the selection of meaningful parameters and the depth of modeling of each component. Currently, many software manufacturers work on a combination of different model abstraction levels (i.e., ↪single input, single output (SISO) systems, network parameter descriptions, finite element models) into a single CAE software with the ability not only to simulate, but also to optimize the model. While this is already possible to a certain amount in commercial software products (e.g., ANSYS™), the ongoing development in these areas will be very useful for the design of haptic systems.

4.1.4 Stage 4: Realization and Verification of Components and System

Based on the optimization, the components can be manufactured and the haptic system can be assembled. Each manufactured component and the complete haptic system should be tested against the requirements, i.e., a verification should be made. Additionally, other design goals such as control stability and transparency (if applicable) should be tested. Due to the above-mentioned interaction analysis (see Sect. 5.2 below for more details), this step ensures that the system generates perceivable haptic signals to the user without any disturbances due to errors. To compare the developed haptic system with others, objective parameters as described in Chap. 13 can be measured.

4.1.5 Stage 5: Validation of the Haptic System

While step 4 ensures that the system is developed correctly with respect to the expected functions and the requirements, this step checks whether the correct system was developed. This is simply made by testing the evaluation criteria defined in the interaction analysis and comparison with other systems with haptic feedback in a user test.

This development process ensures that time-intensive and costly user tests are only conducted in the first and last stages, while all other steps only rely on models and typical engineering tools and process chains. With this detailing of the V-MODEL, the general mechatronic design process is extended in such a way that the interaction with the human user is incorporated in an optimized way in terms of effort and development duration.

4.2 General Design Goals

There are a couple basic goals for the design of haptic systems that can be applied with various extensions to all classes of applications. They do not lead to rigorous requirements, but are helpful to keep all of these in mind when designing an haptic system to ensure a successful product.

Stability Stability in the sense of control engineering should be archived by all haptic systems. It affects the safety of a haptic device as well as the task performance of a haptic system and the interactions performed with it. To ensure stability while improving haptic transparency is the main task of the haptic system control. This is further detailed in Chap. 7.

Haptic Quality To ensure sufficient haptic quality is the second design goal of a haptic system. In general, each system should be able to convey the haptic signals of human–machine interaction without conveying the own mechanical properties to the user. For teleoperation systems, one will find the term *haptic transparency* for this preferable behavior. Analogous to the visual transparency of an ideal window, an ideal haptic teleoperation system will let the user feel exactly the same mechanical properties that are exposed to the manipulator of the teleoperation system. Since physical parts of a haptic system exhibit real physical behavior that cannot be neglected, haptic quality is a control task as well to compensate for this real behavior. It is therefore detailed in Chap. 7.

Usability Since haptics is considered as an interaction modality in this book, all usability considerations of human–machine interfaces should be treated asa design goal. These goals are described in the ISO 9241 standard series[1] and demand *effectiveness* in fulfilling a given task, *efficiency* in handling the system, and *user satisfaction* when working with the system.

Usability has therefore been considered in almost all stages of the development process. This includes the selection of suitable grip configurations that prevent fatigue and allow comfortable use of the system, the definition of clearly distinguishable haptic icons that are not annoying when occurring repeatedly, and the integration of assistive elements such as arm rests. It is advisable to provide for individual adjustment, since this contributes to the usability of a system. This applies to mechanical parts such as adjustable arm rests as well as information-carrying elements such as haptic icons. Methods to assess some of these criteria mentioned are given in Chap. 13 as well as in the standard literature on usability for human–machine interaction, as for example [1].

For the design of haptic systems, the following design principles derived from PREIM's principle for the design of interactive software systems can assist in the development of haptic systems with higher usability [10]:

- Get information about potential users and their tasks;
- Focus on the most important interactions;
- Clarify the interaction options;
- Show system states and make them distinguishable;
- Build an adaptive interface;
- Assist users in developing a mental model, i.e., by consistency of different task primitives;
- Avoid surprising the user;
- Avoid keeping a large amount of information in the user's memory.

4.3 Technical Descriptions of Parts and System Components

Since the design of haptic systems involves several scientific disciplines, one has to deal with different description languages according to the discipline's culture. This section gives a brief introduction into different description languages used in the design of control, kinematics, sensors, and actuators. It is not intended to be sufficient, but to give an insight into the usage and advantages of the different descriptions for components of haptic systems.

[1] The ISO 9241 primarily deals with human–computer interaction in a somewhat limited view of the term "computer" with a strong focus on standard workstations. The general concepts described in the standard series can be transferred to haptics nevertheless, and the ISO 9241-9xx series deals with haptics exclusively.

4.3.1 Single Input, Single Output Descriptions

One of the simplest forms of modeling for systems and components are ↪SISO descriptions. They only consider a single input and a single output with time dependency, i.e., a time-varying force $F(t)$. The description also includes additional constant parameters and derivatives with respect to time of inputs and outputs. If considering a DC motor for example, a SISO description would be the relation between the output torque $M_{\text{out}}(t)$ evoked by a current input $i_{\text{in}}(t)$ as shown in Eq. 4.1.

$$\begin{aligned} M_{\text{out}}(t) &= k_{\text{M}} \cdot i_{\text{in}}(t) \\ \Rightarrow h(t) &= \frac{M_{\text{out}}(t)}{i_{\text{in}}(t)} = k_{\text{M}} \end{aligned} \tag{4.1}$$

The output torque is related to the input current by the transfer function $h(t)$. In this case, the transfer function is just the motor constant k_{M} that is calculated from the strength of the magnetic field, the number of poles and windings, and geometric parameters of the rotor among others. It is normally given in the data sheet of the motor.

SISO descriptions are mostly given in the LAPLACE domain, i.e., a transformation of the time-domain transfer function $h(t)$ into the frequency-domain transfer function $G(s) \bullet\!\!-\!\!\circ\, h(t)$ with the complex LAPLACE operator $s = \sigma + j\omega$. These system descriptions are widely used in control theory to assess the stability and quality of control. However, for the design of complex systems with different components, SISO descriptions have some drawbacks.

- Since only single input and output variables are used, one cannot describe the flow of energy by SISO descriptions accordingly. This is obvious from the above example of a DC motor: Usable output torque will decrease as the revolution speed of the motor increases, since the amount of energy available is limited by the thermal dissipation capabilities of the motor and the counter electromotric force. This behavior cannot be incorporated in Eq. (4.1), since it involves more than one time-dependent input variable.
- When using SISO descriptions for different components that are arranged in a signal and/or energy transmission chain, one has to adjust the interfaces between components accordingly. This complicates the exchange of single components in the transmission chain. Consider an actuator driving a kinematic structure. The exchange of an electrodynamic principle for a piezoelectric principle will require a new SISO description of the kinematic structure, since a input current to the actuator will evoke different kinds of outputs (a force in the case of the electrodynamic principle and an elongation for the piezoelectric principle).

To overcome these disadvantages, one can extend the SISO description to multiple input and multiple output systems (MIMO). For the description of haptic systems, a special class of MIMO systems is advisable, the description based on network parameters as outlined in the following Sect. 4.3.2.

Table 4.1 Analogy between electrical and mechanical network descriptions

Electrical domain			Mechanical domain	
Parameter	Symbol		Parameter	Symbol
Voltage	$\underline{i}$	⇆	Velocity	$\underline{v}$
Current	$\underline{u}$	⇆	Force	$\underline{F}$
Inductivity	L	⇆	Compliance	n
Capacity	C	⇆	Mass	m
Resistance	R	⇆	Viscous damping/friction	$h = \frac{1}{r}$
Impedance	$\underline{Z} = \frac{\underline{u}}{\underline{i}}$	⇆	Admittance (mobility)	$\underline{h} = \frac{1}{\underline{Z}}$
Admittance	$\underline{Y} = \frac{1}{\underline{Z}}$	⇆	Impedance	$\underline{z} = \frac{\underline{F}}{\underline{v}}$

These drawbacks do not necessarily mean that SISO descriptions have no application in the modeling of haptic systems: Despite the usage in control design, they are also useful to describe system parts that are not involved in extensive exchange of energy, but primarily in the exchange of information. Consider a force sensor placed at the tip of the manipulator of a haptic system: While the sensor compliance will effect the transmission of mechanical energy from ↪TCP to the kinematic structure of the manipulator (and should therefore be considered with a more detailed model than a SISO description), the transformation of forces into electrical signals is mainly about information. It is therefore sufficient to use a SISO description for this function of a force sensor.

4.3.2 Network Parameter Description

The description of mechanical, acoustic, fluidic, and electrical systems based on lumped network parameters is based on the similar topology of the differential equations in each of these domains. A system is described by several network elements, which are locally and functionally separated from each other and exchange energy via predefined terminals or ports. To describe the exchange of energy, each considered domain exhibits a flow variable in direct connection with neighboring ports (e.g., current in the electrical domain and force in translational mechanics) and an effort variable (e.g., voltage, respectively, velocity between two arbitrary ports of the network). Table 4.1 gives the mapping of electrical and translational mechanical elements. Historically, there are two analogies between these domains. The one used here depicts physical conditions the best. There is, however, a single incongruent point: The definition of the mechanical impedance is the quotient of flow variable and effort variable.

To couple different domains, lossless transducers are used. Because they are lossless, systems in different domains can be transformed into a single network, which can be simulated with an extensive number of simulation techniques known from electrical engineering like SPICE. The transducers can be divided into two general

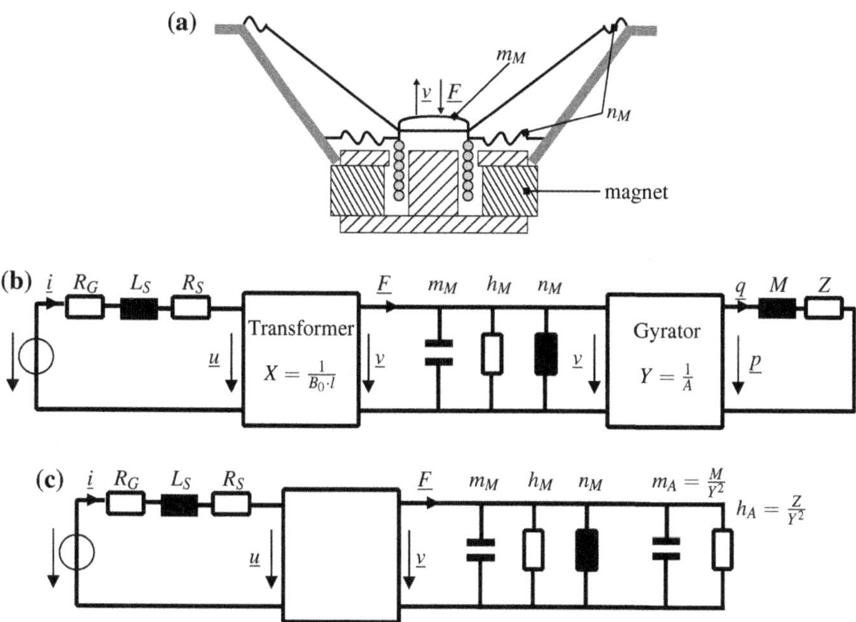

Fig. 4.2 **a** Network model of an electrodynamic loudspeaker. The system consists of an electrical system, the electrodynamic transducer with transformatoric constant X, the mechanical parts of the moving parts, the mechanical–acoustic transducer with gyratory constant Y, and the properties of the acoustic system. **b** shows the corresponding network model and **c** the network model, when acoustic network elements are transformed in equivalent mechanical elements

classes. The first class called *transformer* links the effort variable of domain A with the effort variable of domain B. A typical example for a transformer is a electrodynamic transducer that can be described as shown in Eq. 4.2 with the transformer constant $X = \frac{1}{B_0 \cdot l}$:

$$\begin{pmatrix} \underline{v} \\ \underline{F} \end{pmatrix} = \begin{pmatrix} \frac{1}{B_0 \cdot l} & 0 \\ 0 & B_0 \cdot l \end{pmatrix} \cdot \begin{pmatrix} \underline{u} \\ \underline{i} \end{pmatrix} \tag{4.2}$$

B_0 denotes the magnetic flux density in the air gap of the transducer, and l denotes the length of the electrical conductor in this magnetic field. Further details about these transducers are given in Chap. 9. If different domain networks are transformed into each other by means of a transformer, the network topology stays the same and the transformed elements are weighted with the transformer constant. This is shown in Fig. 4.2 with the example of an electrodynamic loudspeaker and applied to electrodynamic actuators in Sect. 9.2.

The other class of transducers is called *gyrator*, coupling the flow variable from domain A with the effort variable from domain B and vice versa. The coupling is described with the transformer constant Y, and examples (not shown here) include electrostatic actuators and transducers that change mechanical energy into fluidic energy. If different domain networks are transformed, the network topology changes, and series connections become parallel and vice versa. The single elements change as well; for a gyratory transformation between mechanical and electrical domains, an inductor will become a mass and a compliance will turn into a capacitance. A common application for gyratory transformations is the modeling of piezoelectric transducers. This is shown in Sect. 9.3.4.1 in the course of the book.

An advantage of this method is the consideration of influences from other parts in the network, a property that cannot be provided by the representation with SISO transfer functions. On the other side, this method will only work for linear time-invariant systems. Mostly, a linearization around an operating point is made to use network representations of electromechanical systems. Some time dependency can be introduced with switches connecting parts of the network at predefined simulation times. Another constraint is the size of the systems and components modeled by the network parameters. If size and wavelength of the flow and effort variables are in similar dimensions as the system itself, the basic assumption of lumped parameters cannot hold anymore. In that case, distributed forms of lumped parameter networks can be used to incorporate some wave-line transmission properties.

In haptics, network parameters are, for example, used for the description of the mechanical user impedance $\underline{Z}_{\text{user}}$ as shown in Chap. 3 and the condensed description of kinematic structures and the optimization of the mechanical properties of sensors and actuators as shown above. Further information about this method can be found in the work of TILMANNS [14, 15] and LENK ET AL. [7], from which all information in this section was taken.

4.3.3 Finite Element Methods

↪Finite element methods (FEM) are mathematical tools to evaluate ↪partial differential equations (PDE). Since a lot of physical principles are described by partial differential equations, this technique is used throughout engineering to calculate mechanical, thermal, electromagnetic, and acoustic problems [6].

The use of the finite element method requires a discretization of the whole domain, thereby generating several finite elements with finite element nodes as shown in Fig. 4.3. Furthermore, boundary conditions have to be defined for the border of the domain, and external loads and effects are included in these boundary conditions.

Put very simply, FE analysis will run through the following steps: To solve the PDE on the chosen domain, first a partial integration is performed on the differential equations multiplied with a test function. This step leads to a weak formulation of the partial differential equation (also called natural formulation) that incorporates the NEUMANN boundary conditions. Discretization is performed on this natural formulation, leading to a set of PDEs that has to be solved on each single element of the discretized domain. By assuming a certain type of appropriate shape or interpolation function for the PDE on each element, a large but sparse linear matrix is constructed that can be solved with direct or iterative solvers, depending on the size of the matrix.

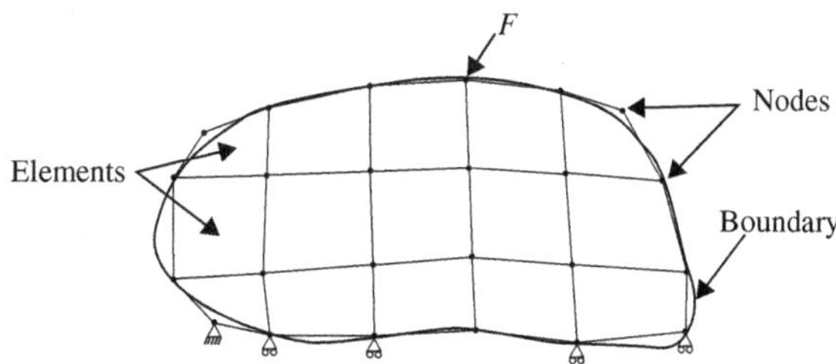

Fig. 4.3 Domain, elements, nodes, and boundary conditions of a sample FEM problem formulation

There are a lot of commercial software products that perform FEM in different engineering fields. They normally include a preprocessor that takes care of discretization, material parameters and boundary conditions, a solver, and a post-processor that turns the solver's results into a meaningful output. For the quality of results of FEM, the choice of the element types depending on geometry of the considered domain and the kind of analysis and mathematical solver used is of great importance.

The advantages of the FE method are the treatment of nonlinear material properties, the application to complex geometries, and the versatile analysis possibilities that include static, transient, and harmonic analysis [19]. The aspect of discretization yields high computational effort, and also spatial resolution of the physical value on investigation.

To overcome the disadvantages of FEM, there are some extensions to the method: The *combined simulation* maps FE results onto network models that are further used in network-based simulations of complex systems [13, 20]. The advantage is the high spatial resolution of the calculation on the required parts only and the resulting higher speed. The data exchange between FE and network model is made by the user. The *coupled simulation* incorporates an automated data exchange between FE and network models at runtime of the simulation. At the moment, many companies work on the integration of this functionality in program packages for FE and network model analysis to allow for multidomain simulation of complex systems.

The application of ↪finite element model (FEM) in haptics can be found in the design of force sensors (see Sect. 10.1), the evaluation of thermal behavior of actuators, and the structural strength of mechanical parts.

4.3.4 Description of Kinematic Structures

A description of the pose, i.e., the position and orientation of a rigid body in space, is a basic requirement to deal with kinematic structures and to optimize their properties. If considering Euclidean space, six coordinates are required to describe the pose of a body. This is normally done by defining a fixed reference frame i with an origin O_i and three orthogonal basis vectors $(\mathbf{x}_i, \mathbf{y}_i, \mathbf{z}_i)$. The pose of a body with respect to the reference frame is described by the differences in position and orientation. The difference in position is also called displacement and describes the change in

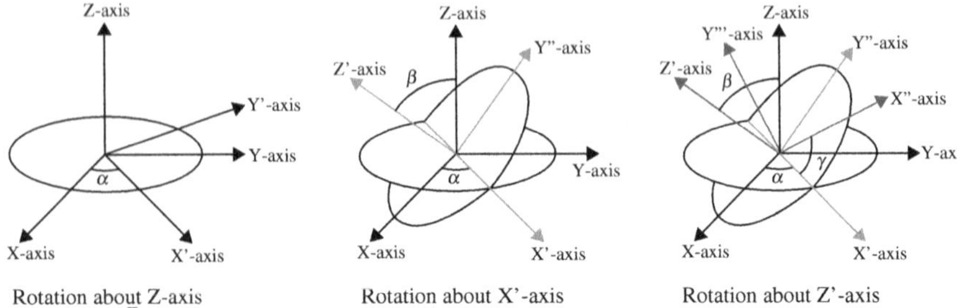

Fig. 4.4 Rotation of a coordinate frame based on Euler angles (α, β, γ) in Z–X–Z order

position of the origin O_j of another coordinate frame j that is fixed to the body. The orientation is described by the angle differences between the two sets of basis vectors $(\mathbf{x}_i, \mathbf{y}_i, \mathbf{z}_i)$ and $(\mathbf{x}_j, \mathbf{y}_j, \mathbf{z}_j)$. This rotation of the coordinate frame j with respect to the reference frame i can be described by the rotation matrix ${}^j\mathbf{R}_i$ as given in Eq. 4.3.

$$
{}^j\mathbf{R}_i = \begin{pmatrix} \mathbf{x}_i \cdot \mathbf{x}_j & \mathbf{y}_i \cdot \mathbf{x}_j & \mathbf{z}_i \cdot \mathbf{x}_j \\ \mathbf{x}_i \cdot \mathbf{y}_j & \mathbf{y}_i \cdot \mathbf{y}_j & \mathbf{z}_i \cdot \mathbf{y}_j \\ \mathbf{x}_i \cdot \mathbf{z}_j & \mathbf{y}_i \cdot \mathbf{z}_j & \mathbf{z}_i \cdot \mathbf{z}_j \end{pmatrix} \tag{4.3}
$$

While the rotation matrix contains nine elements, only three parameters are needed to define the orientation of a body in space. Although there are some mathematical constraints on the elements of ${}^j\mathbf{R}_i$ that ensure the equivalence, several minimal representations of rotations can be used to describe the orientation with less parameters (and therefore less computational effort is required when computing kinematic structures). In this book, only three representations are discussed further, the description by *Euler angles*, *fixed angles*, and *quaternions*.

Euler Angles To minimize the number of elements needed to describe a rotation, the Euler angle notation uses three angles (α, β, γ) that each represents a rotation about the axis of a moving coordinate frame. Since each rotation depends on the prior rotations, the order of rotations has to be given as well. Typical orders are Z–Y–Z and Z–X–Z rotations shown in Fig. 4.4. The description by Euler angles exhibits singularities, when the first and last rotations occur about the same axis. This is a drawback when one has to describe several consecutive rotations and when describing motion, i.e., deriving velocities and accelerations.

Fixed Angles Fixed angle descriptions are basically the same as Euler angle descriptions, and the rotation angles (ψ, θ, ϕ) describe, however, the rotation about the fixed axes of the reference frame. Also known as *yaw* ψ around the $\mathbf{x}_i$-axis, *pitch* θ around the $\mathbf{y}_i$-axis, and *roll* ϕ around the $\mathbf{z}_i$-axis, the fixed angles exhibit the same singularity problem as the Euler angles.

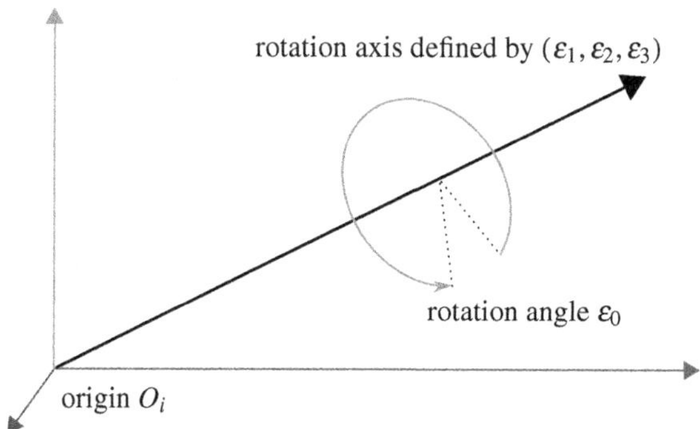

Fig. 4.5 Rotation of a frame defined by a quaternion

Quaternions To overcome this singularity problem, quaternions are used in the description of kinematic structures. Mathematically, also known as Hamilton numbers $\mathbb{H}$, they are an extension of the real number space $\mathbb{R}$. A quaternion ϵ is defined as

$$\epsilon = \epsilon_0 + \epsilon_1 i + \epsilon_2 j + \epsilon_3 k$$

with the scalar components ϵ_0, ϵ_1, ϵ_2, and ϵ_3 and the operators i, j, and k. The operators fulfill the combination rules shown in Eq. (4.4) and therefore allow associative, commutative, and distributive addition as well as associative and distributive multiplication of quaternions.

$$\begin{aligned} &ii = jj = kk = -1 \\ &ij = k, \quad jk = i, \quad ki = j \\ &ji = -k, \quad kj = -i, \quad ik = -j \end{aligned} \tag{4.4}$$

One can imagine a quaternion as the definition of a vector $(\epsilon_1, \epsilon_2, \epsilon_3)$ that defines the axis; the frame is rotated about the scalar part ϵ_0 defining the amount of rotation. This is shown in Fig. 4.5. By dualization, quaternions can be used to describe the complete pose of a body in space, i.e., rotation and displacement. Further forms of kinematic descriptions, for example, the description based on screw theory can be found in [17], on which this section is based primarily and other works like [11, 12].

Recommended Background Reading

[4] K. Janschek: **Mechatronic systems design: methods, models, concepts**. Springer, Heidelberg, 2012.
Design methodologies for mechatronic systems that can also be applied to haptic systems.

[6] M. Kaltenbacher: **Numerical simulation of mechatronic sensors and actuators**. Springer, Heidelberg, 2007.
Broad overview about finite element methods and the application to sensors and actuators.

[7] A. Lenk et al. (Eds.): **Electromechanical Systems in Microtechnology and Mechatronics: Electrical, Mechanical and Acoustic Networks, their Interactions and Applications**. Springer, Heidelberg, 2011.
Introduction to the network element description methodology.

References

1. Boy GA (ed) (2011) The handbook of human–machine interaction—a human-centered approach. Ashgate, Farnham
2. Hart S, Staveland L (1988) Development of NASA-TLX (task load index): results of empirical and theoretical research. Human mental workload 1:139–183. doi:10.1016/S0166-4115(08)62386-9
3. Hatzfeld C (2013) Experimentelle analyse der menschlichen Kraftwahrnehmung als ingenieurtechnische Entwurfsgrundlage für haptische Systeme. Dissertation, Technische Universität Darmstadt. http://tuprints.ulb.tu-darmstadt.de/3392/. München: Dr. Hut Verlag, 2013. ISBN: 978-3-8439-1033-0
4. Janschek K (2012) Mechatronic systems design: methods, models, concepts. Springer, Berlin. ISBN: 978-3-642-17531-2
5. Jones L, Tan H (2013) Application of psychophysical techniques to haptic research. IEEE Trans Haptics 6:268–284. doi:10.1109/TOH.2012.74
6. Kaltenbacher M (2007) Numerical simulation of mechatronic sensors and actuators. Springer, Berlin. ISBN: 978-3-540-71359-3. doi:10.1007/978-3-540-71360-9
7. Lenk A (ed) (2011) Electromechanical systems in microtechnology and mechatronics: electrical, mechanical and acoustic networks, their interactions and applications. Springer, Heidelberg
8. Nattermann R, Anderl R (2013) The W-model-using systems engineering for adaptronics. Procedia Comput Sci 16:937–946. doi:10.1016/j.procs.2013.01.098
9. Pahl G, Wallace K, Blessing L (2007) Engineering design: a systematic approach. Springer, Berlin. ISBN: 978-3540199175
10. Preim B (1999) Entwicklung interaktiver Systeme. Grundlagen, Fallbeispiele und innvoative Anwendungsfelder. Springer, Heidelberg. ISBN: 978-3540656487
11. Sciavicco L (2000) Modelling and control of robot manipulators. Springer, Berlin. ISBN: 978-1-4471-0449-0
12. Selig J (2005) Geometric fundamentals of robotics. Springer, Berlin. ISBN: 978-0-387-27274-0
13. Starke E (2009) Kombinierte Simulation—eine weitere Methode zur Optimierung elektromechanischer Systeme. Dissertation. Technische Universität Dresden.http://tudresden.de/die_tu_dresden/fakultaeten/fakultaet_elektrotechnik_und_informationstechnik/ihm/mst/Veroeffentlichungen_MST/Archiv_MST/mst_2010?fis_type=publikation&fis_id=b1847

14. Tilmans H (1996) Equivalent circuit representation of electromechanical transducers: I. lumped parameter systems. J Micromech Microeng 6: 157. doi:10.1088/0960-1317/6/1/036
15. Tilmans H (1997) Equivalent circuit representation of electromechanical transducers: II. Distributed parameter systems. J Micromech Microeng 7(4) 285. doi:10.1088/0960-1317/7/4/005
16. VDI 2206: Entwicklungsmethodik für mechatronische Systeme. Berlin: Verband Deutscher Ingenieure (VDI), 2004. http://www.vdi.de/technik/fachthemen/produkt-und-prozessgestaltung/fachbereiche/produktentwicklungund-mechatronik/themen/rilis-mechatronische-systeme/richtlinie-vdi-2206-entwicklungsmethodik-fuer-mechatronische-systeme/
17. Waldron K, Schmiedeler J (2008) Kinematics In: Siciliano B, Khatib O (eds) Springer handbook of robotics. Springer, Berlin, pp 9–33. doi:10.1007/978-3-540-30301-5_2
18. Janschek K (2012) Mechatronic systems design: methods, models, concepts. In: Design methodologies for mechatronic systems that can also be applied to haptic systems. Springer, Heidelberg.
19. Kaltenbacher M (2007) Numerical simulation of mechatronic sensors and actuators. In: Broad overview about finite element methods and the application to sensors and actuators. Springer, Berlin
20. Lenk A (ed) (2011) Electromechanical systems in microtechnology and mechatronics: electrical, mechanical and acoustic networks, their interactions and applications. Springer, Heidelberg

Part II
Designing Haptic Systems

In the preceding chapters the focus of the discussion was haptic perception with regard to the human user. In the following chapters, the technical realization of haptic systems will come to the fore. Consequently, the general view will change from a user centralized perspective to a device-specific viewpoint. More definite technological subjects are examined and more practical help for frequent challenges of the design process is offered. The chapters in this part are ordered according to the classic task list to be realized during any technical design process. They start with more general questions concerning the overall system and proceed to specialized questions concerning specific sub-components. The chapters are intentionally ordered in such a way that the ones dealing with questions whose solution spectrum is limited to a great extent are discussed earlier than those which provide more flexible solutions applicable to many situations. The understanding gained as well as the methods for the quantification of haptic perception will still be used for the analysis of the quality of technological solutions.

- Chapter 5: The acquisition of **requirements** for the technical design process is discussed. The design of haptic systems covers a plurality of technological questions. Especially the challenges concerning the high dynamics to be achieved make a systematic approach mandatory for identifying the requirements.
- Chapter 6: After the basic requirements have been identified, a superordinate view of the **structure of the system** to be built is necessary for which a methodology is given in this chapter. The resulting analysis does not only aim at the decision on the control structure of the device, but also defines the technological sub-problems to be addressed during the following design process.
- Chapter 7: Several issues about the **control of haptic systems** are discussed that has to guarantee stability and also haptic quality of new designs.
- Chapter 8: Especially kinaesthetic, but also multidimensional tactile systems have to combine multiple degrees of freedom to fulfill the requirements for certain tasks. This leads to the systematic design of the **kinematics** of the

device. This chapter provides the necessary knowledge on kinematic design and covers specific and sometimes surprising problems of mechanical transfer functions for haptic devices in parallel kinematics.

- Chapter 9: In this most comprehensive chapter of the second part, typical **actuator principles** with respect to their application in haptic devices are discussed. The sections cover all popular actuation principles in an overview, as well as the details of selected principles for a design from scratch.
- Chapter 10: Especially closed-loop controlled haptic devices need **force sensors**. Furthermore, telemanipulation systems—besides simulators—are the second most relevant group of high fidelity haptic devices. A haptic telemanipulator requires force sensors at its end effector, at least. This chapter covers the selection process and the design of force sensors according to the physical principles able to fulfill the specifications of haptic systems. For a complete haptic interaction each system requires a **position measurement**. Technological solutions for this subordinate technological challenge are discussed in this chapter, whereby different positioning- and movement, touch, and imaging-sensors are presented.
- Chapter 11: Typically, haptic devices are interfaced with time discrete systems with digital signal processors, be they controllers for universal **interfaces** or complete simulation systems. This chapter deals with the interfaces between these computing units and gives insight into the performance of different realizations.
- Chapter 12: The most frequent application of complex haptic systems is the perception of **virtual realities** in simulations. For this purpose, it is necessary to calculate the simulated physical properties by a software system running on a programmable device. This chapter deals with a selection of algorithms used for such calculations and gives an overview of the technological challenges. It also provides a terminological basis for mechanics and hardware developers in order to make them understand software engineers. Several available toolkits alleviating the design of corresponding software are presented.
- Chapter 13: Added in the second edition of the book, this chapter deals with measures and methods for the **evaluation** of haptic systems. It covers pure system aspects as well as measures for task-performance and the impact on the user.
- Chapter 14: In this chapter, three task-specific haptic systems and their development and evaluation are described.
- Chapter 15: Final remarks on all previous chapters are made and the most important recommendations for the design of haptic systems are summarized.

Chapter 5
Identification of Requirements

Thorsten A. Kern and Christian Hatzfeld

Abstract In this chapter, the process of requirement definition is described, starting with the definition of the intended application together with the customer. In particular, the derivation of technical parameters from the customers' expectation and useful tools for this step are discussed. Further, the analysis of the intended interaction and the effects on the requirement identification are discussed. To alleviate the identification of requirements, main requirement groups are derived from the intended type of interaction and presented in five technical solution clusters. A review of the relevant standards and guidelines on safety serves as another source of requirements of a haptic system.

5.1 Definition of Application: The Right Questions to Ask

At the beginning of a technical design process, the requirements for the product, which usually are not clear and unambiguous, have to be identified. Frequently, customers formulate wishes and demands, respectively, solutions instead of requirements. A typical example is a task of the kind "to develop a product just like product **P**, but better/cheaper/nicer." If an engineer accepts such an order without getting to the bottom of the original motivation, the project will be doomed to failure. Normally, the original wish of the customer concerning the product has to fulfill two classes of requirements:

The product shall have

- a certain **function**
- in a distinct **technical** and **market-oriented** framework

T.A. Kern (✉)
Continental Automotive GmbH, VDO-Straße 1, 64832 Babenhausen, Germany
e-mail: t.kern@hapticdevices.eu

C. Hatzfeld
Institute of Electromechanical Design, Technische Universität Darmstadt, Merckstr. 25, 64283 Darmstadt, Germany
e-mail: c.hatzfeld@hapticdevices.eu

C. Hatzfeld and T.A. Kern (eds.), *Engineering Haptic Devices*,
Springer Series on Touch and Haptic Systems, DOI 10.1007/978-1-4471-6518-7_5

The formulations of the market-oriented requirements are manifold yet not in the focus of the following analysis (for details of a general systematic product development see [8, 17]). They may be motivated by an existing product **P** to compete with, but usually they are much more comprehensive and cover questions of budget, time frame of development, personal resources and qualifications, and customers to address.

With regard to the technical framework, the customer typically gives just unspecific details. A statement like “a device shall provide a force on a glove” is not a definition of a requirement but already a solution on the basis of existing knowledge on the part of the customer. The complexity of a real technological solution spans from a single actuator to provide, e.g., a vibration to complex kinematics addressing single fingers. Questioning the customer’s original statement, it may even turn out that his intention is, e.g., to simulate the force impression when switching the gears of a clutch in a passenger car. The knowledge about the actual application and following that knowledge about the interaction itself allows the developer a much broader approach, leading to a more optimized technical solution.

5.1.1 Experiments with the Customer

The customer formulates requirements—as mentioned before—typically in an inexact instead of a specific way. Additionally, there is the problem of very unspecific terminology with regard to the design of haptic systems. For the description of haptic sensual impressions, there are numerous adjectives difficult to quantify, such as rough, soft, smooth, gentle, mild, hard, viscous, as well as others derived from substantives such as furry, silky, hairy, watery, and sticky which can be compared to real objects. So what could be more obvious than asking the customer to describe his/her haptic impressions by comparisons?

> Ask the customer to describe the intended haptic impression with reference to objects and items in his/her environment. These items should be easily at hand, e.g., vegetables and fruits which offer a wide spectrum of textures and consistencies for comparison.

Sometimes, the customer first needs to develop a certain understanding of the haptic properties of objects and items. This can best be achieved by his/her directly interacting with them. Examples of haptically extreme objects have to be included in a good sample case, too. The evolving technology of 3D printing allows for a very flexible design of such samples.

Provide a sample case including weights and springs of different size, even marbles, fur, leather, and silk. Depending on the project, add sandpaper of different granularity. Use these items to explain haptic parameters to the customer and help the customer to optimize the description of the product performance expected!

From practical experience, we can recommend also to take spring balances and letter balances or electronic force sensors with you to customer meetings. Frequently, it is possible to attach a handle directly to the items and ask the customer to pull, until a realistic force is reached. This enables customers of non-technical disciplines to quickly get an impression of the necessary torques and forces.

Take mechanical measurement instruments with you to the customer meetings and allow the customer to touch and use them! This gives him/her a good first impression of the necessary force amplitudes.

In order to give a better impression of texture, mechanical workshops may produce patterns of knurls and grooves of different roughness on metals. Alternatively, sandpaper can be used and, by its defined grade of granularity, can provide a standardized scale to a certain extent.

Use existing materials with scales to describe roughness and simulate the impression of texture.

Recently, different toolkits for haptic prototyping are available. They are specific for certain types of applications, for example, cockpit knobs or stylus interaction, as shown in Fig. 5.1. Further examples for Lo-Fi prototyping can be found in [9]. For more sophisticated setups, the usage of a ↪ COTS device and a virtual environment developed with a haptics toolkit (see Sect. 12.4 for a short overview) could be considered.

What Does Not Work

A normal customer without expertise in the area of haptics will not be able to give statements concerning resolutions or dynamics of the haptic sense. This kind of information has to be derived from the type of interaction and the study of the psychophysical knowledge of comparable applications. Therefore, the experience of the developing engineer is still indispensable despite all the systematizations in the technical design process.

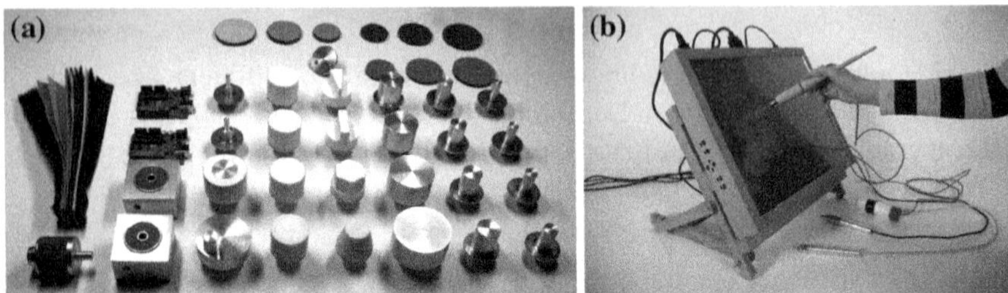

Fig. 5.1 Toolkits for haptic prototyping: **a** Haptic Interface Communication System (HaptICS) to evaluate mechanical characteristics and physical properties of haptic knobs for cockpit use, **b** reflective haptics toolkit for stylus interaction with touchscreens and variable friction. Images courtesy of *Nuilab GmbH*, Schorndorf, www.nuilab.com

Do not confuse the customer by asking questions about the physical resolution! This is necessarily the knowledge of the haptic engineer. However, learn about the dynamics of the interaction and try to assess the application, e.g., by asking about the frame rate of a simulation, or the maximum number of load changes per second of a telemanipulator.

5.1.2 General Design Guidelines

Next to the ideas of the customer and/or user, there are also a number of different guidelines dealing with the design of haptic systems. These guidelines are summarized here briefly, but a close look at the original references is advisable when applicable to the intended haptic system.

Usability and Human–Computer Interaction Guidelines Since all haptic systems are intended to be used as a human–computer interface, the applicable guidelines for these systems are also relevant for the design of haptic systems. As mentioned in Sect. 4.2, the ISO 9241-series deals with usability in general, the ISO 9241-9xx standards specifically address haptic systems and should be considered while working on requirement definitions and beyond. For the use of ↪ COTS devices, MUÑOZ ET AL. [12] introduced a basic guideline for the design of ergonomic haptic interactions that can be useful for these kind of applications.

Design of Haptic Icons The group of BREWSTER works on haptic and auditory icons and did publish several design guidelines for this kind of communication, for example [16].

HCI for Blind Users SJÖSTRÖM [19] developed guidelines for ↪ virtual reality (VR) systems for blind users in addition to existing guidelines for HCI.

Telepresence in Precision Assembly Tasks ACKER [1] investigated the usage of haptic and other feedback in precision assembly tasks.

Presence and Performance in Teleoperation Design factors leading to higher presence and improved performance were investigated by DEML [3, 4]. With a strong focus on human factors, a guideline was developed to optimize the human–machine interface.

Design of VR and Teleoperation Systems Based on a literature review, a design guide for the development of haptic and multimodal interfaces was developed in [14]. The guide selects guidelines based on an interactive front end.

Minimal Invasive Surgery TAVAKOLI ET AL. [20] present the design of a multimodal teleoperated system for minimal invasive surgery and address general questions like control stragies and the effect of time delay.

General Benefits of Haptic Systems Based on a meta-study, NITSCH [15] identified several aspects of haptic feedback on task performance measures. Haptic feedback improves working speed, handling accuracy, and the amount of force exerted in teleoperation and virtual systems. This holds mainly for kinaesthetic force feedback; vibro-tactile feedback predominantly reduces only task completion time.

5.2 Interaction Analysis

Based on the demands of the customer and the clarifications obtained in conversation and experiments, a more technical interaction analysis can be performed. The first goal of this step is a technical description of the user with regard to the intended application. Normally, this will include information about the perception thresholds in the chosen grip configuration, information about the movement capabilities, and the mechanical impedance of the user. Naturally, one will not find fixed values for these parameters, but probably only ranges in the best case. In the worst case, own perception studies and impedance measurements have to be conducted.

The second goal of this step is a definition of suitable evaluation parameters and appropriate testing setups. If a reference system (that has to be improved or equipped with haptic feedback in course of the development) is given, reference values of these parameters should be obtained in this stage of the requirement identification as well.

The following steps are advisable for an interaction analysis that will obtain meaningful information for the following requirement specification as stated in Sect. 5.5. They are based on the works of HATZFELD ET AL. [6, 7].

1. **Task Analysis** Analyze the interaction task as thoroughly as possible. Interaction primitives as described in Sects. 1.2.1 and 2.2 are helpful at this point. Research possible grip configurations suitable for such interactions (see Sect. 3.1.3), if the hand is intended as the primary interface between user and haptic system. Depending on the intended application, other body sites such as the torso, the

back of the hand, or other limbs can be suitable locations for haptic interactions. For ease of reading, the rest of this section will only mention the hand as primary interface without loss of generality with regard to other body sites.
Take the use of tools into account (stylus, gripper, etc.) as well as possible restrictions of the manipulator in a teleoperation scenario (see example below). After this one should have one or more possible interaction configurations that will be able to convey all interaction primitives needed for the intended usage. If one plans to build a teleoperation, comanipulation, or assistive system that adds haptic feedback to interactions that do not already have such, it is worthwhile to discuss if all haptic signals have to be measured, transmitted, and displayed. Sometimes, the display of categorized haptic information (OK/Not OK, Material A/Material B/Material C etc.) could be sufficient in terms of intended usage of the system, facilitates the technical development, and lowers the cost of the final product.
It is advisable to also have a look at some multimodal aspects of the application as well as other environmental parameters: If a visual channel has to be or can be used, special concepts like pseudo-haptic feedback (see Sect. 12.5) can be considered in the design of the system. If the system is to be used in a highly distractive environment, robust communication schemes have to be incorporated or an adjustable feedback mode has to be included. This information will help with formulating the system structure and the detailed requirement list.

2. **Movement Capabilities** Select one or two most promising grip configurations. Based on these, one should define the maximum and comfortable movement spaces of the user and the typical interaction and maximum exertable forces. Section 2.2.4 gives some values for handling forces and velocity, data for typical movement spaces can be found in applicable standards like ISO 7250 or DIN 33402. These are relevant boundaries for the user input kinematics in terms of workspace and structural load. Interaction forces can be further used to define forces on the slave side of a telemanipulation system (as well as input forces in a virtual system that have to be dealt with in the software).

3. **Mechanical Impedance** Research or measure the mechanical impedance of the selected grip configuration. This impedance is relevant for several control issues like stability (local stability of the haptic interface and overall stability in case of teleoperation systems) and haptic transparency as discussed in Sect. 3.2.

4. **Perception Parameters** Research or measure relevant perception parameters for the selected grip or body site configuration. Normally, absolute and different thresholds are needed for an estimation of sensor and actuator resolutions as well as tolerable errors. Based on the intended usage, other perception parameters or other interpretations can be meaningful as well. For example, successiveness limens (SL) and two-point thresholds will affect the design of communication interfaces on all body sites. For an energy-limited system, small JNDs could be beneficial, since they will probably result in a large number of possible transferable information with a small amount of energy.
Keep in mind that force and deflection thresholds can be calculated from

each other using the mechanical impedance according to Eq. (2.9). If possible, obtain data in more than one dimension to facilitate the requirement definition in the intended ↪ DoF. Be sure to check whether there are external conditions that will influence perception favorably for technical development. This could be a maximum contact area or a minimum contact force that will lead to higher perception thresholds for the given contact situation. With means of the system developer, these conditions can be influenced, for example, by the design of the grip or the measurement of a minimum contact force that has to be applied by the user to make the haptic system functional.

5. **Evaluation Criteria** Define suitable evaluation criteria regarding the intended task performance. Chapter 13 gives possible criteria depending on the application class of the haptic system. Despite these measures of task performance, measurements of haptic quality (if applicable) and ergonomic measures can be taken into account. The latter will quantify the cost and benefit of a haptically enhanced system compared to a system without haptics. The definition this early in the development allows for the measurement of reference values and eases the final evaluation, since the intended testing procedure of the haptic system can be incorporated in the design process.

A final decision for a grip configuration can either be made based on the values obtained in this interaction analysis in favor of the technically less demanding option or by conducting user tests considering ergonomic factors such as fatigue and task performance, if this is technically possible (for example with ↪ COTS devices). Obviously, this could involve some iterations of the above-mentioned points. With this structured approach to interaction, a lot of purposeful information is generated for the derivation of requirements. The approach is illustrated with an example in the following.

Example: FLEXMIN Interaction Scheme

The surgical system FLEXMIN is developed to enhance single-port surgery procedures, for example, transanal rectum resection [2] with haptic feedback, additional intracorporal mobility compared to rigid instruments, and a more ergonomic working posture of the surgeon. Task analysis as described above was conducted based on an example rectum resection with commercially available, stiff instruments (TEO system, *Karl Storz*, Tuttlingen, Germany) on an anatomical model. Based on the recordings of the surgeon's movements, system constraints such as workspace, dexterity, instruments, and principal manipulation tasks were identified [11]. This analysis led to the requirements of two manipulators with at least four movement ↪ DoF (positioning in space and rotation along the longitudinal axis) and preferably another DoF for gripping instruments such as scissors or forceps.

Based on additional aspects like request for displaying stiff structures and elements and the available construction space, a parallel kinematic structure was chosen for the

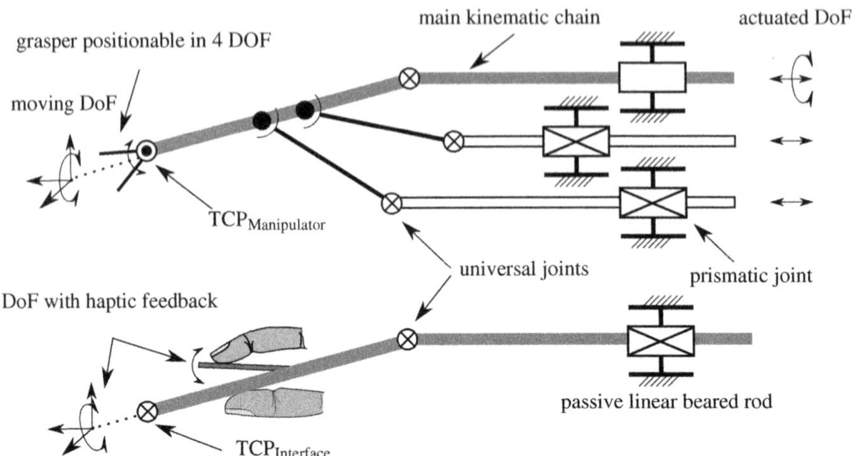

Fig. 5.2 Derivation of the concept of the haptic user interface of FLEXMIN (*lower* part) from the kinematic structure of the intracorporal manipulator (*upper* part). Figure adapted from [13]

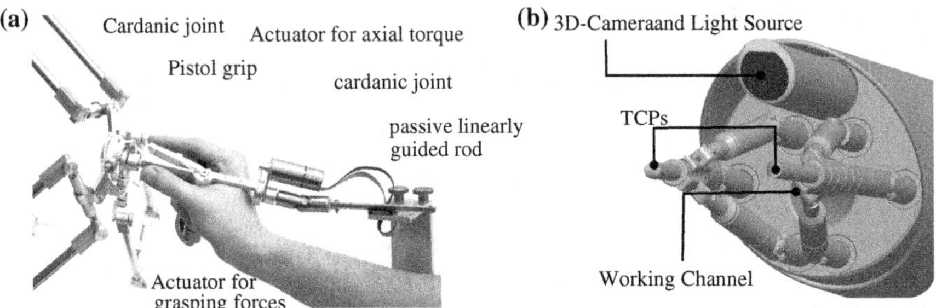

Fig. 5.3 **a** Realization of the haptic user interface of FLEXMIN, **b** Rendering of the intracorporal robot with two manipulator arms, working channel and visual instrumentation. Further information can be found in [10]

intracorporal manipulator already at this point of development [10]. In that case, the ↪ TCP will be at the end of the last part of the lead chain of the parallel mechanism. The movement of this part was chosen as the general form of interaction of the haptic interface used to operate the manipulator [13]. The resulting concept for the haptic interface is shown in Fig 5.2.

Ergonomic considerations about the surgeon handling two of these interfaces led to a passive linear bearing at one end of the main kinematic chain of the user interface. On the other end, a parallel delta kinematic structure was chosen to actuate three DoF of the haptic interface. Additional feedback for the rotatory and the grasping DoF is integrated in the grasping part of the user interface. This is shown in Fig. 5.3.

5.3 Technical Solution Clusters

After interaction analysis and discussion of the customer's expectations toward the haptic system, one should have in-depth knowledge of the intended function of the haptic system. Based on a fundamental description of this function, general types of haptic systems and interactions therewith can be identified. Based on these, this section identifies possible technical realizations and summarizes the necessary questions in clusters of possible applications. The list does not claim to be complete, but is the essence of requirement specifications of dozens of developments achieved during the last few years.

The core of the requirements' identification is the definition of the type of haptic interaction. The first question asked should always refer to the type of interaction with the technical system. Is it a simulation of realistic surroundings, the interaction with physically available objects in terms of telepresence; or is the focus of the interaction on the pure communication of abstract information? In the former cases, the variants are less versatile than in the latter, as described below. In Fig. 5.4, a decision tree for the identification of clusters of questions is sketched. It is recommended to follow the tree from top to bottom to identify the correct application and the corresponding cluster of questions.

Simulation and Telepresence of Objects Does the interaction aim at touching virtual or via telepresence available objects? If this is the case, does the interaction

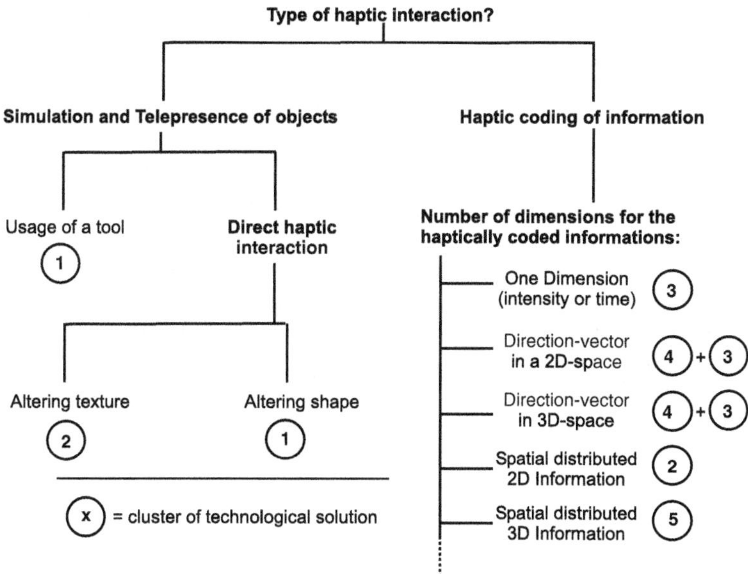

Fig. 5.4 Structure for identifying relevant clusters of questions by analyzing the intended haptic interaction

take place directly via fingers, hands, or skin, or is a mediator, e.g., a tool of the interacting object? Does the user hold a specific tool—a pen, a screw driver, a surgical instrument, a joystick of a plane, in his hands and control one or more other objects with it, or does the user touch a plurality of objects during the interaction with his or her hands? In the case of a tool interaction, the chosen solution can be found in cluster ① "kinaesthetics"; in case of a direct interaction, another detail has to be considered.

Direct Haptic Interaction By touching physical objects, the user can notice the differences between all physical attributes of the volume like mass, elasticity, plasticity and inner friction, and their texture. In the case of interacting with shapes, the questions of cluster ① "kinaesthetics" remain relevant, and in the case of interacting with textures, the questions of cluster ② "surface-tactile" have to be considered. This is not necessarily an alternative decision, however, with the same object interaction, both aspects can be required at the same time or one after the other.

Haptic Coding of Information In the case of abstract, not physical, object-oriented information communication via the haptic sense, the question of the dimension of information becomes relevant:

- Does the interaction include a single event which occurs from time to time (e.g., the call of a mobile phone) or is some permanently active information (e.g., a distance to a destination) haptically communicated? These questions are one-dimensional[1] and covered by cluster ③ "vibro-tactile."
- Is the interaction dominated by directional information coding an orientation in a surface (directional movement) or in a space? In this case, the questions covered by cluster ④ "vibro-directional" are relevant. In such applications time, respectively, distal information is included frequently, also making the questions in cluster ③ to become relevant.
- Does the interaction aim at the communication of data distributed within a two-dimensional information layer, such as geological maps, roadmaps, or texts on a page? In these cases, the questions of cluster ② "surface-tactile" have to be answered.
- In case there is volumetric information—the electrical field of an atomic bonding or medical data sets—to be haptically transmitted, the questions of cluster ⑤ "omnidimensional" are to be considered.

In the following sections, the questions in the clusters are further discussed and some examples are given for the range of possible solutions to the questions aimed at.

[1] As the information includes only one parameter.

5.3.1 Cluster ①: Kinaesthetic

Cluster ① has to be chosen either when an interaction between fingers and shapes happens directly or when the interaction takes place between tool and object. Both cases are technical problems of multidimensional complexity.[2] Each possible dimension movement corresponds to one degree of freedom of the latter technical system. Therefore, the questions to be asked are straightforward and mainly deal with the requirements for these degrees of freedom of tools and users:

- Which degrees of freedom do the tool/movement show? → rotatory, translatory, combinations.[3]
- How large is the volume covered by the movements? → Maximum and minimum values of angles and translations.
- How dynamic is the movement? → Identification of maximum and minimum velocities and accelerations. Usually, this question cannot be answered immediately. A close observation of the intended interaction will help and—as far as possible—measurements of movement velocities of instruments and fingers should be made, e.g., with the aid of videos.
- Which forces/torques happen at each degree of freedom[4]? → Definition of maximum and minimum forces and torques.
- What is the maximum dynamics of the forces and torques? → Bandwidth of forces and torques in frequency range, alternatively maximum slope in time-domain ("from 0 to F_{max} in 0.1 s"). Usually, this question cannot be answered directly. Frequently, measurements are difficult and time-consuming, as an influence of the measurement on the interaction has to be eliminated. Therefore, it is recommended to do an analysis of the interaction itself and the objects the interaction happens with. If it is soft—something typical of surgical simulation, simple viscoelastic models can be used for interpolating the dynamics. The most critical questions with respect to dynamics often address the initial contact with an object, the collision. In this case, especially the stiffness of the object and the velocity of the contact describe the necessary dynamics. But it has to be stated that the resulting high demands are not seldom in conflict with the technical possibilities. In these cases, a split concept based on "events" can be considered, where kinaesthetic clues are transmitted in low-frequency ranges, and highly dynamic clues are coded in pure vibrations (Sects. 11.2.3, 12.5).

---

[2] A tool interaction can be a one-dimensional task, but such an assignment concerning the technical complexity can be regarded as an exception.

[3] In the case of a finger movement, it has to be noted that not necessarily all movement directions have to be equipped with haptic feedback to provide an adequate interaction capability. Frequently, it is even sufficient to provide the grasp movement with haptic feedback, solely.

[4] Frequently, the customer will not be able to specify these values directly. In this case, creative variants of the question should be asked, e.g., by identifying the moving masses, or by taking measurements with one's own tools.

5.3.2 Cluster ②: Surface-Tactile

Haptic texture represents the microstructure of a surface. The lateral movement between this microstructure and the fingertip results in shear forces generating the typical haptic impression of different materials. Haptic bumps on the keyboard keys J and F are a special form of texture. Another variant of texture are Braille letters carrying additional abstract information. Cluster ② has to be chosen when there is a need to present information on any surface via the tactile sense. This can be either coded information on a geological map on a more or less plane surface, but it can also be object-specific features like the material itself. The resulting questions for the technical task are as follows:

- Which body parts perform the interaction? → This trivial question has a significant impact, as the body part selected defines the resolution available on user side and consequently the requirements for the size of the texture-generating elements.
- Is the form of the texture-carrying shape subject to changes? If so, how much and in which areas? → If the shape changes a lot, it is likely that the unit providing the texture information has to be adapted to, e.g., each finger (e.g., as a pin-array or piezoelectric disk), as the fingers will have to be positioned independently of each other. In this case, it may even be necessary to provide a lateral movement between finger and texture unit to generate shear forces in the skin. In case of the shape being fixed, e.g., in the case of a map, a relative movement may happen by the fingers themselves and the texture unit can be designed with less size restrictions.
- How fast does the displayed information change? → Textures change rarely during the simulation of objects and display of maps. This is dramatically different when, e.g., texts or the influences of fluids on textures have to be displayed. The answer to this question has a significant impact on the technical system.
- Which intensity range is covered by the texture? → In the simplest situation, the answer can be given by definite displacements and a resolution in bits. Usually, only qualitative values of the properties of objects for interaction are available. These hints have to be complemented by one's own experiments. With regard to the definition of these requirements, it is important to make sure that the planned spatial addressability and maximum intensity change does not exceed the corresponding resolution of the user. A research on the corresponding psychophysical experiments is highly recommended, as otherwise it may not be possible to transmit the intended information density.

5.3.3 Cluster ③: Vibro-Tactile

Cluster ③ is a solution space for simple one-dimensional technical problems and corresponding questions. It covers independent dimensions of information (e.g., coding an event in a frequency and the importance in the amplitude). In this cluster, distributions of intensity variations and/or time-dependent distributions of single events are filed. Technological solutions are usually vibrational motors or tactors, as being used in mobile phones or game-consoles. But even if the technical solution itself seems quite straightforward, the challenge lies in the coding of information with respect to intensity and time and an appropriate mechanical coupling of the device to the user.

- Which mechanical interface for the transmission of haptic information to the user is planned? → More specifically, is this interface influenced by mechanical limits like housings?
- Which design space is available? → Frequently, vibro-tactile solutions are limited as to the available space at an early stage of the design due to requirements for mobility.
- Which resolution is expected for the planned intensity variation? → The criteria are similar to those of the "surface-tactile" cluster. As the "vibro-tactile" cluster frequently deals with oscillating systems, the dependence of the perception of oscillations on its frequency has to be taken into account. The user's perception is the limiting factor for intensity variations, which themselves are dependent on the mechanical coupling between device and user.

5.3.4 Cluster ④: Vibro-Directional

Vibro-tactile systems code one-dimensional information in the form of intensities. It is obvious that by the combination of multiples of such information sources, directional information can be transmitted. This may happen two-dimensionally in a plane surface, but also three-dimensionally. Cluster ④ deals with such systems. One possible technical solution for directional surface information would be to locate a multitude of active units in the shape of a ring around a body part, e.g., a belt around the belly. The direction is coded in the activity of single elements. This approach can also be transferred to a volumetric vector, whereby in these cases, a large number of units are located on a closed surface, e.g., the upper part of the body. The activity of single elements codes the three-dimensional direction as an origin of a normal vector on this surface. In addition to the questions of cluster ③, this cluster deals with the following questions:

- What is the intended resolution on the surface/in the space? → As well as before dependent on the body surface used, it is likely that the human perception represents the limit for the achievable resolution. The corresponding literature [5, 21] has to be checked carefully before the technical requirements can be met.
- What number of simultaneously displayed vectors is expected? → The fact that users will be able to identify one direction does not guarantee that with a parallel display of two points the user will perform equally well. Simultaneous display of information frequently results in masking-effects hard to be quantified. Experiments and analysis of the intended application are strongly recommended.

- Which frame of reference is used? → The information displayed is usually embedded in a frame of reference, which is not necessarily identical to the user's frame of reference and his or her body. The user may change his position, for example, in a vehicle, which results in a loss of the position of the elements fixed to the body and their orientation in the vehicle. It is necessary to be aware of the active frame of reference (local user-oriented, or vehicle-oriented, or maybe even world-oriented) and to provide measurement equipment for identifying changes in user positions and frame of reference. Additionally, it may become necessary to present a haptic reference signal to the user, which calibrates the user's perception to the frame of reference, e.g., a "north" signal.

5.3.5 Cluster ⑤: Omnidirectional

Cluster ⑤ deals with systems coding real volumetric information. Within such a three-dimensional space, each point either includes intensity information (scalar field) or vector information (vector field). The sources of such data are numerous and frequent, be it medical imaging data, or data of fluid mechanics, of atomic physics, of electrodynamics, or of electromagnetics. Pure systems of haptic interaction with such data are seldom. Frequently, they are combinations of the clusters "kinaesthetic" and "vibro-tactile" for scalar fields, respectively, "kinaesthetic" with six active haptic degrees of freedom for vector fields.[5] Consequently, the specific questions of this cluster add one single aspect to already existing questions of the other clusters:

- Does the intended haptic interaction take place with scalar fields or with vector fields? → For pure vector fields, kinaesthetic systems with the corresponding questions for six active degrees of freedom should be considered. In the case of scalar fields, an analysis of vibro-tactile systems in combination with three-dimensional kinaesthetic systems and the corresponding questions should be considered. Then, the property of the scalar value corresponds to the dynamics of the coded information.

5.3.6 General Requirement Sources

For any development process, there are several questions that always have to be asked. They often refer to the time frame as well as to the resources available for development. For haptic devices, two specific questions have to be focused on, as they can become quite limiting for the design process due to specific properties of haptic devices:

[5] The haptic interaction with objects in a mathematical abstraction always is an interaction with vector fields. In the vectors, forces of surfaces are coded, which themselves are time dependent, e.g., from movements and/or deformations of the objects themselves.

- Which energy sources are available? → It is not a necessary prerequisite that electrical actuators have to be used for haptic devices, especially in the case of telemanipulation systems. The use of pneumatic and hydraulic energy sources, especially for tactile devices, is a real alternative and should be considered.
- The design, how expensive may it be? → The prices of current kinaesthetic haptic systems reach from 200 EUR of mass products to 1.500 EUR of medium-scale products to devices of 25,000 EUR for small series and 100,000 EUR for individual solutions. These prices only partly result from commercial acquisitiveness, but mostly from the technical requirements and the efforts that have to be taken.

Furthermore, safety is a relevant source of requirements for haptic systems. Because of the importance of this issue, it is dealt with in the following section separately.

5.4 Safety Requirements

Since haptic systems are in direct contact with human users, safety has to be considered in the development process. As with usability (see Sect. 4.2), a consideration of safety requirements should be made as early in the development process as possible. Furthermore, certain application areas like medicine require a structured, documented, and sometimes certified process for the design of a product which also has to include a dedicated management of risk and safety issues. In this section, some general safety standards that may be applicable for the design of haptic systems are addressed and some methods for the analysis of risks are given.

5.4.1 Safety Standards

Safety standards are issued by large standard bodies and professional societies such as ↪ International Organization for Standardization (ISO), the national standard organizations, ↪ Institute of Electrical and Electronics Engineers (IEEE), and ↪ International Electrotechnical Commission (IEC). Some relevant standards for the design of haptic systems are listed as follows. Please note that this section will not supersede the study of the relevant standards. For a detailed view on the general contents of the standards, the websites of the standardizing organizations are recommended.

IEC 61508 This standard termed *Functional Safety of Electrical/Electronic/Programmable Electronic Safety-related Systems* defines terms and methods to ensure functional safety, i.e., the ability of a system to stay or assume a safe state, when parts of the system fail. The base principle in this standard is the minimization of risk based on the likelihood of a failure occurrence and the severity of the consequences of the failure. Based on predefined values of these categories, a so-called ↪ Safety Integrity Level (SIL) can be defined, which will impose requirements on the safety measures of the system. It has to be noted that the

IEC 61508 does not only cover the design process of a product, but also the realization and operational phases of the life cycle.
The requirements of functional safety impose large challenges on the whole process of designing technical products and should not be underestimated. The application of the rules are estimated to increase costs from 10 to 20 % in the automotive industry for example [18].

ISO 12100 This standard defines terms and methods for machine and plant safety. It can be considered as detailing the above-mentioned IEC 61508 for the construction of machines, plants, and other equipment. For the design of haptic systems, this standard is probably also useful to assess security requirements for the intended application of the system.

ISO 13485, ISO 14971, IEC 62366, IEC 60601 The ISO 13485 standard defines the requirements on the general design and production management for medical devices, while the ISO 14971 standard deals with the application of risk management tools in the development process of medical devices. One has to note that these standards are a good starting point for devices intended for the European market, but further rules and processes of the ↪ Food and Drug Administration (FDA) have to be considered for products intended for the American market. The IEC 62366 deals with the applicability of usability engineering methods for medical devices. IEC 60601 considers safety and ergonomic requirements on medical devices.

IEEE 830 This standard deals with the requirement specifications of software in general. It can therefore be applied to haptic systems involving considerable amounts of software (as for example haptic training systems). The general principles on requirement definitions (like consistency, traceability, and unambiguousness for example) from this standard can also be applied to the design of technical systems in general.

Since a large number of haptic systems are designed for research purposes and are used in closely controlled environments, safety requirements are often considered secondary. One should note, however, that industry standards as the ones mentioned above resemble the current state of the art and could therefore provide proven solutions to particular problems.

5.4.2 Definition of Safety Requirements from Risk Analysis

As mentioned above, modern safety standards not only define certain requirements (like parameters of electrical grounding or automatic shutdown of certain system parts), but also have an impact on the whole design process. To derive requirements for the haptic system, the following steps are advisable during the design process:

1. Assess the relevant safety standards for the intended application and use of the haptic system. Despite the standards itself, this also includes further regulations and applicable test cases.

2. Define your safety management and development process including project structure, needed certifications, documentation requirements, and the lifecycle management.
3. Conduct a risk analysis and derive technical requirements from the results.

Figure 5.5 shows the general risk management flowchart. Based on a risk identification, a risk assessment is made to evaluate the failure occurrence and severity of the consequences. There are two approaches to identify risks. In a bottom-up approach, possible failures of single components are identified and possible outcomes are evaluated. This approach can be conducted intuitively, mainly based on the engineering experience of the developer or based on a more conservative approach using check lists. On the other hand, a top-down approach can be used by incorporating a ↪ fault tree analysis (FTA). In that case, an unwanted system state or event is analyzed for the possible reasons. This is done consecutively for these reasons until possible failure reasons on component level are reached. In practice, both approaches should be used to identify all possible risks.

For each identified risk, the failure occurrence and the severity of the consequence has to be evaluated. In particular, for hardware components, this is a sometimes hideous task, since some occurrences cannot be calculated easily. Based on these values, a risk graph can be created as shown in Fig. 5.6. Acceptable risks do not require further action, but have to be monitored in the further development process. Risks considered to be in the ↪ as low as reasonably practicable (ALARP) area are considered relevant, but cannot be dealt without an abundant (and therefore not reasonable) effort. Risks in the non-acceptable area have to be analyzed to be at least transferred to the ALARP area. Please note that the definitions of the different axis in the risk graph and the acceptable, ALARP, and non-acceptable area have

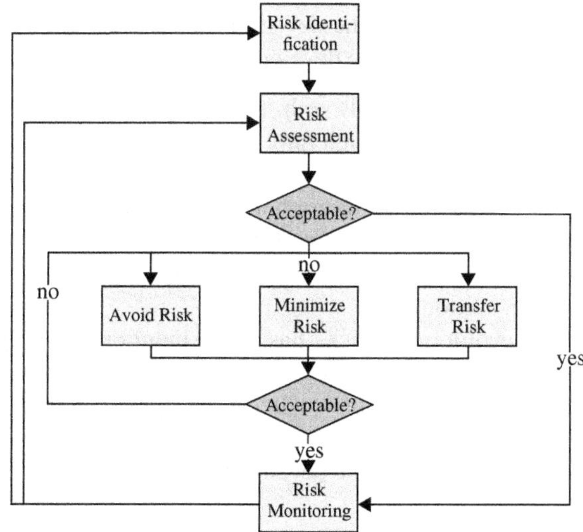

Fig. 5.5 General risk management flowchart

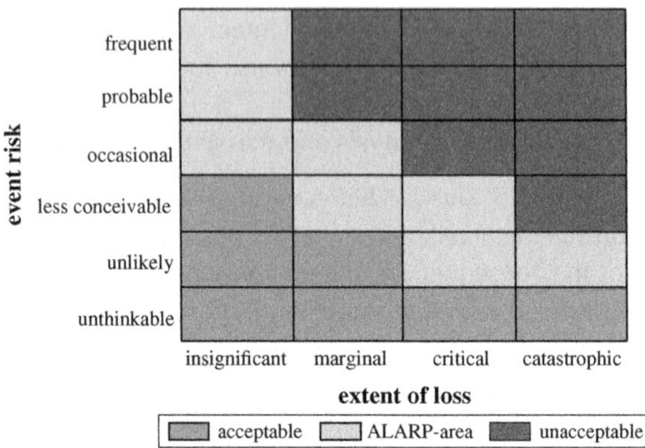

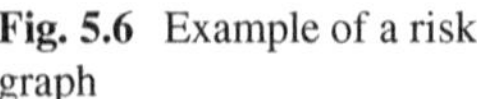
Fig. 5.6 Example of a risk graph

to be defined for each project or system separately based on the above-mentioned standards or company rules.

For each risk, one has three possibilities to deal with the risk, i.e., move it into more acceptable areas of the risk graph:

- First of all, risk can be avoided. If a piezoelectric actuator is used for a tactile application, the user can be exposed to high voltages, if the isolation fails. One can avoid this risk, if no piezoelectric (or other actuator principles) with high-voltage demands are used.
- Secondly, the risk can be minimized. In the above-mentioned example this could be an additional electrically insulating layer with requirements on breakdown voltage, mechanical endurance, and surface texture properties.
- The third possibility is to transfer a risk. This principle is only applicable in a restricted way to the design of haptic interfaces. A possible example would be the assignment of the development of a certain sub-contractor to minimize the technical risks of the development.

After each risk is dealt with, the acceptability has to be evaluated again, i.e., the changes in the risk graph have to be analyzed. Obviously, moving risks into lower risk regions will consume effort and costs. This considerations can lead to ethical dilemmas, when severe harm to humans has to be weighted against financial risks like damage compensations. For this reason, ↪ ALARP classifications for economical reasons are forbidden by ISO 14971 for medical devices starting with the 2013 edition.

The evolution of risks has to be monitored throughout the whole design and production process of a system. If all steps involved are considered, it is obvious that the design of safe systems will have an significant impact on the overall development costs of a haptic system and a thorough knowledge of all components is needed to find possible risks in the development.

Table 5.1 Example of a system specification for a haptic device

R/W	Description	Value	Source/Comment
Especially kinaesthetic-motivated parameters			
R[a]	Number of DOFs	2 × rot., 1 × transl	Shall give an idea of DOFs, name them!
R	Workspace	$100 \times 50 \times 50\,mm^3$	Minimum of workspace to be achieved
W	Maximum workspace	$150 \times 100 \times 100\,mm^3$	Maximum workspace necessary
R	Maximum force in DOF "name"	5 N	
W	Maximum force in DOF "name"	7 N	Always define a range of forces!
R	Minimum force in DOF "name"	0.2 N	
W	Minimum force in DOF "name"	0.1 N	Always define a range of forces!
R	Maximum dynamics (bandwidth) for DOF "name" in a blocked situation	100 Hz	Shows (among other things), for example, the maximum dynamics of the driver electronics
W[b]	Maximum dynamics (bandwidth) for DOF "name" in a blocked situation	200 Hz	Shows the bandwidth the customer dreams of
R	Smallest border frequency when movement is blocked	Static	There may be applications with pure dynamic movements without a static portion. This makes this question interesting
R	Maximum velocity of movement in idle mode	10 mm/s	This is a question regarding security too, as it defines the mechanical energy stored in the system
R	Maximum bandwidth of the velocity change	10 Hz	The change of velocity, which is the acceleration of the system, has a large influence on the energy the system requires
R	Maximum haptic impedance at the output	10 Ns/m	This is an alternative representation to the independent definition of force and velocity for dynamic (but passive) systems!

(continued)

Table 5.1 (continued)

R/W	Description	Value	Source/Comment
R	Minimum haptic impedance at the output	0.01 Ns/m	This is an alternative representation to the independent definition of force and velocity for dynamic (but passive) systems!
R	Smallest position resolution/measurement insecurity for DOF "name"	0.1 mm	Usually measurement of the position is self-evident for haptic interaction
W	Smallest position resolution/measurement insecurity for DOF "name"	0.05 mm	
R	Type of the mechanical interface	Button/pen/none	Is there a handle?
R	Mechanical reference point	Grounded, worn	Has influence on weight, size, and energy
In particular tactile-motivated parameters			
R	Direction(s) of the tactile stimulation	Normal to the skin	An alternative would be lateral stimulation or a combination of both
R	Maximum displacement-amplitude of the tactile elements	1 mm	Is especially relevant for pin-displays, but may be also understood as oscillation-amplitude of vibrational elements
R	Minimum amplitude resolution of displacements	Digital (on/off)	May include several levels for the pin to be moved to
R	Highest density of stimulation	2 mm distance from midpoint to midpoint	Varies extremely in dependency from the chosen skin area in contact
R	Maximum geometrical size of stimulation	2 mm diameter	
R	Maximum frequency range of stimulation	100–300 Hz	Relevant for tactile actuators only, of course
R	Minimum frequency-resolution	1 Hz	For vibro-tactile actuators
R	Maximum force during displacement/stiffness	20 N	Pin-based actuators may not necessarily be stiff. Systems of lower admittance may be used too
R	Connection to the user	Attached to the environment/worn	Necessary to identify, whether there is a relative movement between skin (e.g., finger) and the display
R	Maximum number of fingers simultaneously in contact with the device	1–10	May have an large impact on the design when, for example, full-hand exploration is required

(continued)

Table 5.1 (continued)

R/W	Description	Value	Source/Comment
Digital interface			
R	Minimum resolution of the output data	12 bit	Usually slightly lower than the measurement error of force and position measurement
R	Minimum resolution of the input data	12 bit	Usually slightly larger than the resolution of force and position input data
R	Frequency of the haptic loop[c]	1000 Hz	Should be at least two times, better would be 10 times, larger than the border frequency of the design. Has influence on the perceived stiffness
W	Other interface-requirements	Use USB/firewire...	Typically, the interface to be used is subject to company politics
R	Interface driver	API	As any other hardware, a haptic interface needs an own software driver for abstraction
General parameters			
R	Maximum temperature range for operation	10–50 °C	May become very relevant for actuator principals with little efficiency in extreme environments (automotive)
R	Maximum volume	$500 \cdot 500 \cdot 200\,\text{mm}^3$	Device size
R	Weight	1 kg	In particular relevant if the device is worn. This limit will strongly influence the mechanical energy generated
R	Electrical supply	Battery/110V/ 230V	Very important, devices were spotted on fairs, which ceased to function due to errors made when considering AC voltages of different countries
R	Maximum power	50 W	Primary power consumption including all losses

[a] *R* requirement, *W* wish

[b] The combination of requirements and wishes (R and W) may be used for almost any element of the system specification. It is recommended to make use of this method, but due to clarity in the context of this book, this approach of double questions is aborted here

[c] A "haptic loop" is a complete cycle including the output of the control variable (in case of simulators, this variable was calculated the time step before) and the read operation on the measurement value

5.5 Requirement Specifications of a Haptic System

The defined application together with the assumption from the customer and the interaction analysis will allow to derive individual requirements for the task-specific haptic system. These system requirements should be complemented with applicable safety and other standards to form a detailed requirement list. This list should not only include a clear description of the intended interactions, but also the intended performance measures (see Chap. 13) and as much technical details as possible about the overall system and the included components should be documented. As stated above, the technical solution clusters shown in the preceding Sect. 5.3 will also give possible requirements depending on the intended class of applications.

Table 5.1 gives an example of such a requirement list with the most relevant technical parameters of a haptic system. However, it is meant to be an orientation and has to be adapted to the specific situation by removing obsolete entries and adding application-specific aspects

Additionally, a system specification includes references to other standards and special requirements relating to the product development process. Among others, these are the costs for the individual device, the design process itself, and the number of devices to be manufactured in a certain time frame. Additionally, the time of shipment, visual parameters for the design, and safety-related issues are usually addressed.

Recommended Background Reading

[9] Magnusson, C. & Brewster, S.: **Guidelines for Haptic Lo-Fi Prototyping**, Workshop, 2008.
Proceedings of a workshop conducted during the HaptiMap project with hints and examples for low-fi prototyping of haptic interfaces.

[15] Nitsch, V. & Färber, B.: **A Meta-Analysis of the Effects of Haptic Interfaces on Task Performance with Teleoperation Systems**. In: IEEE Transac tions on Haptics, 2012.
Reviews the effect measures of several evaluation of VR and teleoperation systems. Recommended read for the design of haptic interaction in teleoperation and VR applications.

References

1. Acker A (2011) Anwendungspotential von Telepräsenz-und Teleaktionssystemen für die Präzisionsmontage. Dissertation. Technische Universität München. URL: http://mediatum.ub.tum.de/doc/1007163/1007163.pdf

2. Bhattacharjee HK et al (2011) A novel single-port technique for transanal rectosigmoid resection and colorectal anastomosis on an ex vivo experimental model. Surg endosc 25(6):1844–1857. doi:10.1007/s00464-010-1476-1
3. Deml B (2004) Telepräsenzsysteme: Gestaltung der Mensch-Maschine-Schnittstelle. Dissertation. Universität der Bundeswehr, München, URL: http://d-nb.info/972737340
4. Deml B (2007) Human factors issues on the design of telepresence systems. Presence: Teleoperators Virtual Environ 16(5):471–487. doi:10.1162/pres.16.5.471
5. van Erp J et al. (2004) Vibrotactile waypoint navigation at sea and in the air: two case studies. In: TU-Muenchen (ed) pp 166–173. URL: http://www.eurohaptics.vision.ee.ethz.ch/2004/15f.pdf
6. Hatzfeld C (2013) Experimentelle Analyse der menschlichen Kraftwahrnehmung als ingenieurtechnische Entwurfsgrundlage für haptische Systeme. Dissertation, Technische Universität Darmstadt. http://tuprints.ulb.tu-darmstadt.de/3392/., Dr. Hut Verlag, München. ISBN: 978-3-8439-1033-0
7. Hatzfeld C, Neupert C, Werthschützky R (2013) Systematic consideration of haptic perception in the design of task-specific haptic systems. Biomed Tech 58. doi:10.1515/bmt-2013-4227
8. Janschek K (2012) Mechatronic systems design: methods, models, concepts. Springer, Berlin. ISBN: 978-3- 642-17531-2
9. Magnusson C, Brewster S (2008) Guidelines for haptic Lo-Fi prototyping. conference workshop, NordiCHI. URL: http://www.haptimap.org/organized-events/nordichi-2008-workshop.html
10. Matich S et al (2013) A new 4 DOF parallel kinematic structure for use in a single port robotic instrument with haptic feedback. Biomed Tech 58:1. doi:10.1515/bmt-2013-4403
11. Matich S et al (2013) Teleoperation system with haptic feedback for single insicion surgery—concept and system design. In: CARS proceedings. Heidelberg
12. Muñoz L, Ponsa P, Casals A (2012) Design and development of a guideline for ergonomic haptic interaction. In: Human-computer systems interaction: backgrounds and applications 2, Springer, pp 15–29. doi:10.1007/978-3-642-23172-8_2
13. Neupert C et al (2013) New device for ergonomic control of a surgical Robot with 4 DOF including haptic feedback. Biomed Tech 58. doi:10.1515/bmt-2013-4404
14. Nitsch V (2012) Haptic human-machine interaction in teleoperation systems: implications for the design and effective use of haptic interfaces. Südwestdeutscher Verlag für Hochschulschriften. ISBN: 978-3838132686
15. Nitsch V, Färber B (2012) A meta-analysis of the effects of haptic interfaces on task performance with teleoperation systems. IEEE Trans Haptics 6:387–398. doi:10.1109/ToH.2012.62
16. Oakley I et al (2002) Guidelines for the design of haptic widgets. English. In: People and computers XVI—memorable yet invisible. Springer, Berlin, pp 195–211. doi:10.1007/978-1-4471-0105-5_12
17. Pahl G, Wallace K, Blessing L (2007) Engineering design: a systematic approach. Springer. ISBN: 978-3540199175
18. Schloske A (2010) Funktionale Sicherheit und deren praktische Umsetzung nach IEC 61508 und ISO CD 26262. In: Fachverband Elektronik-Design —FED: Integration und Effizienz–notwendig und möglich: Konferenzband zur 18. FED-Konferenz "Elektronik-Design—Leiterplatten—Baugruppen". URL: http://publica.fraunhofer.de/documents/N-151095.html
19. Sjöström C (2002) Non-visual haptic interaction design-guidelines and applications. Dissertation. Lund University. URL: http://lup.lub.lu.se/record/464997
20. Tavakoli M et al. (2008) Haptics for teleoperated surgical robotic systems. World Scientific Publishing. ISBN: 978-981-281-315-2
21. van Erp J (2005) Vibrotactile spatial acuity on the torso: effects of location and timing parameters. In: Haptic interfaces for virtual environment and teleoperator systems. WHC 2005. First joint Eurohaptics conference and symposium on (2005), pp 80–85. doi:10.1109/WHC.2005.144

Chapter 6
General System Structures

Thorsten A. Kern

Abstract Haptic systems exhibit several basic structures defined by the mechanical inputs and outputs, commonly known as impedance or admittance system structures. This chapter describes these structures in open-loop and closed-loop variants and presents commercial realizations as well as common applications of these structures. Based on the different properties of the structures and the intended application, hints for the selection of a suitable structure are given.

When starting the design of haptic devices, the engineer has to deal with the general structures they can be composed of. Haptic devices of similar functionality can consist of very different modules. There are four big classes of possible system designs:

1. "open-loop admittance controlled systems"
2. "closed-loop admittance controlled systems"
3. "open-loop impedance controlled systems"
4. "closed-loop impedance controlled systems"

Impedance controlled systems are based on the transfer characteristics of a mechanical impedance $\underline{Z} = \frac{F}{\underline{v}}$ and are typical for the structure of many kinaesthetic devices. They generate a force as output and measure a position as input. Admittance controlled systems instead are based on the definition of a mechanical impedance $\underline{Y} = \frac{\underline{v}}{F}$, describing a transfer characteristics with force input and velocity output. These systems generate a position change as haptic feedback and get a force reaction from the user as input source. In the situation of a closed-loop controlled system, this force is measured and used for correction of the position. This analysis can be regarded as analog in the case of torque and angle replacing force and position for rotary systems. Nevertheless, for readability purposes, the following descriptions concentrate on translational movements and devices only.

T.A. Kern (✉)
Continental Automotive GmbH, VDO-Straße 1, 64832 Babenhausen, Germany
e-mail: t.kern@hapticdevices.eu

C. Hatzfeld and T.A. Kern (eds.), *Engineering Haptic Devices*,
Springer Series on Touch and Haptic Systems, DOI 10.1007/978-1-4471-6518-7_6

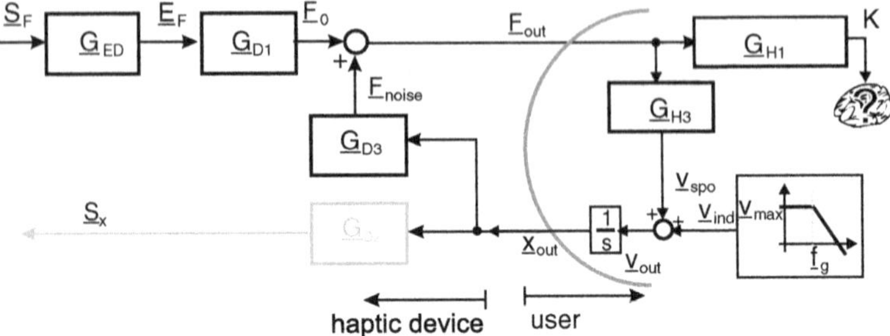

Fig. 6.1 Block diagram of an open-loop impedance controlled haptic system

6.1 Open-Loop Impedance Controlled

Open-loop impedance controlled systems are based on a simple structure (Fig. 6.1). A force signal $\underline{S}_F$ is transferred via a driver $\underline{G}_{\mathrm{ED}}$ into a force-proportional energy form $\underline{E}_F$. This energy is then altered into the output force $\underline{F}_0$ by an actuator $\underline{G}_{\mathrm{D1}}$. This output force interferes with a disturbing force $\underline{F}_{\mathrm{noise}}$. This noise is a result of movements generated by the user $\underline{x}_{\mathrm{out}}$ and the mechanical properties of the kinematic design $\underline{G}_{\mathrm{D3}}$. Typically, such disturbing forces are friction and inertia. The sum of both forces is the actual output force $\underline{F}_{\mathrm{out}}$ of the impedance controller system. Usually, there is an optional part of the system, a sensor $\underline{G}_{\mathrm{D2}}$, which measures the movements and the actual position of the haptic system.

Examples: Universal Haptic Interfaces

Open-loop impedance controlled systems are the most frequently available devices on the market. Due to their simple internal design, a few standard components can be used to build a useful design with adequate haptic feedback, if care is taken for the minimization of friction and masses. Among the cheapest designs available on the market today, the PHANToM Omni, recently renamed as GEOMAGIC Touch, (Fig. 6.2a), connected via Fire-Wire to the control unit, is among the best known. It is frequently used in research projects and for the creative manipulation of 3D data during modeling and design. In the higher price segment, there are numerous products, e.g., the devices of the company *Quanser* (Markham, Ontario, Canada). These devices are usually equipped with a real-time MatLab™ (*The MathWorks*, Natick, MA, USA)-based control station, adding some flexibility to the modifications of the internal data processing by the end customer. The doubled pantograph kinematics of the Haptic Wand (Fig. 6.2b) allow force feedback in up to five degrees of freedom with three translations and two rotations. Although all these devices may be open-loop impedance controlled, the software usually includes simple dynamic models of the mechanical structures. This allows some compensation of inertial and frictional effects of the kinematics based on the continuous measurement of position and velocities.

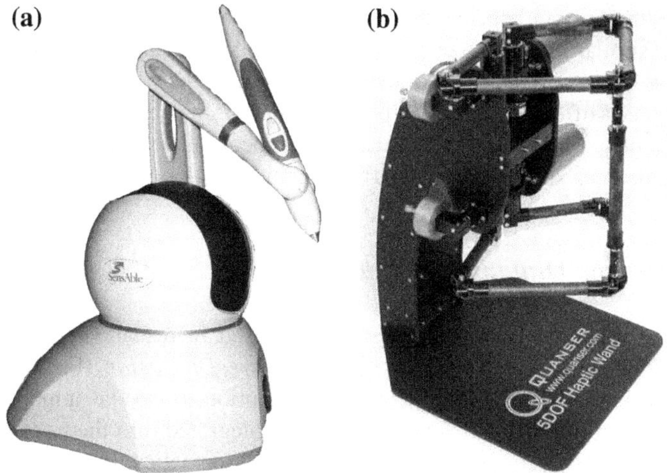

Fig. 6.2 Example of an open-loop impedance controlled system with **a** serial kinematic (GEOMAGIC TOUCH, *3D systems geomagic solutions*), and **b** parallel kinematic (5 DoF HAPTIC WAND, *Quanser*) structure. Images courtesy of *3D systems geomagic solutions*, Morrisville, NC, USA and *Quanser*, Markham, Ontario, Canada, used with permission

6.2 Closed-Loop Impedance Controlled

Closed-loop impedance controlled systems (Fig. 6.3) differ from open-loop impedance controlled systems in such a manner that the output force $\underline{F}_{\text{out}}$ is measured by a force sensor $\underline{G}_{\text{FSense}}$ and is used as a control variable to generate a difference value $\Delta\underline{S}_F$ with the nominal value. An additional component typically is a controller $\underline{G}_{\text{CD}}$ in the control path, optimizing the dynamic properties of the feedback loop. The closed loop makes it possible to compensate the force $\underline{F}_{\text{noise}}$ resulting from the mechanics of the systems. This has two considerable advantages: On the one hand,

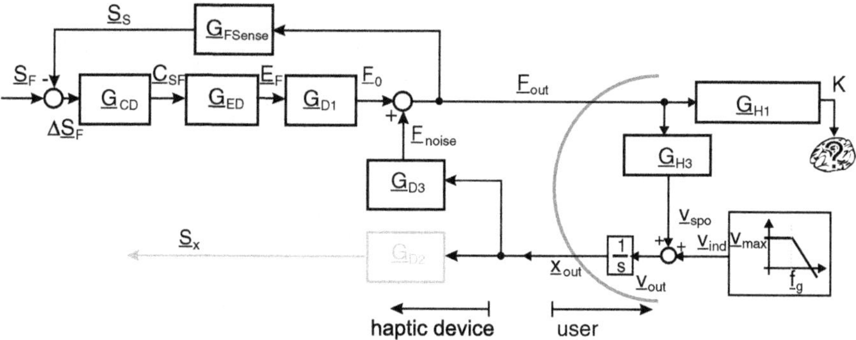

Fig. 6.3 Block diagram of a closed-loop impedance controlled system with force feedback

at idle state, the system behaves in a much less frictional and more dynamic way compared to similar open-loop controlled systems. Additionally, as the closed-loop design allows some compensation of inertia and friction, the whole mechanical setup can be designed stiffer. But it has to be noted that of course part of the maximum output power of actuators will then be used to compensate the frictional force, which makes these devices slightly less powerful than an open-loop design.

Example: Force Dimension Delta Series

Closed-loop impedance controlled systems are used in research projects and as special-purpose machines. The delta series of *ForceDimension* (Fig. 6.4) is one example as it is a commercial system with the option to buy an impedance controlled version. In this variant, force sensors are integrated into the handle, able to measure interaction forces in the directions of the kinematics degrees of freedom. Closed-loop impedance controlled systems are technologically challenging. One the one hand, they have to comply with a minimum of friction and inertia; on the other hand, with little friction, the closed loop tends to become instable, as an energy exchange between user and device may build up. This is why controllers, typically, monitor the passive behavior of the device. Additionally, the force sensor is a cost-intensive element. In case of the delta device, the challenge to minimize moving masses has been faced by a parallel kinematics design.

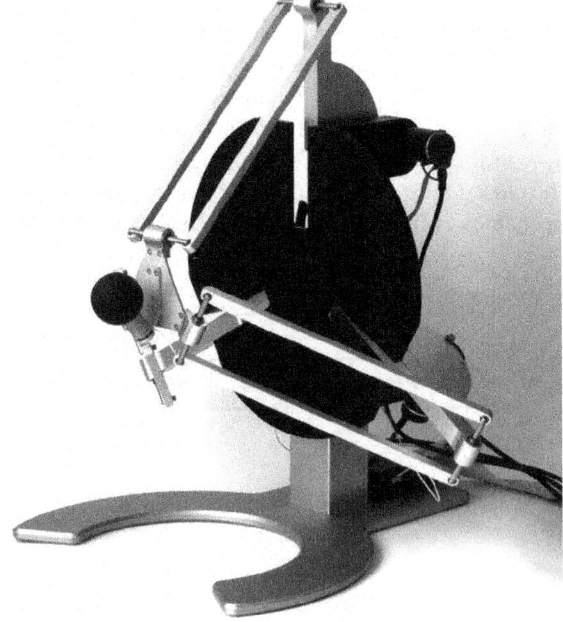

Fig. 6.4 Example of a parallel kinematic closed-loop impedance controlled system (delta3, *ForceDimension*). Image courtesy of *ForceDimension*, Nyon, Switzerland, used with permission

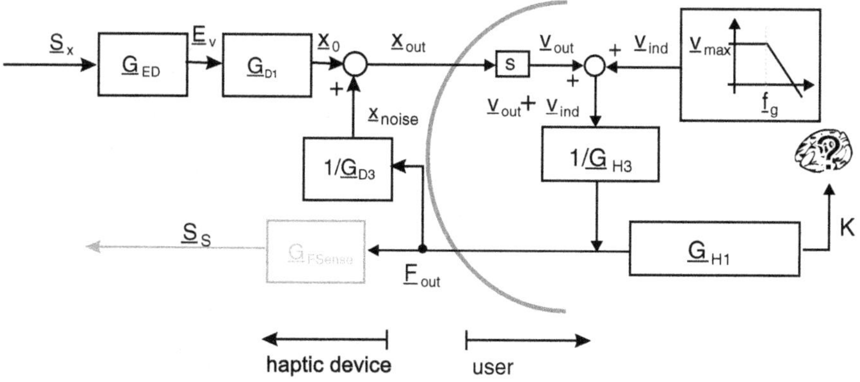

Fig. 6.5 Block diagram of an open-loop admittance controlled haptic system

6.3 Open-Loop Admittance Controlled

Open-loop admittance controlled systems (Fig. 6.5) provide a positional output. Proportionally to the input value $\underline{S}_x$, a control chain with energy converter $\underline{G}_{ED}$ and kinematics $\underline{G}_{D1}$ provides a displacement $\underline{x}_0$. This displacement interferes with a disturbance variable $\underline{x}_{noise}$ which is dependent on the mechanical properties of the kinematics $\underline{G}_{D3}$ and a direct reaction to the user's input $\underline{F}_{out}$. In practice, an open-loop admittance controlled system typically shows a design that allows to neglect the influence of the disturbance variable. Another optional element of open-loop admittance controlled systems is the measurement of the output force with a force sensor $\underline{F}_{Sense}$ without closing the control loop.

Example: Braille Devices

Open-loop admittance controlled systems are used especially in the area of tactile displays. Many tactile displays are based on pin arrays, meaning that they generate spatially distributed information by lifting and lowering pins out of a matrix. These systems' origins are Braille devices (Fig. 6.6) coding letters in a tactile, readable, embossed printing. For actuation of tactile pin-based displays, a variety of actuators are used. There are electrodynamic, electromagnetic, thermal, pneumatic, hydraulic, and piezoelectric actuators and even ultrasonic actuators with transfer media.

6.4 Closed-Loop Admittance Controlled Devices

Closed-loop admittance controlled devices (Fig. 6.7) provide a positional output and a force input to the controlling element identical to impedance controlled devices.

Fig. 6.6 Example of an open-loop admittance controlled system being a Braille row. Image courtesy of Ralf Roletschek, used with permission

The mandatory measurement of the output force $\underline{F}_{\text{out}}$ is used as control variable $\underline{S}_S$ for calculating the difference $\Delta\underline{S}_F$ with the commanding value $\underline{S}_F$. This difference is then fed through the controller $\underline{G}_{\text{CD}}$ into the control circuit. As a result, the displacement $\underline{x}_{\text{out}}$ is adjusted until an aspired force $\underline{F}_{\text{out}}$ is reached.

A variant of a closed-loop admittance controlled device is shown in Fig. 6.8. Closed-loop admittance controlled devices show considerable advantages for many applications requiring large stiffnesses. However, the force sensors $\underline{G}_{\text{FSense}}$ are complex and consequently expensive components, especially when there are numerous degrees of freedom to be controlled. As a variant, the system according to Fig. 6.8 does not use a sensor but just a force-proportional measure, e.g., a current, as control variable. When using a current with electrodynamic actuators, we can identify even the reaction of the user generating an induction as an additional velocity-dependent value.

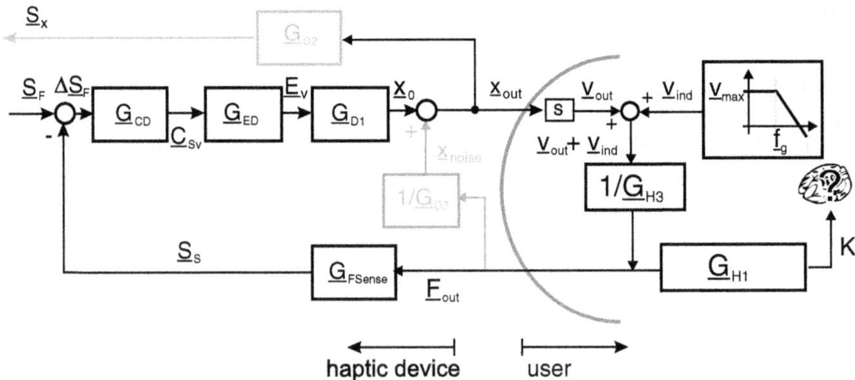

Fig. 6.7 Block diagram of a closed-loop admittance controlled haptic system with force-feedback loop for control

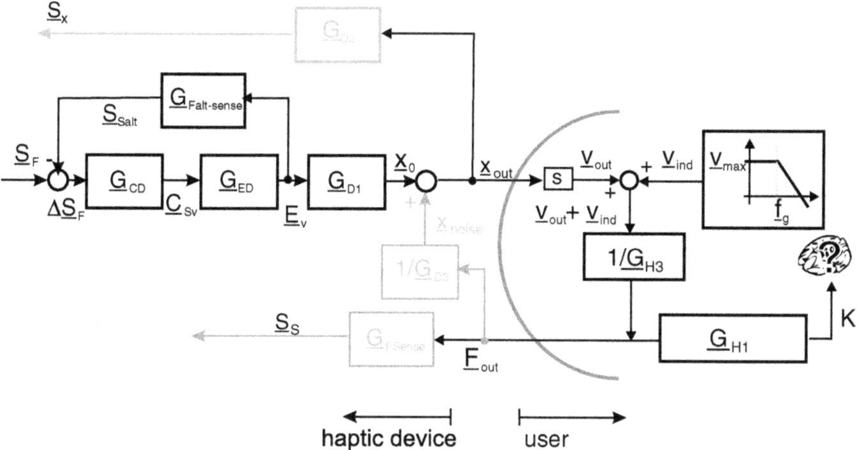

Fig. 6.8 Block diagram of a closed-loop admittance controlled haptic system with a feedback loop measuring an internal force-proportional value

Examples: Universal Haptic Interfaces

At present, closed-loop admittance controlled systems are the preferred approach to provide high stiffnesses with little loss in dynamic properties. The idea to haptically hide the actual mechanical impedance from the user by closing the control loop makes it possible to build serial kinematics with a large workspace. The FCS HapticMaster (Fig. 6.9a) is such a one-meter-high system with three degrees of freedom and a force of up to 100 N. It includes a force sensor at its handle. The axes are controlled by self-locking actuators. The device's dynamics is impressive, despite its size. However, a damping has to be included in the controller for security reasons resulting in a limitation of bandwidth depending on the actual application.

Realizations of the variant of closed-loop admittance controlled devices are the VIRTUOSE systems from *Haption* (Fig. 6.9b). In these devices, the current is measured at electrodynamic electronic commutated actuators and fed back as a control value. The devices show impressive qualities for the display of hard contacts, but have limited capabilities in the simulation of soft interactions, e.g., with tissues. Therefore, the application area of such systems is mainly the area of professional simulation of assembly procedures for manufacture preparation.

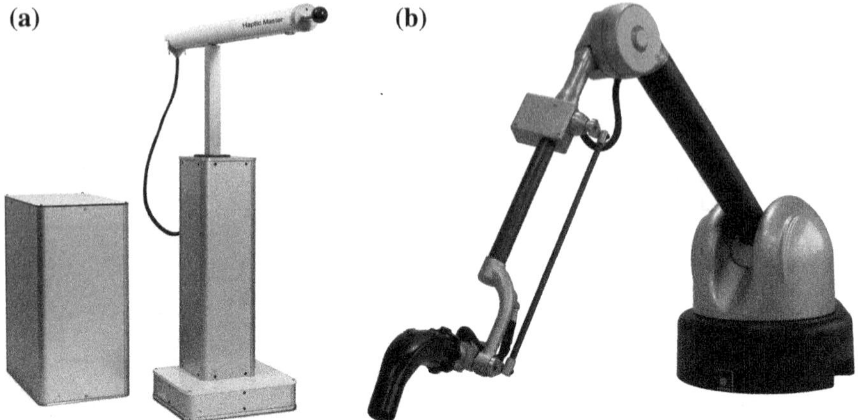

Fig. 6.9 Examples of closed-loop admittance controlled systems in variants with **a** direct force measurement (HAPTICMASTER) and **b** measurement of the actual current (VIRTUOSE 6D35-45). Images courtesy of *Moog FCS*, Nieuw-Vennep, the Netherlands, and *Haption GmbH*, Aachen, Germany, used with permission

6.5 Qualitative Comparison of the Internal Structures of Haptic Systems

As haptic human–machine interaction is based on impedance coupling, it is always a combination of action and reaction, be it via force or position, which has to be analyzed. In fact, without any knowledge about the internal structure of a device, it is impossible to find out whether the system is open-loop impedance controlled, closed-loop impedance controlled, or closed-loop admittance controlled. With experience in the technological borders of the most important parameters such as dynamics and maximum force, an engineer can make a well-founded assumption about the internal structure by simply using the device. But concerning the abstract interface of input and output values, all the devices of the above three classes are absolutely identical to the user as well as to the controlling instance. Despite this fact, the technical realizations of haptic systems differ widely in their concrete technical design. The parameters influencing this design have to be balanced against each other, which are as follows:

- Number of components;
- Maximum impedance to be achieved at slow motion;
- Minimum impedance to be achieved at fast motion;
- Force resolution;
- Impedance of mechanical components (e.g., inertia of kinematics).

These parameters and their mapping onto the technical designs are given qualitatively. In Fig. 6.10, the impedance generated by a device in absolute values and the impedance range covered may be one criterion for the performance of a device.

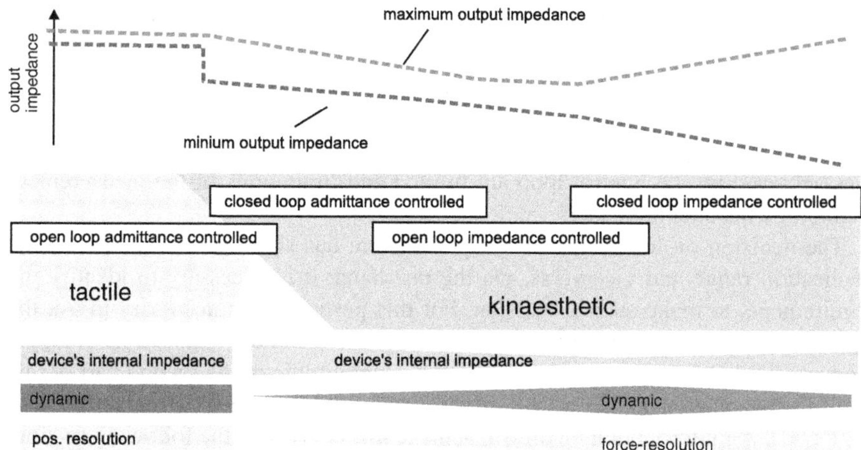

Fig. 6.10 Qualitative comparison of the application areas for different device structures

Analyzing the systems according to this criterion shows that open-loop admittance controlled systems may have high impedance, which shows smaller variability in tighter borders. Closed-loop admittance controlled systems extend these borders by their ability to modulate the impedance due to the feedback loop. Depending on the design, closed-loop admittance controlled systems vary in the width of this modulation. In the lower area of possible, realizable impedances, the open-loop impedance controlled systems follow. They stand out more by their simplicity of design than by the large impedance ranges covered. In comparison with closed-loop admittance controlled systems, they gain some impedance width at the lower border of impedances. In order to be able to equally cover lower as well as higher impedances, the choice should be made of closed-loop impedance controlled systems.

6.5.1 Tactile Devices

Normally, pure open-loop admittance controlled systems are suitable for *tactile devices* only, as with tactile devices, usually there is no active feedback by the user to be measured. The haptic interaction is limited to tensions being coupled to the skin of the user's hands. Such devices show high internal impedance ($\underline{Z}_D$). The dynamics and the resolution concerning the displacement are very high.

6.5.2 Kinaesthetic Devices

Kinaesthetic devices can be built with systems allowing a modulation of the displayed impedances. The closed-loop admittance controlled systems excel due to the possibility to use mechanical components with high impedances. The dynamics of these systems are accordingly low (<100 Hz), and the force resolution is, due to the typical frictions, not trivial when realized. Open-loop impedance controlled systems show a wider dynamic range due to the missing feedback loop with, at the same time, limited dynamic range. Only closed-loop impedance systems allow covering a wide impedance range from the lowest to very high impedances, whereby with increasing requirements of force resolution the dynamics of the maximum velocities achieved by the control loop are limited and limitations of the measurement technology become noticeable.

The decision on the design of a haptic system has significant influence on the application range and vice versa. On the one hand, it is necessary to identify the requirements to make such a decision. For this purpose, it is necessary to ask the right questions *and* to have an insight into possible technical realizations of single elements of the above structures. This is the general topic of the second part of this book. On the other hand, it is necessary to formulate an abstract system description of the device. An introduction on how to achieve this is given in the following section.

6.6 How to Choose a Suitable System Structure

The selection of a suitable system structure is one of the first steps in the design of task-specific haptic systems. Based on the interaction analysis, one should have sufficient insight into the intended interactions between system and user and should be able to decide between a mainly tactile and mainly kinaesthetic device structure. Based on further criteria such as input and output capabilities and the mechanical impedance to be displayed, Fig. 6.11 gives a decision tree for the control structure.

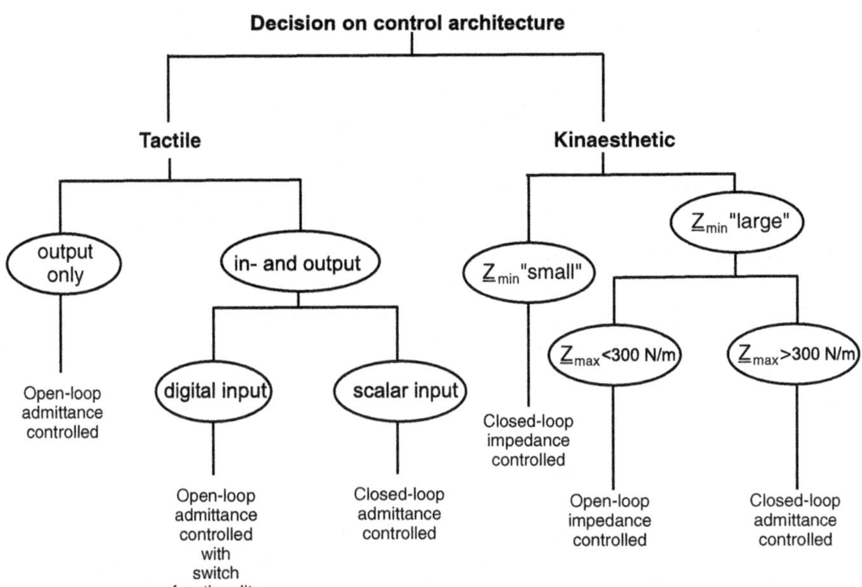

Fig. 6.11 Aid for the decision on the choice of the control structure

In particular, when the application includes an interaction in a multimodal or virtual environment, further additions to the system structure would be wise to consider since they promise a large technical simplification while maintaining haptic quality. This includes the approaches of event-based haptics as well as pseudo-haptics (see Sect. 12.5).

Chapter 7
Control of Haptic Systems

Thomas Opitz and Oliver Meckel

Abstract This chapter reviews some aspects of the control of haptic systems, including advanced forms of technical descriptions, system stability criteria and measures, as well as the design of different control laws. A focus is set on the control of bilateral teleoperation systems including the derivation of control designs that guarantee stability as well as haptic transparency and the handling of time delay in the control loop. The chapter also includes an example for the consideration of thermal properties and non-ideal mechanics in the control of a linear stage made from an EC motor and a ball screw as well as a perception-orientated approach to haptic transparency intended to lower the technical requirements on the control and component design.

The control of technical systems aims at safe and reliable system behavior, and controllable system states. By its depiction as a *system*, the analysis is put on an abstracted level, which allows covering many different technical systems described by their fundamental physics. On this abstracted level, a general analysis of closed-loop control issues is possible using several methods and techniques. The resulting procedures are applicable to a large number of system classes. The main purpose of any depiction and analysis of control systems is to achieve high performance, safe system behavior, and reliable processes. Of course, this also holds for haptic systems. Here, stable system behavior and high transparency are the most important control law design goals. The abstract description used for a closed-loop control analysis starts with the mathematical formulation of the physical principles the system fol-

T. Opitz (✉)
Institute of Electromechanical Design, Technische Universität Darmstadt,
Merckstr. 25, 64283 Darmstadt, Germany
e-mail: t.opitz@hapticdevices.eu

O. Meckel
Wittenstein Motion Control GmbH, Igersheim, Germany
e-mail: o.meckel@hapticdevices.eu

C. Hatzfeld and T.A. Kern (eds.), *Engineering Haptic Devices*,
Springer Series on Touch and Haptic Systems, DOI 10.1007/978-1-4471-6518-7_7

lows. As mentioned above, systems with different physical principles are covered up by similar mathematical methods. The depiction by differential equations or systems of differential equations proves widely usable for the formulation of various system behaviors. Herein, analogies allow transforming this system behavior into the different technical contexts of different systems, provided there exists a definite formulation of the system states that are of interest for closed-loop control analysis. The mathematical formulation of the physical principles of the system, also denoted as modeling, is followed by system analysis including dynamic behavior and its characteristics. With this knowledge, a wide variety of design methods for control systems become applicable. Their main requirements are:

System stability The fundamental requirement for stability in any technical system is the main purpose for closed-loop control design. For haptic systems, stability is the most important criteria to guarantee safe use of the device for the user.

Control quality Tracking behavior of the system states to demanded values, every system is faced with external influences also denoted as disturbances that interfere with the demanded system inputs and disrupt the optimal system behavior. To compensate this negative influence, a control system is designed.

Dynamic behavior and performance In addition to the first two issues, the need for a certain system dynamics completes this requirement list. With a view to haptic systems, the focus lies on the transmitted mechanical impedance which determines the achievable grade of transparency.

Besides the quality of the control result tracking the demanded values, the system behavior within the range of changes from these demanded values is focused. Also, the control effort that needs to achieve a certain control result is to be investigated. The major challenge for closed-loop control law design for haptic systems and other engineering disciplines is to deal with different goals that are often in conflict with each other. Typically, a gained solution is never an optimal one, rather a trade-off between system requirements. In the following Sect. 7.1, basic knowledge of linear and nonlinear system description is given. Section 7.2 gives a short overview of system stability analysis. A recommendation for structuring the control law design process for haptic systems is given in Sect. 7.3. Subsequently, Sect. 7.4 focuses on common system descriptions for haptic systems and shows methods for designing control laws. Closing in Sect. 7.5, a conclusion is presented.

7.1 System Description

A variety of description methods can be applied for mathematical formulation of systems with different physical principles. One of the main distinctions is drawn between methods for the description of linear and nonlinear systems, summarized in the following paragraphs. The description based on single-input–single-output (SISO) systems in the LAPLACE domain was discussed in Sect. 4.3.

7.1.1 Linear State Space Description

Besides the formulation of system characteristics through transfer functions, the description of systems using state space representation in the time domain also allows to deal with arbitrary linear systems. For SISO systems, a description using an nth-order ordinary differential equation is transformable into a set with n first-order ordinary differential equations. In addition to the simplified use of numerical algorithms for solving this set of differential equations, the major advantage is applicability to multi-input–multi-output (MIMO) systems. A correct and systematic model of their coupled system inputs, system states, and system outputs is comparably easy to achieve. In contrast to the system description in the LAPLACE domain by transfer functions $G(s)$, the state space representation formulates the system behavior in the time domain. Two sets of equations are necessary for a complete state space system representation. These are denoted as the *system equation*

$$\dot{\mathbf{x}} = \mathbf{A}\mathbf{x} + \mathbf{B}\mathbf{u} \tag{7.1}$$

and the *output equation*

$$\mathbf{y} = \mathbf{C}\mathbf{x} + \mathbf{D}\mathbf{u}. \tag{7.2}$$

The vectors $\mathbf{u}$ and $\mathbf{y}$ describe the multidimensional system input, respectively, to system output. Vector $\mathbf{x}$ denotes the inner system states.

As an example for state space representation, the second-order mechanical oscillating system as shown in Fig. 7.1 is examined. Assuming the existence of time invariant parameters, the description using second-order differential equation is

$$m\ddot{y} + d\dot{y} + ky = u \tag{7.3}$$

The transformation of the second-order differential Eq. (7.3) into a set of two first-order differential equations is done by choosing the integrator outputs as system states:

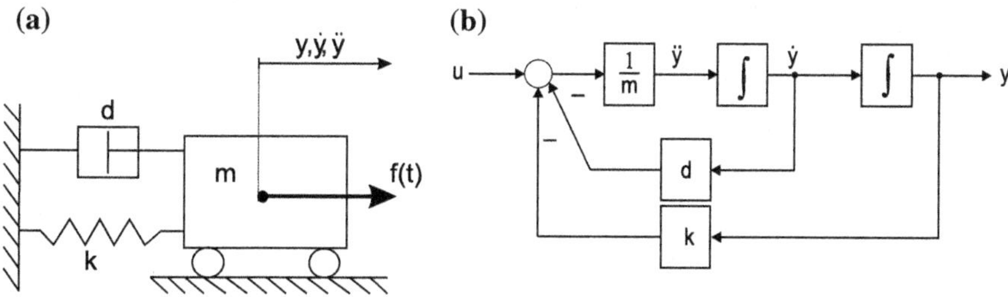

Fig. 7.1 Second-order oscillator **a** scheme, **b** block diagram

$$\begin{aligned} x_1 &= y \Rightarrow \dot{x}_1 = x_2 \\ x_2 &= \dot{y} \Rightarrow \dot{x}_2 = -\frac{k}{m}x_1 - \frac{d}{m}x_2 + \frac{1}{m}u \end{aligned} \tag{7.4}$$

Thus, the system equation for state space representation is as follows:

$$\begin{bmatrix} \dot{x}_1 \\ \dot{x}_2 \end{bmatrix} = \begin{bmatrix} 0 & 1 \\ -\frac{k}{m} & -\frac{d}{m} \end{bmatrix} \begin{bmatrix} x_1 \\ x_2 \end{bmatrix} + \begin{bmatrix} 0 \\ \frac{1}{m} \end{bmatrix} u \tag{7.5}$$

The general form of the system equation is

$$\dot{\mathbf{x}} = \mathbf{A}\mathbf{x} + \mathbf{B}\mathbf{u} \tag{7.6}$$

This set of equations contains the *state space vector* **x**. Its components describe all inner variables of the process that are of interest and that have not been examined explicitly using a formulation by transfer function. The system output is described by the output equation. In the given example as shown in Fig. 7.1, the system output y is equal to the inner state x_1

$$y = x_1 \tag{7.7}$$

which leads to the vector representation of

$$y = \begin{bmatrix} 1 \; 0 \end{bmatrix} \begin{bmatrix} x_1 \\ x_2 \end{bmatrix} \tag{7.8}$$

The general form of the output equation is

$$\mathbf{y} = \mathbf{C}\mathbf{x} + \mathbf{D}\mathbf{u} \tag{7.9}$$

which leads to the general state space representation applicable for single- or multi-input and output systems. The structure of this representation is depicted in Fig. 7.2. Although not mentioned in this example, matrix **D** denotes a direct feedthrough, which occurs in systems whose output signals y are directly affected by the input signals u without any time delay. Thus, these systems show a non-delayed step response. For further explanation on A, B, C, and D, [32] is recommended. Note that in many teleoperation applications where long distances between master device and slave device exist, significant time delays occur.

7.1.2 Nonlinear System Description

A further challenge within the formulation of system behavior is to imply nonlinear effects, especially if a subsequent system analysis and classification is needed. Although a mathematical description of nonlinear system behavior might be found

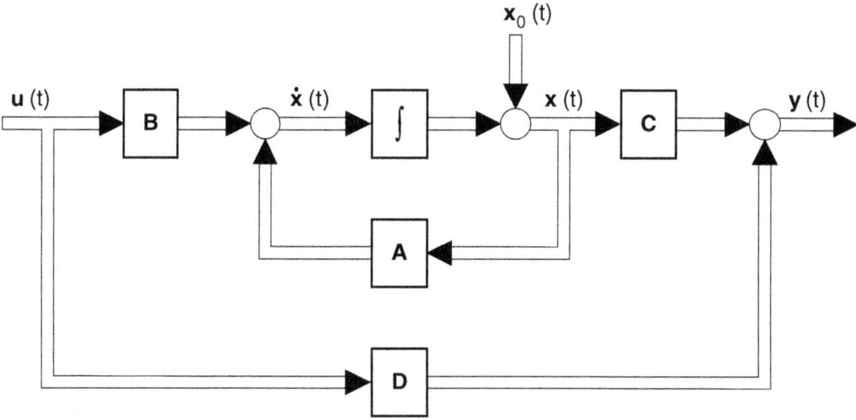

Fig. 7.2 State space description

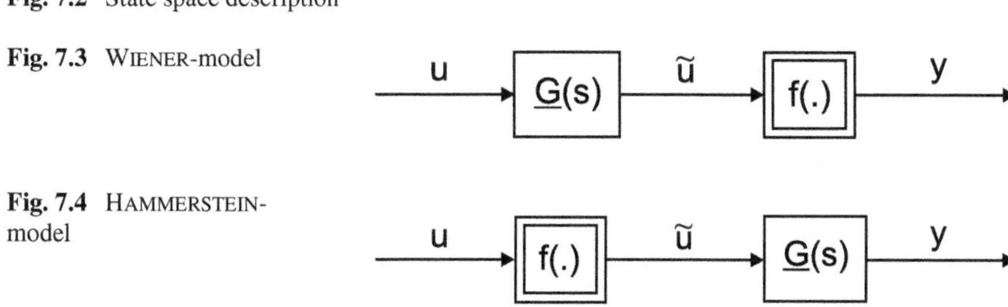

Fig. 7.3 WIENER-model

Fig. 7.4 HAMMERSTEIN-model

fast, the applicability of certain control design methods is an additional problem. Static nonlinearities can be easily described by serial coupling of a static nonlinear and linear dynamic device to be used as a summarized element for closed-loop analysis. Herein, two different models are differentiated. Figure 7.3 shows the block diagram of a linear element with arbitrary subsystem dynamics followed by a static nonlinearity.

This configuration also known as WIENER-model is described as

$$\tilde{u}(s) = \underline{G}(s) \cdot u(s)$$
$$y(s) = f(\tilde{u}(s)).$$

In comparison, Fig. 7.4 shows the configuration of the HAMMERSTEIN-*model* changing the order of the underlying static nonlinearity and the linear dynamic subsystem.

The corresponding mathematical formulation of this model is described as

$$\tilde{u}(s) = f(u(s))$$
$$y(s) = \underline{G}(s) \cdot \tilde{u}(s).$$

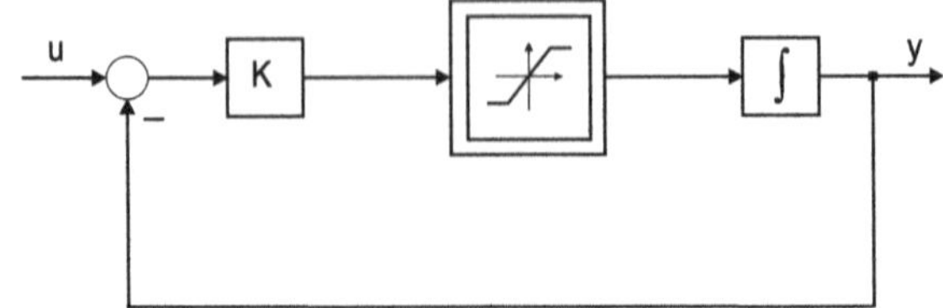

Fig. 7.5 System with internal saturation

More complex structures appear as soon as the dynamic behavior of a system is affected by nonlinearities. Figure 7.5 shows as an example a system with internal saturation. For this configuration, both models cannot be applied as easily as for static nonlinearities, in particular, if a system description is needed usable for certain methods of system analysis and investigation.

Typical examples for systems showing such nonlinear behavior are electrical motors whose torque current characteristic is affected by saturation effects and thus whose torque available for acceleration is limited to a maximum value.

This kind of system behavior is one example of how complicated the process of system modeling may become, as ordinary linear system description methods are not applicable to such a case. Nevertheless, it is necessary to gain a system formulation in which the system behavior and system stability can be investigated successfully. To achieve a system description taking various system nonlinearities into account, it is recommended to set up nonlinear state space descriptions. They offer a wide set of tools applicable to the following investigations. Deriving from Eqs. (7.1) and (7.2), the nonlinear system description for single-, multi-input, and output systems is as follows:

$$\dot{\mathbf{x}} = \mathbf{f}(x, \mathbf{u}, t)$$
$$\mathbf{y} = \mathbf{g}(x, \mathbf{u}, t).$$

This state space description is most flexible to gain a usable mathematical formulation of a systems behavior consisting of static, dynamic, and arbitrarily coupled nonlinearities. In the following, these equations serve as a basis for the examples illustrating concepts of stability and control.

7.2 System Stability

As mentioned above, one of the most important goals of control design is the stabilization of systems or processes during their life cycle, while operative or disabled. Due to the close coupling of haptic systems with a human user via a human–machine interface, safety becomes most relevant. Consequently, the focus of this chapter lies on system stability and its analysis using certain methods applicable to many systems. It must resemble the system behavior correctly and must be aligned with the applied investigation technique. For the investigation of systems, subsystems, close-looped systems, and single- or multi-input output systems, a wide variety of methods exist. The most important ones are introduced in this chapter.

7.2.1 Analysis of Linear System Stability

The stability analysis of linear time invariant systems is easily done by investigation of the system poles or roots derived from the eigenvalue calculation of the system transfer function $G(s)$. The decisive factor is the sign of the real part of these system poles. A negative sign in this real part indicates a stable eigenvalue; a positive sign denotes an unstable eigenvalue. The correspondence to the system stability becomes obvious while looking at the homogeneous part of the solution of the ordinary differential equation describing the system behavior. For example, a system is described as

$$T\dot{y}(t) + y(t) = Ku(t). \tag{7.10}$$

The homogeneous part of the solution $y(t)$ is derived using

$$y_h = e^{\lambda t} \quad \text{with } \lambda = -\frac{1}{T}. \tag{7.11}$$

As it can be seen clearly, the pole $\lambda = -\frac{1}{T}$ has a negative sign only if the time constant T has a positive sign. In this case, the homogeneous part of $y(t)$ disappears for $t \rightarrow \infty$, while it rises beyond each limit exponentially if the pole $\lambda = -\frac{1}{T}$ is unstable. This section will not deal with the basic theoretical background of linear system stability as these are basics of control theory. The focus of this section is the application of certain stability analysis methods. Herein, it will be distinguished between methods for direct stability analysis of a system or subsystem and techniques of closed-loop stability analysis. For direct stability analysis of linear system, investigation of the poles placement in the complex plane is fundamental. Besides the explicit calculation of the system poles or eigenvalues, theROUTH- HURWITZ *criterion* offers to determine the system stability and system pole placement with explicit calculation. In many cases, this simplifies the stability analysis. For the analysis of closed-loop stability, determination of closed-loop pole placement is also a possible approach. Additional methods leave room for further design aspects and extend the basic stability analysis. Well-known examples of such techniques are

- Root locus method,
- NYQUIST's stability criterion.

The applicability of both methods are discussed in the following without looking at the exact derivation.

7.2.1.1 Root Locus Method

The root locus offers the opportunity to investigate pole placement in the complex plane depending on certain invariant system parameters. As example of invariant system parameters, changing time constants or variable system gains might occur. The gain of the open loop is often of interest within the root locus method for closed-loop stability analysis and control design. In Eq. (7.12), G_R denotes the transfer function of the controller and G_S describes the behavior of the system to be controlled.

$$-G_o = G_R G_S \tag{7.12}$$

Using the root locus method, it is possible to apply predefined sketching rules whenever the dependency of the closed-loop pole placement on the open-loop gain K is of interest. The closed-loop transfer function G_g is depicted by Eq. (7.13)

$$G_g = \frac{G_R G_S}{1 + G_R G_S} \tag{7.13}$$

As an example, an integrator system with a second-order delay (IT_2) described by Eq. (7.14)

$$G_S = \frac{1}{s} \cdot \frac{1}{1+s} \cdot \frac{1}{1+4s} \tag{7.14}$$

is examined. The control transfer function is $G_R = K_R$. Thus, we find as open-loop transfer function

$$-G_o = G_R G_S = \frac{K_R}{s(1+s)(1+4s)}. \tag{7.15}$$

Using the sketching rules that can be found in various examples in the literature [33, 43], the root locus graph as shown in Fig. 7.6 is derived. The graph indicates that small gains K_R lead to a stable closed-loop system since all roots have a negative real part. A rising K_R leads to two of the roots crossing the imaginary axis and the closed-loop system becomes unstable. This simplified example proves that this method can be integrated in a control design process, as it delivers a stability analysis of the closed-loop system by only processing an examination on the open-loop system. This issue is also one of the advantages of the NYQUIST stability criterion. Additionally, the definition of the open-loop system is sufficient to derive a stability analysis of the system in a closed-loop arrangement.

7.2.1.2 NYQUIST's Stability Criterion

This section concentrates on the simplified NYQUIST stability criterion investigating the open-loop frequency response described as

$$-G_o(j\omega) = G_R(j\omega) G_S(j\omega).$$

The NYQUIST stability criterion is based on the characteristic correspondence of amplitude and phase of the frequency response. As example, we use the already introduced IT_2 system controlled by a proportional controller $G_R = K_R$. The BODE plot of the frequency response is shown in Fig. 7.7. The stability condition that has

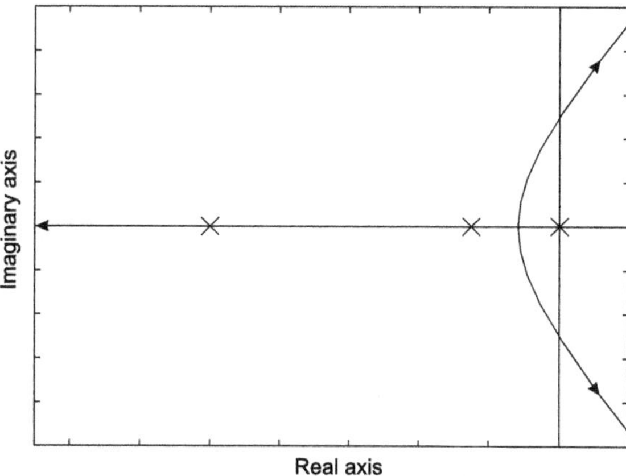

Fig. 7.6 IT_2 root locus

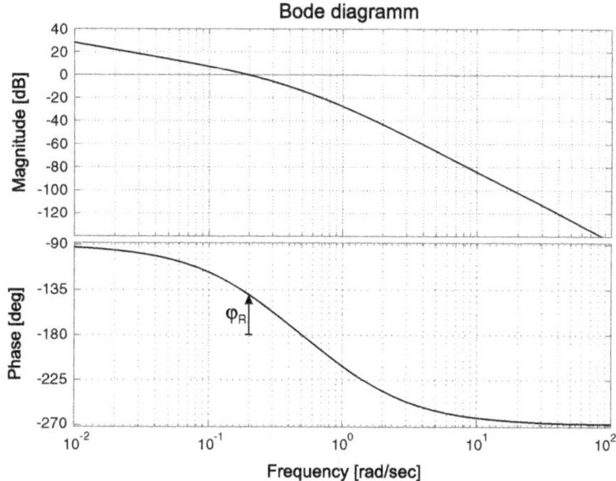

Fig. 7.7 IT_2 frequency response

to be met is given by the phase of the open-loop frequency response, with $\varphi(\omega) > -180°$ in case of the frequency response's amplitude $A(\omega)$ being above 0 dB. As shown in Fig. 7.7, the choice of the controller gain K_R transfers the amplitude graph of the open-loop frequency response vertically without affecting the phase of the open-loop frequency response. For most applications, the specific requirement of a sufficient phase margin φ_R is compulsory. The resulting phase margin is also shown in Fig. 7.7. All such requirements have to be met in the closed-control loop and must be determined to choose the correct control design method. In this example, the examined amplitude and phase of the open-loop frequency response are dependent on the proportional controller gain K_R, which is sufficient to establish system stability including a certain phase margin. More complex control structures such as PI, PIDT_n, or lead lag extend the possibilities for control design to meet further requirements.

This section shows the basic principle of the simplified NYQUIST criterion applicable to stable open-loop systems. For investigation of unstable open-loop systems, the

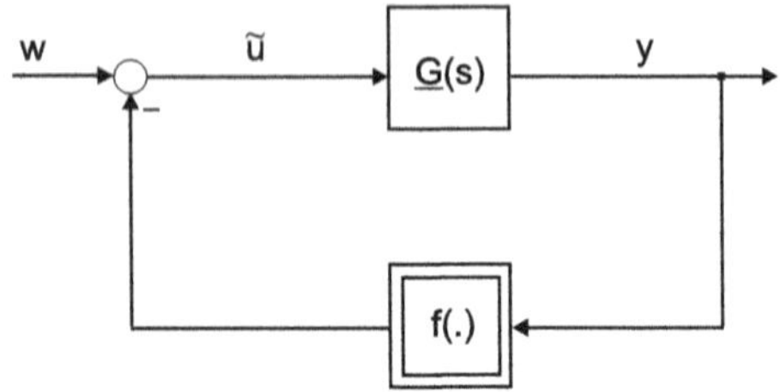

Fig. 7.8 Nonlinear closed-loop system

general form of the NYQUIST criterion must be used, which itself is not introduced in this book. For this basic knowledge, it is recommended to consult [33, 43].

7.2.2 Analysis of Nonlinear System Stability

The application of all previous approaches for the analysis of system stability is limited to linear time invariant systems. Nearly all real systems show nonlinear effects or consist of nonlinear subsystems. One approach to deal with these nonlinear systems is linearization in a fixed working point. All further investigations are focused on this point, and the application of the previously presented methods becomes possible. If these methods are not sufficient, extended techniques for stability analysis of nonlinear systems must be applied. The following are examples of representing completely different approaches:

- Principle of harmonic balance;
- Phase plane analysis;
- POPOV criterion and circle criterion;
- LYAPUNOV's direct method;
- System passivity analysis.

Without dealing with the mathematical background or the exact proof, the principles and application of the chosen techniques are demonstrated. At this point, a complete explanation of this topic is too extensive due to the wide variety of underlying methods. For further detailed explanation, [13–15, 29, 39, 42] are recommended.

7.2.2.1 POPOV Criterion

As preliminary example, the analysis of closed-loop systems can be done applying the POPOV criterion, respectively, the circle criterion. Figure 7.8 shows the block diagram of the corresponding closed-loop structure of the system to be analyzed:

The block diagram consists of a linear transfer function $\underline{G}(s)$ with arbitrary dynamics and static nonlinearity $f(.)$. The state space formulation of $\underline{G}(s)$ is as follows:

$$\dot{\mathbf{x}} = \mathbf{A}\mathbf{x} + \mathbf{B}\tilde{u}$$
$$\mathbf{y} = \mathbf{C}\mathbf{x}$$

Thus, we find for the closed-loop system description:

$$\dot{\mathbf{x}} = \mathbf{A}\mathbf{x} - \mathbf{B}f(y)$$
$$y = \mathbf{C}\mathbf{x}.$$

In case $f(y) = k \cdot y$, this nonlinear system is reduced to a linear system whose stability can be examined with the evaluation of the system's eigenvalues. For an arbitrary nonlinear function $f(y)$, the complexity of the problem is extended. So, the first constraint on $f(y)$ is that it exists only in a determined sector limited by a straight line through the origin with gradient k. Figure 7.9 shows an equivalent example for the nonlinear function $f(y)$. This constraint is depicted by the following equation:

$$0 \leq f(y) \leq ky.$$

The POPOV criterion provides an intuitive handling for the stability analysis of the presented example. The system is asymptotically idle state ($\dot{\mathbf{x}} = \mathbf{x} = \mathbf{0}$) stable if:

- the linear subsystem $\underline{G}(s)$ is asymptotically stable and fully controllable,
- the nonlinear function meets the presented sector condition as shown in Fig. 7.9,
- for an arbitrarily small number $\rho \geq 0$, there exists a positive number α, so that the following inequality is satisfied:

$$\forall \omega \geq 0 \quad \mathrm{Re}[(1 + j\alpha\omega)\underline{G}(j\omega)] + \frac{1}{k} \geq \rho \tag{7.16}$$

Equation (7.16) formulates the condition also known as POPOV *inequality*. With

$$\underline{G}(j\omega) = \mathrm{Re}(\underline{G}(j\omega)) + j\mathrm{Im}(\underline{G}(j\omega)) \tag{7.17}$$

Equation (7.16) leads to

$$\mathrm{Re}(\underline{G}(j\omega)) - \alpha\omega\mathrm{Im}(\underline{G}(j\omega)) + \frac{1}{k} \geq \rho \tag{7.18}$$

With an additional definition of a related transfer function

$$G^* = \mathrm{Re}(\underline{G}(j\omega)) + j\omega\mathrm{Im}(\underline{G}(j\omega)), \tag{7.19}$$

Equation (7.18) states that the plot in the complex plane of $\underline{G}^*$, the so-called POPOV *plot*, has to be located in a sector with an upper limit described as $y = \frac{1}{\alpha}(x + \frac{1}{k})$. Figure 7.10 shows an example for the POPOV plot of a system in the complex plane constrained by the sector condition. The close relation to the NYQUIST criterion for

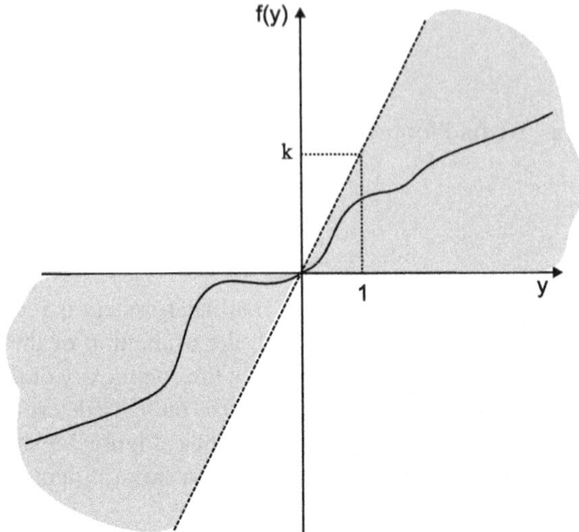

Fig. 7.9 Sector condition

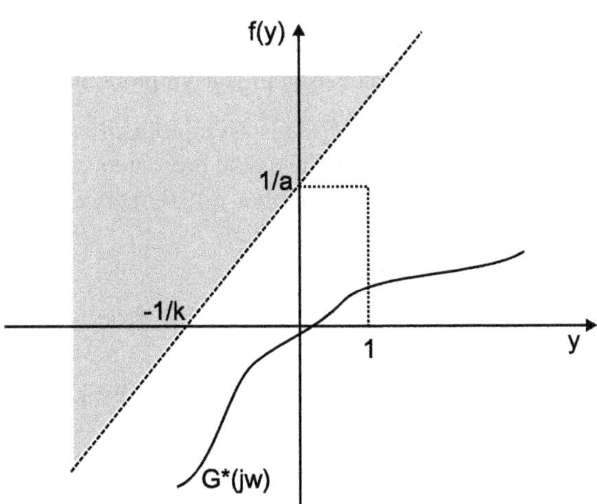

Fig. 7.10 POPOV plot

the stability analysis of linear systems becomes obvious here. While the NYQUIST criterion examines the plot of $\underline{G}(j\omega)$ referred to the critical point $(-1|0)$, the location of the POPOV plot is checked for a sector condition defined by a straight line limit.

The application of the POPOV criterion has the excellent advantage that it is possible to gain a result out of the stability analysis without an exact formulation of the nonlinearity within the system. All constraints for the nonlinear subsystem are restrained to the sector condition and the condition to have memoryless transfer behavior. The most complicated aspect within this kind of analysis is how to formulate the considered system structure in a way that the POPOV criterion can be applied. For completeness, the circle criterion is mentioned whose sector condition is not represented by a straight line, rather

$$k_1 \leq \frac{f(y)}{y} \leq k_2.$$

defines the new sector condition. For additional explanation on these constraints and the application of the circle criterion, it is recommended to consider [29, 39, 42].

7.2.2.2 LYAPUNOV's Direct Method

As a second example for stability analysis of nonlinear systems, the direct method of LYAPUNOV is introduced. The basic principle is that if both linear and nonlinear stable systems tend to a stable steady state, the complete system energy has to be dissipated continuously. Thus, it is possible to gain a result from stability analysis while verifying the characteristics of the function representing the state of energy in the system. LYAPUNOV's direct method generalizes this approach to evaluate the system energy by generation of an artificial scalar function, which can describe not only the energy stored within the considered dynamic system, but is also used as an energy-like function of a dissipative system. Such functions are called LYAPUNOV functions $V(x)$. For the examination of the system stability, the aforementioned state space description of a nonlinear system is used:

$$\dot{x} = f(x, u, t)$$
$$y = g(x, u, t).$$

By the definition of LYAPUNOV's theorem, the equilibrium at the phase plane origin $\dot{\mathbf{x}} = \mathbf{x} = \mathbf{0}$ is globally, asymptotically stable if

1. a positive definite scalar function $V(\mathbf{x})$ with $\mathbf{x}$ as the system state vector exists, meaning that $V(\mathbf{0}) = 0$ and $V(\mathbf{x}) > 0 \, \forall \, \mathbf{x} \neq \mathbf{0}$,
2. $\dot{V}$ is negative definite, meaning $\dot{V}(\mathbf{x}) \leq 0$,
3. $V(\mathbf{x})$ is not limited, meaning $V(\mathbf{x}) \to \infty$ as $\| x \| \to \infty$.

If these conditions are met in a bounded area at the origin only, the system is locally asymptotically stable.

As a clarifying example, the following nonlinear first-order system

$$\dot{x} + fx = 0 \tag{7.20}$$

is evaluated. Herein, $f(x)$ denotes any continuous function of the same sign as its scalar argument x so that $x \cdot fx > 0$ and $f(0) = 0$. Applying this constraint, a LYAPUNOV function candidate can be found described as

$$V = x^2. \tag{7.21}$$

The time derivative of $V(x)$ provides

$$\dot{V} = 2x\dot{x} = -2xf(x). \tag{7.22}$$

Due to the assumed characteristics of $f(x)$, all conditions of LYAPUNOV's direct method are satisfied; thus, the system has globally, asymptotically stable equilibrium at the origin. Although the exact function $f(x)$ is not known, the fact that it exists in the first and third quadrants only is sufficient for $\dot{V}(x)$ to be negative definite. As second example, a MIMO system is examined depicted by its state space formulation

$$\begin{aligned}\dot{x}_1 &= x_2 - x_1(x_1^2 + x_2^2)\\ \dot{x}_2 &= -x_1 - x_2(x_1^2 + x_2^2).\end{aligned}$$

In this example, the system has an equilibrium at the origin too. Consequently, the following LYAPUNOV function candidate can be found:

$$V(x_1, x_2) = x_1^2 + x_2^2. \tag{7.23}$$

Thus, the corresponding time derivative is

$$\dot{V}(x_1, x_2) = 2x_1\dot{x_1} + 2x_2\dot{x_2} = -2(x_1^2 + x_2^2)^2. \tag{7.24}$$

Hence, $V(x_1, x_2)$ is positive definite and $\dot{V}(x_1, x_2)$ is negative definite. Thus, the equilibrium at the origin is globally, asymptotically stable for the system.

A difficult aspect when using the LYAPUNOV direct method is given by how to find LYAPUNOV function candidates. No straight algorithm with a determined solution exists, which is a disadvantage of this method. SLOTINE AND LE [39] propose several structured approaches to gain LYAPUNOV function candidates, namely

- KRASOVSKII's method and
- the variable gradient method.

Besides these, SLOTINE provides additional possibilities to involve the system's physical principles in the procedure for determining of LYAPUNOV function candidates while analyzing more complex nonlinear dynamic systems.

7.2.2.3 Passivity in Dynamic Systems

As another method for the stability analysis of dynamic systems, the passivity formalism is introduced within this section. Functions can be extended to system combinations using LYAPUNOV's direct method and evaluating the dissipation of energy in dynamic systems. The passivity formalism is also based on nonlinear positive definite storage functions $V(\mathbf{x})$ with $V(\mathbf{0} = \mathbf{0})$ representing the overall system energy. The time derivative of this energy determines the system's passivity. As example, the general formulation of a system

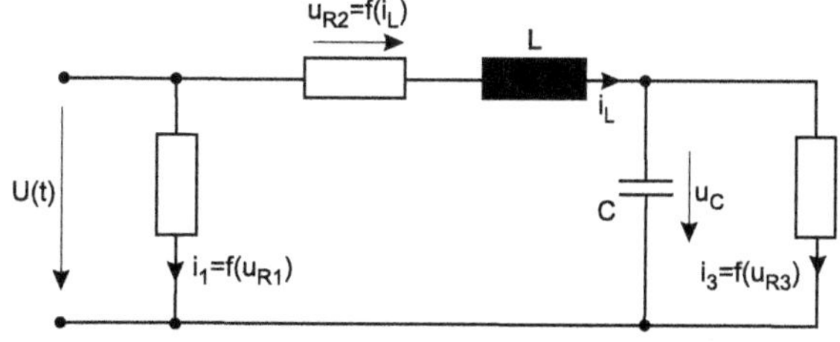

Fig. 7.11 Passivity analysis of an RLC-network

$$\dot{x} = f(x, u, t)$$
$$y = g(x, u, t).$$

is considered. This system is passive concerning the external supply rate $S = \mathbf{y}^T\mathbf{u}$ if the inequality condition

$$\dot{V}(\mathbf{x}) \leq \mathbf{y}^T\mathbf{u} \tag{7.25}$$

is satisfied. KHALIL distinguishes several cases of system passivity depending on certain system characteristics (*Lossless, Input Strictly Passive, Output Strictly Passive, State Strictly Passive, Strictly Passive*) [29]. If a system is passive concerning the *external supply rate S*, it is stable in the sense of LYAPUNOV.

The combination of passive systems using parallel or feedback structures inherits the passivity from its passive subsystems. With the close relation of system passivity to stability in the sense of LYAPUNOV, the examination of system stability is possible by verifying the subsystem's passivity. Based on this evaluation, it can be concluded that the overall system is passive—always with the assumption that a correct system structure was built.

As an illustrating example, the RLC circuit taken from [29] is analyzed in the following. The circuit structure is shown in Fig. 7.11.

The system's state vector is defined as

$$i_L = x_1$$
$$u_C = x_2.$$

The input u represents the supply voltage U, as output y the current i is observed. The resistors are described by the corresponding voltage current characteristics:

$$i_1 = f_1(u_{R1})$$
$$i_3 = f_3(u_{R3})$$

For the resistor that is coupled in series with the inductor, the following behavior is assumed:

$$U_{R2} = f_2(i_L) = f_2(x_1). \tag{7.26}$$

Thus, the nonlinear system is described by the differential equation:

$$\begin{aligned} L\dot{x_1} &= u - f_2(x_1) - x_2 \\ C\dot{x_2} &= x_1 - f_3(x_2) \\ y &= x_1 + f_1(u) \end{aligned}$$

The presented RLC circuit is passive as long as the condition

$$V(\mathbf{x}(t)) - V(\mathbf{x}(0)) \leq \int_0^t u(\tau)y(\tau)\mathrm{d}\tau \tag{7.27}$$

is satisfied. In this example, the energy stored in the system is described by the storage function

$$V(\mathbf{x}(t)) = \frac{1}{2}Lx_1^2 + \frac{1}{2}Cx_2^2. \tag{7.28}$$

Equation 7.27 leads to the condition for passivity:

$$\dot{V}(\mathbf{x}(t), u(t)) \leq u(t)y(t) \tag{7.29}$$

which means that the energy supplied to the system must be equal to or higher than the time derivative of the energy function. Using $V(\mathbf{x})$ in the condition for passivity provides

$$\begin{aligned} \dot{V}(\mathbf{x}, u(t)) &= Lx_1\dot{x}_1 + Cx_2\dot{x}_2 \\ &= x_1(u - f_2(x_1) - x_2) + x_2(x_1 - f_3(x_2)) \\ &= x_1(u - f_2(x_1)) + x_2 f_3(x_2) \\ &= (x_1 + f_1(u))\,u - uf_1(u) - x_1 f_2(x_1) - x_2 f_3(x_2) \\ &= uy - uf_1(u) - x_1 f_2(x_1) - x_2 f_3(x_2) \end{aligned}$$

and finally

$$u(t)y(t) = \dot{V}(\mathbf{x}, u(t)) + uf_1(u) + x_1 f_2(x_1) + x_2 f_3(x_2). \tag{7.30}$$

In case f_1, f_2, and f_3 are passive subsystems, i.e., all functions describing the corresponding characteristics of the resistors exist only in the first and third quadrants, then $\dot{V}(\mathbf{x}, u(t)) \leq u(t)y(t)$ is true; hence, the RLC circuit is passive. Any coupling of this passive system to other passive systems in parallel or feedback structures again results in a passive system. For any passivity analysis and stability evaluation, this method implements a structured procedure and shows high flexibility.

In conclusion, it is necessary to mention that all methods for stability analysis introduced in this section show certain advantages and disadvantages concerning their applicability, information value, and complexity, regardless of whether linear or nonlinear systems are considered. When a stability analysis is expected to be done, the applicability of a specific method should be checked individually. This section gives a brief overview of the introduced methods and techniques and does not claim to be a detailed description due to the limited scope of this section. For further study, the reader is invited to consult the proposed literature.

7.3 Control Law Design for Haptic Systems

As introduced at the beginning of this chapter, control design is a fundamental and necessary aspect within the development of haptic systems. In addition to the techniques for system description and stability analysis, the need for control design and the applicable design rules becomes obvious. For control design of a haptic system, it is especially necessary to deal with several aspects and conditions to be satisfied during the design process. The following sections present several control structures and design schemes to set up a basic knowledge of the toolbox for analytic control design of haptic systems. This also involves some of the already introduced methods for system formulation and stability analysis, as they form the basis for most control design methods.

7.3.1 Structuring of Control Design

As introduced in Chap. 6, various structures of haptic systems exist. Demands on the control of these structures are derived in the following.

Open-loop impedance controlled The user experiences an impression of force that is directly commanded via an open loop based only on a demand value. In Chap. 6, the basic scheme of this structure is shown in Fig. 6.1.

Closed-loop impedance controlled As it can be seen in Fig. 6.3, the user also experiences an impression of force that is fed back to a controller. Here, a specific control design is needed.

Open-loop admittance controlled In this scheme, the user experiences an impression of a defined position. In the open-loop arrangement, this position again is directly commanded based only on a demand value. Figure 6.5 shows the corresponding structure of this haptic scheme.

Closed-loop admittance controlled This last version as depicted in Fig. 6.7 shows its significant difference in the feedback of the force the user applies to the interface. This force is fed back to a demand value. This results in a closed-loop arrangement that incorporates the user and his or her transfer characteristics. Different from the closed-loop impedance controlled scheme, this structure uses a force as demanded value $\underline{S}_F$ compared with the detected $\underline{S}_S$, but the system output is still a position $\underline{x}_{\text{out}}$. This results in the fact that the incorporation of the user into the closed-loop behavior is more complex than in a closed-loop impedance controlled scheme.

All of these structures can be basically implemented in a haptic interaction as shown in Fig. 2.33. From this, all necessary control loops of the overall telemanipulation system become evident:

- On the haptic interface site, a control loop is closed incorporating the user that is valid as long as the user's reaction is fed back to the central interface module for any further data processing or control.
- On the process/environment site also, a closed loop exists if measurable process signals (reactions, disturbances) are fed back to the central interface module for data processing or control.
- Underneath these top-level control loops, various subsystem control loops exist that have a major impact on the overall system too. As an example, each electrical actuator will most likely be embedded in a cascaded control structure with current, speed, and position control.

It becomes obvious that the design of a control system for a telemanipulation system with a haptic interface is complex and versatile. Consequently, a generally valid procedure for control design cannot be given. The control structures must be designed step-by-step involving the following controllers:

1. Design of all controllers for the subsystem actuators;
2. Design of a top-level controller for the haptic interface;
3. Design of a top-level controller for the manipulator/VR environment;
4. Design of the system controller that connects interface and manipulator or VR environment.

This strict separation proposed above might not be the only way to structure the overall system. Depending on the application and functionality, the purposes of the different controller and control levels might be in conflict with each other or might simply overlap. Therefore, it is recommended to set up the underlying system structure and define all applied control schemes corresponding to their required functionality.

While looking at the control of haptic systems, a similar structure can be established. For both the control of the process manipulation and the haptic display or interface, the central interface module will have to generate demand values for force or position that are going to be followed by the controllers underneath. These demand values derive from a calculation predefined by designed control laws. To gain such control laws, a variety of methods and techniques for structural design and optimization can be applied depending on certain requirements. The following sections give an overview of typical requirements to closed-control loop behavior followed by examples for control design.

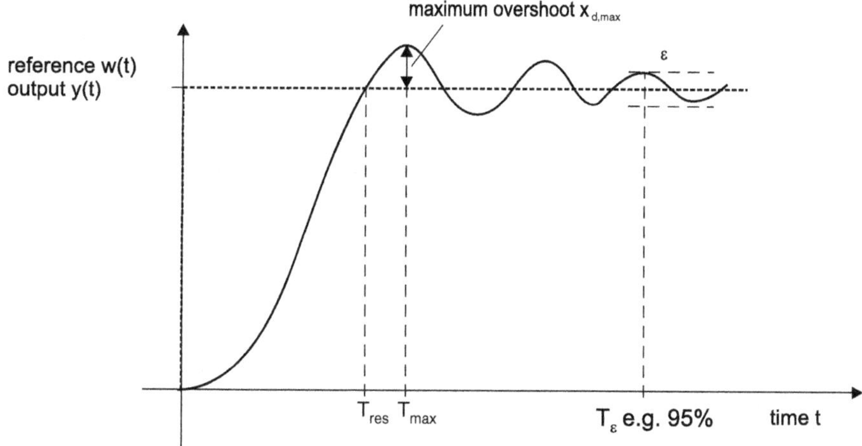

Fig. 7.12 Closed-loop step response requirements

Table 7.1 Parameter for control quality requirements

Parameter	Description
$x_{d,\max}$	Maximum overshoot
$T_{\max}$	Point of time for $x_{d,\max}$
T_ε	Time frame in which the residuum to the demanded value remains within a predefined scope ε
T_{res}	Point of time when the demanded value is reached for the first time

7.3.2 Requirement Definition

Besides the fundamental need for system stability with sufficient stability margins, additional requirements can be set up to achieve a certain system behavior in a closed-loop scheme such as dynamic or precision. A quantitative representation of these requirements can be made by the achievement of certain characteristics of the closed-loop step response.

Figure 7.12 shows the general form of a typical closed-loop step response and its main characteristics. As it can be seen, the demanded value is reached and the basic control requirement is satisfied.

Additional characteristics are discussed and listed in Table 7.1. For all mentioned characteristics, a quantitative definition of certain requirements is possible. For example, the number and amplitude of overshoots shall not extend a defined limit or have a certain frequency spectrum that is of special interest for the control design in haptic systems. As analyzed in Chap. 3, the user's impedance shows a significant frequency range that must not be excited within the control loop of the haptic device. Nevertheless, a certain cut-off frequency has to be reached to establish a good performance of

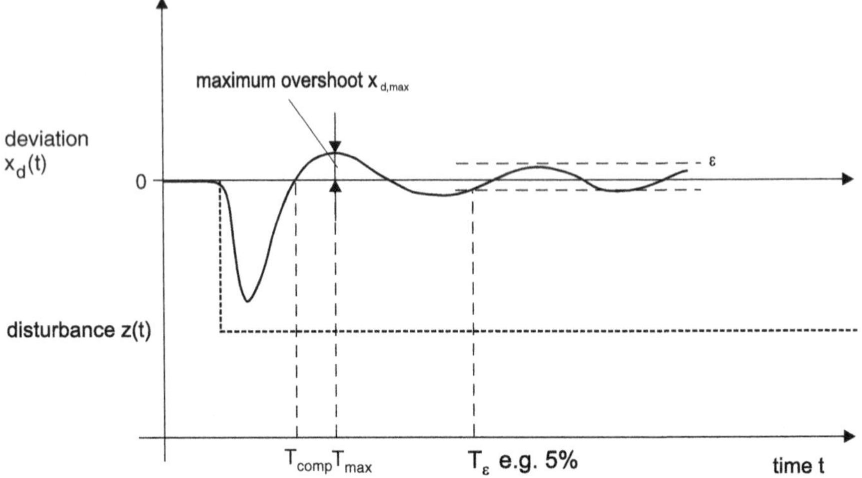

Fig. 7.13 Closed-loop disturbance response requirements

the dynamic behavior. All these issues are valid for requirements to the control design of the process manipulation. In addition to the requirements from the step response due to changes in the setpoint value, it is necessary to formulate requirements concerning the closed-loop system behavior considering disturbances originating from the process. When interpreting the user's reaction as disturbance within the overall system description, a requirement set up for the disturbance reaction of the control loop has to be established. As can be seen in Fig. 7.13, similar characteristics exist to determine the disturbance reaction quantitatively and qualitatively. In most cases, both the step response behavior and the disturbance reaction cannot satisfy all requirements, as they often come into conflict with each other, caused by the limited flexibility of the applied optimization method. Thus, it is recommended to estimate the relevance of step response and disturbance reaction in order to choose an optimization approach that is most beneficial. Although determined quantitatively, it is not possible to use all requirements in a predefined optimization method. In most cases, an adjustment of requirements is necessary to be made, to apply specific control design and optimization methods. As an example, the time T_{res} as depicted above cannot be used directly and must be transferred into a requirement for the closed-loop dynamic characterized by a definite pole placement.

Furthermore, simulation techniques and tests offer iteration within the design procedure to gain an optimal control law. However, this very sufficient way of analyzing system behavior and test designed control laws suggests to forget about the analytic system and control design strategy and switch to a trial an error algorithm.

7.3.3 General Control Law Design

This section presents some possible types of controllers and control structures that might be used in the already discussed control schemes. For optimization of the control parameters, several methods exist, which are introduced here. Depending on the underlying system description, several approaches to set up controllers and control structures are possible. This section presents the classic PID-control, additional control structures, e.g., compensation, state feedback controllers, and observer-based state space control.

7.3.3.1 Classic PID-Control

Maybe, one of the most frequently used controllers is the parallel combination of a proportional (P), an integrating (I) and a derivative (D) controller. This combination is used in several variants including a P-controller, a PI combination, a PD combination, or the complete PID structure. Using the PID structure, all advantages of the individual components are combined. The corresponding controller transfer function is described as

$$\underline{G}_R = K_R \left(1 + \frac{1}{T_N s} + T_V s \right). \tag{7.31}$$

Figure 7.14 shows the equivalent block diagram of a PID controller structure. Adjustable parameters in this controller are the proportional gain K_R, the integrator time constant T_N, and the derivative time T_V.

With optimized parameter adjustment, a wide variety of control tasks can be handled. This configuration offers on the one hand, the high dynamic of the proportional controller and on the other, the integrating component guarantees a high precision step response with a residuum $x_d = 0$ for $t \to \infty$. The derivative finally provides an additional degree of freedom that can be used for a certain pole placement of the closed-loop system.

As major design techniques, the following examples are introduced:

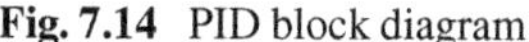

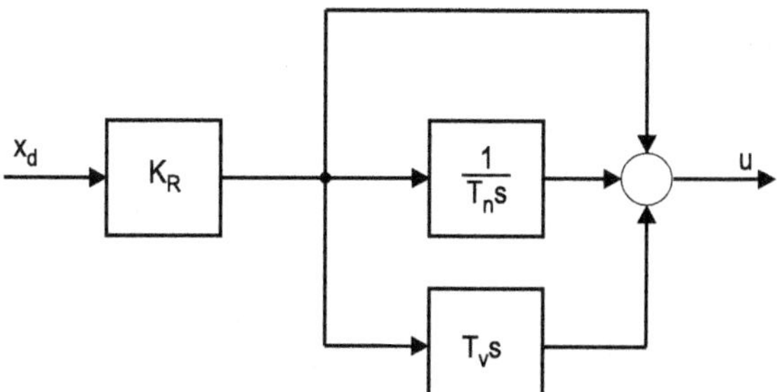

Fig. 7.14 PID block diagram

Root Locus Method This method has its strength by the determined pole placement for the closed-loop system, directly taking into account the dependence on the proportional gain K_R. By a reasonable choice of T_N and T_D, the additional system zeros are influenced that affects directly the resulting shape of the root locus and thus the stability behavior. Besides this, the overall system dynamic can be designed.

Integral Criterion The second method for optimization of the closed-loop system step response or disturbance reaction is the minimization of an integral criterion. The basic procedure for this method is as follows: The tracking error x_d due to changes in the demanded set point or a process disturbance is integrated (and eventually weighted over time). This time integral is minimized by adjusting the controller parameters. In case of convergence of this minimization, the result is a set of optimized controller parameters.

For any additional theoretical background concerning controller optimization, the reader is invited to consult the literature on control theory and control design [32, 33].

7.3.3.2 Additional Control Structures

In addition to the described PID controller, additional control structures extend the influence on the control result without having an impact on the system stability. The following paragraphs present the disturbance compensation and a direct feedforward of auxiliary process variables.

Disturbance Compensation

The basic principle of disturbance compensation assumes that if a disturbance on the process is measurable and its influence is known, this knowledge can be used to establish compensation by corresponding evaluation and processing. Figure 7.15 shows a simplified scheme of this additional control structure.

In this scheme, a disturbance signal is assumed to affect the closed loop via a disturbance z transfer function $\underline{G}_D$. By measuring the disturbance signal and processing, the compensator transfer function $\underline{G}_C$ results in a compensation of the disturbance interference. Assuming an optimal design of the compensator transfer function, this interference caused by the disturbance is completely erased. The optimal design of a corresponding compensator transfer function is depicted as

$$\underline{G}_C = -\frac{\underline{G}_D}{\underline{G}_S}. \tag{7.32}$$

This method assumes that a mathematical and practicable inversion of $\underline{G}_D$ exists. For those cases where this assumption is not valid, the optimal compensator $\underline{G}_K$ must be approximated. Furthermore, Fig. 7.15 states clearly that this additional control

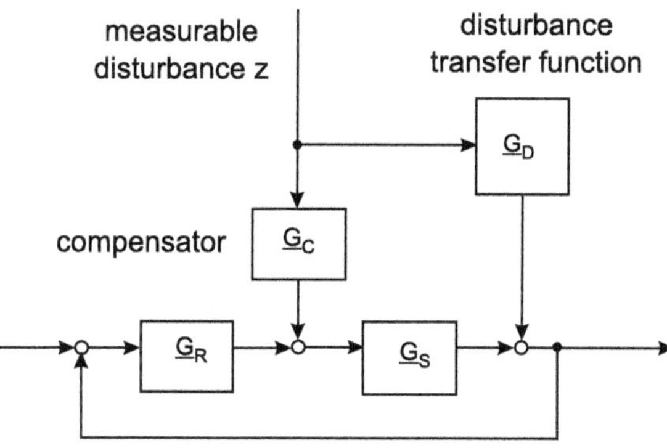

Fig. 7.15 Simplified disturbance compensation

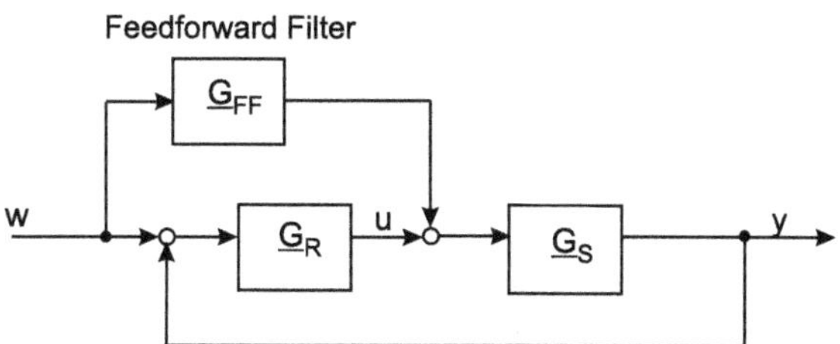

Fig. 7.16 Feedforward of auxiliary input variables

structure does not have any influence on the closed-loop system stability and can be designed independently. Besides, the practicability of the additional effort should be taken into account. This effort will definitively increase just by the sensors to measure the disturbance signals and by the additional costs for realization of the compensator.

Auxiliary Input Feedforward

A structure similar to the disturbance compensation is the *feedforward* of auxiliary input variables. This principle is based on the knowledge of additional process variables that are used to influence the closed-loop system behavior without affecting system stability. Figure 7.16 shows an example of the feedforward of the demanded setpoint w to the controller signal u using a feedforward filter function $\underline{G}_{\mathrm{FF}}$.

7.3.3.3 State Space Control

Corresponding to the techniques for the description of MIMO systems discussed earlier in this chapter, state space control provides additional features to cover the special characteristics within these systems. As described before, MIMO systems are preferably depicted as state space models. Using this mathematical formulation enables the developer to implement a control structure that controls the internal

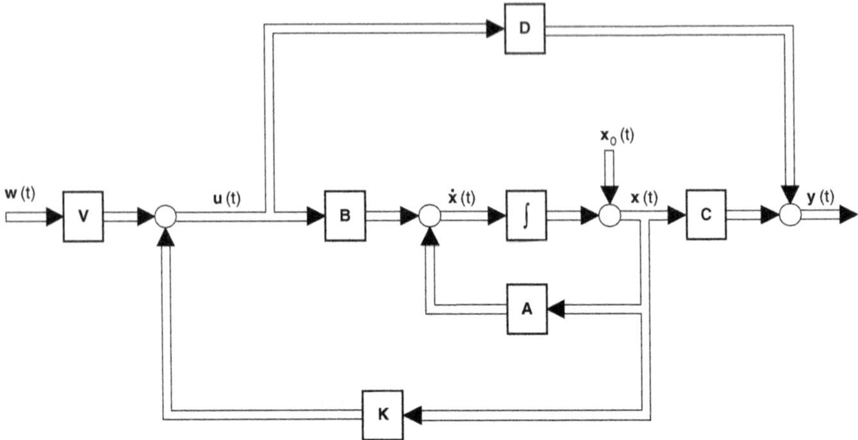

Fig. 7.17 State feedback control

system states to demanded values. The advantage is that the design methods for state space control use an overall approach for control design and optimization instead of a control design step-by-step for each system state. With this approach, it becomes possible to deal with profoundly coupled MIMO systems with high complexity, and design a state space controller simultaneously. This section presents the fundamental state space control structures, which covers the *state feedback control* as well as the *observer-based state space control*. For further detailed procedures as well as design and optimization methods, the reader is referred to [32, 42].

State Feedback Control

As shown in Fig. 7.17, this basic structure for state space control uses a feedback of the system states **x**. Similar to the depiction in Fig. 7.2, the considered system is presented in state space description using matrices **A**, **B**, **C**, and **D**. The system states **x** are fed back gained by the matrix **K** to the vector of the demanded values that were filtered by matrix **V**. The results represent the system input vector u. Both matrices **V** and **K** do not have to be square matrices as a state space description is allowed to implement various dimensions for the state vector, the vector for the demanded values, and the system input vector.

Observer-Based State Space Control

The state space control structure discussed above requires complete knowledge of all system states, which is nothing else but that they have to be measured and processed to be used in the control algorithm. From a practical point of view, this not possible

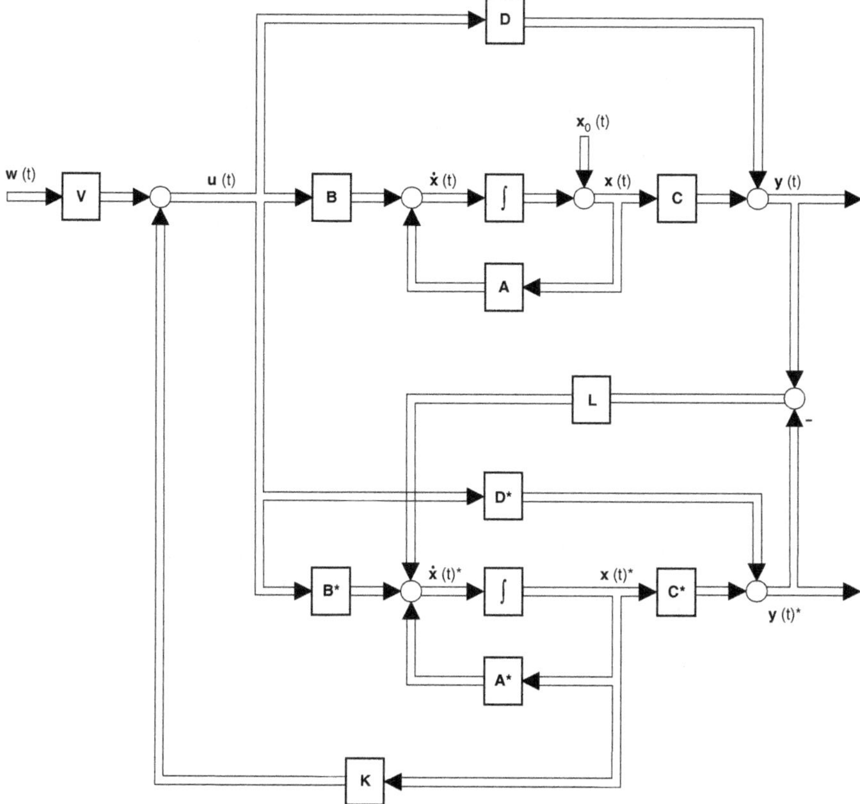

Fig. 7.18 Observer-based state space control

always due to technical limits as well as costs and effort. As a result, the developer is faced with the challenge to establish a state space control without complete knowledge of the system states. As a solution, these system states that cannot be measured due to technical difficulties or significant cost factors estimated using a state space observer structure as shown in Fig. 7.18.

In this structure, a system model is calculated parallel to the real system. As exact as possible, this system model is described by the corresponding parameter matrices $\mathbf{A}^*$, $\mathbf{B}^*$, $\mathbf{C}^*$, and $\mathbf{D}^*$. The model input is also represented by the input vector $\mathbf{u}$. Thus, the model provides an estimation of the real system states $\mathbf{x}^*$ and an estimated system output vector $\mathbf{y}^*$. By comparison of this estimated output vector $\mathbf{y}^*$ with the real output $\mathbf{y}$, which is assumed to be measurable, the estimation error is fed back gained by the matrix $\mathbf{L}$. This results in a correction of the system state estimation $\mathbf{x}^*$. Any estimation error in the system states or the output vector due to varying initial states is corrected and the estimated states $\mathbf{x}^*$ are used to be gained by the equivalent matrix $\mathbf{K}$ and fed back for control.

This structure of an observer-based state space control uses the LUENBERGER observer. In this configuration, all real system states are assumed not to be measurable; thus, the state space control refers to estimated values completely. Practically, the feedback of measurable system states is combined with the observer-based estimation of additional system states. In [32, 42], examples for observer-based state space control structures as well as methods for observer design are discussed in detail.

7.3.4 Example: Cascade Control of a Linear Drive

As an example for the design of a controller, the cascade control of a linear drive buildup of an EC motor and a ball screw is considered in this section based on [27]. The consideration includes nonlinear effects due to friction, temperature change, and a nonlinear degree of efficiency of the ball screw.

A schematic representation of the EC motor is given in Fig. 7.19. in which only one phase is illustrated for simplification. The motor is supplied with voltage u_{DC}. The resistance R and the inductance L represent the stator winding of the motor. The angular speed of the rotor $\underline{\omega}_{\mathrm{M}}$ generates a back electromotive force (back-EMF) u_{EMF}. The mechanical properties of the motor are described by the motor torque $\underline{M}_{\mathrm{e}}$, the load torque $\underline{M}_{\mathrm{L}}$, and the moment of inertia of the rotor J. Mesh analysis yields to the equation for the electrical part of the motor

$$u_{\mathrm{DC}} = Ri + L\frac{\mathrm{d}i}{\mathrm{d}t} + u_{\mathrm{EMF}} \tag{7.33}$$

which can be written in the frequency domain as

$$\underline{U}_{\mathrm{DC}} - \underline{U}_{\mathrm{EMF}} = \underline{I}\,(R + sL) \tag{7.34}$$

The back electromotive force U_{EMF} depends on the angular speed of the rotor ω_M, the back-EMF constant k_e, and the parameter $F(\phi_e)$, which describes the dependence of the back-EMF of the electrical angle ϕ_e.

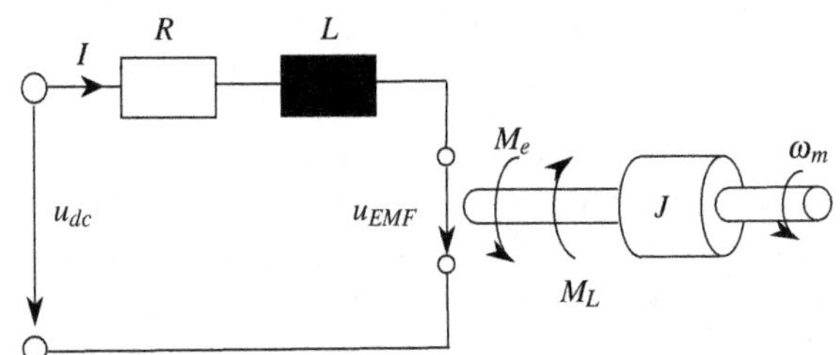

Fig. 7.19 Equivalent circuit of the considered EC motor with attached ball screw to transform rotary into translational movement [27]

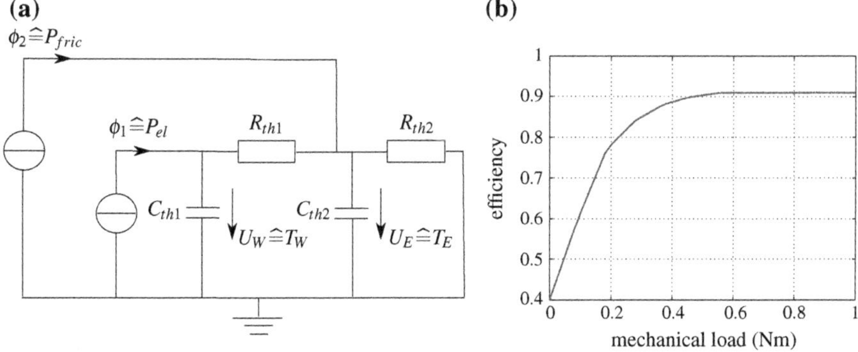

Fig. 7.20 **a** Equivalent thermal circuit of the EC motor, **b** efficiency of the ball screw depending on the mechanical load [27]

$$u_{\text{EMF}} = k_e \omega_{\text{M}} F(\phi_e) \tag{7.35}$$

The motor torque M_{e} generated by the motor current i correlates with the mechanical load M_{L} and the angular acceleration ω_{M} of the rotor with the moment of inertia J. It follows that:

$$M_{\text{e}} = \frac{i \cdot u_{\text{EMF}}}{\omega_{\text{M}}} = i k_{\text{e}} F(\Phi_e) = J \frac{\mathrm{d}\omega_M}{\mathrm{d}t} + M_{\text{L}} \tag{7.36}$$

In the frequency domain, the mechanical properties of the motor are described as

$$M_{\text{e}} - M_{\text{L}} = sJ\omega_{\text{M}}. \tag{7.37}$$

The model takes three different types of nonlinearities into account: friction, temperature change, and a nonlinear efficiency of the ball screw. The friction is modeled as the sum of a static friction K_{F} and a dynamic friction $k_{\text{F}} \cdot \omega_{\text{M}}$. So, the equilibrium of moments of the rotor can now be written as

$$M_{\text{e}} - M_{\text{L}} - K_{\text{F}} = (k_{\text{F}} + sJ)\omega_{\text{M}}. \tag{7.38}$$

The influence of changes in temperature on motor parameters is modeled by a thermal equivalent circuit shown in Fig. 7.20a. The temperature change in the stator winding T_{W} can be determined by

$$\Delta T_{\text{W}} = \frac{R_{\text{th1}} T_{\text{th2}} s + R_{\text{th1}} R_{\text{th2}}}{T_{\text{th1}} T_{\text{th2}} s^2 + (T_{\text{th1}} + T_{\text{th2}}) s} P_{\text{el}} + \frac{R_{\text{th2}}}{T_{\text{th1}} T_{\text{th2}} s^2 + (T_{\text{th1}} + T_{\text{th2}} + R_{\text{th2}} C_{\text{th1}}) s + 1} P_{\text{fric}}. \tag{7.39}$$

with

$$T_{\text{th1}} = R_{\text{th1}} C_{\text{th1}} \quad \text{and} \quad T_{\text{th2}} = R_{\text{th2}} C_{\text{th2}} \tag{7.40}$$

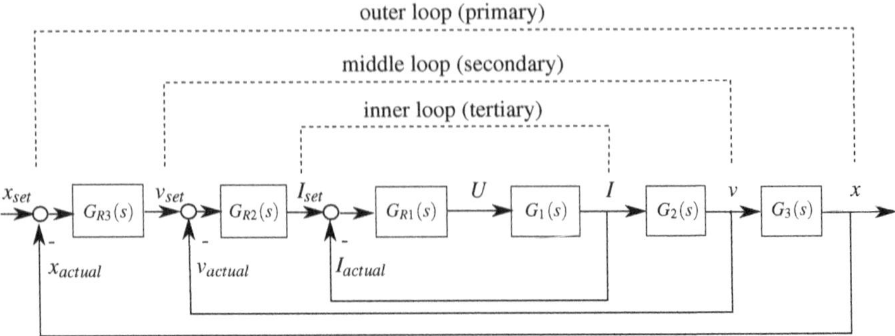

Fig. 7.21 Structure of cascade controller of EC motor [27]

The resulting resistance of the stator winding R_* and the back-EMF constant k_{e*} can be derived with knowledge of the temperature coefficients α_R, α_k from

$$R_* = R(1 + \alpha_R \Delta T_W), \quad k_{e*} = k_e(1 + \alpha_k \Delta T_W). \tag{7.41}$$

The efficiency of the ball screw depends on the mechanical load of the linear drive. Its qualitative characteristics are shown in Fig. 7.20b and can be included in the model as characteristics in a lookup table. The resulting model can be computed for example in Matlab/Simulink and used for simulation and design of a controller. In this example, a cascade controller is chosen (Fig. 7.21). It consists of an inner loop for current control, a middle loop for velocity control, and an outer loop for position control. As controller for the different control loops, P- or PI-controllers are used.

7.4 Control of Teleoperation Systems

In the previous sections, an overview of system description and control aspects in general, which can be used for the design of local and global control laws, was given. The focus of this section lies on special methods used for modeling haptic systems stability analysis of bilateral telemanipulators. In contrast to Sect. 7.3, special tools for the development of control laws are presented here based on the two-port hybrid representation of bilateral telemanipulators (Sect. 7.4.1). Subsequently, in Sect. 7.4.2, a definition of transparency is introduced, which can be used to analyze the performance of a haptic system dependent on the system characteristics and the chosen control law. In Sect. 7.4.3, the general control model for telemanipulators is introduced to close the gap between the closed-loop representation, known from general control theory, and used in Sects. 7.1–7.3, and the two-port hybrid representation. In Sect. 7.4.4, it is shown how a stable and safe operation of the haptic system can be achieved. Furthermore, the design of stable control laws in the presence of time delays are presented in Sect. 7.4.5.

7.4.1 Two-Port Representation

In general, a haptic system is a bilateral telemanipulator, where a user handles a master device to control a slave device which is interacting with an environment. A common representation of a bilateral telemanipulator is the general two-port model as shown in Fig. 7.22.

User and environment are represented by one ports, characterized by their mechanical impedances $\underline{Z}_{\mathrm{H}}$ and $\underline{Z}_{\mathrm{E}}$ as they can be seen as passive elements [28], see Chap. 3. The mechanical impedance $\underline{Z}$ is defined as Eq. (7.42)

$$\underline{Z} = \frac{\underline{F}}{\underline{v}} \tag{7.42}$$

The user manipulates the master device, which controls the slave device. The slave interacts with the environment. The behavior of the telemanipulator is described by its hybrid matrix **H** [16, 37]. So the coupling of user action and interaction with the environment is described by the following hybrid matrix, taking forces and velocities at the master and slave sides and the properties of the haptic system into account.

$$\begin{pmatrix} \underline{F}_{\mathrm{H}} \\ -\underline{v}_{\mathrm{E}} \end{pmatrix} = \begin{pmatrix} \underline{h}_{11} & \underline{h}_{12} \\ \underline{h}_{21} & \underline{h}_{22} \end{pmatrix} \cdot \begin{pmatrix} \underline{v}_{\mathrm{H}} \\ \underline{F}_{\mathrm{E}} \end{pmatrix}. \tag{7.43}$$

In this case, the four h-parameters represent

$$\begin{pmatrix} \underline{F}_{\mathrm{H}} \\ -\underline{v}_{\mathrm{E}} \end{pmatrix} = \begin{pmatrix} \text{Master input impedance} & \text{Backward force gain} \\ \text{Forward velocity gain} & \text{Slave output admittance} \end{pmatrix} \cdot \begin{pmatrix} \underline{v}_{\mathrm{H}} \\ \underline{F}_{\mathrm{E}} \end{pmatrix} \tag{7.44}$$

Please note that the velocity of the slave v_E is taken into account with a negative sign. This is done to fulfill the convention for general two-ports, where the flow is always flowing into a port. The hybrid two-port representation as shown before is often used to determine stability criteria and to describe the performance properties of bilateral telemanipulators. Despite the formulation with force as flow variable (also found in [16, 30], for example), one can also find velocity as flow variable in other two-port descriptions of bilateral telemanipulators [19]. As long as the coupling is defined by the impedance formulation given in Eq. (7.42), both these variants of the two-port descriptions are interchangeable.

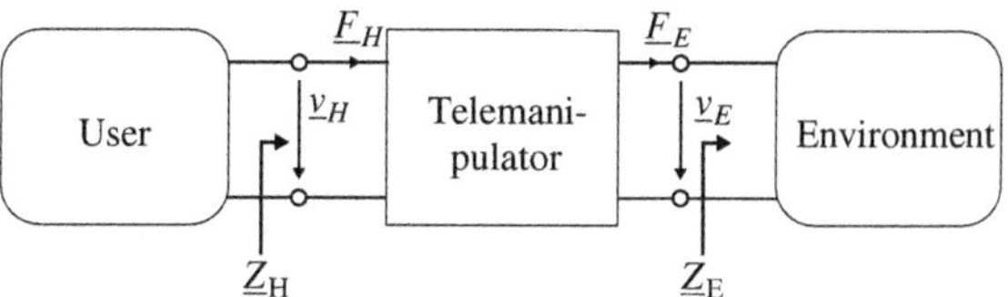

Fig. 7.22 General two-port model of a telemanipulator

7.4.2 Transparency

Besides, system stability performance is an important design criterion in the development of haptic systems. The function of a haptic system is to provide high fidelity force feedback of the contact force at the slave side to the user manipulating the master device of the telemanipulator. One parameter often used to evaluate the haptic sensation presented to the user is transparency. If the user interacts directly with the environment, he experiences a haptic sensation, which is determined by the mechanical impedance $\underline{Z}_{\mathrm{E}}$ of the environment. If the user is coupled to the environment via a telemanipulator system, he experiences a force impression, which is determined by the backward force gain and the mechanical input impedance of the master device. It is desirable that the haptic sensation for the user of the telemanipulator is the same as interacting directly with the environment. Therefore, the telemanipulator has to display the mechanical impedance of the environment $\underline{Z}_{\mathrm{E}}$ at the master device. Assume that $h_{12} = h_{21} = 1$, so there is no scaling of velocity or force. Therefore, the following conditions have to hold to reach full transparency.

$$\underline{F}_{\mathrm{H}} = \underline{F}_{\mathrm{E}} \quad \text{and} \quad \underline{v}_{\mathrm{H}} = \underline{v}_{\mathrm{E}}. \tag{7.45}$$

From this it follows that for perfect transparency [30]

$$\underline{Z}_{\mathrm{H}} = \underline{Z}_{\mathrm{E}} \tag{7.46}$$

Therefore, the force experienced by the user at the master device is

$$\underline{F}_{\mathrm{H}} = \underline{h}_{11}\underline{v}_{\mathrm{H}} + \underline{h}_{12}\underline{F}_{\mathrm{E}}$$

and for velocity at the slave side it holds that

$$-\underline{v}_{\mathrm{E}} = \underline{h}_{21}\underline{v}_{\mathrm{H}} + \underline{h}_{22}\underline{F}_{\mathrm{E}}.$$

Therefore, the mechanical impedance displayed by the master and felt by the user is described as

$$\underline{Z}_{\mathrm{T}} = \frac{\underline{F}_{\mathrm{T}}}{\underline{v}_{\mathrm{T}}} = \frac{\underline{h}_{11}\underline{v}_{\mathrm{H}} + \underline{h}_{12}\underline{F}_{\mathrm{E}}}{\frac{\underline{v}_{\mathrm{E}} - \underline{h}_{22}\underline{F}_{\mathrm{E}}}{\underline{h}_{21}}} \tag{7.47}$$

By analyzing Eq. (7.47), the conditions for perfect transparency can be derived. To achieve perfect transparency, output admittance at the slave side and input impedance at the master side have to be zero. From this it follows that for perfect transparency, in the case of no scaling, the matrix has to be of the form

$$\begin{pmatrix} \underline{F}_{\mathrm{H}} \\ -\underline{v}_{\mathrm{E}} \end{pmatrix} = \begin{pmatrix} 0 & -1 \\ 1 & 0 \end{pmatrix} \cdot \begin{pmatrix} \underline{v}_{\mathrm{H}} \\ \underline{F}_{\mathrm{E}} \end{pmatrix}.$$

It is obvious that perfect transparency is in practice not achievable without further actions taken, due to nonzero input impedance $\underline{h}_{11}$ and output admittance $\underline{h}_{22}$ of the manipulator system. If the input impedance was zero, the user would not feel the mechanical properties of the master device (mass, friction, compliance). An output admittance of zero relates to an ideal stiff slave device.

7.4.2.1 A Perception-Oriented Consideration of Transparency

Christian Hatzfeld, Sebastian Kassner, Carsten Neupert

To obtain a transparent system, the system engineer has two options: Work on the control structure, as described in the following sections or consider the perception capabilities of the human user in the definition of transparency. The latter is the focus of this section, which is based on the detailed elaborations in [22]. It has to be noted that this approach still lacks some experimental evaluation.

Up till now, transparency as defined in Eqs. (7.45) and (7.46) is a binary criterion: A system is either transparent if all conditions are fulfilled, or is not transparent if one of the equalities is not given. Despite this formulation, one can define the absolute transparency error $\underline{e}_{\mathrm{T}}$ according to HEREDIA ET AL. as shown in Eq. 7.48 [25]

$$\underline{e}_{\mathrm{T}} = \underline{Z}_{\mathrm{H}} - \underline{Z}_{\mathrm{E}} \tag{7.48}$$

and the relative transparency error $\underline{e}'_{\mathrm{T}}$ as shown in Eq. (7.49)

$$\underline{e}'_{\mathrm{T}} = \frac{\underline{Z}_{\mathrm{H}} - \underline{Z}_{\mathrm{E}}}{\underline{Z}_{\mathrm{H}}} \tag{7.49}$$

When analyzed along the whole intended dynamic range and in all relevant ↪DoF of the haptic system, Eqs. (7.48) and (7.49) allow for the quantitative comparison of different haptic systems and can give insight into the relevant ranges of frequency that have to be optimized for a more transparent system. They also provide the basis for the integration of perception properties in the assessment of transparency.

From the above-mentioned definitions of transparency [Eqs. (7.45) and (7.46)], one can conclude that $\underline{e}_{\mathrm{T}} = \underline{e}'_{\mathrm{T}} \stackrel{!}{=} 0$ to fulfill the requirement for transparency. On the other hand, it is obvious that a human user will not perceive all possible mechanical impedances, since the perception capabilities are limited as shown in Sect. 2.1. To obtain a quantified range for $\underline{e}_{\mathrm{T}}$ and $\underline{e}'_{\mathrm{T}}$, a thought experiment[1] is conducted in the following [40].

[1] Thought experiments (also *gedankenexperiment*) consider the possible outcomes of a hypothesis without actually performing the experiment, but by applying theoretical considerations. They are conducted when the actual performance of an experiment is not possible or universally valid. Famous thought experiments include Schrödinger's Cat to illustrate quantum indeterminacy.

Experiment Assumptions

The following assumptions are made for the thought experiment about the user and the teleoperation scenario:

1. Linear behavior of haptic perception as discussed in Sect. 2.1.4.2 is assumed, which holds for a wide range of tool-mediated teleoperation scenarios. Superthreshold perception properties like masking are neglected.
2. For each user, there exists a known mechanical impedance $\underline{Z}_{\text{user}}$. This impedance generally depends on external parameters like temperature, contact force as shown in Chap. 3. All of these parameters are assumed to be known and invariant over the course of the experiment. Further, a set of frequency-dependent sensory thresholds for deflection and forces exist. They are labeled as F_θ and d_θ, respectively. Both thresholds can be coupled using the mechanical impedance of the user and $\omega = 2\pi f$ as the angular frequency of the haptic signal as stated in Eq. (7.50) [23].

$$|\underline{Z}_{\text{user}}| = \left| \frac{F_\theta}{j\omega d_\theta} \right| \tag{7.50}$$

3. The user is able to impose an interaction force $\underline{F}_{\text{user,int}}$ or deflection $\underline{d}_{\text{user,int}}$ on the teleoperation system that does not necessarily trigger a sensation event at the contact point. This is, for example, possible by the movement of an arm, while only the fingertips are in contact with the teleoperation system.
4. The teleoperation system is perfectly transparent, i.e., $|\underline{e}_{\text{T}}| = 0$ for all frequencies. The system is able to read and display forces and deflections reproducible below the absolute thresholds of the user.
5. The environment is considered passive for simplification reasons.

Thought Experiment

For the experiment, an impedance type system is assumed, i.e., the user imposes a deflection on the haptic interface of the teleoperation system and interaction forces measured are displayed to the user. First, we assume an environment impedance $\underline{Z}_{\text{E}} < \underline{Z}_{\text{user}}$. Further evaluation leads to Eq. (7.51).

$$\underline{Z}_{\text{E}} = \frac{\underline{F}_{\text{E}}}{j\omega \underline{d}_{\text{E}}} < \frac{\underline{F}_{\text{user}}}{j\omega \underline{d}_{\text{user}}} = \underline{Z}_{\text{user}} \tag{7.51}$$

For an impedance type system, the user can be modeled as a source of deflection or velocity. In that case, the induced deflection of the teleoperation system equals the deflection of the environment $\underline{d}_{\text{user,int}} = \underline{d}_{\text{H}} = \underline{d}_{\text{E}}$. With Eq. (7.51), this leads to $\underline{F}_{\text{H}} = \underline{F}_{\text{E}} < \underline{F}_{\text{user}}$. Assuming that the deflection $\underline{d}_{\text{user,int}}$ imposed by the user is smaller as the user's detection threshold d_θ (assumption no. 3), the resulting amount of force displayed to the user $|\underline{F}_{\text{user}}|$ is smaller than the individual force threshold F_θ according to Eq. (7.50).

This experiment can be extended to admittance-type systems easily. Descriptively, the result can be interpreted as the environment "evading" manipulation, as for example, a slow-moving hand in free air: The arm muscles serve as a deflection source moving the hand, but the interaction forces of the air molecules are too small to be detected.

For large environment impedances, the inequalities above are reversed. In that case, the forces or deflections resulting from the interaction are larger than the detection threshold; the user will feel an interaction with the environment.

Experiment Analysis

One can reason that the user impedance will limit the transparency error function from Eq. (7.49) from the experiment. This is done in such a way that environment impedances lower than the user impedance will be neglected as shown in Eq. (7.52).

$$\underline{e}'_{\text{T}} = \frac{\underline{Z}_{\text{H}} - \max\left(\underline{Z}_{\text{E}}, \underline{Z}_{\text{user}}\right)}{\max\left(\underline{Z}_{\text{E}}, \underline{Z}_{\text{user}}\right)} \tag{7.52}$$

If the user impedance is greater than the environment impedance, the user impedance is used, since the user will not feel any haptic stimuli generated by the lower environment impedance. If the user impedance is smaller than the environment impedance, the environment impedance is used as reference for the transparency error.

Up till now, only absolute detection thresholds were considered that describe the detection properties of haptic perception. In a second step, the discrimination properties will be considered in detail. It is assumed that a system is transparent *enough* for a satisfactory usage, if errors are smaller than the differences that can be detected by the user. This difference can be described in a conservative way by the ↪JND as defined in Sect. 2.1. With that, a limit can be imposed on Eq. (7.52) as given by Eq. (7.53)

$$\underline{e}'_{\text{T}} = \frac{\underline{Z}_{\text{t}} - \max\left(\underline{Z}_{\text{e}}, \underline{Z}_{\text{user}}\right)}{\max\left(\underline{Z}_{\text{e}}, \underline{Z}_{\text{user}}\right)} < c_{\text{JND}(z)} \tag{7.53}$$

This limit $c_{\text{JND}(z)}$ is defined as the JND of an arbitrary mechanical impedance. Although this value is not clearly measurable, it can be either bordered by the JNDs of ideal components like springs, masses, and viscous dampers (see Sect. 2.1 for values) or by the JNDs of forces and deflections (since a change in impedance can be detected if the resulting force or deformation for a fixed imposture of deflection or force, respectively, exceeds the JND). With known values, this leads to a probably sufficient limit of $\left|\underline{e}'_{\text{T}}\right| \leq 3\,\text{dB}$.

With Eq. (7.53), a perception-considering error term of the transparency of haptic teleoperation systems is given. One has to keep in mind the assumptions of the underlying thought experiment and the fact that experimental evaluation of this approach is still the focus of current research activities by the authors.

7.4.3 General Control Model for Teleoperators

In principle, a telemanipulator system can be divided into three different layers as shown in Fig. 7.23. The first layer contains the mechanical, electrical, and local control properties of the master device. The second layer represents the communication channels between the master and slave and therefore eventually occurring time delays. The third layer describes mechanical, electrical, and local control properties of the slave device. As mentioned before, the dynamic behavior of a master and accordingly a slave device (first and third layer) is determined by its mechanical and electrical characteristics. Dependent on the type of actuator used in the master device, respectively, slave device, a distinction is made between impedance and admittance devices. Impedance devices receive a force command and apply a force to their environment. By contrast, admittance devices receive a velocity command and behave as a velocity source interacting with the environment (see Chap. 6).

Customarily, dominant parameters are the mass and friction of the device. Compliance can be minimized by a well-considered mechanical design. In addition, it can be assumed that the dynamic characteristics of the electronic can be disregarded because the mechanical design is dominating the overall performance of the device. A local controller design may extend the usable frequency range of the device and can guarantee a stable operation of the device. In addition, it is possible to change the characteristics of the device from impedance behavior to admittance behavior and vice versa [19].

The second layer describes the characteristics of the communication channel. Significant physical values, which have to be transmitted between master and slave manipulator, are the values for forces and velocities at the master and slave sides.

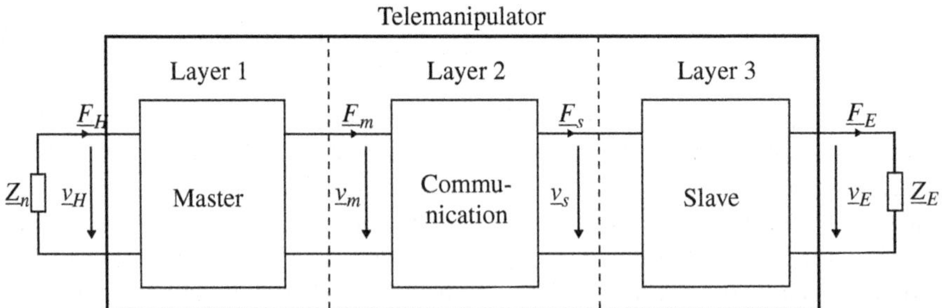

Fig. 7.23 Schematic illustration of a telemanipulator

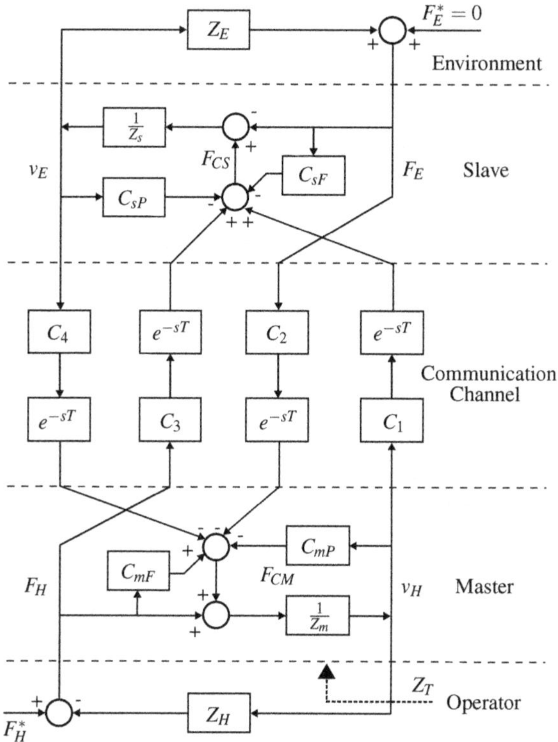

Fig. 7.24 System block diagram of a general telemanipulator in impedance–impedance architecture as shown in [21]

Therefore, telemanipulators exhibit at least two and up to four communication channels for transmitting these values. These communication paths may be afflicted with a significant time delay T, which can cause instability of the whole system.

Figure 7.24 shows the system block diagram of a general four-channel architecture bilateral telemanipulator using impedance actuators for master and slave manipulator, for instance electric motors [21, 30]. In total, there are four possible combinations of impedance and admittance devices, impedance–impedance, impedance–admittance, admittance–impedance, and admittance–admittance.

In this section, the impedance–impedance architecture is used due to its common use because of the high hardware availability. The forces of user and environment $\underline{F}_{\mathrm{H}}$ and $\underline{F}_{\mathrm{E}}$ are independent values. The mechanical impedance of user and environment is described by $\underline{Z}_{\mathrm{H}}$ and $\underline{Z}_{\mathrm{E}}$. The communication layer contains of four transmission elements C_1, C_2, C_3, and C_4 for transmitting the contact forces and velocities $\underline{v}_{\mathrm{H}}, \underline{F}_{\mathrm{E}}$, $\underline{F}_{\mathrm{H}}$, and $\underline{v}_{\mathrm{E}}$ between master and slave sides. $\underline{Z}_{\mathrm{m}}^{-1}$ and $\underline{Z}_{\mathrm{s}}^{-1}$ represent the mechanical admittance of master controller and slave manipulator. In addition, C_{mP} and C_{sP} are local master and slave position controllers and C_{mF} and C_{sF} are local force controllers.

The dynamics of the four-channel architecture are described by the following equations:

$$
\begin{aligned}
\underline{F}_{\mathrm{CM}} &= C_{\mathrm{mF}}\underline{F}_{\mathrm{H}} - C_4 e^{-sT}\underline{v}_{\mathrm{E}} - C_2 e^{-sT}\underline{F}_{\mathrm{E}} - C_{\mathrm{mP}}\underline{v}_{\mathrm{H}} \\
\underline{F}_{\mathrm{CS}} &= C_1 e^{-sT}\underline{v}_{\mathrm{H}} + C_3 e^{-sT}\underline{F}_{\mathrm{H}} - C_{\mathrm{sF}}\underline{F}_{\mathrm{E}} - C_{\mathrm{sP}}\underline{v}_{\mathrm{E}} \\
\underline{Z}_{\mathrm{s}}\underline{v}_{\mathrm{E}} &= \underline{F}_{\mathrm{CS}} - \underline{F}_{\mathrm{E}} \\
\underline{Z}_{\mathrm{m}}\underline{v}_{\mathrm{H}} &= \underline{F}_{\mathrm{CM}} + \underline{F}_{\mathrm{H}}
\end{aligned}
$$

So, the closed-loop dynamics of the telemanipulator are represented by

$$
\left(\underline{Z}_{\mathrm{m}} + C_{\mathrm{mP}}\right) \cdot \underline{v}_{\mathrm{H}} + C_4 e^{-sT}\underline{v}_{\mathrm{E}} = (1 + C_{\mathrm{mF}}) \cdot \underline{F}_{\mathrm{H}} - C_2 e^{-sT}\underline{F}_{\mathrm{E}} \tag{7.54}
$$

$$
-\left(\underline{Z}_{\mathrm{s}} + C_{\mathrm{sP}}\right) \cdot \underline{v}_{\mathrm{E}} + C_1 e^{-sT}\underline{v}_{\mathrm{H}} = (1 + C_{\mathrm{sF}}) \cdot \underline{F}_{\mathrm{E}} - C_3 e^{-sT}\underline{F}_{\mathrm{H}} \tag{7.55}
$$

As presented in Sect. 7.4.1, it is common to describe the dynamics of a telemanipulator by two-port representation. In addition, several stability analysis methods can be applied on two-port model. From Eqs. (7.54) and (7.54) with (7.43), the following h-parameters can be obtained:

$$
\underline{h}_{11} = \frac{(\underline{Z}_{\mathrm{m}} + C_{\mathrm{mP}}) \cdot (\underline{Z}_{\mathrm{s}} + C_{\mathrm{sP}}) + C_1 C_4 e^{-2sT}}{(1 + C_{\mathrm{mF}}) \cdot (\underline{Z}_{\mathrm{s}} + C_{\mathrm{sP}}) - C_3 C_4 e^{-2sT}} \tag{7.56}
$$

$$
\underline{h}_{12} = \frac{C_2(\underline{Z}_{\mathrm{s}} + C_{\mathrm{sP}})e^{-sT} - C_4(1 + C_{\mathrm{sF}})e^{-sT}}{(1 + C_{mF}) \cdot (\underline{Z}_{\mathrm{s}} + C_{\mathrm{sP}}) - C_3 C_4 e^{-2sT}} \tag{7.57}
$$

$$
\underline{h}_{21} = -\frac{C_3(\underline{Z}_{\mathrm{m}} + C_{\mathrm{mP}})e^{-sT} + C_1(1 + C_{\mathrm{mF}})e^{-sT}}{(1 + C_{\mathrm{mF}}) \cdot (\underline{Z}_{\mathrm{s}} + C_{\mathrm{sP}}) - C_3 C_4 e^{-2sT}} \tag{7.58}
$$

$$
\underline{h}_{22} = \frac{(1 + C_{\mathrm{sF}}) \cdot (1 + C_{\mathrm{mF}}) - C_2 C_3 e^{-2sT}}{(1 + C_{\mathrm{mF}}) \cdot (\underline{Z}_{\mathrm{s}} + C_{\mathrm{sP}}) - C_3 C_4 e^{-2sT}} \tag{7.59}
$$

With Eqs. (7.47) and (7.56)–(7.59), the impedance transmitted to the user $\underline{Z}_{\mathrm{T}}$ is given by Eq. (7.60) [19].

$$
\underline{Z}_{\mathrm{T}} = \frac{(\underline{Z}_{\mathrm{m}} + C_{\mathrm{mP}}) \cdot (\underline{Z}_{\mathrm{s}} + C_{\mathrm{sP}}) + C_1 C_4 e^{-2sT} + \left[(1 + C_{sF}) \cdot (\underline{Z}_{\mathrm{M}} + C_{\mathrm{mP}}) + C_1 C_2 e^{-2sT}\right] \cdot \underline{Z}_{\mathrm{E}}}{(1 + C_{\mathrm{mF}}) \cdot (\underline{Z}_{\mathrm{s}} + C_{\mathrm{sP}}) - C_3 C_4 e^{-2sT} + \left[(1 + C_{\mathrm{sF}}) \cdot (1 + C_{\mathrm{mF}}) + C_2 C_3 e^{-2sT}\right] \cdot \underline{Z}_{\mathrm{E}}} \tag{7.60}
$$

Perfect transparency is achievable, if the time delay T is insignificant. The controllers must hold the following conditions, which are known as the transparency-optimized control law [21, 30]:

$$
\begin{aligned}
C_1 &= \underline{Z}_{\,s} + C_{sP} \\
C_2 &= 1 + C_{mF} \\
C_3 &= 1 + C_{sF} \\
C_4 &= -\left(\underline{Z}_{\,m} + C_{mP}\right) \\
C_2, C_3 &\neq 0
\end{aligned} \tag{7.61}
$$

By use of local position and force controllers of master and slave C_{mp}, C_{sp}, C_{mF}, and C_{sF}, a perfect transparency can be achieved with only three communication channels. In this case, the force feedback from slave to master C_2 can be neglected [20, 21].

The most common control architecture is the forward-flow architecture [16] also known as force feedback or position-force architecture [30], which uses the two channels C_1 and C_2. C_3 and C_4 are set to zero. The position, respectively, velocity $\underline{v}_{\,h}$ at the master manipulator is transmitted to the slave. The slave manipulator feeds back the contact forces between manipulator and environment $\underline{F}_{\,e}$. Due to not compensated impedances of master and slave devices, perfect transparency is not achievable by telemanipulator buildup in the basic forward-flow architecture. This architecture has been described and analyzed by many authors [7, 8, 16, 17, 19, 30].

7.4.4 Stability Analysis of Teleoperators

Besides the general stability analysis for dynamic systems from Sect. 7.2, several approaches for stability analysis of haptic devices have been published. Most of them use the two-port representation introduced in Sect. 7.4.1 for stability analysis and controller design and were derived from the classical network theory and communications technology. The following section gives an introduction to the most important of them and also presents methods to guarantee stability of the system under time delay.

7.4.4.1 Passivity

The concept of passivity for dynamic systems is introduced in Sect. 7.2.2. Within this section, the focus is on the application of this concept on the stability analysis of haptic devices. Assume the two-port representation of a telemanipulator as presented in Fig. 7.23. Furthermore, it is assumed that the energy stored in the system at time $t = 0$ is $V(t = 0) = 0$. The power P_{in} at the input of the system at time t is given by the product of the force $F_H(t)$ applied by the user to the master times the master velocity $v_H(t)$.

$$
P_{in} = F_H(t) \cdot v_H(t)
$$

Accordingly, the power P_{out} at the output of the telemanipulator is given by the contact force of the slave $F_{\text{E}}(t)$ manipulating the environment times the velocity of the slave $v_{\text{E}}(t)$

$$P_{\text{out}} = F_{\text{E}}(t) \cdot v_{\text{E}}(t)$$

Thus, the telemanipulator is passive and therefore stable as long as the following inequality is fulfilled:

$$\int_0^t (P_{\text{in}}(\tau) - P_{\text{out}}(\tau)\text{d}\tau) = \int_0^t (F_{\text{H}}(\tau) \cdot v_{\text{H}}(\tau) - F_{\text{E}}(\tau) \cdot v_{\text{E}}(\tau)\text{d}\tau) \geq V(t) \quad (7.62)$$

Alternatively, the criterion can be expressed in the form of the time derivative of Eq. (7.62)

$$F_{\text{H}}(t) \cdot v_{\text{H}}(t) - F_{\text{E}}(t) \cdot v_{\text{E}}(t) \geq V(t) \quad (7.63)$$

From Eq. (7.62), respectively, Eq. (7.63), it can be seen that the telemanipulator must not generate energy to be passive. Thus, an easy method to receive a stable telemanipulator system is to implement higher damping, though it decreases the performance of the system.

Considering the frequency domain passivity of the system can be analyzed using the immitance matrix of the transfer function [7, 8, 10–12, 35–37]. A system is passive and, hence inherently, stable if the immitance matrix $G(s)$ of the n-port network is positive real. The criteria for positive realness of the immitance matrix, which have to be satisfied, are [5, 24]:

1. $G(s)$ has real elements for real s;
2. The elements of $G(s)$ have no poles in $\text{Re}(s) > 0$ and poles on the $j\omega$-axis are simple, such that the associated residue matrix is nonnegative definite Hermitian;
3. For any real value of ω such that no element of $G(j\omega)$ has a pole for this value, $G(jw) + G(jw)$ is nonnegative definite Hermitian.

For real rational $G(s)$, points 1 and 3 may be replaced with

4. $G(s) + G(s)$ is nonnegative definite Hermitian in $\text{Re}(s) > 0$

User and environment can be seen as passive [28]. Therefore, if passivity of the telemanipulator system can be proved, the whole closed loop of user, telemanipulator, and environment can be guaranteed to be passive and hence stable. It has been shown that a robust (passive) control law and transparency are conflicting objectives in the design of telemanipulators [30]. In many cases, the haptic sensation presented to the user can be poor if a fixed damping value is used to guarantee passivity of the telemanipulator. Thus, a new approach using passivity-based control law and improving performance has been done by implementing a passivity observer and passivity controller. The passivity controller increases damping of the system only when needed to guarantee stability. A further benefit of this concept is that no parameter estimation for the dynamic model of the telemanipulator has to be done and if considered, uncertainties can be compensated [18, 38].

7.4.4.2 Absolute Stability Criterion (Llewellyn)

A stability criterion for linear two-ports has been derived by LLEWELLYN [9, 24, 31]. His motivation was the investigation of generalized transmission lines and active networks. Later, several authors have used the criteria formulated by Llewellyn to analyze the stability of telemanipulators or to design control laws for bilateral teleoperation [1, 2, 4, 19]. The criterion is formulated in the frequency domain and it is assumed that the two-port is linear and time invariant, at least locally [3]. A linear two-port is absolute stable if and only if there exists no set of passive terminations for which the system is unstable.

The following criteria provide both necessary and sufficient conditions for absolute stability for linear two-ports.

1. $G(s)$ has no poles in the right half s-plane, only simple poles on the imaginary axis
2. $\mathrm{Re}(g_{11}) > 0, \mathrm{Re}(g_{22}) > 0$
3. $2 \cdot \mathrm{Re}(g_{11}) \cdot \mathrm{Re}(g_{22}) \geq |g_{12}g_{21}| + \mathrm{Re}(g_{12}g_{21}) \quad \forall\omega \geq 0$

Conditions 1 and 2 guarantee passivity of the system when there is no coupling between master and slave. This case occurs when master or slave are free or clamped. Condition 3 guarantees stability if master and slave are coupled.

These criteria may be applied to every type of immitance matrix, thus the impedance matrix, admittance matrix, hybrid matrix, or inverse hybrid matrix. If the criteria are fulfilled for one form of immitance matrix, they are fulfilled for the other three forms as well. A network for which $h_{21} = -h_{12}$, which is the same as $z_{21} = z_{12}$ holds is said to be reciprocal. In this particular case, the tests for passivity and unconditional stability are the same. A passive network will always be absolute stable, but an absolute stable network is not necessarily passive. A two-port that is not unconditional stable is potentially unstable, but this does not mean that it is definitely unstable as shown in Fig. 7.25.

7.4.5 Effects of Time Delay

When master and slave are far apart from each other, communication data have to be transmitted over long distance with significant time delays, which can lead to instabilities unless the bandwidth of signals entering the communication block is severely limited. The reason for this is a non-passive communication block [8], so energy is generated inside the communication block.

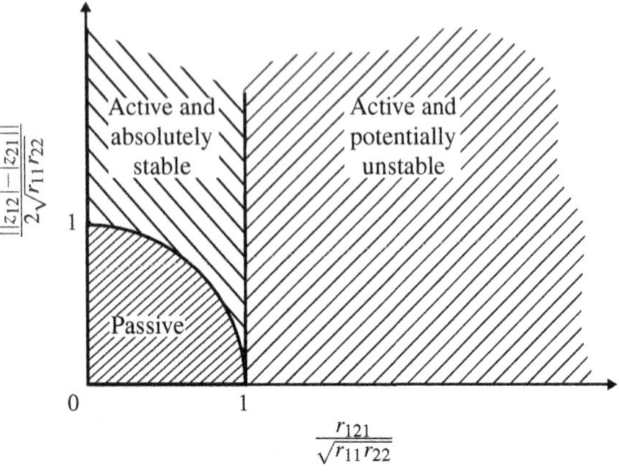

Fig. 7.25 Stability-activity diagram [24]

7.4.5.1 Scattering Theory

ANDERSON [6–8] used the scattering theory to find a stable control law for bilateral teleoperation systems with time delay. Scattering variables were well known from transmission line theory. The scattering operator S maps effort plus flow into effort minus flow and is defined in terms of an incident wave $F(t) + v(t)$ and a reflected wave $F(t) - v(t)$.

$$F(t) - v(t) = S(t)\,(F(t) + v(t))$$

For LTI systems S can be expressed in the frequency domain as follows:

$$F(s) - v(s) = S(s)\,(F(s) + v(s))$$

In the case of a two-port the scattering matrix can be related to the hybrid matrix $\mathbf{H}(s)$ by loop transformation, which leads to

$$S(s) = \begin{pmatrix} 1 & 0 \\ 0 & -1 \end{pmatrix} \cdot (\mathbf{H}(s) - 1)\,(\mathbf{H}(s) + 1)^{-1}$$

To ensure passivity of the system, the reflected wave must not carry higher energy content than the incident wave. Therefore, a system is passive if and only if the norm of its scattering operator $S(s)$ is less than or equal to one [8].

$$\|S(s)\|_\infty \leq 1$$

7.4.5.2 Wave Variables

Wave variables were used by NIEMEYER [35, 36] to design a robust control strategy for bilateral telemanipulation with time delay. It separates the total power flow into two parts, one the power flowing into the system and the other part representing the power flowing out of the system. Later, these two parts are associated with input and output waves. This approach is also valid for nonlinear systems. Assume the two-port shown in Fig. 7.26 using $\dot{x}_m$ and F_e as inputs.

Therefore, the power flow through the two-port can be written as

$$P(t) = \dot{x}_M^T F_T - \dot{x}_s^T F_S = \frac{1}{2} u_M^T u_T - \frac{1}{2} v_M^T v_T + \frac{1}{2} u_S^T u_S - \frac{1}{2} v_S^T v_S.$$

Here, the vectors $\mathbf{u}_M$ and $\mathbf{u}_S$ are input waves, which increase the power flow into the system. Analogous to this $\mathbf{v}_M$ and $\mathbf{v}_S$ are output waves decreasing the power flow into the system. Note that velocity is denoted here as $\dot{x}$. The transformation from the power variables to wave variables is described as

$$\begin{aligned}
u_M &= \frac{1}{\sqrt{2b}}(F_M + b\dot{x}_M) \\
u_S &= \frac{1}{\sqrt{2b}}(F_S - b\dot{x}_S) \\
v_M &= \frac{1}{\sqrt{2b}}(F_M - b\dot{x}_M) \\
v_S &= \frac{1}{\sqrt{2b}}(F_S + b\dot{x}_S)
\end{aligned}$$

The wave impedance b relates velocity to force and represents an opportunity to tune the behavior of the system. Large b values lead to an increased force feedback at the cost of high inertial forces. Small b values lower any unwanted sensations, so fast movement is possible, but also decreases the force impression of contact forces between slave and environment [34]. The wave variables can be inverted to provide the power variables as a function of the wave variables.

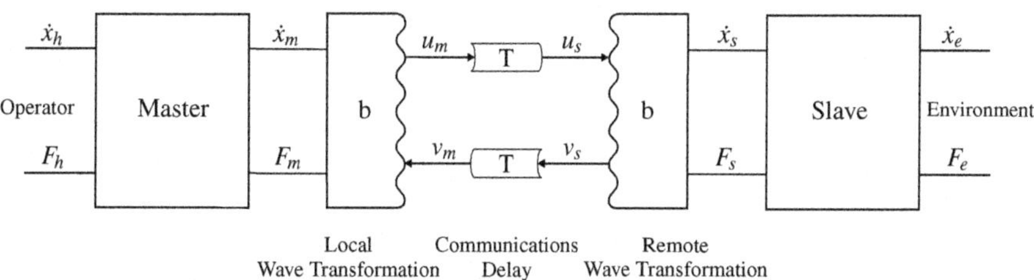

Fig. 7.26 Wave-based teleoperator model

$$
\begin{aligned}
F_M &= \sqrt{\frac{b}{2}}(u_M + v_M) \\
F_S &= \sqrt{\frac{b}{2}}(u_S + v_S) \\
\dot{x}_M &= \frac{1}{\sqrt{2b}}(u_M - \dot{x}_M) \\
\dot{x}_S &= -\frac{1}{\sqrt{2b}}(u_S + v_S)
\end{aligned}
$$

By transmitting the wave variables instead of the power variables, the system remains stable even if the time delay T is not known [35]. Note that when the actual time delay T is reduced to zero, transmitting wave variables is identical to transmitting velocity and force.

7.5 Conclusion

The control design for haptic devices faces the developing engineer with a complex manifold challenge. According to the fundamental requirement, to establish a safe, reliable, and determined influence on all structures, subsystems, or processes, the haptic system composed of an analytical approach for control system design is not negligible anymore. It provides a wide variety of methods and techniques to be able to cover many issues that arise during this design process. This chapter intends to introduce the fundamental theoretical background. It shows several tasks, functions, and aspects the developer will have to focus on, as well as certain methods and techniques that are going to be useful tools for the system's analysis and the process of control design.

Starting with an abstracted view on the overall system, the control design process is based on an investigation and mathematical formulation of the system's behavior. To achieve this, a wide variety of methods exist that can be used for system description depending on the degree of complexity. Besides methods for the description of linear or linearized systems, this chapter introduced techniques for system description to represent nonlinear system behavior. Furthermore, the analysis of MIMO systems is based on the state space description, which is also presented here. All of these techniques on the one hand are aimed at the mathematical representation of the analyzed systems as exactly as possible, on the other hand they need to satisfy the requirement for a system description that further control design procedures are applicable to. These two requirements will lead to a trade-off between establishing an exact system formulation that can be used in analysis and control design procedures without extending the necessary effort unreasonably.

Within system analysis of haptic systems, the overall system stability is the most important aspect that has to be guaranteed and proven to be robust against model uncertainties. The compendium of methods for stability analysis contains techniques applicable to linear or nonlinear system behavior, corresponding to their underlying principles that of course limit the usability. The more complex the mathematical formulation of the system becomes, the higher the effort gets for system analysis. This comes into direct conflict with the fact that a stability analysis of a system with a simplified system description can only provide a proof of stability for this simplified model of the real system. Therefore, the impact of all simplifying assumptions must be evaluated to guarantee the robustness of the system stability.

The actual objective within establishing a control scheme for haptic systems is the final design of controller and control structures that have to be implemented in the system at various levels to perform various functions. Besides the design of applicable controllers or control structures, the optimization of adjustable parameters is also part of this design process. As shown in many examples in the literature on control design, a comprehensive collection of control design techniques and optimization methods exist that enable the developer to cover the emerging challenges and satisfy various requirements within the development of haptic systems as far as automatic control is concerned.

Recommended Background Reading

[19] Hashtrudi-Zaad, K. & Salcudean, S.: **Analysis of control Architectures for Teleoperation Systems with Impedance/Admittance Master and Slave - Manipulators**. In: The International Journal of Robotics Research, SAGE Publications, 2001.
Thorough analysis of different control schemes for impedance and admittance type systems.

[26] Hirche, S. & Buss, M.: **Human perceived transparency with time delay**. In:Manuel Ferre et al. (eds.), Advances in Telerobotics, Springer, 2007.
Analysis of the effects of time delay on transparency and the perception of compliance and mass.

[41] Tavakoli, M.; Patel, R.; Moallem, M. & Aziminejad, A.: **Haptics for Teleoperated Surgical Robotic Systems**. World Scientific Publishing, Shanghai, 2008.
Description and Design of a minimal invasive surgical robot with haptic feedback including an analysis of stability issues and the effect of time delay.

References

1. Adams RJ, Hannaford B (1998) A two-port framework for the design of unconditionally stable haptic interfaces. In: Proceedings of IEEE/RSJ international conference on intelligent robots and systems, 1998, vol 2. IEEE, pp 1254–1259. doi:10.1109/IROS.1998.727471
2. Adams RJ, Hannaford B (2002) Control law design for haptic interfaces to virtual reality. IEEE Trans Control Syst Technol 10(1):3–13. doi:10.1109/87.974333

3. Adams R, Hannaford B (1999) Stable haptic interaction with virtual environments. IEEE Trans Robot Autom, 15(3):465–474. ISSN: hapt. doi:10.1109/70.768179
4. Adams R, Klowden D, Hannaford B (2000) Stable haptic interaction using the Excalibur force display. In: Proceedings of IEEE international conference on robotics and automation, ICRA 2000, vol 1. pp 770–775. doi:10.1109/ROBOT.2000.844144
5. Anderson B (1968) A simplified viewpoint of hyperstability. IEEE Trans Autom Control 13(3):292–294. doi:10.1109/TAC.1968.1098910
6. Anderson R, Spong MW (1988) Hybrid impedance control of robotic manipulators. IEEE J Robot Autom 4(5):549–556. doi:10.1109/56.20440
7. Anderson RJ, Spong MW (1992) Asymptotic stability for force reflecting teleoperators with time delay. Int J Robot Res 11(2):135–149. doi:10.1177/027836499201100204
8. Anderson R, Spong M (1989) Bilateral control of teleoperators with time delay. In: IEEE transactions on automatic control, 34(5):494–501. ISSN, pp 0018–9286, doi:10.1109/9.24201
9. Bolinder E (1975) Survey of some properties of linear networks. In: IRE Transactions on Circuit Theory, 4(3):70–78. ISSN, pp 0096–2007, doi:10.1109/TCT.1957.1086385
10. Colgate J (1993) Robust impedance shaping telemanipulation. IEEE Trans Robot Autom 9(4): 374–384. ISSN, pp 1042–296X, doi:10.1109/70.246049
11. Colgate J, Brown J (1994) Factors affecting the Z-Width of a haptic display. In: Proceedings of IEEE international conference on robotics and automation, 4:3205–3210. doi:10.1109/ROBOT.1994.351077
12. Colgate J, Stanley M, Brown J (1995) Issues in the haptic display of tool use. In: Proceedings of intelligent robots and systems international conference on human robot interaction and cooperative robots, 1995 IEEE/RSJ vol 3. pp 140–145. doi:10.1109/IROS.1995.525875
13. Föllinger O (1970) Nichtlineare regelungen 3. In: Ljapunow-Theorie and Popow-Kriterium. R. Oldenbourg Verlag GmbH
14. Föllinger O (1991) Nichtlineare regelungen 1. Grundlagen and Harmonische Balance. R, Oldenbourg Verlag GmbH. ISBN 3-486-21895-6
15. Föllinger O, (1970) Nichtlineare regelungen. 2. In: Anwendung der Zustandsebene R. Oldenbourg Verlag GmbH, 1970. ISBN: 978-3486218992
16. Hannaford B (1989) A design framework for teleoperators with kinaesthetic feedback. IEEE Trans Robot Autom 5(4):426–434. ISSN: 1042–296X. doi:10.1109/70.88057
17. Hannaford B (1989) Stability and performance tradeoffs in bi-lateral telemanipulation. In: Proceedings of IEEE international conference on robotics and automation 3:1764–1767. doi:10.1109/ROBOT.1989.100230
18. Hannaford B, Ryu J (2002) Time-domain passivity control of haptic interfaces. IEEE Trans Robot Autom 18(1):1–10. doi:10.1109/70.988969
19. Hashtrudi-Zaad K, Salcudean S (2001) Analysis of control architectures for teleoperation systems with impedance/admittance master and slave manipulators. The Int J Robot Res 20(6):419–445. doi:10.1177/02783640122067471
20. Hashtrudi-Zaad K, Salcudean SE (2002) Transparency in time-delayed systems and the effect of local force feedback for transparent teleoperation. IEEE Trans Robot Autom 18(1):108–114. doi:10.1109/70.988981
21. Hashtrudi-Zaad K, Salcudean S (1999) On the use of local force feedback for transparent teleoperation. In: IEEE international conference on robotics and automation, 3:1863–1869 doi:10.1109/ROBOT.1999.770380
22. Hatzfeld C (2013) Experimentelle analyse der menschlichen Kraftwahrnehmung als ingenieurtechnische Entwurfsgrundlage für haptische systeme. Dissertation, Technische Universität München: Dr. Hut Verlag, 2013. Darmstadt. http://tuprints.ulb.tu-darmstadt.de/3392/. ISBN: 978-3-8439-1033-0
23. Hatzfeld C, Werthschützky R (2012) Mechanical impedance as coupling parameter of force and deflection perception: experimental evaluation. In: Isokoski P, Springare J (Eds), Haptics: perception, devices, mobility, and communication. LNCS 7282. Proceedings of the Eurohaptics Conference, Tampere, FIN. Heidelberg: Springer, 2012. doi:10.1007/978-3-642-31401-8_18

24. Haykin S (1970) Active network theory, electrical engineering. Addison-Wesley, Reading ISBN:978-0201026801
25. Heredia E, Rahman T, Kumar V (1996) Adaptive teleoperation transparency based on impedance modeling. Proc SPIE 2901:2–12. doi:10.1117/12.26299
26. Hirche S, Buss M (2007) Human perceived transparency with time delay. In: Manuel Ferre et al. (eds), Advances in telerobotics, Analysis of the effects of time delay on transparency and the perception of compliance and mass, Springer, Berlin
27. Hirsch M (2013) Auswahl und entwurf eines positionsreglers unter Berücksichtigung nichtlinearer Effekte für einen parallelkinematischen Mechanismus. Diploma Thesis. Technische Universität Darmstadt, 2013
28. Hogan N (1989) Controlling impedance at the man/machine interface. In: IEEE International Conference on Robotics and Automation (ICRA). Scottsdale, AZ, USA, 1989, pp 1626–1631. doi:10.1109/ROBOT.1989.100210
29. Khalil HK (2002) Nonlinear systems. Prentice Hall, New Jersey ISBN: 0-130-67389-7
30. Lawrence D (1993) Stability and transparency in bilateral teleoperation. IEEE Trans Robot Autom 9(5):624–637. doi:10.1109/CDC.1992.371336
31. Llewellyn FB (1952) Some fundamental properties of transmission systems. In: Proceedings of the IRE 40(3):271–283. ISSN, pp 0096–8390, doi:10.1109/JRPROC.1952.273783
32. Lunze J (2005) Regelungstechnik 2. Springer, Berlin. ISBN 3-540-22177-8
33. Lunze J (2006) Regelungstechnik 1. Springer, Berlin. ISBN 3-540-28326-9
34. Niemeyer G, Preusche C, Hirzinger G (2008) Telerobotics. In: Siciliano B, Khatib O, (eds) Springer handbook of robotics. Springer, Berlin 2008, pp 741–757. doi:10.1007/978-3-540-30301-5_32
35. Niemeyer G, Slotine J-J (1991) Stable adaptive teleoperation. IEEE J Oceanic Eng 16(1):152–162. ISSN, pp 0364–9059, doi:10.1109/48.64895
36. Niemeyer G, Slotine J-J Towards force-reflecting teleoperation over the internet. In: Proceedings of IEEE international conference on robotics and automation. vol 3. IEEE. 1998, pp 1909–1915. doi:10.1109/ROBOT.1998.680592
37. Raju G, Verghese G, Sheridan T (1989) Design issues in 2-port network models of bilateral remote manipulation. In: IEEE international conference on robotics and automation (ICRA). Scottsdale, AZ, USA, 1989, pp 1316–1321. doi:10.1109/ROBOT.1989.100162
38. Ryu J-H, Kwon D-S, Hannaford B (2004) Stable teleoperation with time-domain passivity control. IEEE Trans Robot Autom 20(2):365–373. doi:10.1109/TRA.2004.824689
39. Slotine J-JE, Li W (1991) Applied nonlinear control. Prentice Hall, New Jersey. ISBN: 0-130-40890-5
40. Sorensen RA (1998) Thought experiments. Oxford University Press, Oxford, GB. ISBN 9780195129137 (Reprint)
41. Tavakoli M et al (2008) Haptics for teleoperated surgical robotic systems. World Scientific Publishing, Singapore. ISBN: 978-981-281-315-2
42. Unbehauen H (2007) Regelungstechnik II: zustandsregelungen, digitale und nichtlineare Regelsysteme Vieweg und Teubner, ISBN: 3-528-83348-3
43. Unbehauen H (2007) Regelungstechnik I. Vieweg und Teubner, ISBN: 3-834-80230-1
44. Hashtrudi-Zaad K Salcudean S (2001) Analysis of control architectures for teleoperation systems with impedance/admittance master and slave manipulators. Int J Robot Res, SAGE Publications, 2001.
45. Hirche S, Buss M (2007) Human perceived transparency with time delay. In: Advances in Telerobotics pp 191–209. doi:10.1007/978-3-540-71364-7_13
46. Tavakoli M, Patel R, Moallem M, Aziminejad A (2008) Haptics for teleoperated surgical robotic systems. Description and design of a minimal invasive surgical robot with haptic feedback including an analysis of stability issues and the effect of time delay, World Scientific Publishing, Shanghai, 2008

Chapter 8
Kinematic Design

Sebastian Kassner

Abstract The kinematic design of haptic interfaces is a crucial step especially when designing interfaces with mainly kinaesthetic feedback (see Sect. 12.2). This is often the case in the context of robotic applications. In these devices, a mechanical mechanism is used to link the user and the feedback generating actuators. Furthermore, the user's input commands are often given by moving a mechanical mechanism, e.g., a joystick.Accordingly, the kinematic design is a crucial aspect for a device with ergonomic design and good haptic transmission. This chapter gives an introduction to the classes of mechanisms and how they are designed.

8.1 Introduction and Classification

Figure 8.1 shows the basic elements of a mechanical mechanism: base platform, rods, joints, and the ↪ TCP. The base platform is that part of a mechanism that is static regarding its motion. This means that for all calculations to be performed to design a mechanism as well as for all calculations executed during the operation of the haptic interface, position, speed, and accelerations are given with respect to the base platform.

In common mechanisms, at least one joint is located in the base platform. Rods and further joints make up a kinematic chain linking the base platform and the ↪ TCP. Kinematic joints used in mechanisms are (Fig. 8.2): revolute (R) joints, prismatic (P) joints, helical (H) joints, universal (U) joints, cylindrical (C) joints, spherical (S) joints, and planar (E) joints.

S. Kassner (✉)
Institute of Electromechanical Design, Technische Universität Darmstadt,
Merckstr. 25, 64283 Darmstadt, Germany
e-mail: s.kassner@hapticdevices.eu

C. Hatzfeld and T.A. Kern (eds.), *Engineering Haptic Devices*,
Springer Series on Touch and Haptic Systems, DOI 10.1007/978-1-4471-6518-7_8

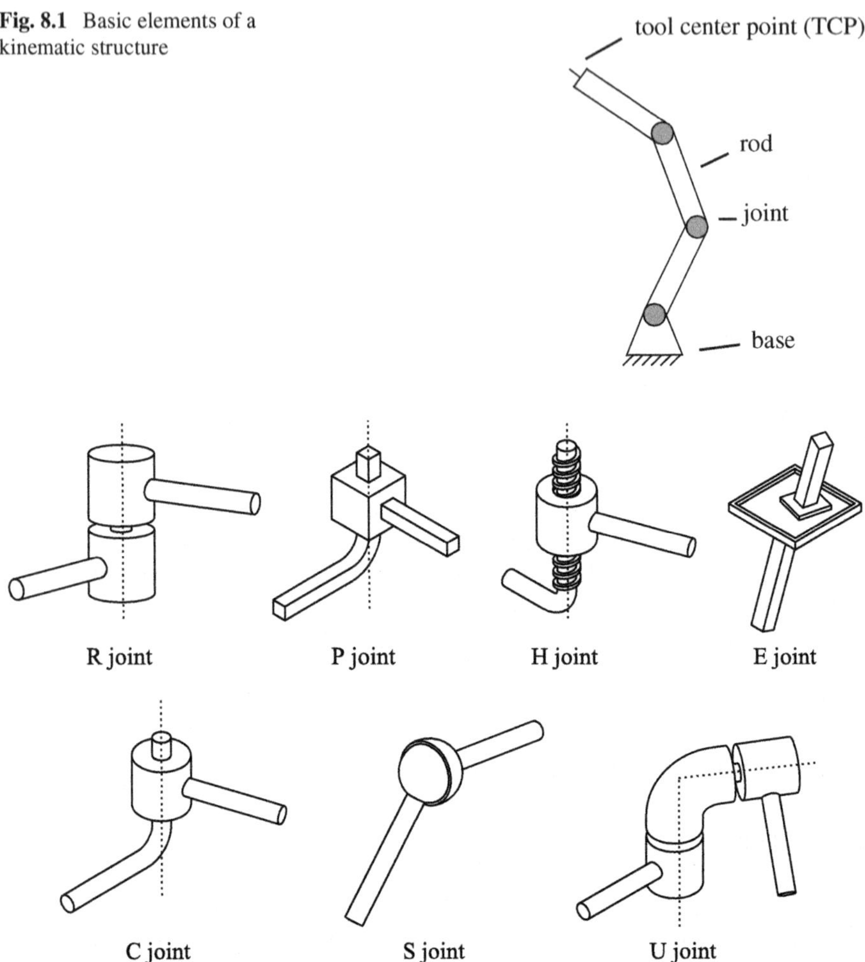

Fig. 8.1 Basic elements of a kinematic structure

Fig. 8.2 Kinematic joints, figure based on [6]

The kinematic chain is a mathematical model to calculate the kinematics of a mechanical system. This is defined as:

> **Definition** *Kinematics* Kinematics is the science of motion of points and bodies in space, characterized by their position, speed, and acceleration. Thereby the external causes of motion (forces and torques) are not taken into account. Motion with respect forces and torques is covered by the science of kinetics.

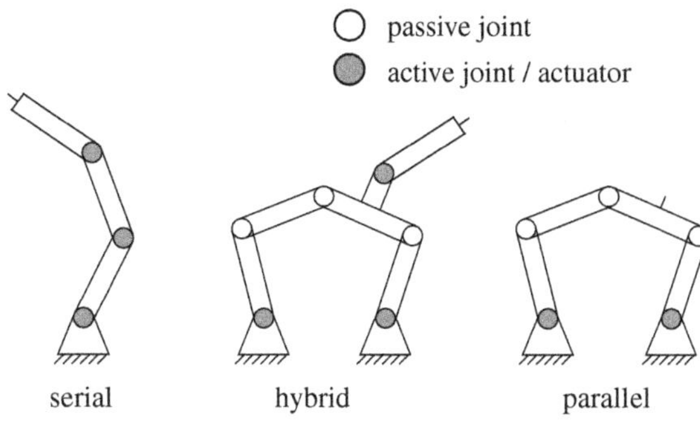

Fig. 8.3 Basic kinematic structures

The ↪ TCP is the point of interaction between mechanism and environment. It is able to move in space with a certain number of ↪ DoF. In the case of haptic interfaces, the haptic feedback and user interaction usually take place via this point.

8.1.1 Classification of Mechanisms

Depending on their kinematic chains mechanical mechanisms are classified as

- serial: open kinematic chains;
- parallel: closed kinematic chains, at least two paths from the base platform to the tool-center point;
- hybrid: combination of serial and parallel mechanisms.

Figure 8.3 shows the three fundamental configurations with typically passive and actively driven joints.

Serial Mechanisms

Serial mechanisms are widely used in all kinds of robotic applications. A classic example is serial assembly robots in an assembly line production of motor vehicles. Purely serial mechanisms include no passive joints. All actuators are in serial order within one single kinematic chain.

The advantages of serial mechanisms are their simple design and their relatively large workspace. They are furthermore easy to control, especially in positioning tasks. This is mainly due to the serial sequence of joints and rods allowing the application of mathematical step-by-step transformations. An established method is the DANAVIT-HARTENBERG transformation [4], which is not covered in this chapter.

The major drawback of a serial mechanism is its dynamic behavior. Since a load is carried by a single kinematic chain, serial mechanisms usually have lower structural stiffness with respect to their own weight. Additionally, the dynamic behavior is restricted by the comparatively high masses. One reason for this effect is the mass

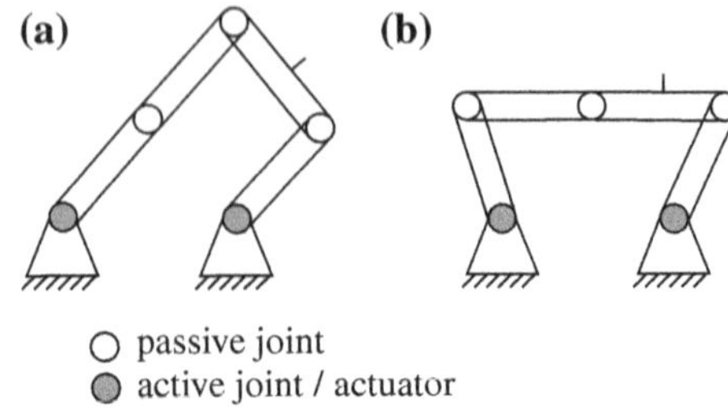

Fig. 8.4 Singular positions: **a** First kind. **b** Second kind

of the rods to gain a high structural stiffness. A second reason is the weight of the actuators within the mechanism. Every actuator has to accelerate all following actuators in the kinematic chain.

Parallel Mechanisms

This is the main advantage of parallel structures. They comprise actuators that are fixed to the base platform or that are only moved slightly in space. The load on a ↪ TCP is distributed into several kinematic chains. This allows the design of lightweight, yet stiff mechanical structures. This leads to a dynamic transmission behavior with high cutoff frequencies and thus a more transparent transmission of the haptic feedback. Because of these properties, parallel mechanisms are of high significance in the design of haptic interfaces.

On the other hand, parallel mechanism has a small workspace in comparison with serial structures. The parallel mechanism's kinematic is often mathematically more complex and usually nonlinear. Furthermore, the transmission behavior changes with respect to the mechanism's position. Thus, it is directional and anisotropic throughout the workspace.

Parallel mechanisms have special positions that have to be taken into account when designing a haptic interface: singular positions. These positions occur when at least two rods of the mechanism are aligned. One distinguishes two types of singular positions (Fig. 8.4):

Singularity of the first kind The actuator's motion is not transmitted to the ↪ TCP any longer. This position typically occurs at the edge of the mechanism's workspace.

Singularity of the second kind The actuator's force or torque is not transmitted to the ↪ TCP any longer and the ↪ TCP can carry no load. The mechanism is jammed regarding its actuators. This position typically occurs within the mechanism's workspace.

If a mechanism approaches a singular position its transmission or gear ratio changes quickly until the mechanism is locked in the singular position. In the singular position, the mechanism's degrees of freedom change in an undesirable way. The mechanism runs into danger of damage or cannot be controlled any longer. Thus, possible singular positions have to be analyzed thoroughly during the design of a parallel haptic device. During operation, they have to be avoided by all means. How singular positions can be identified mathematically is discussed in Sect. 8.3.

If a serial, parallel, or hybrid mechanism is suitable for the design of a haptic interface, it should be decided on a case-by-case basis. All are used in haptic applications.

8.2 Design Step 1: Topological Synthesis—Defining the Mechanism's Structure

The topological synthesis is the first step in designing a haptic interface. It leads to the basic configuration of joints, rods, and actuators. While the basic structure of the haptic interface is defined in this step, the topological synthesis has be carried out thoroughly.

Topological synthesis should be based on an analysis of the specific task and the following issues must be addressed:

- *degrees of freedom*: In how many ↪ DoFs should the user interact with the haptic interface? Which ↪ DoFs are required (e.g., one pure rotatory as in a jog wheel, three to mimic spatial interaction or even six to display three translations and three rotations)?
- *adaption of existing structures*: Should the device adapt the structure of the task (e.g., a controlled robot) or of the user (e.g., the user's finger or arm)?
- *workspace*: How large is the desired workspace, the ↪ TCP has to move in? Are there any restrictions (e.g., areas of the workspace that should not be accessed)?
- *mobility*: Is the haptic interface designed as a device standing on a fixed place, e.g., on a table or is designed as a portable device?

The analysis of these requirements lays the foundation for the design of an easy-to-use and ergonomic haptic interface, which will be accepted by the user.

8.2.1 Synthesis of Serial Mechanisms

A serial mechanism is not less nor more than a sequence of rods and actuators, whereas the actuators can be regarded as driven joints. Whether the actuators are linear or rotary is of no importance for the complexity of the kinematic problem. For the workspace and the orientation of the tool-center-point, however, this aspect is of highest importance. A spacial serial mechanism with three rotatory drives changes the orientation of its ↪ TCP all over its workspace. If it is not intended to generate a torque as output to the user, the handle attached to this serial mechanism has to be

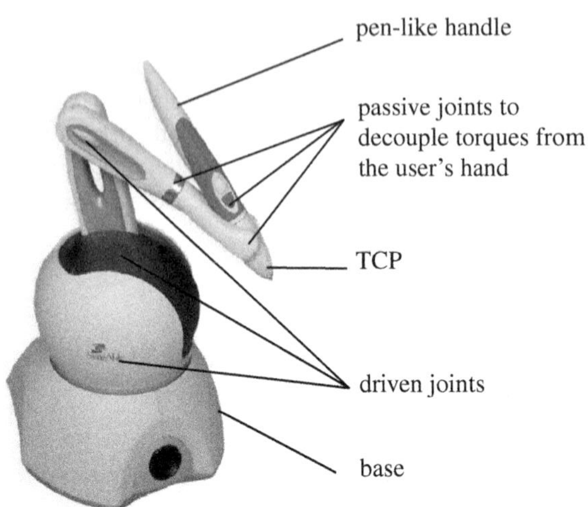

Fig. 8.5 The GEOMAGIC TOUCH haptic device is an example for a spacial working serial kinematic haptic device. The hand is decoupled from rotational movements by passive joints. Thus, no torques are induced to the hand

equipped with a passive universal joint. Such a realization as haptic device can be found in Fig. 8.5. Torques are decoupled from the hand. The handle does not have to be placed exactly in the ↪ TCP, as the moments are eliminated by the passive joints. Force vectors can be displaced arbitrarily within space. As a result, the hand experiences the same forces as the ↪ TCP.

8.2.2 Synthesis of Parallel Mechanisms

The synthesis of a parallel mechanism in general is a less intuitive process than the synthesis of a serial mechanism.

Since a parallel structure comprises several kinematic chains, the first step is to determine the required number of kinematic chains with respect to the desired degrees of freedom of the mechanism. This can be done using the ratio of the number of chains k and the degrees of freedom F of the mechanism leading to the degree of parallelism (see [4])

$$P_g = \frac{k}{F}. \tag{8.1}$$

A mechanism is partially parallel for $P_g < 1$, fully parallel for $P_g = 1$, and highly parallel for $P_g > 1$. Assuming the most common case of fully parallel mechanisms, this results in a number of kinematic chains—or "legs" of the mechanism—which is equal to the desired number of the mechanism's ↪ DoF.

The next step is to determine the number of joints ↪ DoF in the mechanism. This is done using the GRUEBLER-KUTZBACH-CHEBYCHEFF mobility criterion

$$F = \lambda \cdot (n - g - 1) + \sum_{i=1}^{g} f_i - f_{id} + s \tag{8.2}$$

with

F mechanism's ↪ DoF
n number of rods
g number of joints
f_i ↪ DoF of the ith joint
f_{id} number of identical links
s number of constrains
λ factor with $\lambda = 3$ for planar and $\lambda = 6$ for spatial mechanisms

An identical link is given for example when a rod has universal joints at both its ends. The rod will be able to rotate around its axis, without violating any constraints. Another example is two coaxial-oriented linear joints.

Constraints appear whenever conditions have to be fulfilled to enable the movement. If five joint axes have to be parallel to a 6th axis to enable a movement, then $s = 5$. Another example for a passive condition is two driving rods that have to be placed in parallel to enable a motion.

Applying Eq. (8.2) at this stage of design is usually not possible in the given form since the number of rods n and joints g is not known yet. However, it is possible to link $n - g$ with the already known number of kinematic chains k via

$$n = g - k + 2. \tag{8.3}$$

Assuming a spatial mechanism ($\lambda = 6$) then leads to

$$\sum_{i=1}^{g} f_i = F + 6 \cdot (k - 1). \tag{8.4}$$

for the sum of the joint ↪ DoFs that have to be distributed in the mechanism.

8.2.3 Special Case: Parallel Mechanisms with Pure Translational Motion

An important task of many haptic interfaces is the displaying of three-dimensional spatial sensation. An example is interaction with a pen-like tool where only forces in (x, y, z) should be displayed to the user.

A special class of 3-↪ DoF parallel mechanisms is used for these applications: ↪ Translational Parallel Machines (TPM). A ↪ TPM is a mechanism whose ↪ TCP can only move in three Cartesian coordinates (x, y, z). This is achieved by kinematic chains which block one or more rotatory ↪ DoFs of the ↪ TCP and are able to perform translational motion in all directions.

According to CARRICATO [1, 2], two restrictions have to be fulfilled to ensure a parallel kinematic mechanism with pure translational motion:

- ball joints shall not be used
- the rotatory axis of rotatory joints shall not be parallel to the axis of a degree of freedom which should be constrained

Neglecting overdetermined configurations, this results in so-called T_5-mechanisms, each comprising four or five rotatory joints. Each joint constrains the rotation of the ↪ TCP about one axis, defined by the unity vector $\mathbf{n}_i$ $(i = 1, 2, 3)$. To constrain a rotation about $\mathbf{n}_i$, all rotatory axes of a chain are orientated perpendicularly to $\mathbf{n}_i$.

There are two ways to design a T_5 mechanism. The first type is made of three T_5' chains, each having

- two rotatory joints following each other, with axis parallel to the unity vector $\mathbf{w}_{1i}$;
- two rotatory joints following each other, with axis parallel to the unity vector $\mathbf{w}_{2i}$ but not parallel to $\mathbf{w}_{1i}$;
- a prismatic joint at an arbitrary position in the chain or a fifth rotatory joint, parallel to one of its contiguous joints.

The second type is made of three T_5'' chains, having

- two rotatory joints with axis parallel to the unity vector $\mathbf{w}_{1i}$;
- two rotatory joints located between the first two rotatory joints with axis parallel to $\mathbf{w}_{2i}$ but not parallel to $\mathbf{w}_{1i}$;
- a prismatic joint at an arbitrary position in the chain or a fifth rotatory joint, parallel to one of its contiguous joints.

Figure 8.6 shows examples of these ↪ TPM chains with only 1-↪ DoF joints. This does not mean that ↪ TPMs are restricted to 1-↪ DoF. For instance can two adjoining and perpendicular rotatory joints be concentrated as a universal joint?

An important distinction between T_5' and T_5'' chains which is taken into account during design is the position of singular positions: Whereas in a T_5' mechanism, singular positions only occur at the edge of the workspace, T_5'' mechanisms can have singular positions within the workspace as well. Since a mechanism cannot pass through a singular position, this can lead to a split and therefore restricted usable workspace.

An exemplary topology synthesis is shown in the following example. However, one should keep in mind that topology synthesis is a process that requires some experience and cannot only be executed by application of straightforward design rules!

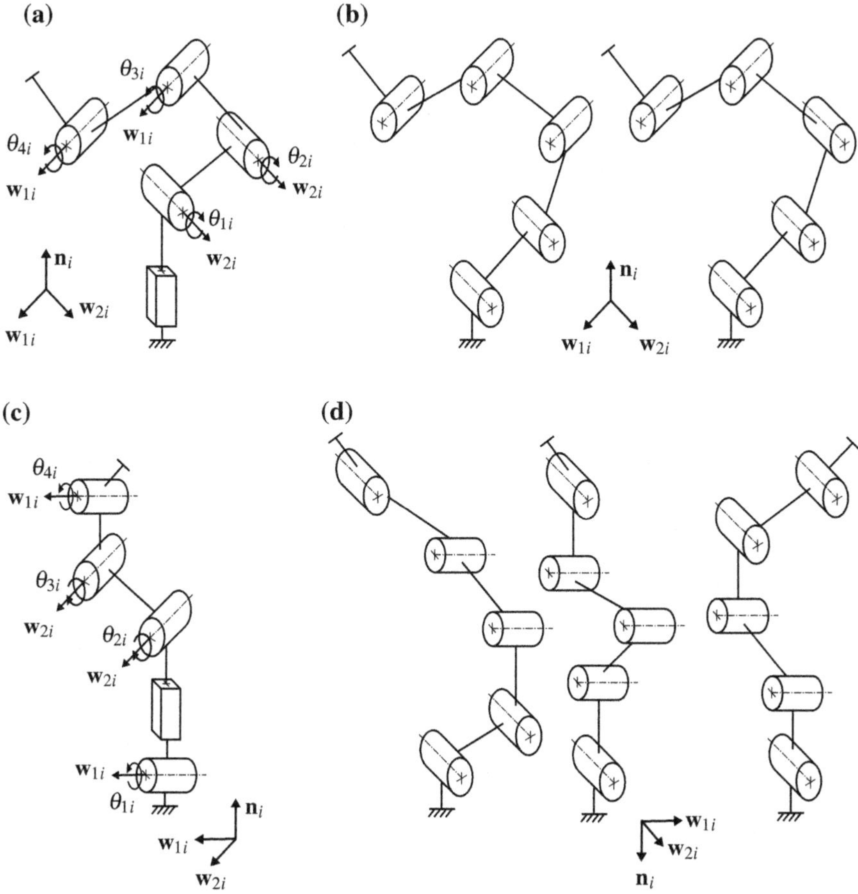

Fig. 8.6 Examples for ↪ TPM chains with five joint ↪ DoFs (based on [1])

8.2.4 Example: The DELTA Mechanism

One of the most common topologies to display spatial interaction is the parallel DELTA mechanism. Due to its relevance in the field of haptic interfaces, it is used as an example for topological synthesis.

Let us assume the design goal of a parallel kinematic haptic interface for spatial interaction in (x, y, z). Thus, a mechanism with three degrees of freedom is required. Using Eq. (8.1) for a fully parallel mechanism ($P_g = 1$) on $F = 3$ haptic degrees of freedom leads to a mechanism with $k = 3$ kinematic chains or legs.

In a second step, we have to determine the required joint degrees of freedom using GRUEBLER's formula (Eq. 8.4). This leads to the sum of $\sum_i f_i = 15$ joint degrees of freedom. Regarding an equal behavior in all spatial directions, it is self-evident to

Table 8.1 Topologies for 3-↪ DoF mechanisms with five ↪ DoF in each kinematic chain

Joints per chain	Topologies
1 × 1 DoF, 2 × 2 DoF	UUP, UPU, PUU, UUR, URU, RUU, CUP, CPU, CUR, CRU, RCU, UCP, UPC, PCU, UCR, URC, RUC, CCP, CPC, PCC, CCR, CRC, RCC
2 × 1 DoF, 1 × 3 DoF	SPP, SRR, SPR, SRP, PSP, RSP, PSR, RSR, PPS, RRS, RPS, PRS
3 × 1 DoF, 1 × 2 DoF	RRRU, RRUR, RURR, URRR, RRPU, RRUP, RURP, URRP, RPRU, RPUR, RUPR, URPR, PRRU, PRUR, PURR, UPRR, RPPU, RPUP, RUPP, URPP, PRPU, PRUP, PURP, UPRP, PPPU, PPUP, PUPP, UPPP, RRRC, RRCR, RCRR, CRRR, RRPC, RRCP, RCRP, CRRP, RPRC, RPCR, RCPR, CRPR, PRRC, PRCR, PCRR, CPRR, RPPC, RPCP, RCPP, PRPC, PCRP, PPPC, PPCP, PCPP, CPPP
5 × 1 DoF	32 iterations of P- and R-joints

distribute the 15 joint degrees of freedom with five degrees in each leg. This leads to the topologies in Table 8.1. The topologies are denominated according to the joints in one leg, e.g., a UUP mechanism comprises two universal and one prismatic joint.

The selection of an appropriate topology then can be carried on by a systematic reduction of the 3 ↪ DoF topologies in Table 8.1. The reduction is based on the following criteria:

- *Functionality as a* ↪ TPM: Criteria like the number of R-joints or the existence of an S-joint eliminate a large number of topologies.
- *Position of actors*: Rotatory, linear, or piston actors (e.g., in a hydraulic system) act as R-, P-, or C-joints. When having topologies with a U-joint attached to the base platform, this would lead to actors located within the kinematic chain. The required acceleration to move the actors' relatively high masses then would inhibit the dynamic advantages of a parallel mechanism to have the fullest effect.
- *Number of joints*: A concentration of two R-joints into one U-joint and an R- and P-joint into a C-joint, respectively, simplifies the mechanism's geometry and thereby its kinematic equations.

Table 8.2 shows the eliminated topologies. The remaining configurations are: UPU, PUU, CUR, CRU, RUU, and RUC.

Looking carefully at these topologies in Fig. 8.7 one recognizes that only RUU and RUC have rotatory joint attached to the base platform. Thus, these are the only two topologies that can be reasonably driven by a rotatory electrical motor. What makes the RUU- or DELTA mechanism special is that there are only joints with rotatory degrees of freedom within the kinematic chains. All forces and torques are converted into rotatory motion and there is no chance for the mechanism to cant. From a ↪ TPM point of view, the RUU/DELTA is a T_5'' mechanism having singular positions within the workspace. This has to be considered when dimensioning the mechanism.

The RUU/DELTA was introduced in 1988 by CLAVEL [3]. Besides acting as spatial haptic interface, the mechanism is widely used for robotic applications (e.g., pick-and-place tasks). Two popular examples are shown in Fig. 8.8.

Table 8.2 Eliminated topologies, sorted by the number of 1-, 2-, and 3-DoF joints in each leg

Elimination criterion	5 × 1 DoF	3 × 1 DoF, 1 × 2 DoF	2 × 1 DoF, 1 × 3 DoF	1 × 1 DoF, 2 × 2 DoF
No TPM	RRRPP, RRPRR, RRPPR, RRPPP, RPRRR, RPRRP, RPRPR, RPRPP, RPPRR, RPPRP, RPPPR, RPPPP, PRRRR, PRRRP, PRRPR, PRRPP, PRPRR, PRPRP, PRPPR, PPRRP, PPRPR, PPRPP, PPPRR, PPPRP, PPPPR, PPPPP, PRPPP	RPPU, RPUP, RUPP, URPU, PURP, UPRP, PPPU, PPUP, PUPP, UPPP, RRPC, RRCP, RCRP, CRRP, RPRC, RPCR, RCPR, CRPR, PRRC, PRCR, PCRR, CPRR, RPPC, RPCP, RCPP, CRPP, PRPC, PRCP, PCRP, CPRP, PPPC, PPCP, PCPP, CPPP	SPP, SRR, SPR, SRP, PSP, RSP, PSR, RSR, PPS, RRS, RPS, PRS	CUP, CPU, RCU, UCP, UPC, PCU, UCR, CCP, CPC, PCC, CCR, CRC, RCC
High number of joints	RRRRR, RRRRP, RRRPR, RRPRP, PPRRR	RRRU, RRUR, RURR, RRPU, RRUP, RURP, RPRU, RPUR, RUPR, PRRU, PRUR, PURR, UPRR, RRRC, RRCR RCRR, CRRR		
Base joint cannot be used as an actor		URRR, URRP, URPR		UUP, UUR, URU, URC

In these devices with mainly kinaesthetic feedback, a mechanical mechanism is used to link the user and the feedback generating actuators. Furthermore, the user's input commands are often given by moving a mechanical mechanism.

8.3 Design Step 2: Kinematic Equations

The kinematics of a mechanical mechanism describe its motion by means of position and orientation, speed, acceleration, and—of special importance for haptic interfaces—force and torques. The kinematic equations relate those measures at the input and output of a mechanism, typically at the base platform and the ↪ TCP. In other words: the kinematics represent the gearing properties of a mechanism.

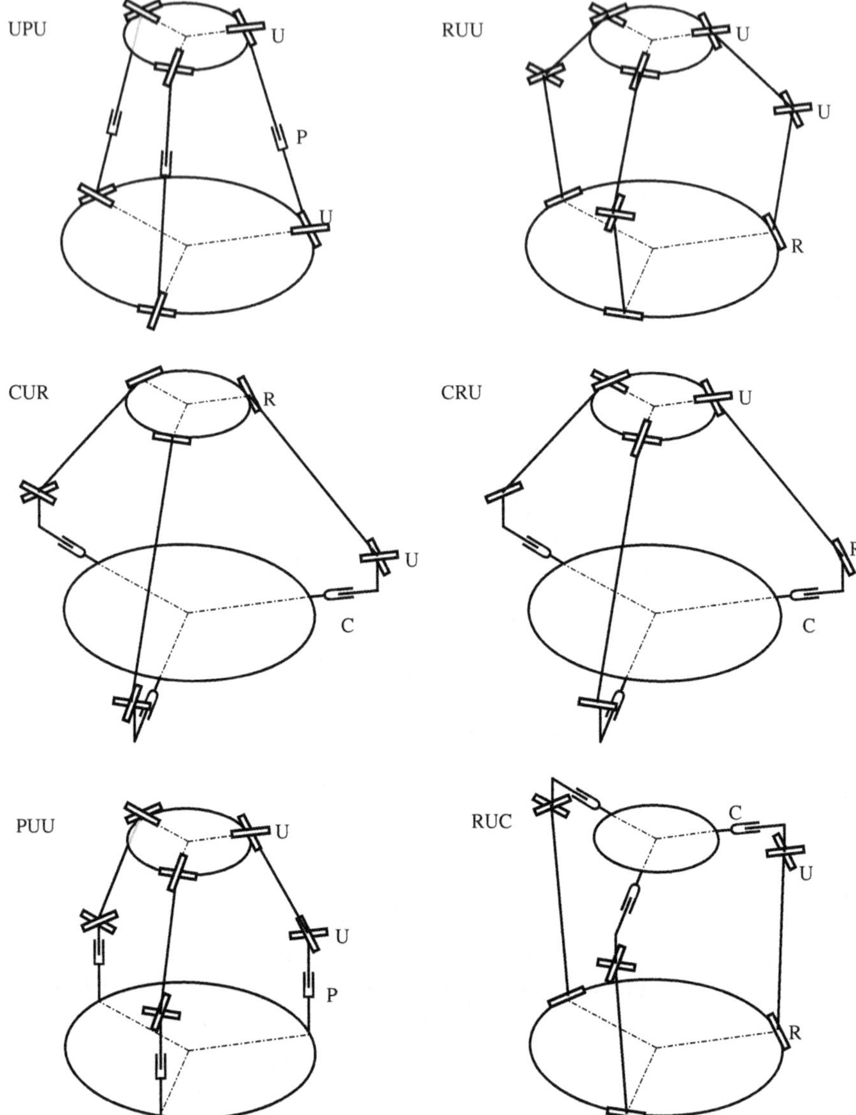

Fig. 8.7 Possible translational topologies with 3 DoF

They are of equal importance in the design and operation of a haptic device. This chapter gives an introduction to the basics of kinematic equations.

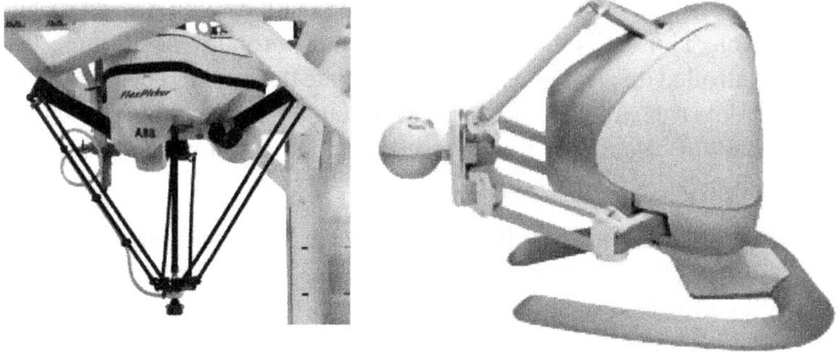

Fig. 8.8 *Left* FLEX PICKER for pick-and-place tasks (*source* ABB); *right* the FALCON as a 3-DoF haptic device (*source* Novint Technologies Inc.)

8.3.1 Kinematics: Basic Equations for Design and Operation

The transmission of motion in a mechanism can be described in two directions: from the actors to the ↪ TCP and vice versa. This leads to two basic kinds of kinematic equations: the *direct kinematic problem* or forward kinematics and the *inverse kinematic problem* or inverse kinematics.

Direct Kinematic Problem

The direct kinematic problems give a vector $\mathbf{x} = (x_1, x_2, ..., x_m)$ of ↪ TCP coordinates (position and orientation) with respect to a vector $\mathbf{q} = (q_1, q_2, ..., q_n)$ of actor coordinates by

$$\mathbf{x} = f(\mathbf{q}). \tag{8.5}$$

In contrast to serial mechanisms, for parallel mechanisms, the direct kinematic problem can only be solved numerically. However, there are exceptions as can bee seen later.

An important application of the direct kinematic problem is the calculation of an input command in impedance-controlled device. The users move the device, the mechanism's joint is detected and based; thereon the mechanism's ↪ TCP position is derived by Eq. (8.5).

Inverse Kinematic Problem

In the opposite direction, the transformation from ↪ TCP coordinates is given by the inverse kinematic problem

$$\mathbf{q} = f^{-1}(\mathbf{x}). \tag{8.6}$$

Equation (8.6) is used to determine required actor positions with respect to a desired ↪ TCP position. Thus, it is the essential equation in robotic positioning tasks. In admittance-controlled displays, it is used to calculate the required evasive movement in order to regulate a desired contact force between the user and the haptic interface.

The procedure of calculating the inverse kinematic problem can be split into the following three steps:

1. Formulation of closed vector chains for each leg, starting at the reference coordinate system enclosing the ↪ TCP coordinate system and going back to the reference coordinate system.
2. Splitting the vector chains in all—Cartesian—movement directions of the individual leg.
3. Solving the resulting system of equations according to the ↪ TCP coordinates.

In the design process, inverse kinematic problem the inverse kinematic problem can be used to derive the haptic interface's workspace, the space in which the user can operate the haptic device. This can be done using the fact that a point $\mathbf{x}$ is within the workspace if it yields a real solution for Eq. (8.6).

JACOBIan Matrix

In both the direct and inverse kinematic problems, the vectors $\mathbf{x}$ and $\mathbf{q}$ are linked via the mechanism's gearing properties. These properties are represented by the JACOBIan matrix $\mathbf{J}$. For the mechanism's kinematics, the JACOBIan matrix represents the transmission matrix of the first order. It carries all information regarding dimensions and transmission properties. $\mathbf{J}$ is defined by the partial derivative of the direct kinematic problem Eq. (8.5) with respect to the actor or joint coordinates $\mathbf{q}$.

From a mathematical perspective, the transformation is a mapping of the differentiable function $f : \mathbb{R}^n \to \mathbb{R}^m$, $n = 1 \dots 6$, $m = 1 \dots 6$ via a $n \times m$ matrix. It is

$$\mathbf{J}(\mathbf{q}) = \frac{\partial f}{\partial \mathbf{q}^{\mathrm{T}}} = \begin{pmatrix} \frac{\partial f_1}{\partial q_1} & \cdots & \frac{\partial f_1}{\partial q_n} \\ \vdots & \ddots & \vdots \\ \frac{\partial f_m}{\partial q_1} & \cdots & \frac{\partial f_m}{\partial q_n} \end{pmatrix}. \tag{8.7}$$

By its derivatives, the JACOBIan matrix gives correlation of speeds between actor and ↪ TCP coordinates.

Using the JACOBIan matrix the direct kinematic problem is expressed as

$$\mathrm{d}\mathbf{x} = \mathbf{J} \cdot \mathrm{d}\mathbf{q} \tag{8.8}$$

where the inverse kinematic problem is given by

$$\mathrm{d}\mathbf{q} = \mathbf{J}^{-1} \cdot \mathrm{d}\mathbf{x}. \tag{8.9}$$

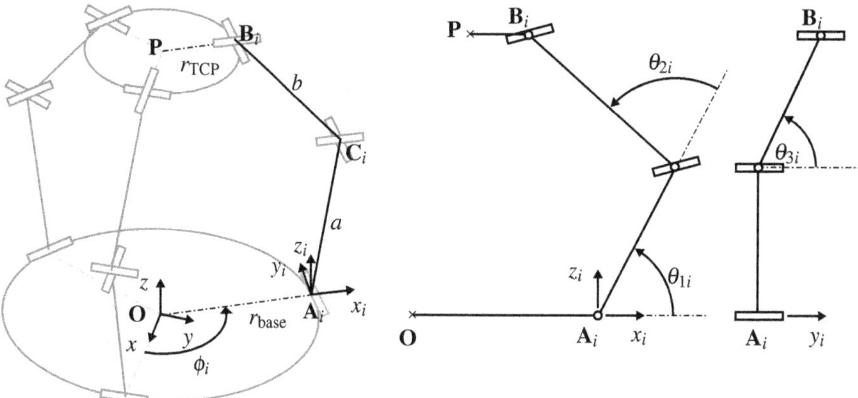

Fig. 8.9 Geometry of the DELTA mechanism (based on [9])

For the design and operation of haptic interfaces a third equation of high importance is the transformation of forces and torques by a mechanism. Again the JACOBIan matrix can be used. To display a force or torque vector $\mathbf{F}$ the required actor forces and torques $\boldsymbol{\tau}$ are given by

$$\boldsymbol{\tau} = \mathbf{J}^{\mathrm{T}} \cdot \mathbf{F}. \tag{8.10}$$

8.3.2 Example: The DELTA Mechanism

Figure 8.9 shows the dimensions and angles necessary to derive the mechanism's kinematic equations. It is desired to express all these equations with respect to the world coordinate system (WKS) in the middle of the base platform.

The x axis points toward the first leg. By means of simplification, we introduce a local coordinate system (x_i, y_i, z_i) in the start point $\mathbf{A}_i$ of the ith leg. The local coordinate system is rotated by $\phi_i = (i-1) \cdot 120$, $i = 1, 2, 3$ with respect to the WKS.

8.3.2.1 Direct Kinematic Problem

As mentioned above, the direct kinematic problem in general cannot be solved for parallel kinematic mechanisms. In case of the DELTA mechanism, it is different. Here the method of trilateration can be applied. This approach is based on the fact that—if looking at one leg—all points $\mathbf{B}_i$ are on the surface of a sphere with radius b and the center point $\mathbf{C}_i$. The surface is given by the sphere equation

$$\left(x - x_{C_i}\right)^2 + \left(y - y_{C_i}\right)^2 + \left(z - z_{C_i}\right)^2 = b^2 \tag{8.11}$$

with the center coordinates $\left(x_{C_i}, y_{C_i}, z_{C_i}\right)$ of the sphere. Assuming a leg's start point $\mathbf{A}_i'$ not in the distance r_{Basis} from the basis' origin but with the distance $(r_{\text{base}} - r_{\text{TCP}})$, all sphere surfaces of the three legs intersect in the point $\mathbf{P}$. With respect to the WKS, the assumed center point $\mathbf{C}_i'$ of the sphere is at

$$\mathbf{C}_i' = \begin{pmatrix} \cos(-\phi_i) & \sin(-\phi_i) & 0 \\ -\sin(-\phi_i) & \cos(-\phi_i) & 0 \\ 0 & 0 & 1 \end{pmatrix} \cdot \left[\begin{pmatrix} a \cdot \cos\theta_{1i} \\ 0 \\ a \cdot \sin\theta_{1i} \end{pmatrix} + \begin{pmatrix} r_{\text{base}} - r_{\text{TCP}} \\ 0 \\ 0 \end{pmatrix} \right] \tag{8.12}$$

The position of the ↪ TCP center point $\mathbf{P} = (x_P \; y_P \; z_P)^{\mathrm{T}}$, which is the solution to the direct kinematic problem, can be derived by the solution of the three sphere equations with the center points $\mathbf{C}_i'$

$$\left(x_P - x_{C_i'}\right)^2 + \left(y_P - y_{C_i'}\right)^2 + \left(z_P - z_{C_i'}\right)^2 = b^2 \tag{8.13}$$

Since the lower rod in point $\mathbf{A}_i'$, respectively, $\mathbf{A}_i$ rotates solely around the axis x_i axis, point $\mathbf{C}_i'$ is always within the x_i/z_i plane and for the y coordinate of $\mathbf{C}_i'$ one can write

$$y_{C_i'} = 0. \tag{8.14}$$

With respect to the WKS, the assumed center of the assumed sphere $\mathbf{C}_i'$ hence is given as

$$\mathbf{C}_i' = \underbrace{\begin{pmatrix} \cos(-\phi_i) & \sin(-\phi_i) & 0 \\ -\sin(-\phi_i) & \cos(-\phi_i) & 0 \\ 0 & 0 & 1 \end{pmatrix}}_{\text{rotational matrix}} \cdot \left[\begin{pmatrix} a \cdot \cos\theta_{1i} \\ 0 \\ a \cdot \sin\theta_{1i} \end{pmatrix} + \begin{pmatrix} r_{\text{base}} - r_{\text{TCP}} \\ 0 \\ 0 \end{pmatrix} \right] \tag{8.15}$$

The rotational matrix maps the coordinates of the local coordinate system in $\mathbf{A}_i$ respectively $\mathbf{A}_i'$ into the WKS. The rotation is done clockwise by $-\phi_i$.

This means that the equation system of the three sphere equations has to be solved:

$$\left(x_P - x_{C_1'}\right)^2 + \left(y_P - y_{C_1'}\right)^2 + \left(z_P - z_{C_1'}\right)^2 = b^2 \tag{8.16}$$

$$\left(x_P - x_{C_2'}\right)^2 + \left(y_P - y_{C_2'}\right)^2 + \left(z_P - z_{C_2'}\right)^2 = b^2 \tag{8.17}$$

$$\left(x_P - x_{C_3'}\right)^2 + \left(y_P - y_{C_3'}\right)^2 + \left(z_P - z_{C_3'}\right)^2 = b^2 \tag{8.18}$$

from (8.16)–(8.17) we get

$$x_P\left(-2x_{C_1'}+2x_{C_2'}\right)+y_P\cdot 2y_{C_2'}+z_P\left(-2z_{C_1'}+2z_{C_2'}\right)$$
$$=-x_{C_1'}^2+x_{C_2'}^2+y_{C_2'}^2-z_{C_1'}^2+z_{C_2'}^2, \tag{8.19}$$

from (8.16)–(8.18) we get

$$x_P\left(-2x_{C_1'}+2x_{C_3'}\right)+y_P\cdot 2y_{C_3'}+z_P\left(-2z_{C_1'}+2z_{C_3'}\right)$$
$$=-x_{C_1'}^2+x_{C_3'}^2+y_{C_3'}^2-z_{C_1'}^2+z_{C_3'}^2 \tag{8.20}$$

and from (8.17)–(8.18) we get

$$x_P\left(-2x_{C_2'}+2x_{C_3'}\right)+y_P\left(-2y_{C_2'}+2y_{C_3'}\right)+z_P\left(-2z_{C_2'}+2z_{C_3'}\right)$$
$$=-x_{C_2'}^2+x_{C_3'}^2-y_{C_2'}^2+y_{C_3'}^2-z_{C_2'}^2+z_{C_3'}^2 \tag{8.21}$$

The solution yields to two points of intersection of the spheres, whereas only one solution is geometrically meaningful. The calculation of $\mathbf{P}=(x_P\ y_P\ z_P)^{\mathrm{T}}$ should be computer-assisted , e.g., by using Mathematica®.

Inverse Kinematic Problem

The DELTA mechanism is especially known from impedance-controlled devices. In the mode of operation, the inverse kinematic problem is not needed. However, it is a useful tool in the design process to determine the available workspace which is shown later. Furthermore, it provides an effective way to determine the JACOBIan matrix of DELTA mechanism.

As mentioned above a standard approach to determine a mechanism's inverse kinematic is using closed vector chains. In the case on hand, this can be done via

$$\overrightarrow{\mathbf{OP}}+\overrightarrow{\mathbf{PB}_i}=\overrightarrow{\mathbf{OA}_i}+\overrightarrow{\mathbf{A}_i\mathbf{C}_i}+\overrightarrow{\mathbf{C}_i\mathbf{B}_i}. \tag{8.22}$$

This leads to the coordinates of the point $\mathbf{B}_i$ with

$$\begin{pmatrix} x_{B_i}\\ y_{B_i}\\ z_{B_i}\end{pmatrix}=\begin{pmatrix} a\cdot\cos\theta_{1i}+b\cdot\sin\theta_{3i}\cdot\cos(\theta_{1i}+\theta_{2i})\\ b\cdot\cos\theta_{3i}\\ a\cdot\sin\theta_{1i}+b\cdot\sin\theta_{3i}\cdot\sin(\theta_{1i}+\theta_{2i})\end{pmatrix} \tag{8.23}$$

Since we want to derive the mechanism's base angles with respect to the $\hookrightarrow$ TCP position in point $\mathbf{P}$, we can use the relation between $\mathbf{P}$ and $\mathbf{B}_i$ which is given by

$$\begin{pmatrix} x_{B_i}\\ y_{B_i}\\ z_{B_i}\end{pmatrix}=\begin{pmatrix}\cos\phi_i & \sin\phi_i & 0\\ -\sin\phi_i & \cos\phi_i & 0\\ 0 & 0 & 1\end{pmatrix}\cdot\begin{pmatrix} x_P\\ y_P\\ z_P\end{pmatrix}+\begin{pmatrix} r_{\mathrm{TCP}}-r_{\mathrm{base}}\\ 0\\ 0\end{pmatrix}. \tag{8.24}$$

By solving Eq. (8.23), we can determine the solution for the inverse kinmatic equations as given by TSAI [9] as

$$\theta_{3i} = \arccos \frac{y_{B_i}}{b} \quad (8.25)$$

$$\theta_{2i} = \arccos \frac{x_{B_i}^2 + y_{B_i}^2 + z_{B_i}^2 - a^2 - b^2}{2ab \sin \theta_{3i}} \quad (8.26)$$

$$\theta_{1i} = \arctan \frac{x_{B_i} - b \sin \theta_{3i} \cos(\theta_{1i} + \theta_{2i})}{z_{B_i} - b \sin \theta_{3i} \sin(\theta_{1i} + \theta_{2i})} \quad (8.27)$$

Especially Eq. (8.25) gives the angles of joints attached to the base platform with respect to the ↪ TCP position, and therefore, the solution for the inverse kinematic problem. Furthermore, the inverse JACOBIan matrix

$$\mathbf{J}^{-1} = \begin{pmatrix} j_{11} & j_{12} & j_{13} \\ j_{21} & j_{22} & j_{23} \\ j_{31} & j_{32} & j_{33} \end{pmatrix} \quad (8.28)$$

comprises the matrix elements [9]

$$j_{i1} = \frac{\cos(\theta_{1i} + \theta_{2i}) \sin \theta_{3i} \cos \phi_i - \cos \theta_{3i} \sin \phi_i}{a \sin \theta_{2i} \sin \theta_{3i}} \quad (8.29)$$

$$j_{i2} = \frac{\cos(\theta_{1i} + \theta_{2i}) \sin \theta_{3i} \sin \phi_i + \cos \theta_{3i} \sin \phi_i}{a \sin \theta_{2i} \sin \theta_{3i}} \quad (8.30)$$

$$j_{i3} = \frac{\sin(\theta_{1i} + \theta_{2i})}{a \sin \theta_{2i}}. \quad (8.31)$$

This closed-form solution provides an effective way to calculate the DELTA mechanism's JACOBIan matrix during operation.

8.4 Design Step 3: Dimensioning

The design step of dimensioning covers the optimization and determination of all designable lengths and angles within a topology that has been defined in step 1. Since especially the dimensioning of parallel kinematic mechanisms is a rather complex procedure, the following section focuses on this class of mechanisms.

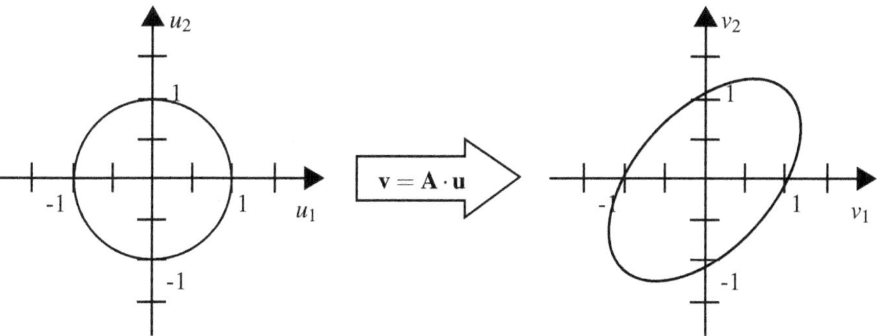

Fig. 8.10 Linear mapping of a vector, Example based on [5]

8.4.1 Isotropy and Singular Positions

As discussed in Sect. 8.1, parallel kinematic has a rather complex transmission behavior. Especially two effects have to be taken into account: isotropy and singular positions.

The key to analyze these effects in the design process is again based on properties of the JACOBIan matrix. A key performance index which is derived from the JACOBIan matrix properties is the condition number κ. It is introduced in the following section.

The Conditioning Number

The kinematic transmission behavior is rated by the singular values σ_i of the inverse JACOBIan matrix $\mathbf{J}^{-1}$. In general the singular values of a matrix $\mathbf{A}$ are defined as

$$\sigma_i(\mathbf{A}) = \sqrt{\lambda_i\left(\mathbf{A}^{\mathrm{T}}\mathbf{A}\right)}. \tag{8.32}$$

They are a measure for the distortion of the general linear projection of the vector $\mathbf{u}$ to $\mathbf{v}$ via

$$\mathbf{v} = \mathbf{A} \cdot \mathbf{u} \tag{8.33}$$

Figure 8.10 shows an example for this kind of distortion in a two-dimensional projection.

The role of the singular values can be shown by GOLUB's method of singular value decomposition. It is based on the fact that each complex $m \times n$-matrix $\mathbf{A}$ with the rank r can be fractioned in the product

$$\mathbf{A} = \mathbf{U} \cdot \mathbf{\Sigma} \cdot \mathbf{V}^* \tag{8.34}$$

of the unitary $m \times m$-matrix $\mathbf{U}$ and the adjoint matrix $\mathbf{V}^*$ of the $n \times n$-matrix $\mathbf{V}$. $\mathbf{\Sigma}$ is a $m \times n$- diagonal matrix with

$$\mathbf{\Sigma} = \left(\begin{array}{ccc|c} \sigma_1 & & & \vdots \\ & \ddots & & \cdots 0 \cdots \\ & & \sigma_r & \vdots \\ \hline & \vdots & & \\ \cdots & 0 & \cdots & \cdots 0 \cdots \\ & \vdots & & \vdots \end{array}\right) \tag{8.35}$$

and $\sigma_1 \geq \cdots \geq \sigma_r > 0$. In the linear projection, $\mathbf{U}$ and $\mathbf{V}^*$ act as rotations and $\mathbf{\Sigma}$ as elongation and compression of the ellipse from Fig. 8.10. σ_{min} and σ_{max} quantity the minimal and maximal amplifications of the vector $\mathbf{u}$.

A measure to rate the distortion is the conditioning number

$$\kappa = \frac{\sigma_{\text{max}}\left(\mathbf{J}^{-1}\right)}{\sigma_{\text{min}}\left(\mathbf{J}^{-1}\right)} \tag{8.36}$$

as the ratio of the two maximal singular values σ_{max} and σ_{min} of $\mathbf{J}^{-1}$. Thus, the conditioning number κ is a measure for the equal amplification $\mathbf{u}$ in all spatial directions. As a function of the JACOBIan matrix κ changes with respect to the mechanism's position. The conditioning number can reach values from $\frac{1}{\kappa} = 0 \ldots 1$.

Isotropic Transmission and Singular Positions

The goal of kinematic design is a highly isotropic transmission. From the distortion properties of singular values, the design target of

$$\frac{1}{\kappa} = 1 \tag{8.37}$$

for an isotropic transmission can be derived. On the other hand, one has to avoid singular behavior with

$$\frac{1}{\kappa} = 0. \tag{8.38}$$

In singular positions, the rank of the JACOBIan matrix decreases. This means that the transformation equations are no longer independent of each other. Practically, this leads to the loss of one or several controllable degrees of freedom.

For the two introduced kinds of singular positions (see Fig. 8.4), the loss of rank is characterized as

1. Singularity of the first kind: $\det(\mathbf{J}) = 0$
2. Singularity of the second kind: $\det\left(\mathbf{J}^{-1}\right) = 0$

For all conclusions drawn from the JACOBIan matrix, one has to take care of the used definition. In this book, we use the definition as in Eq. (8.7), but also the inverse definition

$$\mathbf{J}_{\text{alternative}}(\mathbf{x}) = \begin{pmatrix} \frac{\partial q_1}{\partial x_1} & \cdots & \frac{\partial q_1}{\partial x_n} \\ \vdots & \ddots & \vdots \\ \frac{\partial q_m}{\partial x_1} & \cdots & \frac{\partial q_m}{\partial x_n} \end{pmatrix} = \mathbf{J}^{-1} \tag{8.39}$$

is possible and used in the literature. However, given the fact that for singular values of a matrix $\mathbf{A}$ applies

$$\sigma(\mathbf{A}) = \frac{1}{\sigma\left(\mathbf{A}^{-1}\right)} \tag{8.40}$$

the derived conditions from the JACOBIan matrix can be transferred into each other.

Transmission of Force and Speed

Besides the analysis of isotropy the second aspect one has to take care of in the design process is the transmission of force and speed.

1. *Transmission of force*: To limit the maximal required force and torque and thereby limit also the size of the used actuators, it is important reach a good transmission of forces and torques even in cases of a disadvantageous scaling σ_i. From the transposed Eq. (8.10)

$$\mathbf{F} = \mathbf{J}^{\text{-T}} \cdot \boldsymbol{\tau} \tag{8.41}$$

   we can derive the criteria

$$\sigma_{\min}\left(\mathbf{J}^{\text{-T}}\right) \rightarrow \max \tag{8.42}$$

   and with $\sigma(\mathbf{A}) = \sigma\left(\mathbf{A}^{\mathrm{T}}\right)$ we get

$$\sigma_{\min}\left(\mathbf{J}^{-1}\right) \rightarrow \max \tag{8.43}$$

   as a design criterion for the force transmission.

   Therefore, we have to maximize the speed transmission for the most disadvantageous spatial direction. With the JACOBIan matrix, a measure for speed transmission

$$\dot{\mathbf{x}} = \mathbf{J} \cdot \dot{\mathbf{q}} \tag{8.44}$$

   we accordingly get

$$\sigma_{\min}(\mathbf{J}) \rightarrow \max \tag{8.45}$$

Table 8.3 Summarization of the experiments

Design aspect	Criterion
Force transmission	$\sigma_{\min}(\mathbf{J}) \to \max$
No singular positions	$\sigma_{\min}(\mathbf{J}) \to \max$
High stiffness	$\sigma_{\min}(\mathbf{J}) \to \max$
Speed transmission	$\sigma_{\max}(\mathbf{J}) \to \min$
Isotropy	$\frac{\sigma_{\min}(\mathbf{J})}{\sigma_{\max}(\mathbf{J})} \to \max$

as a design goal. Analogous to the design goal for the force transmission with Eq. (8.40) from

$$\begin{aligned}\sigma_{\min}(\mathbf{J}) &= \min\{\sigma_1(\mathbf{J}), \ldots, \sigma_r(\mathbf{J})\} \\ &= \frac{1}{\max\left\{\frac{1}{\sigma_1(\mathbf{J})}, \ldots, \frac{1}{\sigma_r(\mathbf{J})}\right\}} \\ &= \frac{1}{\max\left\{\sigma_1\left(\mathbf{J}^{-1}\right), \ldots, \sigma_r\left(\mathbf{J}^{-1}\right)\right\}} \\ &= \frac{1}{\sigma_{\max}\left(\mathbf{J}^{-1}\right)}\end{aligned} \tag{8.46}$$

we can derive the criterion

$$\sigma_{\max}\left(\mathbf{J}^{-1}\right) \to \min. \tag{8.47}$$

Table 8.3 sums up the various design aspects.

Looking at the definition of the conditioning number in Eq. (8.36), we see that the requirements for $\sigma_{\min}\left(\mathbf{J}^{-1}\right) \to \max$ are $\sigma_{\max}\left(\mathbf{J}^{-1}\right) \to \min$ are covered by one single value. Therein both requirements are weighted equally. Thus, the conditioning number $1/\kappa$ covers the evaluation of isotropy and of force and torque, respectively, speed at the same time.

One major drawback of Eq. (8.36) is that it rates the mechanism for JACOBIan matrix or position. The pure optimization of $1/\kappa$ would in fact lead to one single position where the mechanism reaches high isotropy. However, one cannot draw the conclusion that the whole workspace in total has an optimized transmission behavior.

What is needed is a measure to rate $1/\kappa$ of a whole workspace. This measure is provided by the global conditioning index (e.g., MERLET [7])

$$\nu = \frac{\int_W \frac{1}{\kappa} \mathrm{d}W}{\int_W \mathrm{d}W}. \tag{8.48}$$

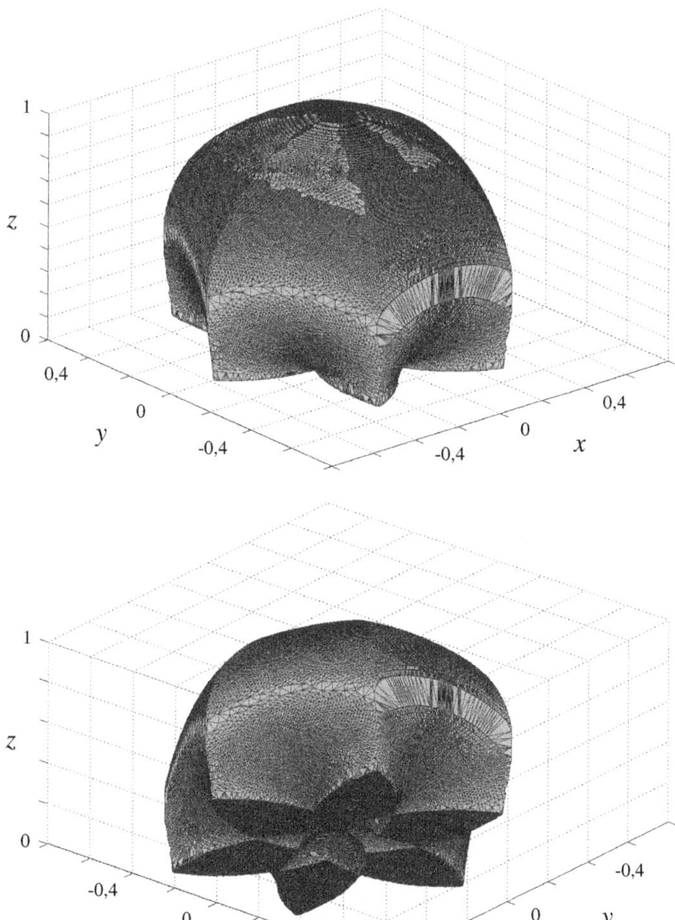

Fig. 8.11 Hull of DELTA's workspace from two angles

In a computer-assisted algorithm, it can be programmed in discrete form as

$$\nu = \frac{\mathrm{d}W \cdot \sum_n \frac{1}{\kappa}}{n \cdot \mathrm{d}W} \tag{8.49}$$

with n as the number of sampling points in the workspace and dW the size of a discretized voxel.

8.4.2 Example: The DELTA Mechanism

The parameters to be designed are the length of the rods a and b and the radii of the ↪ TCP platform r_{TCP} and of the base platform r_{base}. The design goal is a mechanism with a workspace of suitable size and isotropic transmission behavior.

This is done by an algorithm which is executed in two steps for each set of $a, b, r_{\mathrm{TCP}}, r_{\mathrm{base}}$

1. determine the workspace,
2. evaluate the global conditioning index.

The workspace is defined by the set of possible ↪ TCP positions. As mentioned above, a point is within this set if a real solution for the inverse kinematic problem exists. Using this criterion, the workspace can be determined using an algorithm that solves the inverse kinematic equations (8.25), (8.26), (8.27) pointwise in space. Simultaneously, the size of resulting workspace is determined by the number of points or voxels which fulfills the inverse kinematic equation. The shape of the DELTA's workspace obtained by this method is shown in Fig. 8.11.

To rate the mechanism's isotropy in each workspace, we use the global conditioning index as in Eq. (8.49). Figure 8.12 shows ν, calculated for rod lengths $a = 0.2 \ldots 0.8$. Cubic spaces have been discretized with voxels of $\mathrm{d}W = 0.001$. The result shows a global maximum at $a = 0.46$ and $b = 0.54$, respectively. This ratio yields a mechanism with a maximal global conditioning index.

A similar analysis shows that the variations of r_{TCP} and r_{base} have rather small influence on the global conditioning index and regarding the size of the workspace $r_{\mathrm{TCP}} = r_{\mathrm{base}} = 1$ turns out to be a good choice.

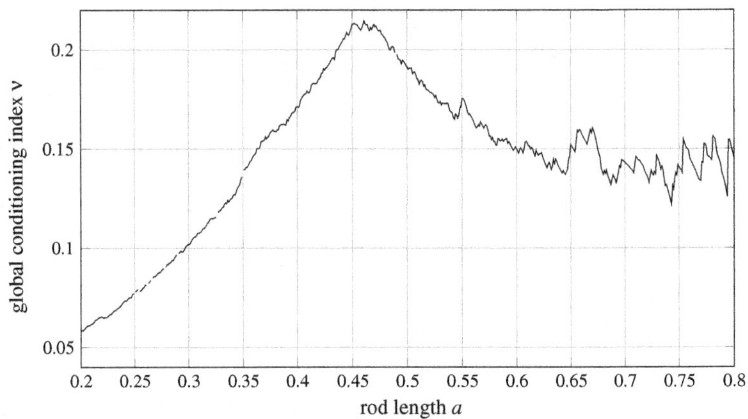

Fig. 8.12 Global conditioning index ν for varied rod lengths a and $b = 1 - a$ with $r_{\mathrm{TCP}} = r_{\mathrm{base}} = 1$ (all lengths without dimensions)

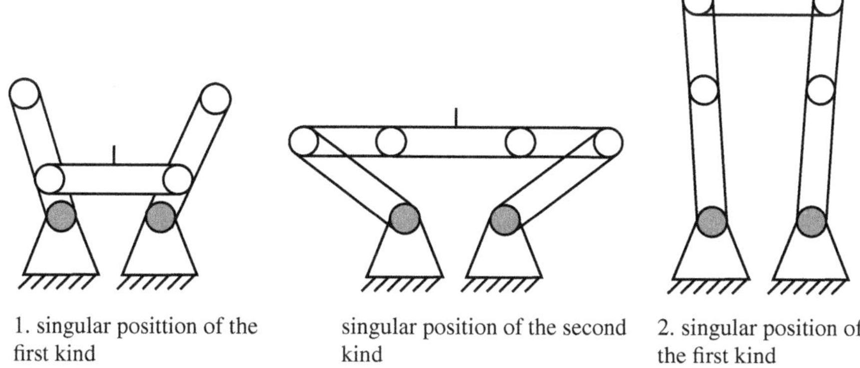

Fig. 8.13 Singular positions

As mentioned in the first part of the example in Sect. 8.2.4, we have to keep in mind that the RUU/DELTA mechanism is a T_5'' mechanism having singular positions within the workspace. Moving the ↪ TCP through its workspace this leads to the singular positions as shown in Fig. 8.13. Since a mechanism cannot cross singular positions—or in the case of a haptic device provide a feedback in singular positions—the workspace is divided by the second kind of singular position. The device is only operated in one part of the theoretically available workspace. In the shown case, this would be in the part above the second kind of singular position.

Recommended Background Reading

[6] Kong, X. & Gosselin, C.: **Synthesis of Parallel Mechanisms**. Springer, Germany, 2007.
Comprehensive description of the design of parallel kinematic mechanisms.

[7] Merlet, J.P.: **Parallel robots**. Springer, Netherlands, 2009.
Standard textbook on parallel mechanisms with a great variety of examples.

[8] Tsai, L.: **Robot analysis: the mechanics of serial and parallel manipulators**. John Wiley & Sons, Netherlands, 1999.
Standard textbook on parallel and serial mechanisms with deep discussion of mathematical backgrounds.

References

1. Carricato M, Parenti-Castelli V (2002) Singularity-free fully-isotropic translational parallel mechanisms. Int J Robot Res 21(2):161–174. doi:10.1177/027836402760475360

2. Carricato M, Parenti-Castelli V (2003) A family of 3-DOF translational parallel manipulators. J Mech Des 125:302. doi:10.1115/1.1563635
3. Clavel R (1988) DELTA, a fast robot with parallel geometry. In: 18th international symposium on industrial robots. pp 91–100
4. Husty M et al (1997) Kinematik und Robotik. Springer, Berlin. ISBN: 978-3-642-59029-0
5. Kirchner J (2001) Mehrkriterielle Optimierung von Parallelkinematiken. Dissertation. Technische Universität Chemnitz. ISBN: 3928921673
6. Kong X, Gosselin C (2007) Type synthesis of parallel mechanisms, vol 33. Springer Tracts in Advanced Robotics, Berlin
7. Merlet J (2006) Parallel robots. Springer, New York. ISBN 978-1402003851
8. Tsai L-W (1999) Robot analysis: the mechanics of serial and parallel manipulators. Wiley, New York. ISBN: 978-0471325932
9. Tsai L, Joshi S (2002) Kinematic analysis of 3-DOF position mechanisms for use in hybrid kinematic machines. J Mech Des 124:245–253. doi:10.1115/1.1468860

Chapter 9
Actuator Design

Henry Haus, Thorsten A. Kern, Marc Matysek and Stephanie Sindlinger

Abstract Actuators are the most important elements of every haptic device, as their selection, respectively, their design influences the quality of the haptic impression significantly. This chapter deals with frequently used actuators structured based on their physical working principle. It focuses on the electrodynamic, electromagnetic, electrostatic, and piezoelectric actuation principle. Each actuator type is discussed as to its most important physical basics, with examples of their dimensioning, and one or more applications given. Other rarely used actuation principles are mentioned in several examples. The previous chapters were focused on the basics of control-engineering and kinematic design. They covered topics of structuring and fundamental character. This and the following chapters deal with the design of technical components as parts of haptic devices. Experience teaches us that actuators for haptic applications can rarely be found "off-the-shelf". Their requirements always include some outstanding features in rotational frequency, power density, working point, or geometry. These specialties make it necessary and helpful for their applicants to be aware of the capabilities and possibilities of modifications of existing actuators. Hence, this chapter addresses both groups of readers: users who want to choose a certain actuator and the mechanical engineer who intends to design a specific actuator for a certain device from scratch.

H. Haus (✉)
Institute of Electromechanical Design, Technische Universität Darmstadt,
Merckstr. 25, 64283 Darmstadt, Germany
e-mail: h.haus@hapticdevices.eu

T.A. Kern
Continental Automotive GmbH, VDO-Straße 1, 64832 Babenhausen, Germany
e-mail: t.kern@hapticdevices.eu

M. Matysek
Continental Automotive GmbH, Babenhausen, Germany
e-mail: m.matysek@hapticdevices.eu

S. Sindlinger
Roche Diagnostics GmbH, Mannheim, Germany
e-mail: s.sindlinger@hapticdevices.eu

C. Hatzfeld and T.A. Kern (eds.), *Engineering Haptic Devices*,
Springer Series on Touch and Haptic Systems, DOI 10.1007/978-1-4471-6518-7_9

9.1 General Facts About Actuator Design

Thorsten A. Kern

Before a final selection of actuators is made, the appropriate kinematics and the control-engineering structure, according to the previous chapters, should have been fixed. However, in order to handle these questions in a reasonable way, some basic understanding of actuators is mandatory. In particular, the available energy densities, forces, and displacements should be estimated correctly for the intended haptic application. This section provides some suggestions and guidelines to help and preselect appropriate actuators based on the typical requirements.

9.1.1 Overview of Actuator Principles

There are a certain number of approaches to transform an arbitrary energy source into mechanical energy. Each of these ways is one actuation principle. The best-known and most frequently used principles are as follows:

Electrodynamic principle A force, so-called LORENTZ force, acting upon a conductor conducting a current.

Electromagnetic principle A force, acting upon a magnetic circuit to minimize the enclosed energy.

Piezoelectric principle A force, acting upon the atomic structure of a crystal and deforming it.

Capacitive principle A force, resulting from charges trying to minimize the energy in a capacitor.

Magnetorheological principle Viscosity change within a fluid resulting from particles trying to minimize the energy contained within a magnetic circuit.

Electrochemical principle Displacement of or pressure within a closed system, whereby a substance emits or bounds a gas and consequently changes its volume due to the application of electrical energy.

Thermal principle Change of length of a material due to controlled temperature changes, making use of the material's coefficient of thermal expansion.

Shape-memory alloy Sudden change of an object's shape made of special material due to relatively small temperature changes ($\approx$500 °C). The object transforms into a root shape embossed during manufacture by the application of high temperatures ($\approx$1,000 °C).

Each of these principles is used in different embodiments. They mainly differ in the exact effective direction of, e.g., a force vector[1] or a building principle.[2] As a

---

[1] The electromagnet principle for instance is divided into magnetic actuators and actuators based on the reluctance principle; the piezoelectric principle is subdivided into three versions depending on the relative orientation of electrical field and movement direction.

[2] E.g., resonance drives versus direct drives.

consequence, a widespread terminology exists for naming actuators. The major terms are given as follows:

Electric motor The most general term of all. It may describe any electromechanic transformer. However, in most cases, it refers to an actuator rotating continuously whose currents are commutated (mechanically or electronically), or which is equipped with a multiphase alternating current unit. Typically, it is a synchronous motor, a drive with a rotor moving synchronously to the rotating electromagnetic field. In a more general understanding, the term includes hysteresis motors and squirrel-cage rotors. The latter however has not yet reached any relevance for haptic systems, not even in very exceptional cases.

EC-motor Specific embodiment of the synchronous motor and common to haptic applications. Motor based on the electromagnetic or electrodynamic principle with an electronic control unit for the rotating field (electronic-commutated, electronically commutated).

DC-motor Another specific form of a synchronous motor and used among haptic applications because of its cheapness and simplicity. This is an actuator based on the electromagnetic or electrodynamic principle with a mechanical control unit for rotating field using switching contacts (mechanically commutated).

Resonance actuator Generic term for a whole class of actuators with different actuation principles. The term describes an actuator containing one component which is driven in its mechanical resonance mode (or nearby its resonance mode). Typically, parts of this component make an elliptic oscillation driving a second component in small steps via frictional coupling. As a result of the high frequency, the movement of the second component seems uninterrupted. The term is most frequently applied to piezoelectric actuators.

Ultrasonic actuator Resonance actuator performing steps at a frequency within ultrasonic ranges (>15 kHz). These actuators are built almost always based on the piezoelectric principle.

Voice-coil actuator Translational drive based on the electrodynamic principle. Mainly consisting of a conductor wrapped around a cylinder. The cylinder itself is placed in a magnetic circuit, resulting in a Lorentz force when a current is applied to the conductor. There are two major embodiments, one with a "moving coil", another variant with a "moving magnet".

Shaker Another form of a voice-coil actuator with an elastic suspension of the coil. When current is applied to the coil, an equilibrium condition between the suspension's spring and the Lorentz force is achieved at a specific displacement. Actuators based on this structure are frequently used for fast and dynamic movements of masses for vibration testing (this is where its name comes from).

Plunger-type magnet Actuator based on the electromagnetic principle. A rod made of ferromagnetic material is pulled into a magnetic circuit equipped with a coil. These actuators have nonlinear force–displacement characteristics.

Stepper motor Generic term for all actuation principles moving forward step-by-step. In contrast to the resonance drives, no component of a stepper motor is driven in resonance mode. Their step frequency is below any resonance of the system and may vary. These actuators may even be used in a "microstep mode", interpolating movement between so-called "full-steps", which are original to their mechanical design. The term is most frequently used for rotatory drives, for those working based on the reluctance principle or another electromagnetic actuation principle especially.

Pneumatic and Hydraulic These actuation principles do not have a direct electric input value. They transform pressure and volume flow into displacement and force. The media for pressure transmission is air in case of pneumatics and a fluid, typically oil, in case of hydraulics. Usually, the pressure itself is generated via actuators, e.g., electrical actuators attached to a compressor, and controlled via electrical valves.

Bending actuator Actuator with an active layer, frequently made of piezoelectric material—attached to a passive mechanical substrate. By actuating the active layer, mechanical tensions between this layer and the substrate build up, resulting in a bending movement of the whole actuator.

Piezoelectric stack A larger number of piezoelectric layers mechanically connected in series. Small displacements of each layer sum up to a large usable displacement of the whole actuator.

Piezoelectric motor/drive Generic term for all drives based on the piezoelectric principle. It frequently refers to drives moving a rotor or translator with frictional coupling. However, this movement does not need to happen in resonance mode.

Capacitive actuator Actuator based on the capacitive principle and frequently used in microtechnology. Usually, equipped with a comb-like structure of electrode pairs, generating forces in millinewton range with micrometers of displacement.

Shape-memory wire Wire on the basis of shape-memory alloys capable to shorten in the range of percents ($\approx 8\,\%$ of its total length) when changing its temperature (e.g., by controlling a current flowing through the wire. The current heats up the wire based on its thermal loss at the wire's electrical resistance).

Surface-wave actuators Generic term for a group of actuators generating high-frequency waves in mechanical structures or exciting the resonance modes of structures. This actuator is frequently based on piezoelectric principles and has been used for the generation of haptic textures for some years.

Each of the above actuation principles can be found in tactile and/or kinaesthetic systems. To simplify the decision process for a new design, all actuators can be grouped into classes. Most of the physical working principles can be grouped either into "self-locking" or "free-wheeling" systems. These groups are identical to:

- Positional sources (x), respectively, angular sources (α)
- Force sources (F), respectively, torque sources (M)

Table 9.1 Typical application areas for actuator principles in haptic systems

Control type		Admittance		Impedance	
Type	Actuator	Closed-l	Open-l	Open-l	Closed-l
Rotatory	Electric motor[a]	X	X	(X)[b]	–
Rotatory and translatory	EC-motor	–	–	X	X
Rotatory and translatory	DC-motor	–	–	X	X
Rotatory and translatory	Resonance actuator	X	X	(X)	–
Rotatory and translatory	Ultrasonic actuator	X	X	(X)	–
Translatory	Voice coil	–	–	X	X
Translatory	Shaker	X	X	–	–
Translatory	Plunger-type magnet	X	–	–	–
Rotatory (and translatory)	Stepper motor	X	X	–	–
Translatory (and rotatory)	Pneumatic	(X)	X	–	–
Translatory (and rotatory)	Hydraulic	–	X	–	–
Translatory	Bending actuator	–	X	–	–
Translatory	Piezo-stack	(X)	X	–	–
Translatory and rotatory	Piezo-actuator	X	X	X	–
Translatory	Capacitive	–	(X)	–	–
Translatory	Shape-memory	–	(X)	–	–
Translatory	Surface wave	–	(X)	–	–

X is frequently used by many groups and even commercialized
(X) Some designs, especially in research
– Very rare to almost none, and if it is used, it is only in the context of research
Type: Gives an idea about which actuator design (translatory or rotatory) is used more often. If the actuator is unusual but does exist, the marker is set into brackets
Annotations
[a]In the meaning of a mechanically commutated drive with a power between 10–100 W
[b]By high-frequency vibrations of the commutation

According to the basic structures of haptic systems (Chap. 6), it is likely that both classes are used within different haptic systems. The correlation between basic structures of haptic systems and actuators is depicted in Table 9.1. This table shows a tendency toward typical applications; however, by adding mechanical elements (springs, dampers), it is possible to use any actuator for any basic structure of haptic systems.

9.1.2 Actuator Selection Aid Based on Its Dynamics

Different actuator designs according to the same physical principle still cover wide performance ranges regarding their achievable displacements or forces. Based on the author's experience, these properties are put in relation to the dynamical ranges relevant for haptic applications. In Fig. 9.1, the most important actuation principles

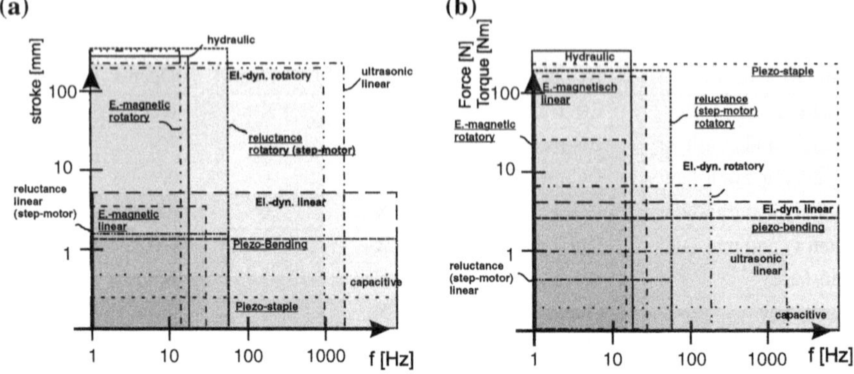

Fig. 9.1 Order of the actuator principles according to the achievable displacement (**a**) and forces respectively torques (**b**) in dependency from their dynamics. Further information can be found in [52]

are visualized in squares scaled according to their achievable displacements (a)[3] and typical output forces and torques (b). The area covered by a specific version of an actuator is typically smaller than the area shown here. The diagram should be read in such a way that, e.g., for haptic applications, electromagnetic linear actuators exist, providing a displacement up to 5 mm at ≈50 Hz. These designs are not necessarily the same actuators that are able to provide ≈200 N, as with electromagnetic systems, the effectively available force increases with smaller maximum displacement (Sect. 9.4). The diagrams in Fig. 9.1 visualize the bandwidths of realization possibilities according to a certain actuation principle and document the preferred dynamic range of their application. Using the diagrams, we have to keep in mind that the borders are fluent and have to be regarded in the context of the application and the actuator's individual design.

9.1.3 Gears

In general machine engineering, the use of gears is a matter of choice for adapting actuators to their load and vice versa. Gears are available in many variants. A simple lever can be a gear; complex kinematics according to Chap. 8 are a strongly nonlinear gear. For haptic applications, specialized gear designs are discussed for specific actuation principles in the corresponding chapters. However, there is one general aspect of the application of gears with relevance to the design of haptic systems, which has to be discussed independently: the scaling of impedances is only in the context of research.

[3] For continuously rotating principles, all displacements are regarded as unlimited.

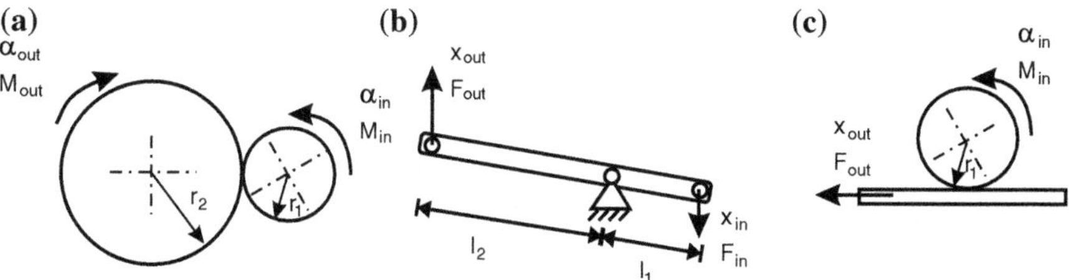

Fig. 9.2 Simple gear design with wheels (**a**), a lever (**b**) and a cable, rope, or belt (**c**)

There is no principal objection to the use of gears for the design of haptic systems. Each gear (Fig. 9.2) be it rotatory/rotatory (gearwheel or frictional wheel), translatory/translatory (lever with small displacements), rotatory/translatory (rope/cable/capstan) has a transmission "tr." This transmission ratio scales forces and torques neglecting loss due to friction according to

$$\frac{F_{\text{out}}}{F_{\text{in}}} = \text{tr} = \frac{l_2}{l_1}, \tag{9.1}$$

$$\frac{M_{\text{out}}}{M_{\text{in}}} = \text{tr} = \frac{r_2}{r_1}, \tag{9.2}$$

$$\frac{F_{\text{out}}}{M_{\text{in}}} = \text{tr} = \frac{1}{2\pi\, r_1}, \tag{9.3}$$

and displacements resp. angles according to

$$\frac{x_{\text{in}}}{x_{\text{out}}} = \text{tr} = \frac{l_2}{l_1}, \tag{9.4}$$

$$\frac{\alpha_{\text{in}}}{\alpha_{\text{out}}} = \text{tr} = \frac{r_2}{r_1}, \tag{9.5}$$

$$\frac{\alpha_{\text{in}}}{x_{\text{out}}} = \text{tr} = \frac{1}{2\pi\, r_1}. \tag{9.6}$$

The velocities and angular velocities scale analogously to the differential of the above equations. Assuming the impedance of the actuator $\underline{Z}_{\text{transl}} = \frac{\underline{F}}{\underline{v}}$ resp. $\underline{Z}_{\text{rot}} = \frac{\underline{M}}{\underline{\alpha}'}$, one consequence of the load condition of a driven impedance $\underline{Z}_{\text{out}}$ from the perspective of the motor is

$$\underline{Z}_{\text{transl}} = \frac{\underline{F}_{\text{in}}}{\underline{v}_{\text{in}}} = \frac{\underline{F}_{\text{out}}}{\underline{v}_{\text{out}}} \frac{1}{\text{tr}^2} = \underline{Z}_{\text{transl out}} \frac{1}{\text{tr}^2} \tag{9.7}$$

$$\underline{Z}_{\text{rot}} = \frac{\underline{M}_{\text{in}}}{\underline{\alpha}'} = \frac{\underline{M}_{\text{out}}}{\underline{\alpha}'} \frac{1}{\text{tr}^2} = \underline{Z}_{\text{rot out}} \frac{1}{\text{tr}^2} \tag{9.8}$$

The transmission ratio tr is quadratic for the calculation of impedances. From the perspective of an actuator, the driven impedance of a system gets small with a gear with a transmission ratio larger than one. This is favorable for the actuating system (and the original reason for the application of gears). For haptic applications, especially for impedance controlled ones, the opposite case has to be considered. In an idle situation and with a transmission ratio larger than one,[4] the perceived mechanical impedance of a system $\underline{Z}_{\text{out}}$ increases to the power of two with the transmission ratio. Another aspect makes this fact even more critical, as the increase in output force changes only in a linear way with the transmission ratio, whereas a motor's unwanted moment of inertia is felt to increase quadratically. This effect is obvious to anyone who has tried to rotate a gear motor with a high transmission ratio (e.g., tr >100) at its output. The inertia and the internal frictions upscaled by the gear are identical to a self-locking of the actuator.

As a consequence, the use of gears with force-controlled haptic systems makes sense only for transmission ratios of 1–20 (with emphasis on the lower transmission ratios between 3 and 6). For higher transmission ratios, designs according to Fig. 9.2c and Eq. (9.6) based on ropes or belts have proved valid. They are used in many commercial systems, as with the aid of the definition $\text{tr} = \frac{1}{2\pi\, r_1}$ and the included factor 2π, a comparably high transmission ratio can be achieved easily. In combination with rotatory actuators (typically EC-drives) with low internal impedances, this design shows impressive dynamic properties. Figure 9.3 shows an example for the application of such a gear to drive a delta mechanism [102].

Recently, a new type of gear came into view of several researchers [140]. The twisted-string actuator (TSA) is based on a relatively small motor with large rotation speed that twists a string or a set of strings. Because of the twisting, the strings contract and provide pulling forces in the range of several ten newtons that can be transferred via bowden cables. Applications include exoskeletons as presented in [103] and other devices that are weight sensitive.

Some advice may be given here out of practical experience: Wheel-based gears are applicable for haptic systems but tend to generate unsteady and waving output torques due to toothing. Careful mechanical design can reduce this unsteadiness. The mechanical backlash should be minimized (which is typically accompanied by an increase in friction), for example, by material combinations with at least one soft material. At least one gear should have spur/straight gearing, whereas the other can keep involute gearing.

[4] Which is the normal case, as typically the fast movement of an actuator is transmitted into a slower movement of the mechanism.

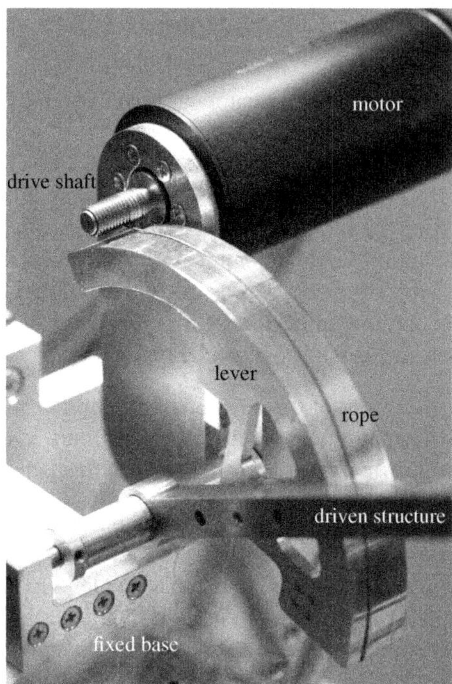

Fig. 9.3 Rope-based gear as widely used in haptic interfaces. The driven structure is connected to a lever on which the driving rope is running. The driving rope is wound around the driving shaft of the motor. The number of revolutions around the shaft is determined by the amount of torque to be transmitted via the gear; threads are used to minimize friction and wearout between individual turns of the driving rope

9.2 Electrodynamic Actuators

Thorsten A. Kern

Electrodynamic actuators are the most frequently used type of drives for haptic applications. This popularity is a result of the direct proportion between their output value (force or torque) from their input values (the electrical current). In case of kinaesthetic applications, they are typically used as open-loop controlled force sources. But even with tactile applications, these very dynamic actuators are frequently applied for oscillating excitements of skin areas. They are equally often used rotatory and translatory. Depending on the design either the electrical coil or the magnet is the moving component of the actuator. This section gives a short introduction to the mathematical basics of electrodynamic systems. Later, some design variants are discussed in detail. The final subsection deals with the drive electronics necessary to control electrodynamic systems.

9.2.1 The Electrodynamic Effect and Its Influencing Variables

Electrodynamic actuators are based on the LORENTZ force

$$\mathbf{F}_{\text{Lorentz}} = \mathbf{i} \cdot l \times \mathbf{B}, \tag{9.9}$$

acting upon moving charges in magnetic fields. The LORENTZ force is dependent on the current **i**, the magnetic induction **B** such as the length of the conductor l, which is typically formed as a coil. This section deals with optimization of each parameter for the maximization of the generated output force F_{Lorentz}. Any electrodynamic actuator is made of three components:

- generator of the magnetic field (coil or most frequently a permanent magnet);
- magnetic flux conductor (iron circuit, magnetic core);
- electrical conductor (frequently formed as coil or a more complex winding).

After a quick look, a recommendation for the maximization of the output force could be to simply increase the current **i** in the conductor. However, with the given limited available space for the conductor's length l (coil's cross section) and a flux density **B** with an upper border (0.8–1.4 T), the effectiveness of this change has to be put into question. This can be shown with a simple calculation example.

9.2.1.1 Efficiency Factor of Electrodynamic Actuators

For example, a straightforward design of an electrodynamic actuator similar to the actor in Fig. 9.4 is analyzed. It contains a wound coil with a permanent magnet in a

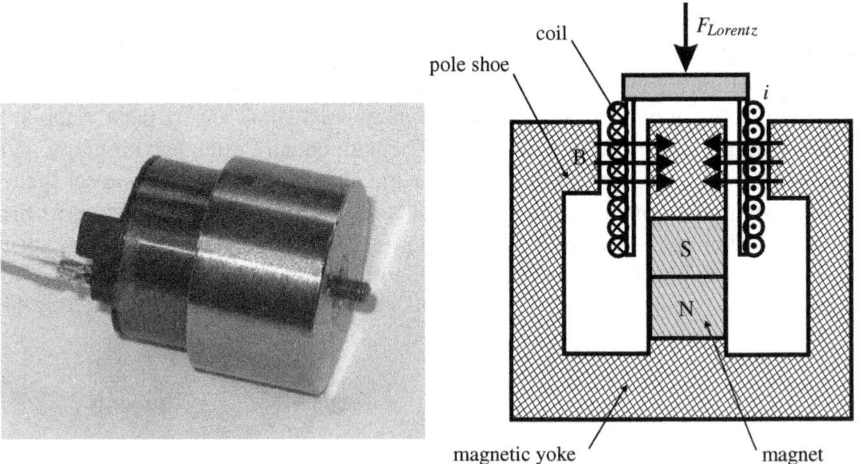

Fig. 9.4 Moving-coil actuator and corresponding functional elements

ferromagnetic core. The electrical power loss P_{el} of this electrodynamic system is generated mainly in a small moving coil with a pure ohmic resistance $R_{coil} = 3.5\,\Omega$ and a nominal current $i = 0.78$ A:

$$P_{el} = R_{coil}\, i^2 = 3.5\,\Omega \cdot 0.78\,\mathrm{A}^2 = 2.13\,\mathrm{W}. \tag{9.10}$$

With this electrical power loss, at flux density B = 1.2 T, with an orthogonal conductor orientation, and a conductor length within the air-gap $l = 1.58$ m, the actuator itself generates the force

$$F_{Lorentz} = i\, l\, B = 0.78\,\mathrm{A} \cdot 1.58\,\mathrm{m} \cdot 1.2\,\mathrm{T} = 1.48\,\mathrm{N}\,. \tag{9.11}$$

Assuming the system is driven in idle mode—working against the coil's own mass of $m = 8.8$ g only—being accelerated from idleness, and performing a displacement of $x = 10$ mm, above electrical power P_{el} is needed for a period of

$$t = \sqrt{2\frac{x}{a}} = \sqrt{2\frac{xm}{F}} = 0.011\,\mathrm{s} \tag{9.12}$$

seconds. The electrical energy loss sums up to

$$W_{el} = P_{el} \cdot t = 23,4\,\mathrm{mJ}. \tag{9.13}$$

This gives an efficiency factor of $\frac{W_{mech}}{W_{el}+W_{mech}} = 38\,\%$ for idle mode and continuous acceleration. Assuming now that the same actuator shall generate a force of 1 N against a fingertip for a period of, e.g., two seconds, an electrical power of $W_{el} = 2.13\,\mathrm{W} \cdot 2\mathrm{s} = 4.26\,\mathrm{J}$ is needed. This would be identical to an efficiency factor well below 1 %. And indeed, the efficiency factor of electrodynamic actuators in haptic applications lies in the area of low percentages due to the common requirement to generate static forces without much movement. This simple calculation points to one major challenge with electrodynamic actuators: The electrical power loss due to heat transmission extends the mechanically generated power by far. Consequently, during the design of electrodynamic actuators, attention has to be paid to the optimization of power within the given actuator volume and to the thermal management of energy losses.

9.2.1.2 Minimization of Power Loss

Typical designs of electrodynamic actuators either have a winded conductor, which by itself is self-supportive, or which is wound on a coil carrier (Fig. 9.5). The available space for the electrical coil within the homogeneous magnetic flux is limited (A_{Coil}). The number of coil turns $N_{Conductor}$ is also limited, too, within this area due to the cross-sectional surface; a single turn needs $A_{Conductor}$. This cross-sectional surface

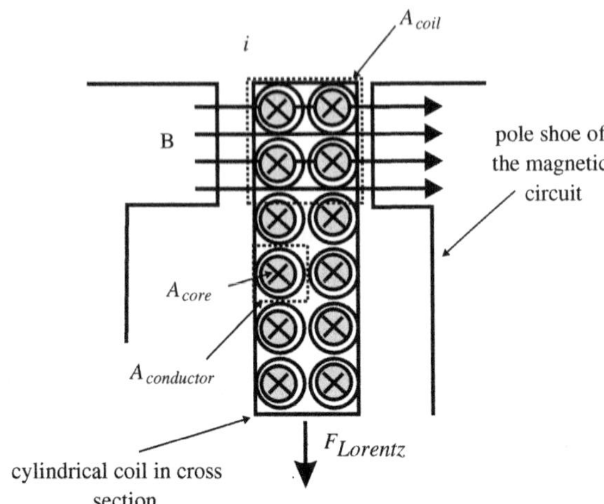

Fig. 9.5 Cross section through a cylindrical electrodynamic actuator based on the moving-coil principle

is typically more than the actual cross section of the conductor, as the winding will have gaps in between single turns [Eq. (9.15)]. Additionally, the actual conducting core with the cross-sectional surface A_{Core} will be smaller than the cross section of the conductor itself due to its isolation. Both parameters describing the geometrical losses in cross sections are available in tables of technical handbooks [95] and are assumed as factors $k \geq 1$ according to Eq. (9.14). The length l of the conductor can be easily calculated by multiplication with the number of turns and the mean circumference *Circ* of the coil [Eq. (9.16)].

The choice of the conductor's diameter influences the resistance of the coil via the conducting area A_{Core} . The specific length-based resistance R_{spezf} of a conductor is given according to Eq. (9.17). Big conducting diameters with large cross sections A_{Core} allow coils to conduct high currents at low voltages but—due to the limited volume available—few windings. Small diameters limit the necessary currents at high voltages and carry more windings. By a careful choice of wire diameter, the winding can be adjusted as a load to the corresponding source to drain the maximum available power.

The power loss P_{Loss} [Eq. (9.18)] acceptable within a given winding is limited. This limit is defined by the generated heat able to dissipate. The technical solutions are dependent on the time of continuous operation, the thermal capacity resulting from the volume of the actuator and the materials it consists of, and a potential cooling system. A calculation of heat transmission is specific to the technical solution and cannot be solved in general within this book. However, the dependency of LORENTZ-force on power loss can be formulated as

$$A_{\text{Conductor}} = k \cdot A_{\text{Core}} \tag{9.14}$$

$$N_{\text{Conductor}} = \frac{A_{\text{Coil}}}{A_{\text{Conductor}}} \tag{9.15}$$

$$l_{\text{Conductor}} = N_{\text{Conductor}} \cdot \text{Circ} \tag{9.16}$$

$$R_{\text{spezf.}} = \frac{l_{\text{Conductor}}\, \rho}{A_{\text{Conductor}}} \tag{9.17}$$

$$P_{\text{Loss}} = i^2 \cdot R_{\text{Coil}} \tag{9.18}$$

From Eq. (9.18) follows

$$i = \sqrt{\frac{P_{\text{Loss}}}{R_{\text{Coil}}}} \tag{9.19}$$

With Eq. (9.17) there is

$$i = \sqrt{\frac{P_{\text{Loss}}\, A_{\text{Core}}}{\rho\, l_{\text{Conductor}}}} \tag{9.20}$$

put into Eq. (9.9) (keeping the direction of current flow $\mathbf{e}_i$) there is

$$F_{\text{Lorenz}} = \sqrt{\frac{P_{\text{Loss}}\, A_{\text{Core}}\, l_{\text{Conductor}}}{\rho}}\, \mathbf{e}_i \times \mathbf{B} \tag{9.21}$$

by considering Eqs. (9.15)–(9.16) the result is

$$F_{\text{Lorenz}} = \sqrt{\frac{P_{\text{Loss}}\, A_{\text{Coil}}\, N\, \text{Circ}}{\rho\, k}}\, \mathbf{e}_i \times \mathbf{B} \tag{9.22}$$

Equations (9.15)–(9.18) put into Eq. (9.9) give a precise picture of the influence values on the LORENTZ force [Eq. (9.22)]. The level of Lorentz force is given by the power loss P_{Loss} acceptable within the coil. If there is room for modifications to the geometrical design of the actuator, the cross-sectional area of the coil and the circumference of the winding should be maximized. Additionally, a choice of alternative materials (e.g., alloy instead of copper) may minimize the electrical resistance. Furthermore, the filling factor k should be reduced. One approach could be the use of wires with a rectangular cross-section to avoid empty spaces between the single turns.

The question for the maximum current itself is only relevant in combination with the voltage available and in the context of adjusting the electrical load to a specific electrical source. In this case for i_{Source} and u_{Source}, the corresponding coil resistance has to be chosen based on Eq. (9.23).

$$
\begin{aligned}
P_{\text{Source}} &= u_{\text{Source}} \cdot i_{\text{Source}} = i_{\text{Source}}^2 \cdot R_{\text{Coil}} \\
R_{\text{Coil}} &= \frac{P_{\text{Source}}}{i_{\text{Source}}^2}
\end{aligned}
\tag{9.23}
$$

Surprisingly, from the perspective of a realistic design, an increase in current is not necessarily the preferred option to increase the LORENTZ force according to Eq. (9.22). The possibility to optimize P_{Loss} by adding cooling or to analyze the temporal pattern of on- and off-times is much more relevant. Additionally, the flux density **B** has—compared to all other influence factors—quadratically higher influence on the maximum force.

9.2.1.3 Maximization of the Magnetic Flux Density

For the optimization of electrodynamic actuators, a maximization of the flux density **B** is necessary within the area where the conducting coils are located. This place is called air-gap and resembles an interruption of the otherwise closed ferromagnetic core conducting the magnetic flux. The heights of the magnetic flux density are influenced by

- the choice of ferromagnetic material for the magnetic core,
- the field winding/exciter winding of the static magnetic field, and
- the geometrical design of the magnetic core.

In the context of this book, some basic design criteria for magnetic circuits are given. For an advanced discussion and optimization, process source [66] is recommended.

Basics for the Calculation of Magnetic Circuits

Calculating magnetic circuits show several parallels to the calculation of electrical networks. As shown in Table 9.2 several analogies between electrical and magnetic variables can be defined.

The direct analogy to the magnetic flux ϕ is the electrical charge Q. For the application of equations, however, it is helpful to regard electrical currents I as a counterpart to the magnetic flux. Please note that this is an aid for thinking and not a mathematical reality, although it is very common. The actual direct analogy for the current I would be a time-dependent magnetic flux $\frac{\mathrm{d}\phi}{\mathrm{d}t}$, which is usually not defined with an own variable name. The exception with this model is the magnetomotive force Θ, which resembles the sum of all magnetic voltages V identical to a rotation within an electrical network. Nevertheless, it is treated differently, as many applications require generating a magnetomotive force Θ by a certain number of winding turns N and a current I, often referred to as ampere turns. The coupling between field and flux variables is given by the permittivity ϵ in case of electrical values and by the permeability μ in case of magnetic values. It is obvious that the field constants

Table 9.2 Analogies between electric and magnetic values

Description	Electric	Magnetic
Flux Differential flux	Charge Q ($C = \text{As}$) $I = \frac{dQ}{dt}$ (A)	Magnetic flux ϕ (Vs)
Flux value	Dielectric flux density D (C/m^2) $Q = \int_A \mathbf{D}\mathrm{d}\mathbf{A}$ Current density J (A/m^2) $I = \int_A \mathbf{J}\mathrm{d}\mathbf{A}$	Flux density B ($\text{T} = \text{Vs/m}^2$) $\phi = \int_A \mathbf{B}\mathrm{d}\mathbf{A}$
Electromagnetic coupling formerly:	Voltage U (V) Electromotive force	Flux/ampere turns Θ (A) Magnetomotive force
Induction laws	$U = -N\frac{d\phi}{dt}$	$\Theta = N\frac{dQ}{dt}$ $\Theta = N\,I$ (N = turns)
Field values	El. field strength E (V/m)	Magn. field strength H (A/m)
Differential values	Voltage U (V) $U = \int_a^b \mathbf{E}\mathrm{d}\mathbf{s}$	Magnetic voltage V (A) $V = \int_a^b \mathbf{H}\mathrm{d}\mathbf{l}$
Mesh equations	$U_{\text{ges}} = \sum_i U_i$	$\Theta = \sum_i V_i$
Resistances	El. resistance R (Ω) $R = \frac{U}{I}$	Magn. resistance R_m (A/Vs) reluctance $R_m = \frac{V}{\phi}$
Coupling factors	Permittivity $\epsilon = \epsilon_0\,\epsilon_r$ ($\epsilon_0 = 8{,}854 \times 10^{-12}$ C/Vm)	Permeability $\mu = \mu_0\,\mu_r$ ($\mu_0 = 1{,}256 \times 10^{-6}$ Vs/Am)
Coupling between field and flux values	$\mathbf{D} = \epsilon\,\mathbf{E}$	$\mathbf{B} = \mu\,\mathbf{H}$
Power (W)	$P_{\text{el}} = U \cdot I$	
Energy (J)	$W_{\text{el}} = P_{\text{el}}\,t$	$W_{\text{mag}} = \phi\,V$ $W_{\text{mag}} = \sum_n H_n\,l_n \cdot B_n\,A_n$

ϵ_0 differ from μ_0 by the factor 10^6. This is the main reason for the electromagnetic effect being the preferred physical realization of actuators in macroscopic systems.[5]

However, there is another specialty with the field constants. The electrical permittivity can be regarded as constant (Sect. 9.5) even for complex actuator designs and can be approximated as linear around an operating point. The permeability μ_r of typical flux-conducting materials, however, shows a strong nonlinear relationship; the materials are reaching saturation. The level of magnetic flux has to be limited to prevent saturation effects in the design of magnetic core.

[5] In micromechanical systems, the energy density relative to the volume becomes more important. The manufacture of miniaturized plates for capacitive actuators is easier to realize with batch processes than the manufacture of miniaturized magnetic circuits.

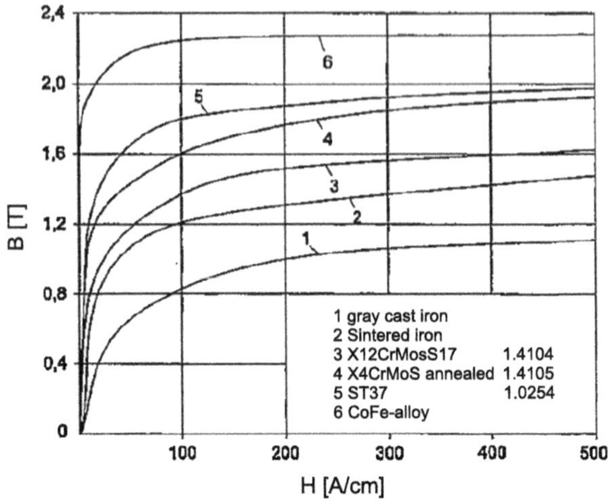

Fig. 9.6 Saturation curve of typical magnetic materials [66]

Magnetic Circuits

For maximization of the magnetic flux density, it is necessary to either analyze the magnetic circuit mathematically, analytically, and/or do a numerical simulation of it. For the simulation of magnetic fields, common CAD and FEM products are available.[6] For classification of the mathematical problem, three solution levels exist: stationary, quasistationary, and dynamic magnetic fields. With stationary magnetic fields, there is no time-dependent change in the magnetic circuit. A steady state of flux density is assumed. With quasistationary field, the induction considered results from changes within the current generating the magnetic field or a linearized change within the geometry of the magnetic circuit (e.g., a movement of an anchor). Dynamic magnetic fields consider additional effects covering the dynamic properties of moving mechanical components up to the change of the geometry of the magnetic circuit and the air-gaps during operation. Dealing with electrodynamic actuators, the analysis of static magnetic circuits is sufficient for first dimensioning. The relevant dynamic drawbacks for electrodynamic actuators are presented in Sect. 9.2.1.4.

There are two principal possibilities to generate magnetic flux densities within the volume of the conducting coil:

1. Generation via winded conductors with another coil (exciter winding);
2. Generation via a permanent magnet.

Both approaches show specific pros and cons: With a winded conductor, the flux density $B = \mu\,(NI - H_{Fe}\,l_{Fe})$ can be raised without any theoretical limit. In practical

---

[6] From the very beginning, several free or open software projects are available for electrical and magnetic field simulation, e.g., for rotatory or planar systems, a program by David Meeker named "FEMM".

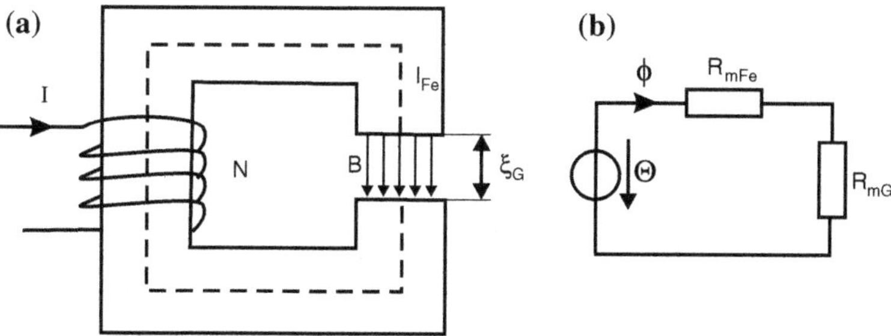

Fig. 9.7 Magnetic field generation B via a current-conducting coil with N turns (**a**) and derived equivalent circuit representation (**b**)

applications, the flux-conducting material will reach saturation (Fig. 9.6) actually limiting the achievable maximum flux density. Additionally, the ohmic resistance of the winding will generate electrical power losses, which will have to be dissipated in addition to the losses resulting from the electrodynamic principle itself (Sect. 9.2.1.1). Abandoning any flux-conducting material and using exciter windings with extremely low electrical resistance, extraordinarily high field densities can be reached.[7] Till date, such a technological effort for haptic devices is yet to be made.

Building a magnetic circuit with a permanent magnet, the practical border for the flux density is given by the remanence flux density B_r of the magnetic material. Such a magnet can be compared to a source providing a certain magnetic power. The flux density—being the relevant quality for electrodynamic actuators—is not independent of the magnetic load attached to the permanent magnet. Additionally, the relevant properties of the magnetic material are temperature dependent, and wrong use of specific magnet materials may harm its magnetic properties.[8]

Nevertheless, modern permanent magnetic materials made of "rare earths" are the preferred source to generate static magnetic fields for electrodynamic actuators. The following section gives some basics on the calculation for simple magnetic circuits. In extension to what is shown here, a more precise analytical calculation is possible [66]. However, it is recommended to use simulation tools early within the design process. Leakage fields are a great challenge in the design of magnetic circuits. Especially, beginners should develop a feeling for the look of these fields with the aid of simulation tools.

---

[7] MRI systems for medical imaging generate field densities of 2 T and more within air-gaps of up to 1 m diameter by use of supraconducting coils and almost no magnetic circuit at all.

[8] E.g., when removing AlNiCo magnets out of their magnetic circuit after magnetization, they may drop below their coercive field strength actually losing performance.

Direct Current Magnetic Field

Figure 9.7a shows a magnetic circuit of iron with a cross section A and an air-gap of length ξ_G ($G =$ Gap). The magnetic circuit has a winding with N turns conducting a current I. The medium length of the magnetic circuit is l_{Fe}. For calculation, the circuit can be transformed into a magnetic equivalent network (Fig. 9.7b). According to the analogies defined in Table 9.2, the magnetic induction generates a magnetomotive force Θ as a differential value. In combination with two magnetic resistances of the iron circuit R_{mFe} and the air-gap R_{mG}, a magnetic flux ϕ can be identified.

For the calculation of flux density B in the air-gap, it is assumed that this magnetic flux ϕ is identical to the flux within the iron part of the circuit. Leakage fields are disregarded in this example.[9]

$$B = \frac{\phi}{A}$$

The magnetic resistance of materials and surfaces is dependent on the geometry and can be found in special tables [66]. For magnetic resistance of a cylinder of length l and diameter d, a resistance according to Eq. (9.24) is given.

$$R_m = \frac{4\,l}{\mu\,\pi\,d^2} \tag{9.24}$$

For the magnetic circuit, the magnetic resistances $R_{m\text{Fe}}$ and $R_{m\text{G}}$ can be regarded as known or at least calculable. The magnetic flux is given by

$$\phi = \frac{\Theta}{R_{m\text{Fe}} + R_{m\text{G}}}, \tag{9.25}$$

and the flux density by

$$B = \frac{\Theta}{(R_{m\text{Fe}} + R_{m\text{G}})\,A}. \tag{9.26}$$

Using this procedure, a clever approximation of the magnetic resistances of any complex network of magnetic circuits can be made. In this specific case of a simple horseshoe-formed magnet, an alternative approach can be chosen. Assuming that the magnetic flux density in the air-gap is identical to the flux density in the iron (no leakage fields, see above) the flux density B is given by:

$$B = \mu_0 \mu_r\, H \tag{9.27}$$

Assuming that μ_r is given either as a factor or with a characteristic curve (like in Fig. 9.6) only the magnetomotive force Θ within the iron has to be calculated. With

[9] Considering leakage fields would be identical to a parallel connection of additional magnetic resistors to the resistance of the air-gap.

Table 9.3 Magnetic properties of permanent magnet materials [66]

Material	B_r (T)	H_{cB} (kA/m)	(BH)$_{max}$ (kJ/m^3)
AlNiCo (isotropic)	0.5–0.9	10–100	3–20
AlNiCo (anisotropic)	0.8–1.3	50–150	30–70
Hard ferrite (isotropic)	0.2–0.25	120–140	7–9
Hard ferrite (anisotropic)	0.36–0.41	170–270	25–32
SmCo (anisotropic)	0.8–1.12	650–820	160–260
NdFeB (anisotropic)	1.0–1.47	790–1100	200–415

$$\Theta = H_{\text{Fe}}\, l_{\text{Fe}} + H_{\text{G}}\, \xi_G = \frac{B}{\mu_0 \mu_r}\, l_{\text{Fe}} + \frac{B}{\mu_0}\, \xi_G \tag{9.28}$$

the flux density

$$B = \Theta \, \frac{1}{\frac{l_{\text{Fe}}}{\mu_0 \mu_r} + \frac{\xi_G}{\mu_0}}, \tag{9.29}$$

results and can be written down immediately.

Permanent Magnets Generating the Magnetic Field

As stated earlier, the typical approach to generate the magnetic field within an electrodynamic actuator is the choice of a permanent magnet. Permanent magnets cannot simply be regarded as flux or field sources. Therefore, some basic understanding of magnet technology is necessary.

As a simple approach, a magnet is a source of energy that is proportional to the volume of the magnet. Magnets are made of different magnetic materials (Table 9.3) differing in the maximum achievable flux density [remanence flux density B_r), maximum field strengths (coercive field strength H_{cB} and H_{cJ})], and their energy density BH_{max}, such as the temperature coefficient. Additionally, identical materials are differentiated based on being isotropic or anisotropic. With isotropic magnets, the substance is made of homogeneous material, which can be magnetized in one preferred direction. With anisotropic material, a magnetic powder is mixed with a binding material (e.g., epoxy) and formed via a casting or injection-molding process. The latter approach enables almost unlimited freedom for the magnet's geometry and a large influence on the pole distribution on the magnet. However, anisotropic magnets are characterized by slightly worse characteristic values in energy density such as maximum field strengths and flux densities.

Figure 9.8 shows the second quadrant of the $B-H$-characteristic curve (only this quadrant is relevant for an application of a magnet within an actuator) of different magnetic materials. The remanence flux density B_r equals the flux density with short-circuit pole shoes (a magnet surrounded by ideal iron as magnetic circuit). When there is an air-gap within the magnetic circuit (or even by the magnetic resistance of the real magnetic circuit material itself), a magnetic field strength H appears as a load. As a reaction, an operation point is reached, which is shown here as an example on a curve of NdFeB for a flux density of about 200 kA/m. The actually available flux density at the poles is decreased accordingly. As electrodynamic actuators for haptic applications face high requirements according to their energy density, there are almost no alternatives to the usage of magnet materials based on rare earths (NdFeB, SmCo). This is very accommodating for the design of the magnetic circuit, as nonlinear effects near the coercive field strength such as with AlNiCo or Barium ferrite are of no relevance[10]. Rare earth magnets allow an approximation of their B/H-curve with a linear equation, providing a nice relationship for their magnetic resistance (Fig. 9.9c):

$$R_{\text{Mag}} = \frac{V}{\phi} = \frac{H_c\, l_{\text{Mag}}}{B_r\, A} \tag{9.30}$$

With this knowledge, the magnetic circuit of Fig. 9.9a and the corresponding equivalent circuit (Fig. 9.9b) can be calculated identical to an electrically excited magnetic circuit.

The flux density within the iron is once again given by

$$B = \frac{\phi}{A} \tag{9.31}$$

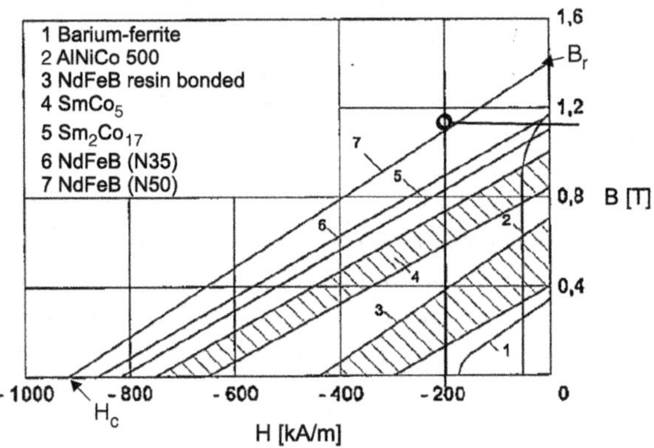

Fig. 9.8 Demagnetization curves of different permanent magnet materials [66]

[10] The small coercive field strength of these materials, e.g., results in the effect, which a magnet magnetized within a magnetic circuit does not reach its flux density anymore once removed and even after reassembly into the circuit again. This happens due to the temporary increase in the air-gap, which is identical to an increase in the magnetic load to the magnet beyond the coercive field strength. Additionally, the temperature dependency of the coercive field strength and of the remanence flux density is critical. Temperatures just below freezing point may result in a demagnetization of the magnet.

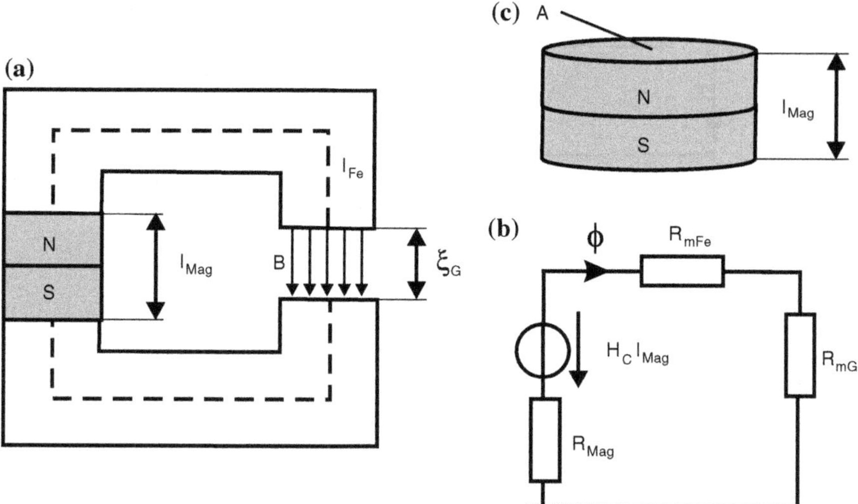

Fig. 9.9 Magnetic field generation B via permanent magnets (**a**), derived equivalent circuit (**b**), and dimensions of the magnet (**c**)

For the given magnetic circuit, the resistances $R_{m\text{Fe}}$ and $R_{m\text{G}}$ are assumed as known or calculable. From Eq. (9.30), the magnetic resistance of the permanent magnet is known. The source within the equivalent circuit is defined by the coercive field strength and the length of the magnets $H_c\, l_{\text{Mag}}$. These considerations result in

$$\phi = \frac{H_c\, l_{\text{Mag}}}{R_{m\text{Fe}} + R_{m\text{G}} + R_{\text{Mag}}}, \tag{9.32}$$

and the flux density

$$B = \frac{H_c\, l_{\text{Mag}}}{(R_{m\text{Fe}} + R_{m\text{G}} + R_{\text{Mag}})\, A}. \tag{9.33}$$

Slightly rearranged and R_{Mag} included gives

$$B = \frac{B_r\, H_c\, \frac{l_{\text{Mag}}}{A}}{(R_{m\text{Fe}} + R_{m\text{G}})\, B_R + H_c\, \frac{l_{\text{Mag}}}{A}}. \tag{9.34}$$

Equation (9.34) states by the factor $B_r\, H_c\, \frac{l_{\text{Mag}}}{A}$ that it is frequently helpful for achieving a maximum flux density B in the air-gap to increase the length of a magnet with at the same time minimized cross-sectional area of the magnetic circuit, which is of course limited by the working distance within the air-gap and the saturation field strengths of the magnetic circuit.

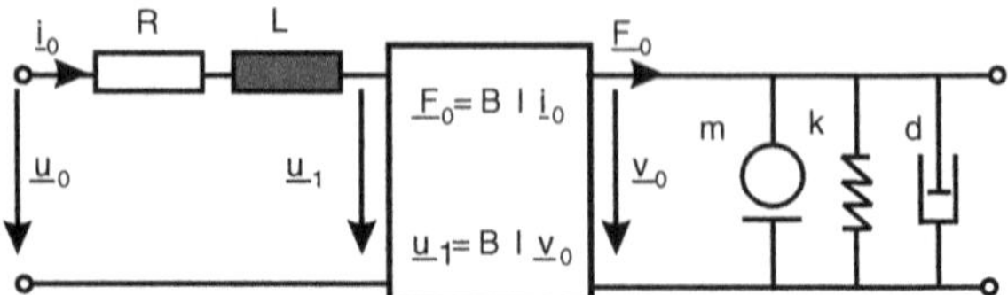

Fig. 9.10 Electrical and mechanical equivalent circuit of an electrodynamic actuator as a transformer

9.2.1.4 Additional Effects in Electrodynamic Actuator

To do a complete characterization of an electrodynamic actuator, there are at least three more effects, whose influences are sketched in the following sections.

Induction

For a complete description of an electrodynamic actuator besides the geometrical design of its magnetic circuit and the mechanical design of its winding and considerations concerning electrical power losses, its other dynamic electrical properties have to be considered. For this analysis, the electrodynamic actuator is regarded as a bipolar transformator (Fig. 9.10).

A current $\underline{i}_0$ generates via the proportional constant $B\,l$ a force $\underline{F}_0$, which moves the mechanical loads attached to the actuator. The movement itself results in a velocity $\underline{v}_0$, which is transformed via the induction law and the proportional constant to an induced voltage $\underline{u}_1$. By measurement of $\underline{u}_1$ and a current source, the rotational velocity or the movement velocity v can be measured, with a voltage source the measurement of $\underline{i}_0$ provides a force- or torque-proportional signal. This is the approach taken by the variant of admittance-controlled devices as a control value (see Sect. 6.4).

The induction itself is a measurable effect, but should not be overestimated. Typically, electrodynamic actuators are used within haptic systems as direct drives at small rotational or translational velocities. Typical coupling factors with rotatory drives are—depending on the size of the actuators—in an area between 100 and 10 $\frac{\text{revolutions}}{s\,V}$. At a rotational speed, which is already fast for direct drives of 10 Hz, induced voltage amplitude $|\underline{u}_1|$ of 0.1 to 1 V can be achieved. This is around 1 to 5 % of the control voltage's amplitude.

Electrical Time Constant

Another aspect resulting from the model according to Fig. 9.10 is the electrical transfer characteristics. Typical inductances L of electrodynamic actuators lie in the area of 0.1–2 mH. The ohmic resistance of the windings largely depends on the actual design, but as a rule of thumb values between 10 and 100 Ω can be assumed. The step response of the electrical transfer system $\frac{\underline{i}_0}{\underline{u}_0}$ shows a time constant $\tau = \frac{L}{R} = 10$ to $30\,\mu$s and lies within a frequency range $\gg$10 kHz, which is clearly above the relevant dynamic area of haptics.

Field Response

A factor that cannot so easily be neglected when using electrodynamic actuators for high forces is the feedback of the magnetic field generated by the electromagnetic winding on the static magnetic field. Taking the actuator from the example at the beginning (Fig. 9.4), positive currents generate a field of opposite direction to the field generated by the magnet. This influence can be considered by substitution of both field sources. Depending on the direction of current, this field either enforces or weakens the static field. With awkward dimensioning, this can result in a directional variance of the actuator properties. The problem is not the potential damage to the magnet, modern magnetic materials are sufficiently stable, but a variation in the magnetic flux density available within the air-gap. An intended application of this effect within an actuator can be found in an example in Fig. 9.11.

An detailed discussion of electrodynamic actuators based on concentrated elements can be found in [77].

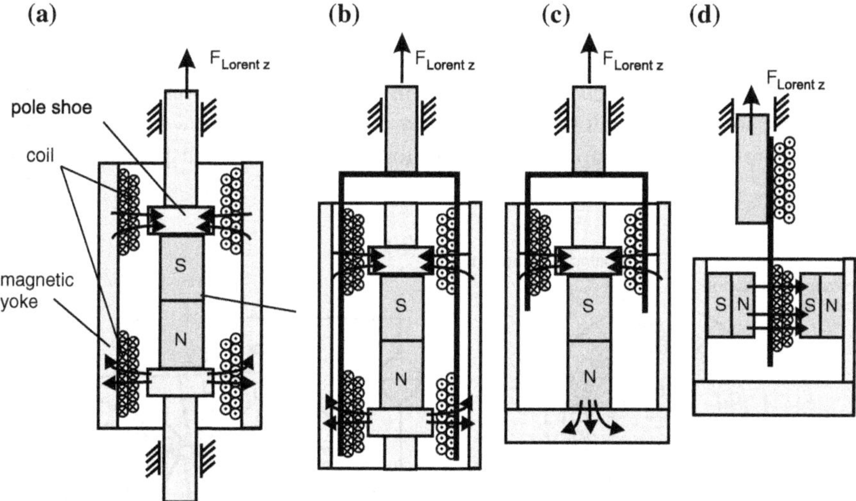

Fig. 9.11 Variants of electrodynamic actuators for translational movement with moving magnets (**a**), moving coils (**b**), as plunger type (**c**), and as flat coil (**d**)

9.2.2 Actual Actuator Design

As stated earlier, electrodynamic actuators are composed of three basic components: coil/winding, magnetic circuit, and magnetic exciter. The following section describes a procedure for the design of electrodynamic actuators based on these basic components. As the common principle for excitation, a permanent magnet is assumed.

9.2.2.1 Actuator Topology

The most fundamental question for the design of an electrodynamic actuator is its topology. Usually, it is known whether the system shall perform rotary or translatory movements. Later, the components' magnetic circuit, the location of magnets, pole shoes, and the coil itself can be varied systematically. A few common structures are shown in Fig. 9.12 for translational actuators and in Fig. 9.11 for rotatory actuators. For the design of electrodynamic actuators in any case, the question to be asked should be whether the coil or the magnetic circuit moves. By this variation, apparently complex geometrical arrangements can be simplified drastically. Anyway, it has to be considered that a moving magnet has more mass and can typically be moved less dynamically than a coil. On the other hand, there is no contact or commutating problem to be solved with non-moving windings.

Moving Coils

Electrodynamic actuators according to the principle of moving coils with a fixed magnetic circuit are named "moving coil" in case of linear movement and "ironless rotor" in case of rotatory actuator. They always combine few moving masses and as a result high dynamics. The translatory version shows displacements of a

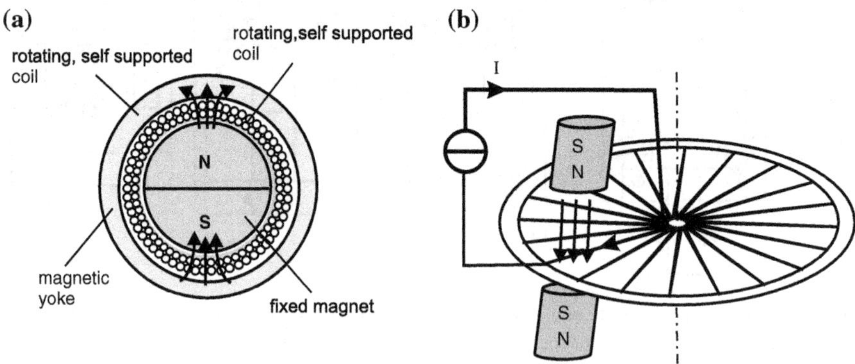

Fig. 9.12 Variants of electrodynamic actuators for rotatory movements with self-supportive winding (**a**), and with disk winding (**b**)

few millimeters and is used especially within audio applications as loudspeaker. Actuators according to the principle of "moving coils" have two disadvantages:

- As the coil is moving, the electrical contact is subject to mechanical stress. With high displacements especially, the contact has to be mechanically robust.
- If there is the idea to design moving coils as pure force sources with large displacements, always only a small area of the conducting coil is within the air-gap and therefore contributes to force generation. With large displacements, moving coils show an even lower efficiency factor. This can be compensated by switching the active coil areas, which again results in the necessity to have more contacts.

A similar situation happens with rotatory systems. Based on the electrodynamic principle, there are two types of windings applicable to rotatory servo-systems: the *Faulhaber* and the *Maxon* winding of the manufacturers with identical names. These actuators are also known as "iron-less" motors. Both winding principles allow the manufacture of self-supportive coils. A diagonal placement of conductors and a baking process after winding generates a structure sufficiently stable for centrifugal forces during operation. The baked coils are connected to the rotating axis via a disk. The complete rotor (Fig. 9.13) is built of these three components. By the very small inertia of the rotor, such actuators show impressive dynamic properties. The geometrical design allows placing the tubular winding around a fixed, diametrally magnetized magnet. This enables another volume reduction compared to conventional actuators as its housing has to close the magnetic circuit only instead of providing additional space for magnets.

Within self-supportive winding, there are areas of parallel-lying conductors combined to poles.[11] With moving coils, there is always the need for a specialized contactor, either via contact rings, or electronic commutation or via mechanical switching.

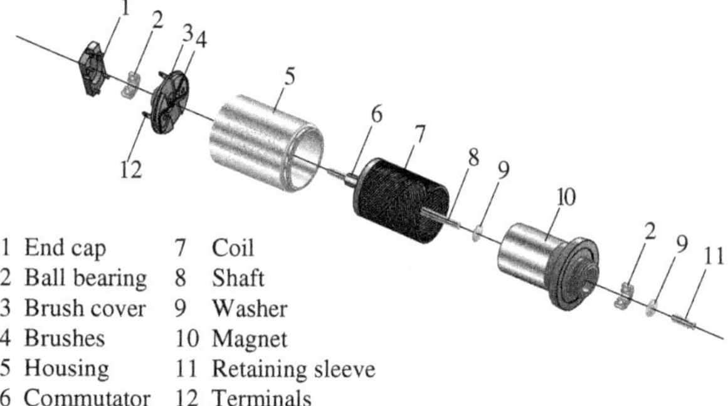

Fig. 9.13 Design of an electrodynamic actuator with self-supportive winding based on the FAULHABER principle. Picture courtesy of *Dr. Fritz Faulhaber GmbH*, Schöneich, Germany

[11] The *Faulhaber* and the *Maxon* excel by a very clever winding technique. On a rotating cylinder, respectively, a flatly pressed rectangular winding pole can be combined by contacting closely located areas of an otherwise continuous wire.

Depending on the number of poles, all coils are contacted at several points. In case of mechanical switching, these contacts are placed on the axis of the rotor and connected via brushes with the fixed part of the actuator named "stator". This design enables a continuous movement of the rotor, whereas a change in the current flow is made purely mechanical by sliding the brushes on the contact areas of the poles on the axis. This mechanical commutation is a switching procedure with an inductance placed in parallel.

As such, an actuator can be connected directly to a direct current source, known as "DC-drives". As stated in Sect. 9.1, the term "DC-drive" is not only limited to actuators according to the electrodynamic principle but is also frequently applied to actuators following the electromagnetic principle (Sect. 9.4).

Moving Magnet

In case of translatory (Fig. 9.11a) systems, actuators based on the principle of a moving magnet are designed to provide large displacements with compact windings. The moving part of the actuator is composed almost completely of magnetic material. The polarity direction of this material may vary in its exact orientation. Actuators according to this principle are able to provide large power but are expensive due to the quantity of magnet material necessary. Additionally, the moving magnet is heavy; the dynamics of the actuator is therefore smaller than in the case of a moving coil. In case of a rotatory system, a design with moving magnet is comparable to a design with moving coil. Figure 9.14 shows such a drive. The windings fixed to the stator are placed around a diametral magnetized magnet. It rotates on an axis, which frequently additionally moves the magnetic circuit, too. Providing the right current feed to the coil the orientation of the rotor has to be measured. For this purpose, sensors based on the Hall effect or optical code-wheels are used.

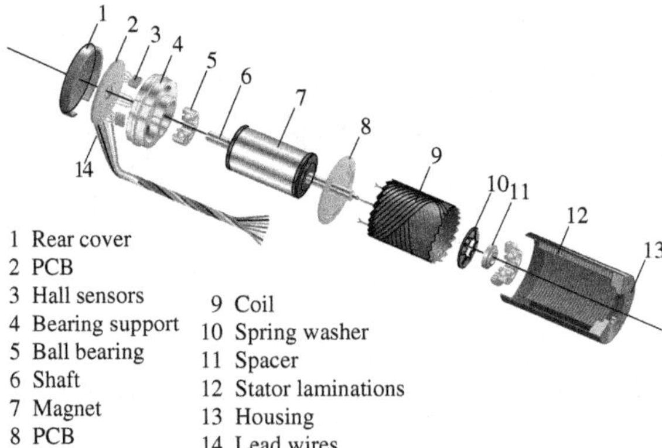

Fig. 9.14 Components of an EC-drive. Pictures courtesy of *Dr. Fritz Faulhaber GmbH*, Schöneich, Germany.

Electrodynamic actuators with moving magnet are known as EC-drives (electronic-commutated). This term is not exclusive to electrodynamic actuators, as there are also electronic-commutated electromagnetic drives. EC-drives—whether electrodynamic or electromagnetic—combined with the corresponding driver electronics are frequently known as servo-drives. Typically, a servo-drive is an actuator able to follow a predefined movement path. Servo-drives are rarely used for haptic devices. However, the use of EC-drives for haptic application is frequent, but they are equipped with specialized driver electronics.

9.2.2.2 Commutation in the Context of Haptic Systems

The necessary commutation of current for rotating actuators has huge influence on the quality of force output with respect to torque output.

Mechanically Commutating Actuators

With mechanically commutating actuators, the current flow is interrupted suddenly. Two effects of switching contacts appear: The voltage at the contact point increases, sparks may become visible—an effect called electrical brush sparking. Additionally, the remaining current flow induces a current within the switched-off part of the winding that itself results in a measurable torque. Depending on the size of the motor, this torque can be felt when interacting with a haptic system and has to be considered in the design.

The current and torque changes can be reduced by he inclusion of resistors and capacitors into the coil. However, this results in high masses of the rotor and worse dynamic properties. Besides, a full compensation is impossible. Nevertheless, mechanically commutating actuators are in use for inexpensive haptic systems. The GEOMAGIC TOUCH from *geomagic* and the FALCON from *Novint* use such actuators.

Electronic-Commutated Electrodynamic Actuators

Electronically commutated electrodynamic actuators differ from mechanically commutated actuators by the measurement technology used as a basis for switching currents. There are four typical designs for this technology:

- In sensor-less designs (Fig. 9.15a), an induced voltage is measured within a coil. At zero-crossing point, one pole is excited with a voltage after an interpolated 30o phase delay dependent on the actual revolution speed of the rotor. In combination with measurement of the inductance followed by a switched voltage, a continuous rotation with batch-wise excitation is realized. This procedure cannot be applied to low rotation speeds, as the induced voltage becomes too low, and accordingly, the switching point can hardly be interpolated. Additionally, the concept of using

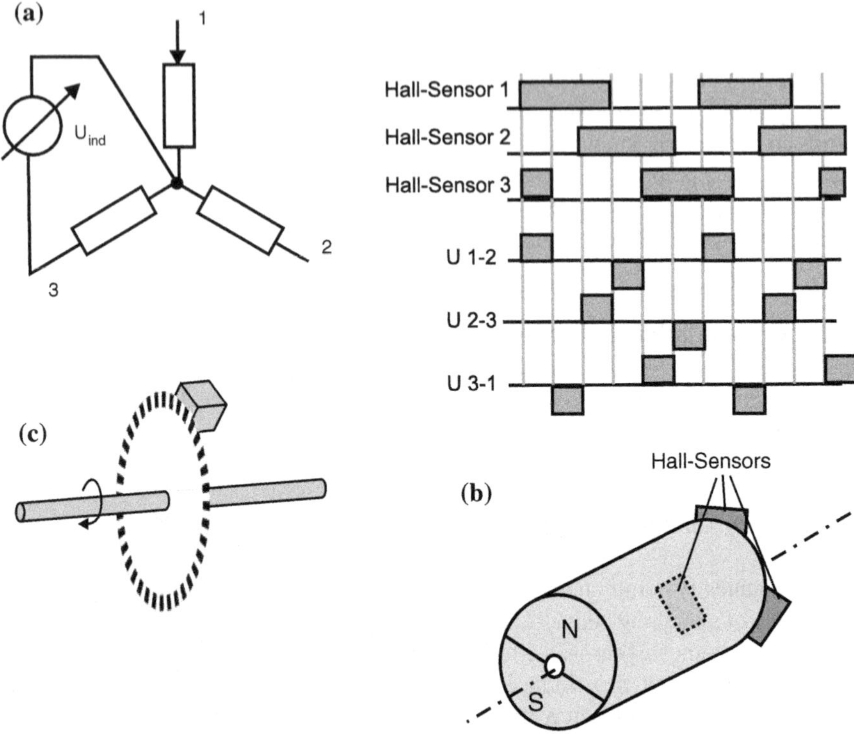

Fig. 9.15 Technologies for different commutation methods: sensorless (**a**), block-commutation (**b**) and optical code-wheel (**c**)

one to two coils for torque generation results in high torque variations at the output of up to 20 %, making this approach not useful for haptic systems.

- Block-commutating procedures (Fig. 9.15 b) are based on the use of simple hall switches or field plates for position detection of the rotor. Three sensors located at 120° angular phase shift allow the detection of six different rotor positions. Reducing positioning information to six orientations per revolution makes this approach equally inappropriate for haptic applications, as the torque varies in a range of >15 % for one revolution.
- Sinus-commutating procedures with analog hall sensors are based on the measurement of the rotor position by at least two sensors. They are placed with an angle of either 120° or 90° at the front of the rotor. They provide voltages in angular phase shift according to their geometrical position. By analyzing the polarity and the absolute height of the voltages, absolute positioning information can be obtained and used for commutating the windings. If the phase lag between both sensor signals is identical to the phase lag between the poles of the winding, a direct control of current drivers can be performed without the need for digitization or a specific calculation step.

- Sinus commutating with digital code-wheels (Fig. 9.15c) are based on the measurement of rotor position by use of, usually optical, code disks. By reflective or transmissive measurement, the rotor position is sampled with high resolution. This relative positioning information can be used for position measurement after initial calibration. Depending on the code-wheels resolution, a smooth sinusoid commutation can be achieved with this method.

The sinus-commutating methods are the preferred solutions used for haptic applications due to the little torque variations and their applicability for slow revolution speeds typical of direct drives.

9.2.3 Actuator Electronics

Electrodynamic actuators require specific electrical circuits. In the following section, the general requirements on these electronics are formulated.

9.2.3.1 Driver Electronics

Driver electronics are electrical circuits transforming a signal of low power (several volts, some milliampere) into a voltage or current level appropriate to drive an actuator. For electrodynamic actuators in haptic applications, driver electronics have to provide a current in a dynamic range from static up to several kilohertz. This section describes general concepts and approaches for such circuits.

Topology of Electric Sources

Driver electronics for actuators—independently from the actuation principle they are used for—are classified according to the flow of electrical energy (Fig. 9.16). There are four classes of driver electronics:

- One-quadrant controllers are capable of generating positive output currents and voltages. An actuator driven by them is able to move in one direction. These controllers use only the first quadrant according to Fig. 9.16a.
- Switched one-quadrant controllers are capable of a direction change by input of a logical signal. They work within the first and third quadrants based on Fig. 9.16a. The switching point is a nonlinear step in their characteristic curve.
- Real two-quadrant controllers are capable of providing a characteristic curve, which is steady around the zero point. They function in the first and third quadrants based on Fig. 9.16a, but are not capable to conduct currents and voltages with opposite directions.
- Four-quadrant controllers function within all four quadrants of Fig. 9.16a. They are able to control currents and voltages in any combination of directions. Four-

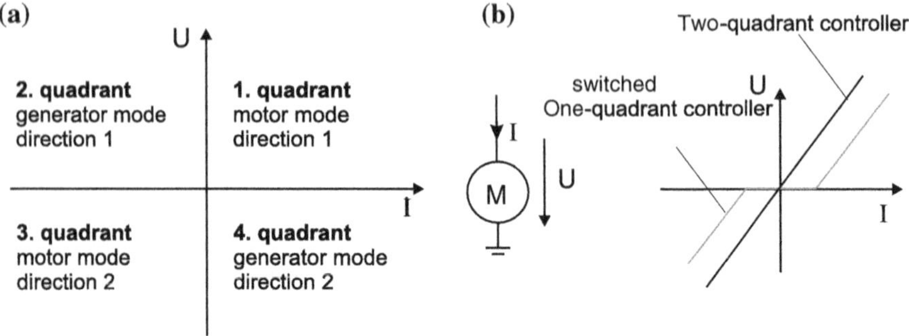

Fig. 9.16 **a** Visualization of the four quadrants of an electric driver, formed by the directions of current and voltage. **b** characteristic curves of a two-quadrant controller and a switched one-quadrant controller

quadrant controllers allow energy recovery by induced current to an energy storage, which is especially relevant for mobile applications.

For haptic application, the switched one-quadrant controller is frequently met, as many haptic systems do not have the necessity to control the device near the voltage- or current-zero point. However, for systems with high dynamics and low impedance, the two-quadrant and the four-quadrant controller are relevant, as the unsteadiness near the zero point is perceivable with high-quality applications.

Pulse-Width Modulation and H-Bridges

With the exception of some telemanipulators, the sources controlling the actuators are always digital processors. As actuators need an analog voltage or current to generate forces and torques, some transformer between digital signals and analog control value is necessary. There are two typical realizations of these transformers:

1. Use of a digital–analog converter (D/A-converter)
2. Use of a ↪ Pulse-Width Modulation (PWM)

The use of D/A-converters as external components or integrated within a micro-controller is not covered further in this book, as it is, if necessary to use, extremely simple. It just requires some additional efforts in circuit layout. The latter results in it being not used much for control of actuators.

With electrodynamic actuators, the method of choice is driver electronics based on PWM (Fig. 9.17a). With the PWM, a digital output of a controller is switched with a high frequency (>10 kHz[12]). The period of the PWM is given by the frequency. The program controls the duty cycle between on- and off-times. Typically, one byte

[12] Typical frequencies lie between 20 and 50 kHz. However, especially within automotive technology for driving LEDs, PWMs for current drivers with frequencies below 1 kHz are in application. Frequencies within this range are not applicable to haptic devices, as the switching in the control

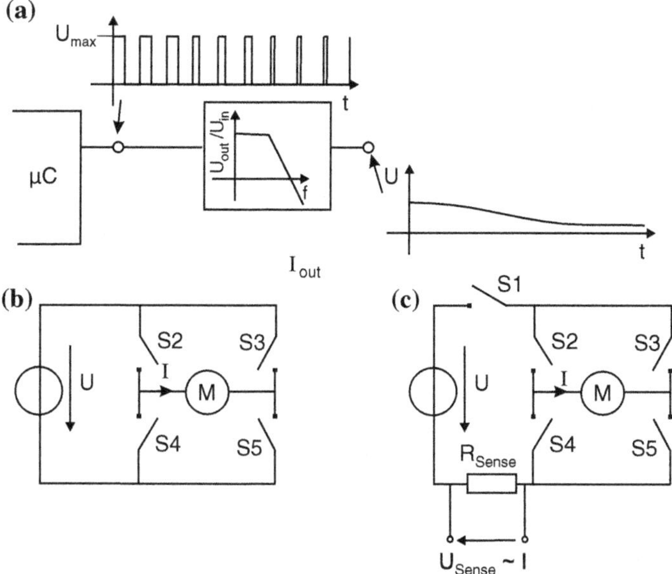

Fig. 9.17 Principle of pulse-width modulation (PWM) at a digital μC-output (**a**), H-bride circuit principle (**b**), and extended H-bridge with PWM (S1) and current measurement at (R_{Sense}) (**c**)

is available to provide a resolution of 256 steps within this period. After filtering the PWM, either via an electrical low-pass or via the mechanical transfer characteristics of an actuator, a smoothed output signal becomes available.

Pulse-width modulation is frequently used in combination with H-bridges (Fig.9.17b). The term H-bridge results from the H-like shape of the motor surrounded by four switches. The H-bridge provides two operation modes for two directions of movement and two operation modes for braking. If according to Fig. 9.17b, the two switches S2 and S5 are on, current I will flow through the motor in positive direction. If instead switches S3 and S4 are switched on, current I will flow through the motor in negative direction. One additional digital signal acting on the H-bridge will change the direction of movement of the motor. This is the typical procedure with switched 1-quadrant controllers. Additional switching states are given by switching the groups S2 and S3, respectively, S4 and S5. Both states result in short circuit of the actuator and stops its movement. Other states like simultaneously switching S2 and S4, respectively, S3 and S5 results in short circuit of the supply voltage, typically destroying the integrated circuit of the driver.

To combine the H-bridge with a PWM, either switch groups S2 and S5 can be switched according to the timing of the PWM, or additional switches S1 (Fig. 9.17c)

value may be transmitted by the actuator and will therefore be perceivable especially in static conditions. Typical device designs show mechanical low-pass characteristics even at frequencies in the area of 200 Hz. However, due to the sensitivity of tactile perception in an area of 100–200 Hz, increased attention has to be paid to any switched signal within the transmission chain.

can be placed in series to the H-bridge modulating the supply voltage U. In practical realization, the latter is the preferred design, as the timing of the switches S2 to S5 is critical to prevent likely short circuits in the supply voltage. The effort to perform this timing between the switching is usually higher than the cost of another switch in series. The practical realization of H-bridges is done via field-effect transistors. The discrete design of H-bridges is possible, but not easy. In particular, the timing between switching events, the prevention of short circuits, and the protection of the electronics against induced currents is not trivial. There are numerous integrated circuits available on the market that include appropriate protective circuitry and provide only a minimum of necessary control lines. The ICs L6205 (2A), L293 (2.8A), and VNH 35P30 (30A) are some examples common in test-bed developments. For EC-drives, there are specific ICs performing the timing for the field-effect transistors and reducing the number of necessary PWMs from the microcontroller. The IR213xx series switches three channels with one external half-bridge per channel built from N-MOS transistors with a secure timing for the switching events.

The PWM described above with an H-bridge equals a controlled voltage source. For electrodynamic systems, such a control is frequently sufficient to generate an acceptable haptic perception. Nevertheless, for highly dynamic haptic systems, a counter-induction (Sect. 9.2.1.4) due to movement has to be expected, resulting in a variation in the current within the coils generating an uncontrolled change in the LORENTZ force. Additionally, the power loss within the coils (Sect. 9.2.1.1) may increase the actuator's internal temperature resulting in a change in conductivity in the conductor's material. The increasing resistance with increasing temperatures of the conductor results in less current flow at a constant voltage source. An electrodynamic actuator made of copper as conductive material generates a reduced force when operated. With higher requirements on the quality of haptic output, a controlled current should be considered. In case of a PWM, a resistor with low resistance (R_{Sense} in Fig. 9.17c) has to be integrated, generating a current-proportional voltage U_{Sense}, which itself can be measured with an A/D input of the controller. The control circuit is closed within the microcontroller. However, the A/D-transformation and the closing of the control circuit can be challenging for state-of-the-art electronics with highly dynamic systems with border frequencies of some kilohertz. Therefore, analog circuits should be considered for closed-loop current controls too.

Analog Current Sources

Analog current sources are—to make it simple—controlled resistors within the current path of the actuator. Their resistance is dynamically adjusted to provide the wished current flow. Identical to classical resistors, analog current sources transform the energy that is not used within the actuator into heat. Consequently, in comparison with the switched H-bridges, they are generating a lot of power loss. By use of a discrete current control (Fig. 9.18a), analog current sources for almost any output currents can be built by the choice of one to two field-effect transistor (FET). For heat dissipation, they are required to be attached to adequate cooling elements. There

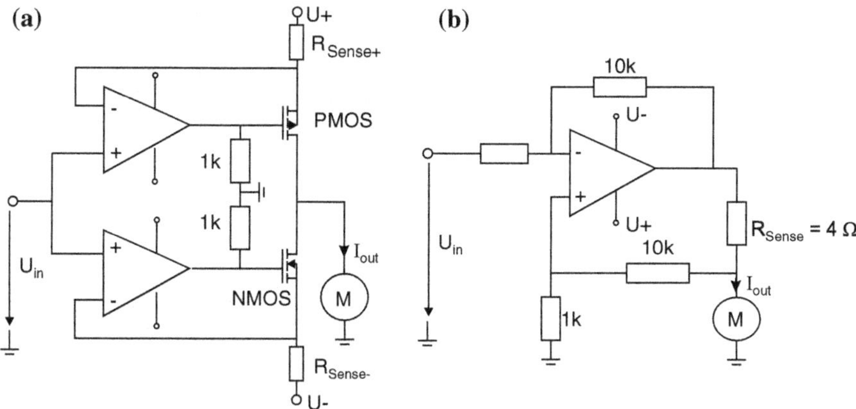

Fig. 9.18 Discrete closed-loop current control [122] (**a**), and closed-loop current control with a power-operational amplifier (**b**)

are only little requirements on the operational amplifiers themselves. They control the FET within its linear range proportional to the current-proportional voltage generated at R_{Sense}. Depending on the quadrant used within operational mode (1 or 3), either the N-MOS transistors or the P-MOS transistor is conductive. An alternative to such discrete designs is the use of power amplifiers (e.g., LM675, Fig. 9.18b). It contains fewer components and is therefore less dangerous to make errors. Realized as non-inverting or inverting operational amplifier with a resistor for measurement R_{Sense}, they can be regarded as a voltage-controlled current source.

9.2.3.2 Monitoring Temperature

Resulting from the low efficiency factor and the high dissipative energy from electrodynamic actuators, it is useful to monitor the temperature nearby the coils. Instead of including a measuring resistor PT100 nearby the coil, another approach monitors the electrical resistance of the windings themselves. Depending on the material of the windings (e.g., cooper, Cu), the conductivity changes proportional to the coil's temperature. With copper, this factor is 0.39 % per Kelvin temperature change. As any driver electronics, either works with a known and controlled voltage or current, measurement of the other component immediately provides all information to calculate resistance and consequently the actual coil temperature.

Fig. 9.19 Electrodynamic tactor by *Audiological Engineering Inc.* with a frequency range from 100 to 800 Hz

9.2.4 Examples for Electrodynamic Actuators in Haptic Devices

Electrodynamic actuators are most frequently used as force and torque sources within kinaesthetic systems. In particular, EC-drives can be found in products of *Quanser*, *ForceDimension*, *Immersion*, and *SensAble/geomagic*. Mechanically commutated electrodynamic actuators are used within less expensive devices, like the GEOMAGIC TOUCH or the NOVINT FALCON. For tactile applications, electrodynamic actuators appear only with linear moving coils or magnets as oscillation sources. The possibility to control frequency and amplitude independently from each other makes them interesting for tactors. Tactors (Fig. 9.19) are small, disk-like actuators that can be integrated in clothes or mobile devices to transmit information via tactile stimulation in small areas of the skin.

9.2.4.1 Cross-Coil System as Rotary Actuator

Besides self-supportive coils, electrodynamic actuators according to the design of cross-coils are one possibility to generate defined torques. Continental VDO developed a haptic rotary actuator device being a central control element for automotive applications (Fig. 9.20). It contains a diametral magnetized NdFeB-magnet. The magnet is surrounded by a magnetic circuit. The field lines reach from the magnet to the magnetic circuit. The coils surround the magnet in an angular phase of 90°, and the electrodynamic active winding section lies in the air-gap between magnetic circuit and magnet. The rotary position control is made via two hall sensors placed in a 90° position. The actuator is able to generate a ripple-free torque of $\approx$25 m Nm at a geometrical diameter of 50 mm, which is additionally increased by an attached gear to $\approx$100 m Nm torque output.

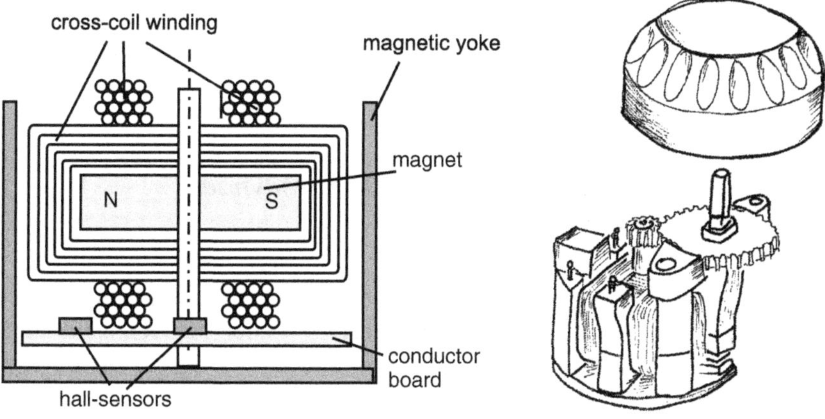

Fig. 9.20 Electrodynamic cross-coil system with moving magnet as haptic rotary actuator

9.2.4.2 Reconfigurable Keypad: HapKeys

The electrodynamic linear actuators building the basis of this device are equipped with friction-type bearings and moving magnets with pole shoes within cylindrically wound fixed coils as shown in Fig. 9.21. The coils have an inner diameter of 5.5 mm and an outer diameter of 8 mm. The magnetic circuit is decoupled from other nearby

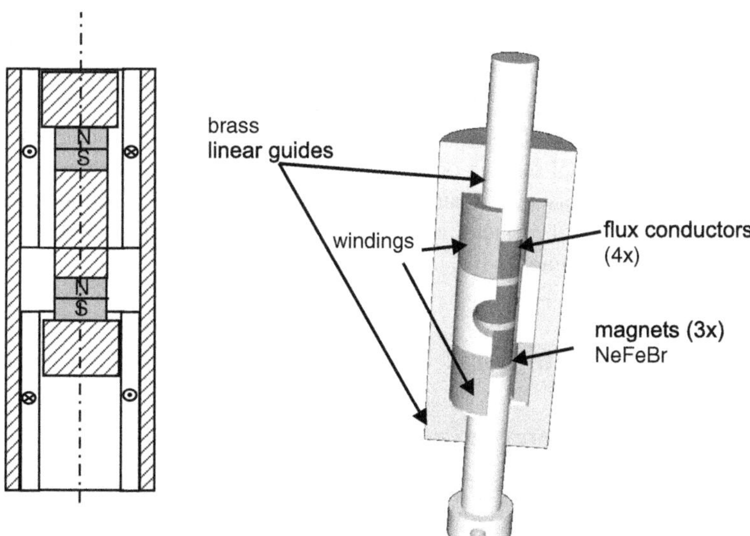

Fig. 9.21 Electrodynamic linear actuator with moving magnet [25]

elements within the actuator array. It is made of a tube with a wall thickness of 0.7 mm of a cobalt–iron alloy with very high saturation flux density. Each actuator is able to generate 1 N in continuous operation mode.

9.2.5 Conclusion About the Design of Electrodynamic Actuators

Electrodynamic actuators are the preferred actuators used for kinaesthetic impedance-controlled haptic devices due to their proportional correlation between the control value "current" and the output values "force" or "torque." The market of DC- and EC-drives offers a wide variety of solutions, making it possible to find a good compromise between haptic quality and price for many applications. Most suppliers of such components offer advice on how to dimension and select a specific model based on the mechanical, electrical, and thermal properties as for example shown in [26].

If there are special requirements to be fulfilled, the design, development, and start of operation of special electrodynamic actuator variants are quite easy. The challenges in thermal and magnetic design are manageable, as long as some basic considerations are not forgotten. The examples of special haptic systems seen in the above section prove this impressively. Just driver electronics applicable to haptic systems and its requirements are still an exceptional component within the catalogs of manufacturers of automation technology. They must either be paid expensively or be built by oneself. Therefore, commercial manufacturers of haptic devices, e.g., *Quanser*, offer their haptic-applicable driver electronics independent of the own systems for sale.

For design of low-impedance haptic systems, currently no alternative to electrodynamic systems is at hand. Other actuation principles that are discussed within this book need a closed-loop control to overcome their inner friction and nonlinear force/torque transmission. This always requires some kind of measurement technology such as additional sensors or the measurement of inner actuator states. The efforts connected with this are still an advantage for electrodynamic actuators, which is gained by a low efficiency factor and as a consequence, the relatively low energy density per actuator volume.

9.3 Piezoelectric Actuators

Stephanie Sindlinger, Marc Matysek

Next to the frequently found electrodynamic actuators, the past few years' piezoelectric actuators were used for a number of device designs. Their dynamic properties in resonance mode especially allow an application for haptics, which is different from the common positioning application they are used for. As variable impedance, a wide spectrum of stiffnesses can be realized. The following chapter gives the calculation basics for the design of piezoelectric actuators. It describes the design variants and their application in haptic systems. Besides specific designs for tactile and kinaesthetic devices, approaches for the control of the actuators and tools for their dimensioning are presented.

9.3.1 The Piezoelectric Effect

The piezoelectric effect was first discovered by JACQUES and PIERRE CURIE. The term is derived from the Greek word "piedein–piezo" = "to press" [61].

Figure 9.22 shows a scheme of a quartz crystal (chemical: SiO2). With force acting on the crystal mechanical displacements of the charge centers can be observed within the structure, resulting in microscopic dipoles within its elementary cells. All microscopic dipoles sum up to a macroscopic measurable voltage. This effect is called "reciprocal piezoelectric effect." It can be reversed to the "direct piezoelectric effect." If a voltage is applied on a piezoelectric material, a mechanical deformation happens along the crystal's orientation, which is proportional to the field strength in the material [3].

Piezoelectric materials are anisotropic—direction dependent—in their properties. Consequently, the effect depends on the direction of the electrical field applied, and on the angle between the direction of the intended movement and the plane of polarization. For the description of these anisotropic properties, the directions are labeled with indices. The index is defined by a Cartesian space with the axes numbered 1, 2, and 3. The plane of polarization of the piezoelectric material is typically orientated on direction 3. The shear at the axes is labeled with indices 4, 5, and 6.

Among all possible combinations, there are three major effects (Fig. 9.23), commonly used for piezoelectric applications: longitudinal, transversal, and shear effect.

The *longitudinal effect* acts in the same direction as the applied field and the corresponding field strength E_3. As a consequence, the resulting mechanical tensions T_3 and strains S_3 also lie within plane 3. With the *transversal effect*, mechanical

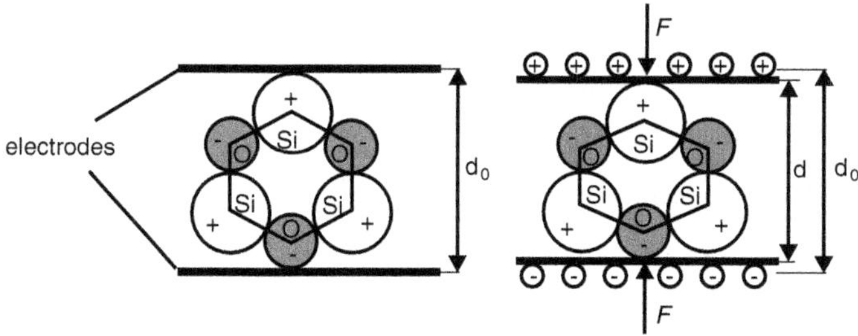

Fig. 9.22 Crystal structure of quartz in initial state and under pressure

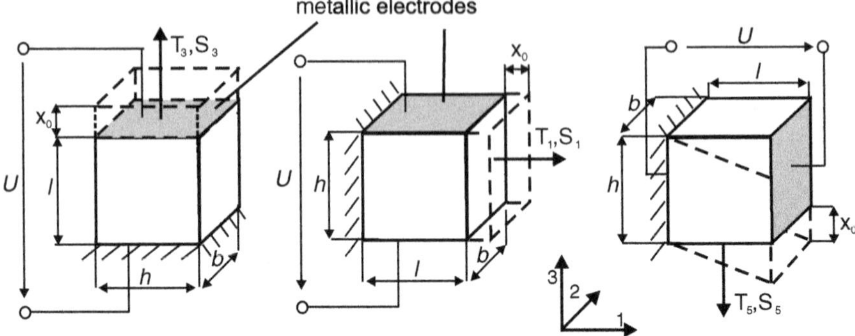

Fig. 9.23 Effects during applied voltage: longitudinal effect (*left*), transversal effect (*center*), shear effect (*right*)

actions show normal to the electrical field. As a result from a voltage U_3 with the electrical field strength E_3, the mechanical tensions T_1 and strains S_1 appear. The *shear effect* happens with electrical voltage U applied along plane 1 orthogonal to the polarization plane. The resulting mechanical tensions appear tangential to the polarization, in the direction of shear, and are labeled with the directional index 5.

9.3.1.1 Basic Piezoelectric Equations

The piezoelectric effect can be described most easily by state equations:

$$P = e \cdot T \tag{9.35}$$

and

$$S = d \cdot E \tag{9.36}$$

with

P = Direction of polarization (in C/m^2)
S = Deformation (non-dimensional)
E = Electrical field strength (in V/m)
T = Mechanical tension (in N/m^2)

The piezoelectric coefficients are

- the piezoelectric coefficient of tension (also: coefficient of force) e (reaction of the mechanical tension on the electrical field)

$$e_{ij,k} = \frac{\partial T_{ij}}{\partial E_k}\partial \tag{9.37}$$

- and the piezoelectric coefficient of strain (also: coefficient of charge) d (reaction of the strain on the electrical field)

$$d_{ij,k} = \frac{\partial \epsilon_{ij}}{\partial E_k} \partial \tag{9.38}$$

The correlation of both piezoelectric coefficients is defined by the elastic constants C_{ijlm}

$$e_{ij,k} = \sum_{lm} \left(C_{ijlm} \cdot d_{lm,k} \right) \tag{9.39}$$

Usually, the tensors shown in the equation above are noted as matrix. In this format, matrices result in six components identical to the defined axes. The matrix shown below describes the concatenation of the dielectrical displacement D, the mechanical strain S, the mechanical tension T, and the electrical field strength E.

	T_1	T_2	T_3	T_4	T_5	T_6	E_1	E_1	E_3
D_1	0	0	0	0	d_{15}	0	ϵ_{11}	0	0
D_2	0	0	0	d_{15}	0	0	0	ϵ_{11}	0
D_3	d_{31}	d_{31}	d_{33}	0	0	0	0	0	ϵ_{11}
S_1	s_{11}	s_{12}	s_{13}	0	0	0	0	0	d_{31}
S_2	s_{12}	s_{11}	s_{13}	0	0	0	0	0	d_{31}
S_3	s_{13}	s_{13}	s_{33}	0	0	0	0	0	d_{33}
S_4	0	0	0	s_{44}	0	0	0	d_{15}	0
S_5	0	0	0	0	s_{44}	0	d_{15}	0	0
S_6	0	0	0	0	0	$2(s_{11} - s_{12})$	0	0	0

This matrix can be simplified for the specific cases of a longitudinal and a transversal actuator. For a longitudinal actuator with electrical contact in direction 3, the following equations are the result:

$$D_3 = \epsilon_{33}^T E_3 + d_{31} T_1 \tag{9.40}$$

$$S_3 = d_{31} E_3 + s_{11}^E T_1. \tag{9.41}$$

Accordingly for a transversal actuator, the correlation

$$D_3 = \epsilon_{33}^T E_3 + d_{33} T_3 \tag{9.42}$$

$$S_3 = d_{33} E_3 + s_{33}^E T_3 \tag{9.43}$$

with becomes valid.

D_3 = dielectric displacement in C/m^2 D = 0: open-ended
E_3 = field strength in V/m   E = 0: short cut
$S_1, S_3 = \Delta L/L$ = strains, dimensionless $S = 0$: mech. short cut
T_1, T_3 = mechanical tensions N/m^2 $T = 0$: idle mode
ϵ_{33}^T = relative dielectricity constant at mechanical tension = 0
d_{31}, d_{33} = piezoelectric charge constant in C/N
s_{11}^E, s_{33}^E = elasticity constant at field strength = 0

Therefore, the calculation of piezoelectric coefficients simplifies into some handy equations: The charge constant d can be calculated for the electrical short circuit—which is $E = 0$—to

$$d_{E=0} = \frac{D}{T} \tag{9.44}$$

and for the mechanical idle situation—which is $T = 0$—to

$$d_{T=0} = \frac{S}{E}. \tag{9.45}$$

The piezoelectric tension constant is defined as

$$g = \frac{d}{\epsilon^T}. \tag{9.46}$$

The coupling factor k is given by Eq. (9.47). It is a quantity for the energy transformation and consequently for the strength of the piezoelectric effect. It is used for comparison among different piezoelectric materials. However, note that it is not identical to the efficiency factor, as it does not include any energy losses.

$$k = \frac{\text{converted energy}}{\text{absorbed energy}}. \tag{9.47}$$

A complete description of the piezoelectric effect, a continuative mathematical discussion, and values for piezoelectric constants can be found in the literature, such as [16, 55, 77].

9.3.1.2 Piezoelectric Materials

Till 1944, the piezoelectric effect was observed within monocrystals only. These were quartz, turmalin, lithiumniobat, potassium- and ammonium-hydrogen phosphate (KDP, ADP), and potassium sodium tartrate [3]. With all these materials, the direction of the spontaneous polarization was given by the direction of the crystal lattice [61]. The most frequently used material was quartz.

The development of polarization methods made it possible to retrospectively polarize ceramics by the application of a constant exterior electrical field in 1946. By this approach, "piezoelectric ceramics" (also "piezoceramics") were invented. By this development of polycrystalline materials with piezoelectric properties, the whole group of piezoelectric materials obtained increased attention and technical significance. Today, the most frequently used materials are barium titanate ($BaTiO_3$) or lead zirconate titanate (PZT). C 82 is a piezoelectric ceramic suitable for actuator design due to its high *k-factor*. However, like all piezoelectric ceramic materials, it shows reduced long-term stability compared to quartz. Additionally, it has a pyro-electric effect, which is a charge increase due to temperature changes in the material [77]. Since the 1960s, the semicrystalline synthetic material polyvinylidene fluoride (PVDF) is known. Compared to the materials mentioned before, PVDF excels by its high elasticity and reduced thickness (6–9 μm).

Table 9.4 shows different piezoelectric materials with their specific values.

Looking at these values, PZT is the most suitable for actuator design due to its high coupling factor with large piezoelectric charge modulus and still a high Curie temperature. The Curie temperature represents the temperature at which the piezoelectric properties from the corresponding material are lost permanently. The value of the Curie temperature depends on the material (Table 9.4).

Table 9.4 Selection of piezoelectric materials with characteristic values [77]

Constant	Unit	Quartz	PZT-4	PZT-5a	C 82	PVDF
d_{33}	$10^{-12}\frac{\text{m}}{\text{V}}$	2.3	289	374	540	−27
d_{31}		−2.3	−123	−171	−260	20
e_{33}	$\frac{A\cdot s}{\text{m}^2}$	0.181	15.1	15.8	28.1	108
e_{31}		−0.181	−5.2	−5.4	−15.4	–
s_{33}^E	$10^{-12}\frac{\text{m}^2}{\text{N}}$	12.78	15.4	18.8	19.2	–
s_{11}^E		12.78	12.3	16.4	16.9	–
c_{33}^E	$10^{10}\frac{\text{N}}{\text{m}^2}$	7.83	6.5	5.3	5.2	–
c_{11}^E		7.83	8.1	6.1	5.9	–
$\frac{\epsilon_{33}^T}{\epsilon_0}; \frac{\epsilon_{33}^S}{\epsilon_0}$	–	4.68; 4.68	1,300; 635	1,730; 960	3,400; –	12; 12
$\frac{\epsilon_{11}^T}{\epsilon_0}; \frac{\epsilon_{11}^S}{\epsilon_0}$	–	4.52; 4.41	1,475; 730	1,700; 830	3,100; –	–
k_{33}	–	0.1	0.7	0.71	0.72	0.20
k_{31}	–	–	0.33	0.34	0.36	0.15
ϑ_{Curie}	$\frac{1}{°\text{C}}$	575	328	365	190	80
ρ	$\frac{\text{kg}}{\text{m}^{-3}}$	2,660	7,500	7,500	7,400	1,790

9.3.2 Designs and Properties of Piezoelectric Actuators

Actuators using the piezoelectric effect are members of the group of solid actuators (also: solid-state actuators). The transformation from electrical into mechanical energy happens without any moving parts, resulting in fast reaction time and high dynamics compared to other actuation principles. Additionally, piezoelectric actuators have high durability. The thickness changes are smaller compared to other actuation principles, although the generated forces are higher.

9.3.2.1 Basic Piezoelectric Actuator Designs

Depending on the application, different designs may be used. One may require a large displacement; another one may require self-locking or high stiffness. The most frequently used actuator types are bending actuators and actuator staples. A schematic sketch of each design is given in Fig. 9.24a, c.

Stacked actuator are based on the longitudinal piezoelectric effect. For this purpose, several ceramic layers of opposite polarity are stapled above each other. In between each layer contact, electrodes are located for electrical control. A staple is

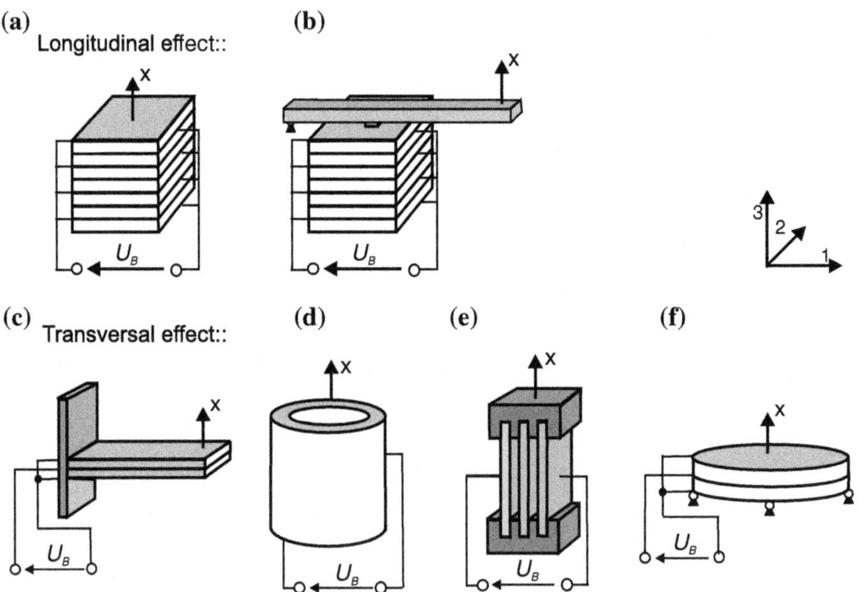

Fig. 9.24 Piezoelectric transducers separated by longitudinal and transversal effect: longitudinal effect: **a** stack, **b** stack with lever transformation, change of length: $x = d_{33} \cdot U_B$ transversal effect: **c** bending actuator, **d** cone, **e** band, **f** bending disk, change of length: $x = -d_{31} \cdot U_B$. Further information can be found in [61]

able to generate high static forces up to several 10 kN. The achievable displacement of 200 μm is low compared to other piezoelectric designs. By use of levers Fig. 9.24, (b) the displacement can be significantly increased (see Fig. 9.24b). Voltages of several 100 V are necessary to drive a piezoelectric actuation staple.

Bending actuators are based on the transversal piezoelectric effect. Designed according to the so-called bimorph principle—with two active layers—they are used in applications requiring large displacements. The transversal effect is characterized by comparably low controlling voltages [3, 61]. These electrical properties and large displacements can be achieved by thin ceramic layers in the direction of the electrical fields, and an appropriate geometrical design. Other geometrical designs using the transversal effect are tubular actuators, film actuators, or bending disks, Fig. 9.24d–f. Due to their geometry, they equal staple actuators in their mechanical and electrical characteristics. The achievable displacements of 50 μm are comparably low, whereas the achievable forces excel bending actuators at several orders of magnitude.

The use of the *shear effect* is uncommon in actuator design. This is somewhat surprising as it shows charge modulus and coupling factor, which is twice as much as the transversal effect. Additionally, it is possible to increase the elongation x_0 in idle mode (displacement without any load) by optimization of the length to thickness (l/h) ratio. However, the clamping force F_k of the actuator is not influenced by these parameters.

Table 9.5 summarizes the properties of different geometrical designs. Typical displacements, actuator forces, and control voltages are shown.

9.3.2.2 Selection of Special Designs for Piezoelectric Actuators

Besides the standard designs shown above, several variations of other geometrical designs exist. In this section, examples of an ultrasonic drive with resonator, oscillatory/traveling waves actuators, and piezoelectric stepper motors are discussed.

Table 9.5 Properties of typical piezoelectric actuator designs based on [61]

Standard designs	Stack	Stack with lever	Bending actuator	Tape actuator	Tubus	Bending disks
Actuator displacements (μm)	20–200	≤1.000	≤1.000	≤50	≤50	≤500
Actuating forces (N)	≤30.000	≤3.500	≤5	≤1000	≤1000	≤40
Control voltages (V)	60–200 200–500 500–1000	60–200 200–500 500–1000	10–400	60–500	120–1000	10–500

Ultrasonic actuators are differentiated according to resonators with bar-like geometry and rotatory ring geometry.

Ultrasonic Actuators with Circular Resonators

As mentioned before, besides actuators providing standing waves, another group of actuators based on traveling waves exists. The traveling wave actuators known best are circular in their design. The first actuator according to this principle was built in 1973 by SASHIDA [125]. Traveling waves actuators count to the group of ultrasonic actuators as their control frequencies typically lie between 20 and 100 kHz.

This section is reduced to the presentation of ring-shaped traveling wave actuators with a bending wave. Other design variants for linear traveling wave actuators can be found in the corresponding literature [29, 46, 47].

Figure 9.25 shows an actuator's stator made of piezoelectric elements. They have an alternating polarization all around the ring. The stator itself carries notches actually enabling the formation of the rotating traveling wave.

Each point on the surface of the stator performs a local elliptic movement (trajectory). This movement is sketched schematically in Fig. 9.25. These individual elliptic movements overlie to a continuous wave on the stator. With frictional coupling, this movement is transferred on the rotor, resulting in a rotation. The contact between stator exists with the same number of contact points anytime during operation.

The movement equation of the traveling wave actuator is given by

$$u(x, t) = A\cos(kx - \omega t) \tag{9.48}$$

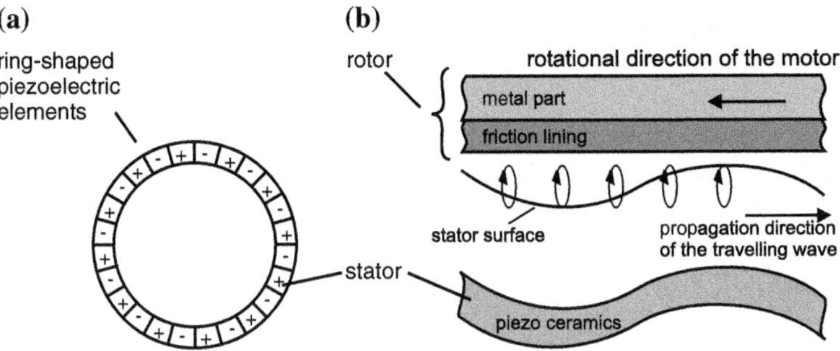

Fig. 9.25 Piezoelectric traveling wave motor: **a** stator disk with piezoelectric elements. **b** schematic view of the functionality of a ring-shaped piezoelectric traveling wave motor [61]

By reshaping it, the following form results in

$$u(x,t) = A(\cos(kx))(\cos(wt)) + A(cos(kx - \pi/2))(cos(kx + \pi/2)) \quad (9.49)$$

The second term of Eq. (9.49) includes important information for the control of traveling wave actuators. A traveling wave can be generated by two standing waves being spatially and timely different. Within typical realization, the spatial difference of $x_0 = \lambda/4$ is chosen, and a time phase lag of $\Phi_0 = \pi/2$. The use of two standing waves is the only practical possibility for generating a traveling wave. The direction of rotor movement can be switched by changing the phase lag from $+\pi/2$ to $-\pi/2$ [43, 49, 50, 125].

Figure 9.26 shows the practical realization of a traveling wave motor. The advantages of a traveling wave motor are the high torques possible to achieve at low rotational speeds. It has low overall size and is of little weight. This enables a thin design as shown in Fig. 9.26 . In passive mode, the traveling mode motor has high locking torque of several Nm.

Other advantages are given by the good control capabilities, high dynamics, and robustness against electromagnetic noise such as the silent movement [65]. Typical applications of traveling wave actuators are autofocus functions in cameras.

Piezoelectric Stepper Motors

Another interesting design can be found with the actuator PI Nexline. It combines the longitudinal effect with the piezoelectric shear effect, resulting in a piezoelectric stepper motor.

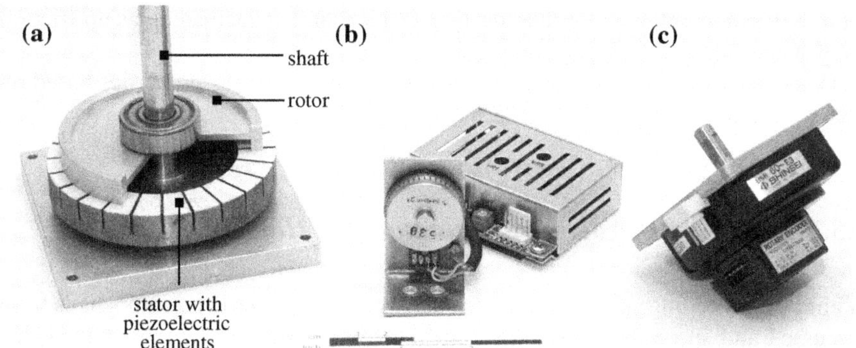

Fig. 9.26 Realization of a traveling wave motor. **a** cross-sectional model with functional parts, **b** motor model USR30 with a maximum speed of 300 rpm and a nominal torque of 0.05 Nm with driver D6030, **c** model USR60 with attached rotary encoder, maximum speed of 150 rpm and a nominal torque of 0.05 Nm. Scale shown is cm (*top*) and inch (*bottom*), valid for all parts of the figure. All examples by *Shinsei Corporation*, Tokyo, JP

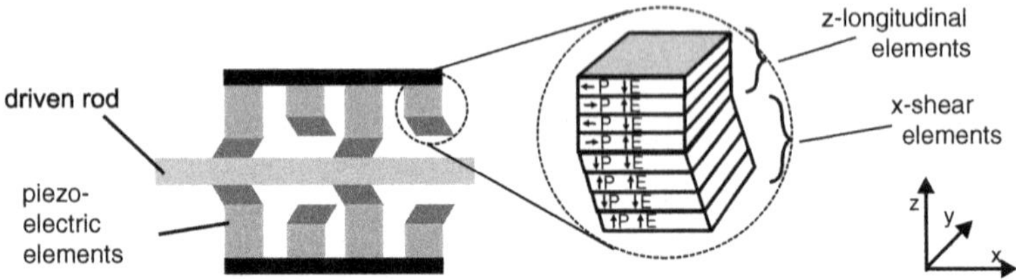

Fig. 9.27 Piezoelectric stepper motor using the shear effect and the longitudinal effect [101]

The principal design is sketched in Fig. 9.27. The movement of the motor is similar to the inchworm principle. Drive and release phases of the piezoelectric elements produce a linear movement of the driven rod. The piezoelectric longitudinal elements generate the clamping force in *z*-direction, the shear elements rotated by 90° a translational movement in *y*-direction is also possible.

The advantage of this design is given by the high positioning resolution. Over the whole displacement of 20 mm, a resolution of 0.5 nm can be achieved. The stepping frequency is given by—dependent on the control—up to 100 Hz and enables, depending on its maximum step-width, velocities of up to 1 mm/s. The step-width can be chosen continuously between 5 nm and 8 μm. The intended position can be achieved either closed loop or open loop controlled. For closed-loop control, a linear encoder has to be added to the motor. In open-loop control, the resolution can be increased to 0.03 nm in a high-resolution dithering mode.

The actuator can generate push and pull forces of 400 N maximum. The self-locking reaches up to 600 N. The typical driving voltage is 250 V. The specifications given above are based on the actuator N-215.00 Nexline® of the company *Physik Instrumente (PI) GmbH & Co. KG*, Karlsruhe, Germany [101]. Besides the impressive forces and positioning resolutions that can be achieved, these actuators have high durability compared to other designs of piezoelectric actuators, as no friction happens between moving parts and stator.

OLSSON ET AL. presented a haptic glove to display stiffness properties based on such an actuator as shown in Fig. 9.28 [96].

9.3.3 Design of Piezoelectric Actuators for Haptic Systems

Within the preceding section, the basic designs of piezoelectric actuators have been discussed and special variants were shown. This section transfers this knowledge about the design of piezoelectric actuators focusing on haptic applications now.

First of all, the principal approach for designing is sketched. Hints are given about designs appropriate for such applications. Later, three tools for practical engineering are shown: description via electromechanic networks, analytical formulations, and finite-element simulations.

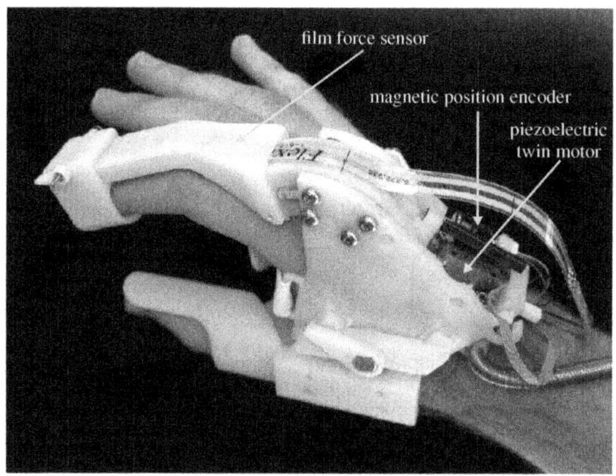

Fig. 9.28 Hand exoskeleton for the display of stiffness parameters. Forces exerted by the user are recorded with thin force sensors and the actuator position is tracked magnetically. Figure taken from [96]

9.3.4 Procedure for the Design of Piezoelectric Actuators

Figure 9.29 gives the general procedure for the design of piezoelectric actuators.

The choice of a general design based on those shown in the previous section is largely dependent on the intended application. For further orientation, Fig. 9.30 shows a decision tree for classifying the own application.

The following section describes the appropriate designs for specific application classes according to this scheme. The list has to be regarded as a point for orientation, but it does not claim to be complete. The creativity of an engineer will be able to find and realize other and innovative solutions besides those mentioned here.

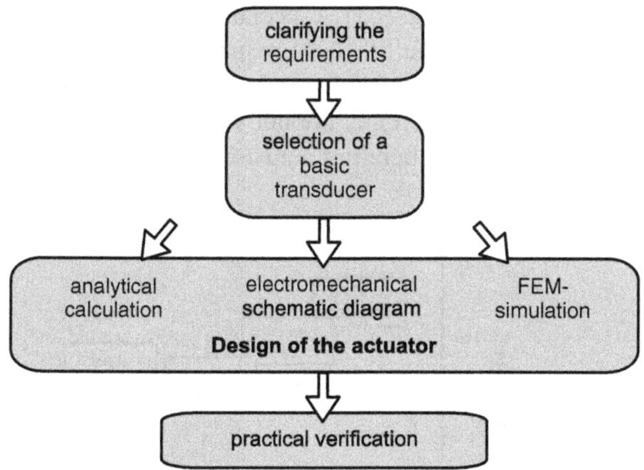

Fig. 9.29 Procedure of designing piezoelectric actuators

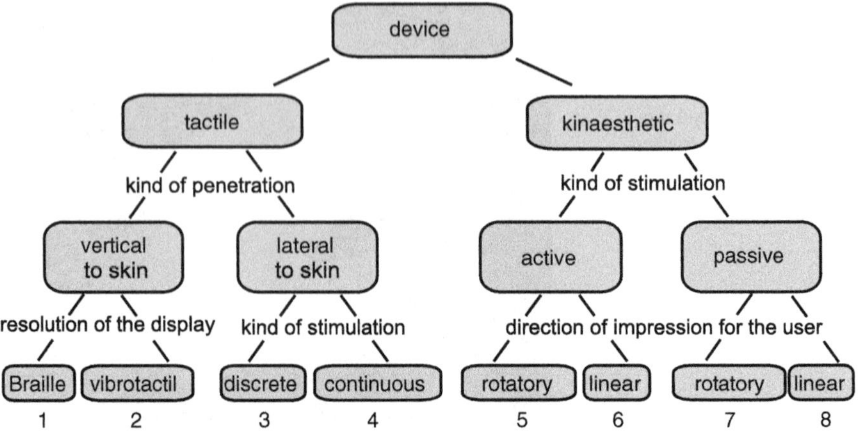

Fig. 9.30 Decision tree for the selection of a type of piezoelectric actuator

Nevertheless, especially for the design of tactile devices, some basic advice can be given based on the classification in Fig. 9.30:

1. Braille displays have to act against the finger's force. At typical resolutions, this requires forces in the area of mN at displacement of around 100 μm. The requirements on the dynamics are in the lower range of several Hertz. The smallest resolution of a pixel has to be in the area of $1 \times 1\,\text{mm}^2$ and is defined by the lowest resolution at the finger's tip. Looking at the force–amplitude diagram of Fig. 9.31 bending actuators fit well to these requirements.
2. In comparison with Braille displays, there are vibrotactile displays, which need higher frequencies and smaller displacements and forces to present a static shape to the user. With the diagram in Fig. 9.31, especially bending disk or staple actuators would be appropriate to fulfill these requirements, although these are overpowered concerning the achievable forces.

3, 4 Such displays are subject to current research and are not yet applied in broad range. Their design is typically based on bending actuators, as shear on the

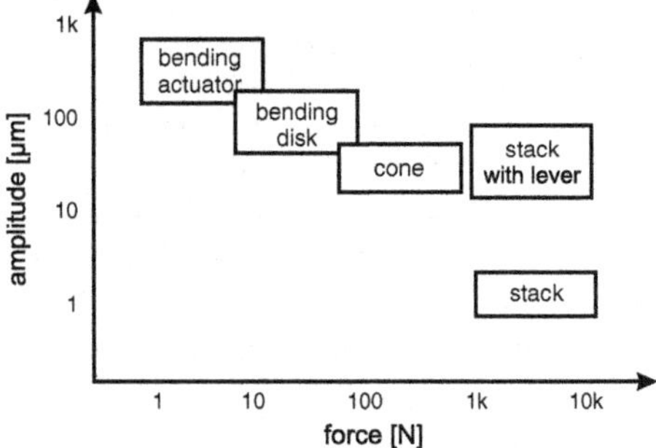

Fig. 9.31 Force–amplitude diagram for classification of the piezoelectric actuating types

skin requires less force than forces acting normal to the skin by generating a comparable perception.

5–8 In contrast to tactile displays, the actuator selection for kinaesthetic systems is more dependent on the actual application. Forces, displacement, and degrees of freedom influenced by the kinematics alter in wide ranges. Additionally, the actuator's volume may be a criterion for selection. Figure 9.31 gives an overview of piezoelectric actuation principles and has to be interpreted based on the specific kinaesthetic problem at hand. Generally speaking, ultrasonic piezoelectric actuators are usually a matter of choice for kinaesthetic devices, although they have to be combined with a closed-loop admittance control.

Figure 9.31 gives an overview of piezoelectric actuation principles and has to be interpreted according to the specific kinaesthetic problem at hand. Generally speaking, ultrasonic piezoelectric actuators are usually a matter of choice for kinaesthetic devices, although they have to be combined with a closed-loop admittance control.

Additional reference for actuator selection found in Sect. 9.3.2 are suitable for haptic applications, but still need some care in their use due to high voltages applied and their sensitivity on mechanical damage. This effort is often rewarded by piezoelectric actuation principles, which can be combined to completely new actuators. The only thing required is the creativity of the engineer.

After choosing the general actuator, the design process follows. For this purpose, three different methods are available, which are presented in the following and discussed with their pros and cons. In addition, some hints on further references are given.

9.3.4.1 Methods and Tools for the Design Process

There are three different engineering tools for the design of piezoelectric actuators:

- Description via the aid of electromechanical concentrated networks
- Analytical descriptions
- Finite-element simulations

Description via the Aid of Electromechanical Concentrated Networks

The piezoelectric basic equations from Sect. 9.3.1.1 are the basis for the formulation of the electromechanic equivalent circuit of a piezoelectric converter.

The piezoelectric actuator can be visualized as an electromechanical circuit. Figure 9.32 shows the converter with a gyratory coupling (a), alternatively a transformatory coupling (b) is also possible. For gyratory coupling, Eqs. (9.50)–(9.53) summarize the correlations for calculation of values for the concentrated elements. They are derived from the constants e, c, ϵ such as the actuator's dimensions l and A [77].

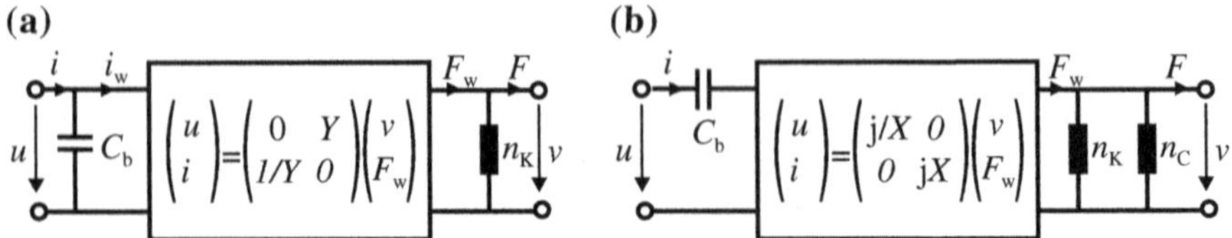

Fig. 9.32 Piezoelectric actuator as a electromechanical schematic diagram **a** gyratory and **b** transformatory combination [77]

$$C_b = \epsilon \cdot \frac{A}{l} = (\epsilon - d^2 \cdot c)\frac{A}{l} \quad \text{with} \quad v = 0 \tag{9.50}$$

$$n_K = \frac{1}{C} \cdot \frac{l}{A} = s \cdot \frac{A}{l} \qquad \text{with} \quad U = 0 \tag{9.51}$$

$$Y = \frac{1}{e} \cdot \frac{l}{A} = \frac{s}{d} \cdot \frac{l}{A} \tag{9.52}$$

$$k^2 = \frac{e^2}{\epsilon \cdot c} = \frac{d^2}{\epsilon \cdot c} \tag{9.53}$$

With the piezoelectric force constants

$$e = d \cdot c = \frac{d}{s} \tag{9.54}$$

This makes for the transformatory coupling

$$X = \frac{1}{\omega C_b \cdot Y} \quad \text{and} \quad n_C = Y^2 \cdot C_b \tag{9.55}$$

Figure 9.33 shows the sketch of an element Δx taken out of a piezoelectric bimorph bending actuator (dimensions $\Delta l \times \Delta h \times \Delta b$) as a electromechanical equivalent circuit.

It is:

$$C_b = 4\epsilon_{33}^T \left(1 - k_L^2\right) \frac{b \cdot \Delta x}{h}$$

$$\Delta n_{RK} \approx 12\, s_{11}^E \frac{(\Delta x)^3}{b \cdot h^3}$$

$$\frac{1}{Y} = \frac{1}{2} \frac{d_{31}}{s_{11}^E} \frac{b \cdot h}{\Delta x}$$

The piezoelectric lossless converter couples the electrical with the mechanical rotatory coordinates first, which are torque M and angular velocity Ω. To calculate the force F and the velocity v, additional transformatory coupling between rotatory and translatory mechanical network has to be introduced. As a result, the complete description of a subelement Δx of a bimorph is given in a ten-pole equivalent circuit.

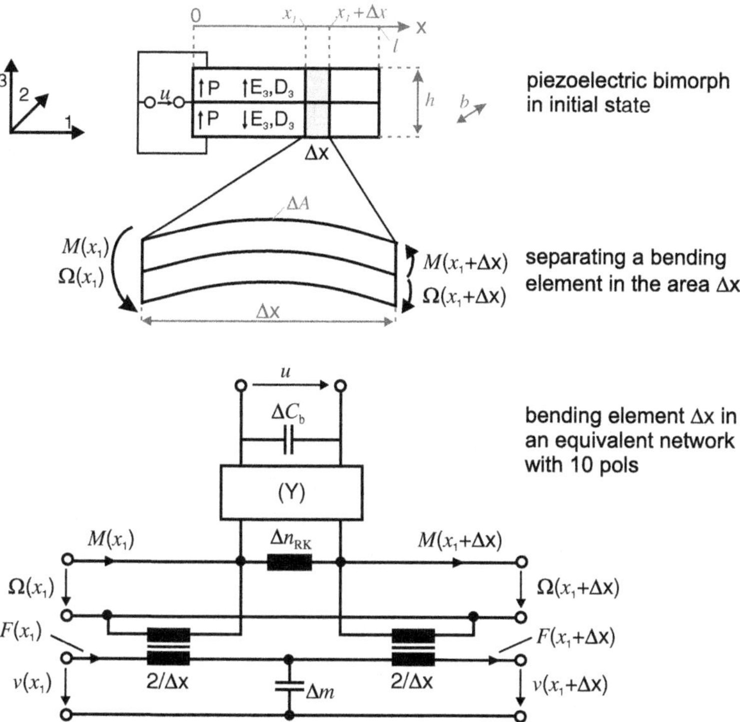

Fig. 9.33 Piezoelectric bimorph bending element in a electromechanical schematic view in quasi-static state [77]

Analytical Calculations

A first approach for the design of piezoelectric actuators is given by the application of analytical equations. The advantage of analytical equations lies in the descriptive visualization of physical interdependencies. The influence of different parameters on a target value can be derived directly from the equations. This enables high flexibility in the variation of dimensions and material properties. Additionally, the processing power for the solution of equations can—compared to simulations—be neglected.

A disadvantage of analytical solution results from the fact that they can only be applied to simple and frequently only symmetrical geometrical designs. Although already limited to such designs, even simple geometries may result in complex mathematical descriptions requiring a large theoretical background for their solution.

The following gives the relevant literature for familiarization with specific analytical challenges faced with during the design of piezoelectric actuators:

- Compelling and complete works on the design of piezoelectric actuators are [124–127].
- The theory of the piezoelectric effect and piezoelectric elements are discussed in [16, 55, 56].

- The mathematical description of traveling wave actuators can be found in [108, 141].
- The contact behavior between stator and rotor with piezoelectric multilayer bending actuators is analyzed in [42, 118, 133, 142].
- In the static and dynamic behavior of multilayer beam bending actuators is described.
- The description of the mechanical oscillations for resonance shapes is discussed in [8, 27, 30, 112, 114, 135] elaborately.

Finite-Element Simulation

The application of both approaches given before is limited to some limited geometrical designs. In reality, complex geometrical structures are much more frequent that cannot be solved with analytical solutions or mechanical networks. Such structures can be analyzed according to the method of finite-element simulation (FEM).

For design of piezoelectric actuators, the use of coupled systems is relevant. One example of a FEM simulation for a piezoelectric traveling wave motor is shown in Fig. 9.34.

9.3.5 Piezoelectric Actuators in Haptic Systems

Piezoelectric actuators are among the most frequently used actuation principles in haptic systems. The designs shown before can be optimized and adapted for a reasonable number of applications. One of the most important reasons for their use is their effectiveness at a very small required space, which is identical to a high power density. To classify the realized haptic systems, a division into tactile and kinaesthetic is done for the following sections.

9.3.5.1 Piezoelectric Actuators for Tactile Systems

For design of tactile systems, the application area is of major importance. The bandwidth ranges from macroscopic table-top devices, which may be used for embossed printings in Braille being placed below a PC keyboard, up to highly integrated sys-

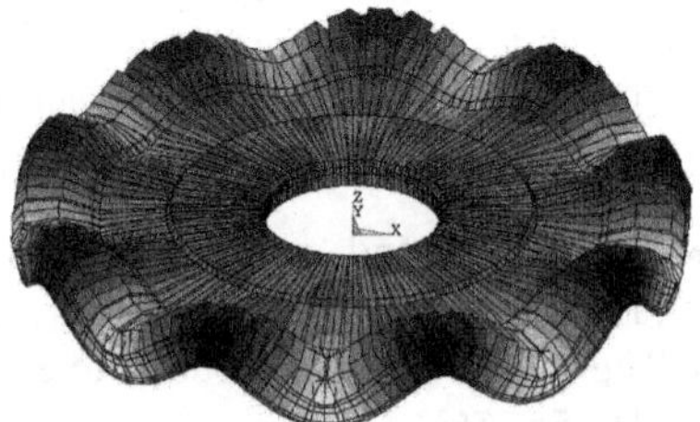

Fig. 9.34 Example FEM simulation of the oscillation shape of the stator of a piezoelectric traveling wave motor (view highly exaggerated)

tems, which may be used for mobile applications. In particular for the latter use, the requirements on volume, reliable, and silent operation, but also on low weight and energy consumption are enormous. The following examples are structured into two subgroups. Each of them addresses one of two directions of the penetration of the skin: lateral and normal.

Tactile Displays with Normal Stimulation

Braille Devices

A Braille character is encoded by a dot pattern formed by embossed points on a flat surface. By touching this pattern made of eight dots (two columns with four rows of dots each), combinations of up to 256 characters can be addressed. Since the 1970s reading tablets for visually handicapped people have been developed who are capable to present these characters with a 2×4 matrix of pins. The most important technical requirements are a maximum stroke of $0.1-1$ mm and a counter-force of 200 mN. Early in this development, electromagnetic drives have been replaced by piezoelectric bimorph bending actuators. These actuators enable a thinner design, are more silent during operation, and are faster. At typical operating voltages of $\pm 100-200$ V and at nominal current of 300 mA, they additionally need less energy than the electromagnetic actuators used before. Figure 9.35 shows the typical design of a Braille character driven by a piezoelectric bimorph actuator. A disadvantage of this system is the high price, as for 40 characters with eight elements each, altogether 320 bending actuators are needed. Additionally, they still require a large volume as the bending elements have to show a length of several centimeters to provide the required displacements. This group of tactile devices is part of the shape-building devices. The statically deflected pins enable the user to detect the displayed symbol.

In the HYPERBRAILLE project, a two-dimensional graphics-enabled display for blind computer users based on piezoelectric bending actuators is realized. The pin matrix of the portable tablet display consists of 60 rows with 120 pins each to present objects such as text blocks, tables, menus, geometric drawings, and other elements of a graphical user interface. The array is an assembly of modules that integrate 10

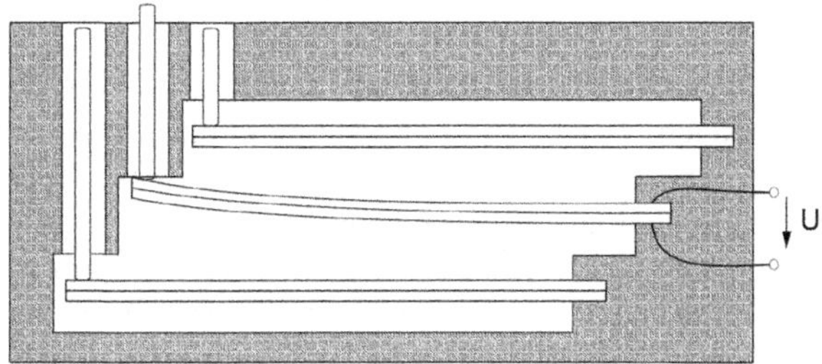

Fig. 9.35 Schematic setup of a Braille row with piezoelectric bending actuators

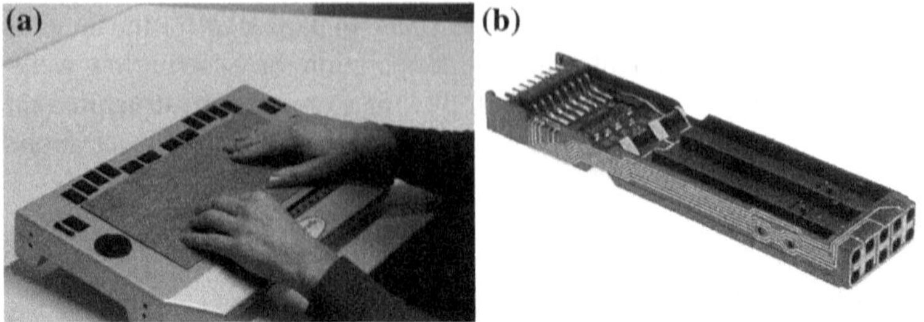

Fig. 9.36 HyperBraille display: whole device (**a**) [54], and single actuator module (**b**) [87]. Pictures courtesy of *metec AG*, Stuttgart, Germany

pins, spaced at intervals of 2.5 mm as shown in Fig. 9.36. The benders raise the pins above the plate by 0.7 mm [130].

Vibrotactile Devices

With vibrotactile devices, the user does not detect the displacement of the skin's surface but the skin itself is put into oscillations. At smaller amplitude, the sensation is similar to the static elongation. The general design of vibrotactile displays equals an extension of the Braille character to an $N \times N$ matrix, which is actuated dynamically. The tactile image generated is not perceived by the penetration depth but by the amplitude of the oscillation [58]. Another impact factor is the oscillation frequency, as the tactile perception depends extremely on the frequency. With the knowledge of these inter-dependencies, optimized tactile displays can be built generating a well-perceivable stimulation of the receptors. Important for displays according to this approach is a large surface, as movements performed by the own finger disturb the perception of the patterns.

The TEXTURE EXPLORER presented in [57] is designed as a vibrating 2×5 pin array. It is used for research on the perception of tactile stimulation, such as the overlay of tactile stimulation with force feedback. The surfaces displayed change according to their geometry and roughness within the technical limits of the device. The contact pins have a size of $0.5 \times 0.5\,\text{mm}^2$ with a point-to-point distance of 3 mm. Each pin is actuated separately by a bimorph bending actuator at a driving voltage of 100 V and a frequency of 250 Hz. The maximum displacement of these pins with respect to surface level is 22 μm and can be resolved down to 1 μm resolution.

An even more elaborate system is based on 100 individually controlled pins [116]. It can be actuated dynamically in a frequency range of 20–400 Hz. Figure 9.37 shows a schematic sketch. Twenty piezoelectric bimorph bending actuators (PZT-5H, *Morgan Matrinic, Inc.*) in five different layers one above the other are located in a circuit around the stimulation area. Each bending actuator carries one stimulation pin, which is placed 1 mm above the surface in idle state. The pins have a diameter of 0.6 mm and are located equally spaced at 1 mm distance. At a maximum voltage of $\pm$ 85 V, a displacement of $\pm$ 50 μm is achieved. A circuit of equally high passive pins is located around the touch area to mark the borders of the active display.

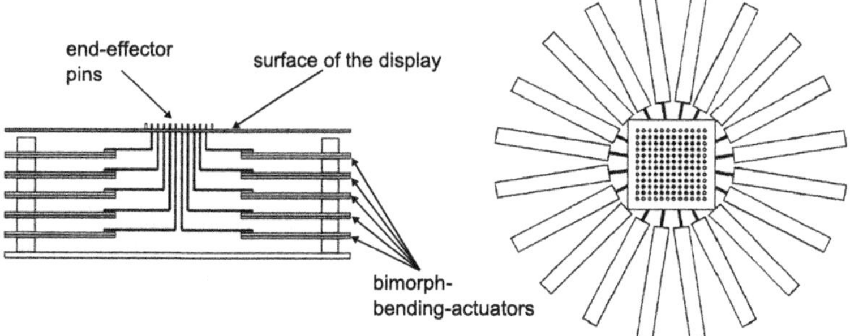

Fig. 9.37 Schematic setup of the 100 pin array [116]

Another even more compelling system can be found in [75] (Fig. 9.38). The very compact 5×6 array is able to provide static and dynamic frequencies of up to 500 Hz. Once again, piezoelectric bending actuators are used to achieve a displacement of 700 μm. However, the locking force is quite low with a maximum of 60 mN.

Ubi-Pen

The "Ubi-Pen" is one of the highest integrated tactile systems. Inside of a pen, both components, a spinning disk motor and a tactile display, are assembled. The design of the tactile display is based on the "TULA35" ultrasonic linear drive (*Piezoelectric Technology Co.*, Seoul, South Korea). A schematic sketch of the design is given in Fig. 9.39. The actuator is made of a driving component, a rod, and the moving part. The two piezoelectric ceramic disks are set to oscillation resulting in the rod oscillating upwards and downwards. The resulting movement is elliptical. To move upwards, the following procedure is applied: in the faster downward movement, the inertial force is excelled by the speed of the operation and the element remains in the upper position. Whereas in the upwards movement, the speed is controlled slow enough to carry the moving part up by the frictional coupling between moving

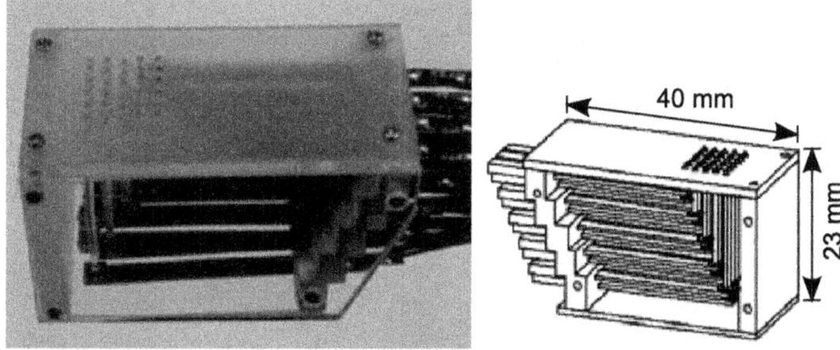

Fig. 9.38 Schematic setup of the 5×6 pin array [75]

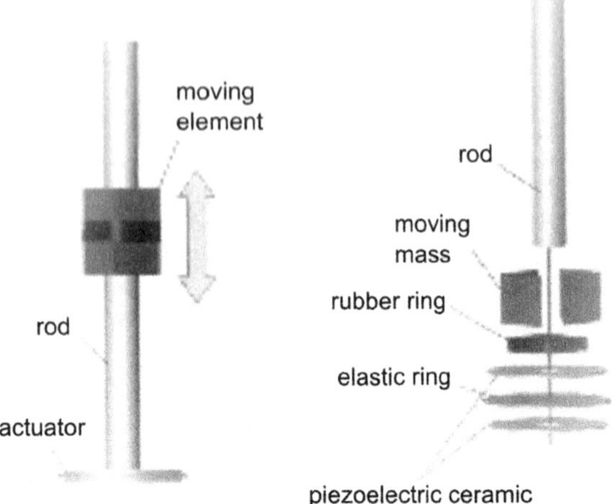

Fig. 9.39 Schematic setup of the ultrasonic motor "TULA35" [74]

element and central rod. The actuator disks have a diameter of 4 mm and height of 0.5 mm. The rod has a length of 15 mm and a diameter of 1 mm. It can be used as contact pin directly. The actuator's blocking force is larger than 200 mN and at a control frequency of 45 kHz velocities of 20 mm/s can be reached.

Figure 9.40 shows the design of a 3 x 3 pin array. In particular, the very small size of the design is remarkable: All outer dimensions have a length of 12 mm. The pins are distributed in a matrix of 3 mm. On the area of 1.44 cm^2, nine separate drives are located. To achieve such a high actuator density, the lengths of the rods have to be different, allowing the moving parts to be placed directly next to each other. If this unit is placed at the upper border of the display, all pins move in, respectively, out of the plane. The weight of the whole unit is 2.5 g. When the maximum displacement of 1 mm is used, a bandwidth of up to 20 Hz can be achieved.

The integration in a pen including another vibratory motor at its tip is shown in Fig. 9.41. This additional drive is used to simulate the contact of the pen with a surface. The whole pen weighs 15 g.

Fig. 9.40 Tactile 3 × 3 pin array [75]

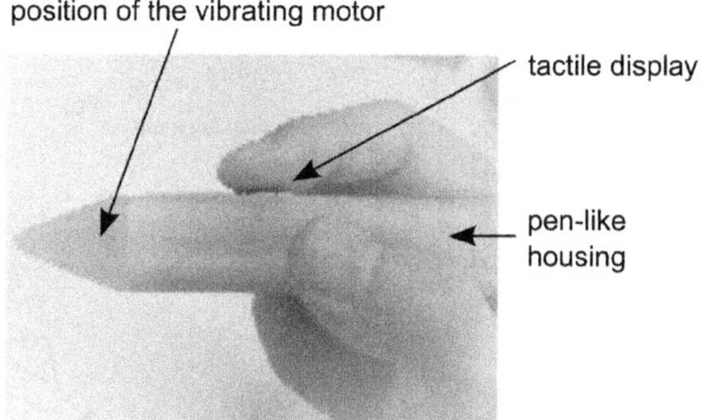

Fig. 9.41 Prototype of the "Ubi-Pen" [74]

The Ubi-Pen provides surface structures such as roughness and barriers or other extreme bumpy surfaces. To realize this, vibrations of the pins are superimposed with the vibratory motor. If the pen is in contact with a touch-sensitive surface (touch panel), the shown graphical image may be displayed in its grayscale values by the pins of the tactile display. The system has been used for a number of tests for recognition of information displayed in different tactile modalities [73]. The results are good with a mean recognition rate of 80 % with untrained users.

Tactile Displays with Lateral Stimulation

Discrete Stimulation

The concept of discrete stimulation is based on an excitation lateral to the skin's surface ("laterotactile display") [45]. Figure 9.42a shows the schematic sketch of a one-dimensional array of actuators. An activation of a piezoelectric element results in its elongation and a deformation of the passive contact comb (crown). If the skin of a touching finger is stretched by this movement, a contact point is perceived. Part (b) of Fig. 9.42 shows a two-dimensional display on the right. With the extension from 1D to 2D array, it is necessary to consider the more complex movement patterns resulting from it. A deeper analysis of the capabilities of such a system and the application for the exploration of larger areas can be found in [79, 134], proving the applicability of a laterotactile display as a virtual 6-point Braille device and to render surface properties. Further tests prove the generated tactile impression very realistic, it can hardly be distinguished from a moving pin below the finger surface.

Continuous Stimulation

The transfer from the discrete points to a piezoelectric traveling wave drive is shown in [12]. The touching finger faces a closed and continuous surface. Due to this design, the tactile display itself becomes less sensitive in its performance to movements of the finger. With the contact surface beyond the skin being excited as a standing wave, the user perceives a surface texture. With relative movement between wave and finger

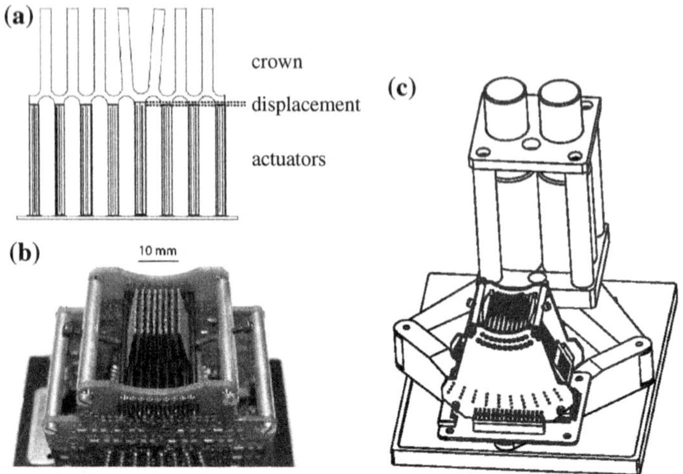

Fig. 9.42 **a** 1D array of the "laterotactile display" [45]. **b** 2D STReSS^2 display [79], **c** integration of an optimized version of the 2D display mounted on a Pantograph haptic device to simulate texture on large surfaces [134]

even a roughness can be perceived. By modifying the shape of the traveling wave, a simulation of a touch force perceivable by the moving finger can be achieved. Figure 9.43 shows the schematic sketch of the contact between finger and traveling wave, such as the corresponding movement direction.

In a test-bed application [12], the stator of a traveling actuator USR60 from *Shinsei* has been used. This actuator provides a typical exploration speed by its tangential wave speed of 15 cm/s and forces up to $\approx$2 N. This system enables to generate continuous and braking impressions by the change of wave shapes. An additional modulation of the ultrasonic signals with low frequency periodical signal generates the sensation of surface roughness. Actual research is performed on the design of linear ultrasonic traveling wave displays.

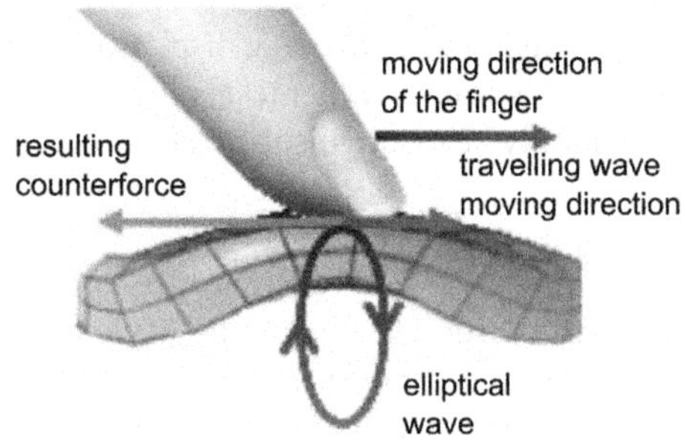

Fig. 9.43 Contact area of finger and traveling wave [12]

9.3.5.2 Piezoelectric Actuators for Kinaesthetic Systems

Piezoelectric actuators used in kinaesthetic systems are usually part of active systems (in a control-engineering sense). The user interacts with forces and torques generated by the actuator. A classic example is a rotational knob actuated by a traveling wave motor. With passive systems, the actuator is used as a switching element, which is able to consume power from an actuator or user in either time-discrete or continuous operation. Examples are breaks and clutches.

Active Kinaesthetic Systems

Piezoelectric traveling wave actuators show a high torque to mass ratio compared to other electrical actuation principles. They are predestined for use in applications with high torque at small rotational velocity, as they do not need additional bearing or other transmission ratios. Kinaesthetic systems require exactly these properties. A very simple design can be used to build a haptic knob: A rotationally mounted plate with a handle attached for the user is pressed on the stator of a traveling wave piezoelectric motor. A schematic sketch of the critical part is shown in Fig. 9.25. Due to the properties specific to the traveling wave motor, the rotational speed of the rotor can be adjusted easily by increasing the wave's amplitude w at the stator. As this actuation principle is based on a mechanical resonance mode, it is actuated and controlled with frequencies nearby its resonance. Coincidentally, this is the most challenging part of the design, as the piezoelectric components show nonlinear behavior at the mechanical stator–rotor interface. Hence, the procedures for its control and electronics have a large influence on its performance.

Figure 9.44 shows the speed versus torque characteristics for different wave amplitudes w of the actuator [39]. The torque is highly dependent on the actual rotational speed and amplitude. By monitoring the speed and controlling the phase and wave amplitude, the system can be closed-loop controlled to a linear torque–displacement

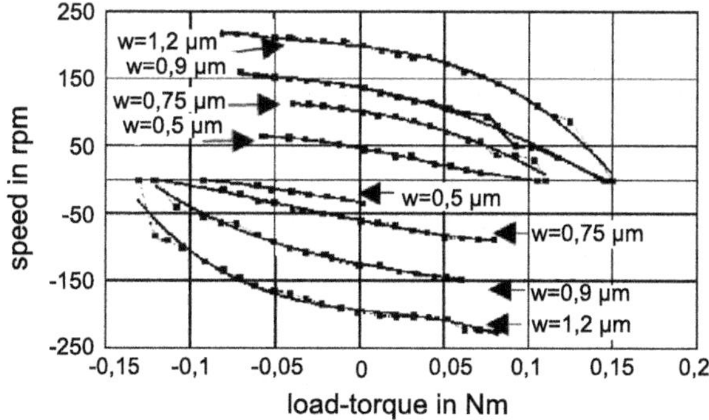

Fig. 9.44 Speed/load torque characteristics for different wave amplitudes. Figure taken from [39]

characteristic. In this mode, a maximum torque of about 120 mNm can be achieved with this example. A deeper discussion of the phase control for a piezoelectric traveling wave motor according to this example is given in [38]. A specialized design of such a device is used in neurological research for application with magneto-resonance tomography [31]. For closed-loop control, the admittance of the device the torque has to be measured. In the specific application, near a large magnetic field, glass fibers are used, measuring the reflection at a bending polymer body. Preventing disturbance from the device on the switching gradients of MRI and vice versa, the device is designed from specific non-conductive materials. It was based on a traveling wave motor in a special MR-compatible version of the "URS60" from *Shinsei*.

Hybrid Systems

Another class of kinaesthetic systems are so-called hybrid systems. If there is need for generating a wide bandwidth of forces and elongations with a single device, there is no actuator fulfilling all requirements alone. Due to this reason, several hybrid systems are designed with two (or more) components complementing each other. A typical example is the combination of a dynamic drive with a brake; the latter is used to provide large blocking torques. As seen in the above section, the closed-loop control of a traveling wave motor's impedance is a challenging task. A simple approach to avoid this problem is the combination of a traveling wave actuator with a clutch. The main difference between a traveling wave actuator and other types of actuators is given by its property to rotate with a load-independent velocity. Providing a certain torque, this system is accompanied with a clutch. Other designs add a differential gear or a break. Such a system is presented in [21]. If the system experiences a mechanical load, the operation of the break is sufficient to increase friction or even block the system completely. Consequently, the system provides the whole dynamic range of the traveling motor in active operation, whereas passive operation is improved significantly by the break. Due to the simple mechanical design and the reduction of energy consumption, such systems are suitable for mobile applications as well.

Passive Kinaesthetic Systems

Objects can be levitated by standing acoustic waves. To achieve this, an ultrasonic source has to provide a standing wave first. Within the pressure nodes of the wave, a potential builds up, attracting a body placed nearby the node. The size of the object is important for the effect to take place, as with a large size, the influence on the next node may be too high. A system based on this principle is described in [72]. It shows the design of an exoskeleton in the form of a glove with external mechanical guides and joints. The joints are made of piezoelectric clutches with a schematic design shown in Fig. 9.45. In their original state, both disks are pressed together by an external spring generating a holding torque. With the vibrator being actuated, the levitation mode is achieved between rotor and stator creating a gap h. This reduces the friction drastically allowing to make both disks almost free of each other.

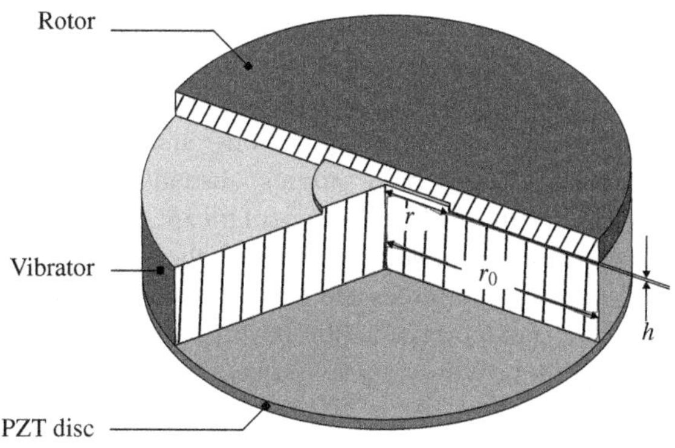

Fig. 9.45 Schematic setup of the "levitation clutch" presented in [72]

9.3.5.3 Summary

Tactile systems are distinguished according to their direction of movement. With a normal movement into the direction of the skin, additional differences are made between passive systems simulating a more or less static surface by their pins, and active systems—so-called vibrotactile systems—providing information by a dynamic excitement of the skin's surface. The user integrates this information into a static perception. The advantages of this approach are given by the reduced requirements on the necessary force and displacements, as the dynamical perception of oscillations is higher than the perception of static slow-changing movements. When the display is not fixed to the finger, however, its fast movements will be a problem. With a static display in a fixed frame the user is able to repeatedly touch the display, increasing the dynamics of the haptic impression by own movements. With a dynamic display, this interaction does not work as well anymore as periods of oscillations from the vibrating elements are lost.

Another alternative is tactile systems with lateral movement of the skin. With appropriate control, the human can be "fooled" to feel punctual deformations analogous to an impression of a normal penetration. Systems with a closed surface are comfortable variants of such displays, but their dynamic control is demanding for finger positions moving across larger surfaces. Typically, today's solutions show smaller contact areas than with other variants, as the actuator elements cannot be placed as close together as necessary.

Kinaesthetic (force feedback) systems can be distinguished in active and passive systems in a control-engineering sense. Active systems are able to generate counterforces and supportive forces. The spectrum of movements is only limited by the degrees-of-freedom achieved by the mechanical design. A stable control for active system tends to become very elaborate due to the required measurement technology and complex control algorithm of sufficient speed. As with all active systems, a danger that remains is: an error in the functionality of the system may harm the user. This danger increases for piezoelectric actuators, as the available forces and torques are high. Passive systems with breaks and clutches enable the user to feel the resistance against their own movement as reactive forces. These designs are simpler to build and less dangerous by definition of passivity. The general disadvantages of passive systems can be found in their high reaction times, change in their mechanical properties in long-time applications, and their comparably large volume. Hybrid systems combining both variants—usually including another actuation principle—may enlarge the application area of piezoelectric actuators. Although the mechanical design increases in volume and size, the requirements on control may become less and large holding forces and torques can be achieved with low power consumption. From the standpoint of haptic quality, they are one of the best actuator solutions for rotating knobs with variable torque/angle characteristics available today.

9.4 Electromagnetic Actuators

Thorsten A. Kern

Electromagnetic actuators are the most frequently used actuator type within the general automation industry. Due to their simple manufacture and assembly, they are a matter of choice. Additionally, they do not necessarily need a permanent magnet, and their robustness against exterior influences is high. They are used within coffee machines, water pumps, and for redirecting paper flow within office printers. But nevertheless, their applicability for haptic devices, especially kinaesthetic devices, is limited. Their main fields of application are tactile systems. This can be reasoned by several special characteristics of the electromagnetic principle. This chapter gives the theoretical basics for electromagnetic actuators. Technical realizations are explained with examples. First, the general topology and later the specific designs are shown. The chapter closes with some examples of haptic applications of electromagnetic actuators.

9.4.1 Magnetic Energy

The source responsible for the movement of a magnetic drive is the magnetic energy. It is stored within the flux-conducting components of the drive. These components are given by the magnetic core (compare Sect. 9.2.1.3) and the air-gap, such as all leakage fields, which are neglected for the following analysis. It can be seen from Table 9.2 that stored magnetic energy is given by the products of fluxes and magnetic voltages in each element of the magnetic circuit:

$$W_{\mathrm{mag}} = \sum_n H_n\, l_n \cdot B_n\, A_n \tag{9.56}$$

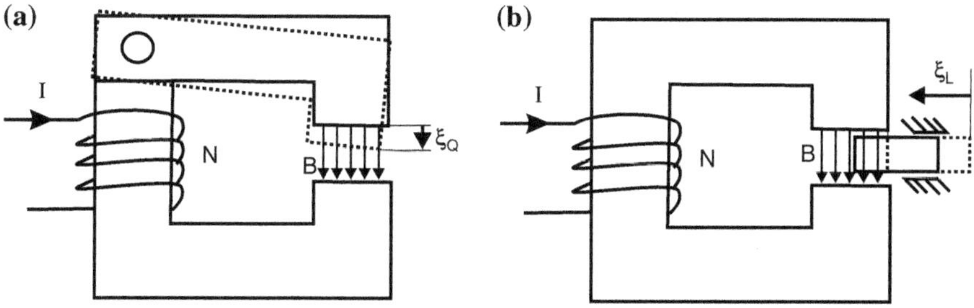

Fig. 9.46 Electromagnetic transversal effect (**a**) and longitudinal effect (**b**)

As every other system does, the magnetic circuit tries to minimize its inner energy.[13] Concentrating on electromagnetic actuators, the minimization of energy almost always refers to reduction in the air-gap's magnetic resistance R_{mG}. For this purpose, two effects may be used, which can be found within electrostatic for electrical fields too (Sect. 9.5):

- Electromagnetic longitudinal effect (Fig. 9.46a) (also: reluctance effect)
- Electromagnetic transversal effect (Fig. 9.46b)

The forces, respectively, torques generated with the individual effects are the derivations of the energy according to the corresponding direction,

$$\mathbf{F}_{\xi} = \frac{\mathrm{d}\, W_{\mathrm{mag}}}{d\xi}, \tag{9.57}$$

being equal to a force in the direction of the change of the air-gap

$$\mathbf{F}_{\xi} = -\frac{1}{2}\phi^2 \frac{\mathrm{d}\, R_{\mathrm{mG}}}{\mathrm{d}\xi}. \tag{9.58}$$

9.4.1.1 Example: Transversal Effect

The magnetic resistance of an arbitrary homogeneous element of length l between two boundary surfaces (Fig. 9.47a) with the surface A is calculated as

$$R_{\mathrm{m}} = \frac{l}{\mu\, A}. \tag{9.59}$$

This gives the stored energy W_{mag} within the magnetic resistance:

[13] Minimizing potential energy is the basis for movements in all actuator principles. Actuators may therefore be characterized as "assemblies aiming at the minimization of their inner energy."

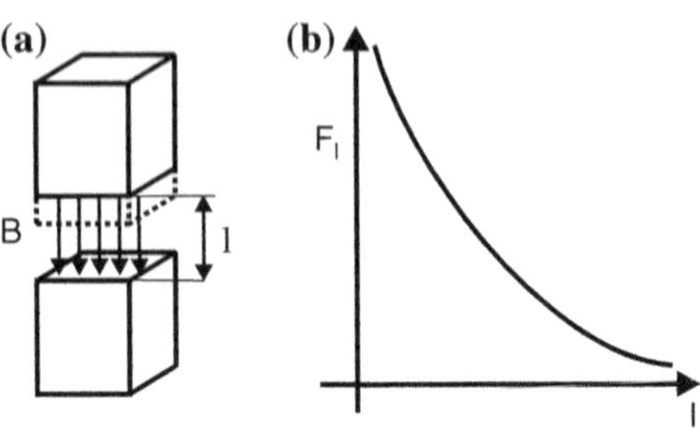

Fig. 9.47 Electromagnetic transversal effect in the air-gap (**a**) and with a qualitative force plot (**b**)

$$W_{\mathrm{mag}} = (B\,A)^2\,\frac{l}{\mu\,A}. \tag{9.60}$$

The flux density B is dependent on the length of the material. Assuming that the magnetic core contains one material, only the magnetomotive force Θ is calculated as

$$\Theta = \frac{B}{\mu l} = NI, \tag{9.61}$$

which gives

$$B = NI\,\frac{\mu}{l}. \tag{9.62}$$

This equation used to replace the flux density in Eq. (9.60), with several variables canceled, finally results in the magnetic energy

$$W_{\mathrm{mag}} = (NI)^2 A\mu\,\frac{1}{l}. \tag{9.63}$$

With the assumption about the magnetic energy concentrating within the air-gap—which is identical to the assumption that the magnetic core does not have any relevant magnetic resistance—the approximation of the force for the transversal effect in the direction of l can be formulated as

$$F_l = -\frac{1}{2}(NI)^2\,A\mu\,\frac{1}{l^2}. \tag{9.64}$$

The force shows an anti-proportional quadratic coherence (Fig. 9.47b) to the distance l. The closer the poles become, the higher the force attracting the poles increases.

9.4.1.2 Example: Longitudinal Effect, Reluctance Effect

The same calculation can be repeated for the longitudinal effect. Assuming that the surface A from Eq. (9.63) is rectangular and its edge lengths are given by a and b,

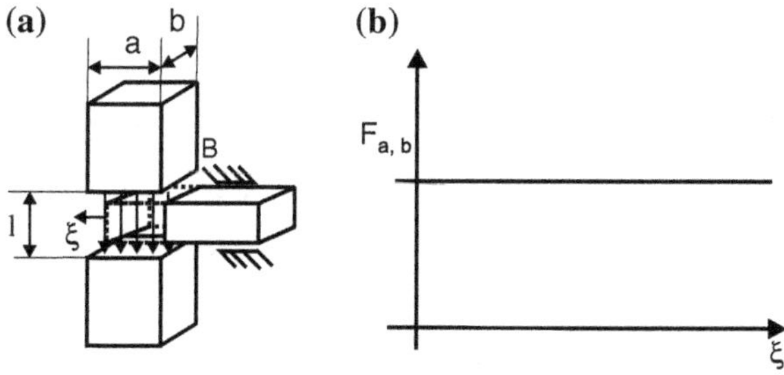

Fig. 9.48 Electromagnetic longitudinal effect in the air-gap (**a**) and as qualitative force plot (**b**)

and further assuming that a flux-conducting material is inserted into direction a, the forces in longitudinal direction can be calculated as

$$F_a = (NI)^2 \, b\mu \, \frac{1}{l}, \tag{9.65}$$

and in direction b as

$$F_b = (NI)^2 \, a\mu \, \frac{1}{l}. \tag{9.66}$$

The reluctance effect is—in contrast to the transversal effect—linear (Fig. 9.48b). The force is dependent on the length of the mobbing material's edge only. Consequently, the stored energy within the magnetic circuit is necessary for the design of an electromagnetic actuator. The above examples have the quality of rough estimations. They are sufficient to evaluate the applicability of an actuation principle—no more, no less. The magnetic networks sufficient for a complete dimensioning should contain effects with magnetic leakage fields and the core's magnetic resistance. Therefore, it is necessary to further deal with the design of magnetic circuits and their calculation.

9.4.2 Design of Magnetic Circuits

The basic interdependencies for the design of magnetic circuits have already been discussed in Sect. 9.2.1.3 in the context of electrodynamic actuators. Taken from the approach of longitudinal and transversal effect, several basic shapes (Fig. 9.49) can be derived applicable to electromagnetic actuators. In contrast to electrodynamic actuators, the geometrical design of the air-gap within electromagnetic actuators is freer. There is no need to guide an electrical conductor within the air-gap anymore. Besides the designs shown in Fig. 9.49, there are numerous other geometrical variants. For example, all shapes can be transferred to a rotational-symmetrical design around one axis. Additional windings and even permanent magnets can be added. There are however two limits to their design:

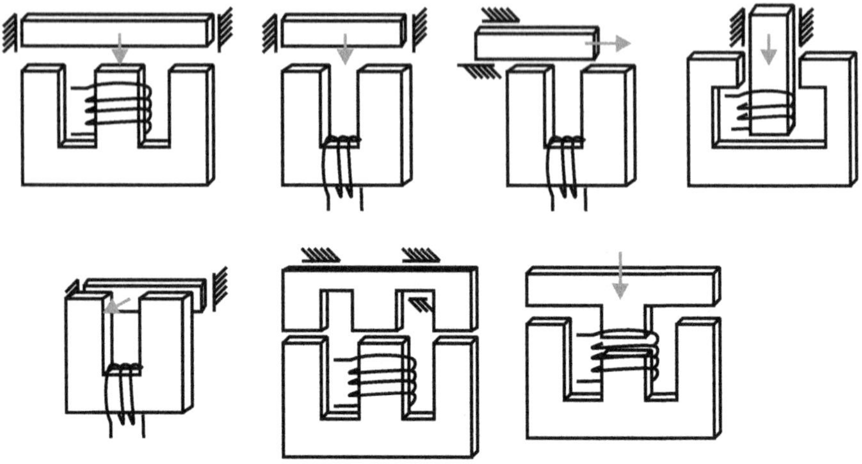

Fig. 9.49 Basic shapes of electromagnetic actuators

- A sufficient cross section of the flux-conducting elements has to be guaranteed to prevent the material from getting into saturation.
- A sufficient distance between flux-conducting elements has to be kept to prevent a magnetic leakage-field via the air.

9.4.2.1 Cross Section Surface Area: Rough Estimation

The calculation of the cross section surface area for dimensioning the magnetic core is simple. A common, easily available and within precision engineering and prototype-design gladly used material is steel ST37. The B/H-characteristic curve with its saturation is given in Fig. 9.8. For this example, we choose a reasonable flux density of 1.2 T. This equals a field intensity of $H \approx 1{,}000$ A/m. Within the air-gap a flux density of 1 T should be achieved. The magnetic flux within the air-gap is given as

$$\phi = A_{\mathrm{G}}\, B_{\mathrm{G}}. \tag{9.67}$$

As the magnetic flux is conducted completely via the magnetic core—neglecting leakage fields and other side bypasses—the relation

$$A_{\mathrm{Iron}}\, B_{\mathrm{Iron}} = A_{\mathrm{G}}\, B_{\mathrm{G}}, \tag{9.68}$$

is given, and consequently with the concrete values from above:

$$\frac{A_{\mathrm{Iron}}}{A_{\mathrm{G}}} = \frac{B_{\mathrm{G}}}{B_{\mathrm{Iron}}} = 0,833\,. \tag{9.69}$$

At its tightest point, the magnetic core may have 83 % of the cross section of the air-gap. More surface of the cross section results in lower field intensities, which should be aimed at if geometrically possible. Please note that $A_G \leq A_{Iron}$ is with almost all technical realization, as the boundary surface of the magnetic core is always one pole of the air-gap.

9.4.2.2 Magnetic Energy in the Magnetic Core and Air-Gap

In the above examples, the assumption that the energy stored within the magnetic core is clearly less than the energy within the air-gap should now be checked for validity. Calculating the magnetic resistance of an arbitrary element

$$R_m = \frac{l}{\mu A}, \tag{9.70}$$

the relation of two elements with identical length and cross section scales via the magnetic resistance with permeability μ:

$$\frac{R_{m1}}{R_{m2}} = \frac{\mu_2}{\mu_1} \tag{9.71}$$

The permeability $\mu_r = \frac{B}{H \mu_0}$ is given by the relation between flux density versus field strength relatively to the magnetic constant. It is nonlinear (Fig. 9.50) for all typical flux-conducting materials within the flux-density areas relevant for actuator design between 0.5 and 2 T. It is identical to the inverse gradient of the curves given in diagram 9.8. The maximum permeability values are given in tables frequently, but refer to field strengths within the material only. They range from 6,000 for pure iron over 10,000 for nickel alloys up to 150,000 for special soft-magnetic materials.

Mechanical processing of flux-conducting materials and the resulting thermal changes within its microstructure will result in considerable degradation of its magnetic properties. This change can be restored by an annealing process.

Generally speaking, however, even outside an optimum value for flux density the stored energy within typical materials is always several orders of magnitudes below the energy within the air-gap. This legitimates to neglect this energy portion

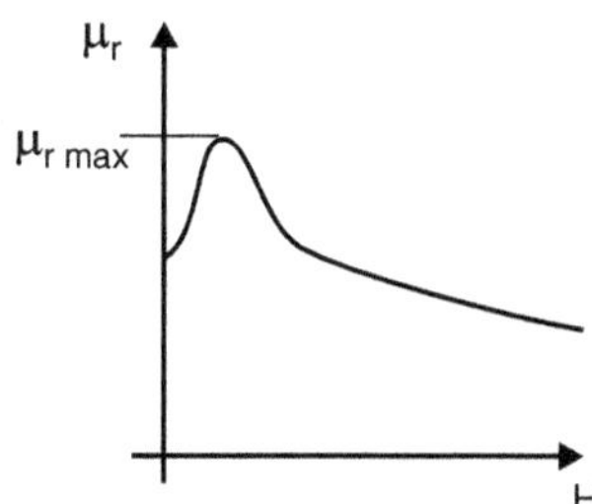

Fig. 9.50 Qualitative change permeability for common flux-conducting materials

for the rough estimations in actuator design, but also shows that there is potential within the optimization of electromagnetic actuators. This potential can be used by the application of FEM-software, which is typically available as module for CAD software.[14]

9.4.2.3 Permanent Magnets in Electromagnetic Actuators

Permanent magnets do not differ significantly in their properties from coils conducting DC current. They generate a polarized field, which—in combination with another field—provide attraction or repulsion. For calculating combined magnetic circuits, a first approach can be taken by substituting the sources within the magnetic equivalent circuit (neglecting saturation effects). The calculation is analogous to the methods mentioned in this chapter on electrodynamic actuators (Sect. 9.2.1.3). A permanent magnet within a circuit either allows

- the realization of a state held without any current applied
- or switching between two states with just one winding powered.

A good example for a currentless held state [66] shows the calculation of a polarized magnetic clamp (Fig. 9.51). With non-active winding, the flux is guided through the upper anchor and held securely. With active coil, the magnetic flux via the upper anchor is compensated. The magnetic bypass above the coil prevents the permanent magnet from being depolarized by a counter field beyond the kink in the B/H-curve.

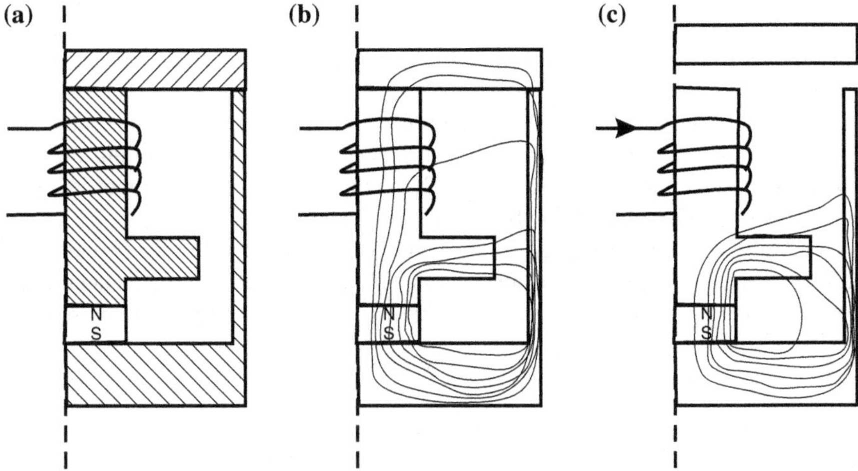

Fig. 9.51 Permanent magnet in the magnetic circuit in shape (**a**), field lines with inactive coil (**b**) and field lines with active coil (**c**), releasing the anchor

[14] Or as free software, e.g., the tool "FEMM" from David Meeker

9.4.3 *Examples for Electromagnetic Actuators*

Electromagnetic actuators are available in many variants. The following section gives typical designs for each principle and shows the corresponding commercial products. Knowledge of the designs will help to understand the freedom in the design of electromagnetic circuits more broadly.

9.4.3.1 Claw-Pole Stepper Motor

The electromagnetic claw-pole stepper motor (Fig. 9.52) is one of the most frequently used rotatory actuation principles. These actuators are made of two stamped metal sheets (1, 2) with the poles—the so-called claws—bent by 90° to the interior of the motor. The metal sheets are the magnetic core for conducting the flux of one coil each (3). The permanent magnet rotor (4) with a pole subdivision equalizing the claw pattern orientates to the claws in currentless state. In stepper mode, the coils are powered subsequently, resulting in a combined attraction and repulsion of the rotor. The control of the coil currents may happen either by simply switching them or by a microstep mode with different current levels being interpolated between discrete steps. The latter generates stable states for the rotor not only in the positions of the claws but also in between.

Claw-pole stepper motors are available with varying number of poles, different numbers of phases, and varying loads to be driven. As a result of the permanent magnet, they show a large holding torque with respect to their size. The frequency of single steps may reach up to 1 kHz for fast movements. By counting the steps, the position of the rotor can be detected. Step losses—the fact that no mechanical step happens after a control signal—are not very likely with a carefully designed power chain. Claw-pole stepper motors are the working horses in the electrical automation technology.

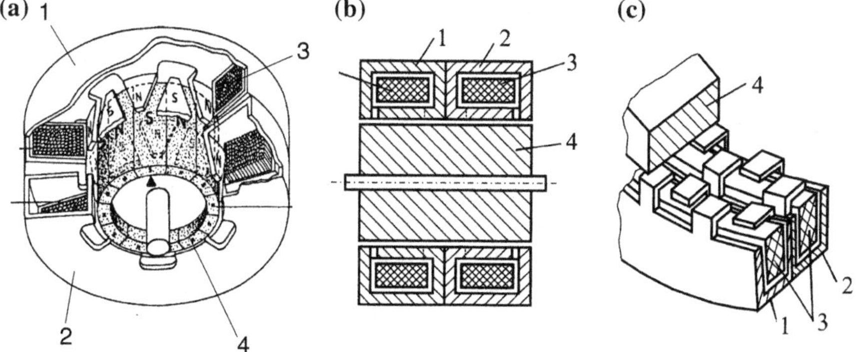

Fig. 9.52 Two-phase stepper motor made of stamped metal sheets and with a permanent magnet rotor in a 3D-sketch (**a**), cross section (**b**), and with details of the claw-poles (**c**). (Figure based on Kallenbach et al.)

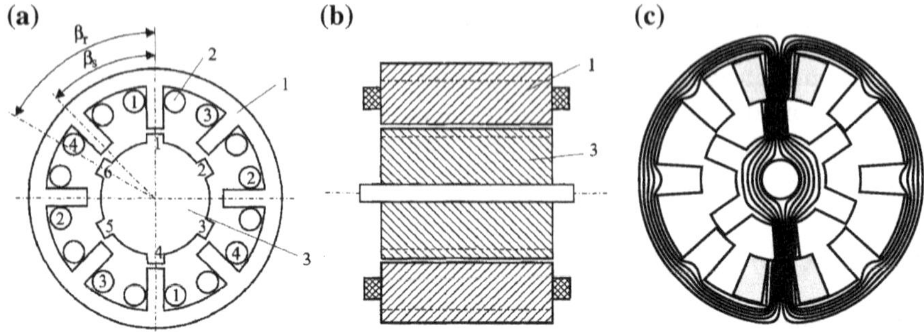

Fig. 9.53 Switched reluctance drive with pole- and coil-layout (**a**), in cross section (**b**), and with flux lines of the magnetic excitation (**c**). Figure based on [65]

9.4.3.2 Reluctance Drives

The rotatory reluctance drives (Fig. 9.53) are based on the electromagnetic longitudinal effect. By clever switching of the windings (2), it is possible to keep the rotor (3) in continuous movement with minimal torque ripples. To make this possible, the rotor has to have fewer poles than the stator. The rotor's pole angle β_r is larger than the pole angle β_s. Reluctance drives can also be used as stepper motors by the integration of permanent magnets. Generally speaking, it excels by high robustness of the components and a high efficiency factor with—for electromagnetic drives—comparably little torque ripples.

9.4.3.3 Electromagnetic Brakes

Electromagnetic brakes (Fig. 9.54) are based on the transversal effect. They make use of the high force increase at electromagnetic attraction to generate friction on a spinning disk (1). For this purpose, usually rotational-symmetrical flux-conducting magnetic cores (2) are combined with embedded coils (3). The frontal area of the magnetic core and braking disk (1) itself is coated with a special layer to reduce abrasion and influence positively the reproducibility of the generated torque. The current/torque characteristic of electromagnetic brakes is strongly nonlinear. On one hand, this is the result of the quadratic proportion between force and current of the electromagnetic transversal effect, and on the other hand, it is also a result of the friction pairing. Nevertheless, they are used in haptic devices for simulation of "hard contacts" and stop positions. A broad application for haptic devices is nevertheless not visible. This is likely a result of their limits in reproducibility of the generated torque, the resulting complex control of the current, and the fact that they can only be used as a break (passive) and not for active actuation.

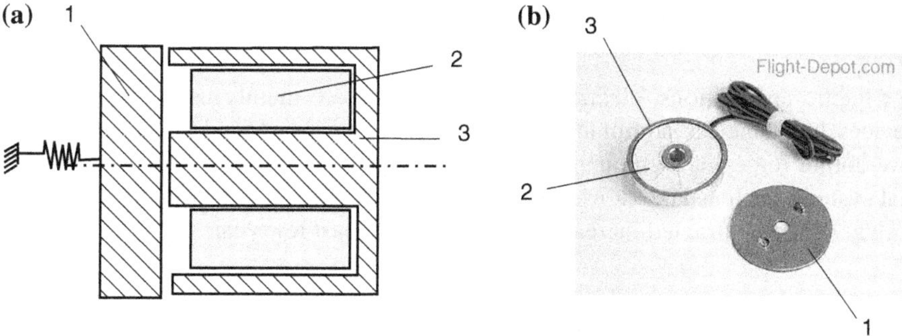

Fig. 9.54 Electromagnetic brake in cross section (**a**) and as technical realization for an airplane model (**b**)

9.4.3.4 Plunger-Type Magnet

Electromagnetic plunger-type magnets (Fig. 9.55) are frequently based on the electromagnetic transversal effect. Their main uses are switching and control applications, requiring actuation in specific states. With a magnetic core frequently made of bended iron steel sheets, (2) a coil-induced (3) flux is guided to a central anchor, which itself is attracted by a yoke (4). The geometry of the yoke influences significantly the characteristic curve of the plunger-type magnet. By varying its geometry, a linearization of the force–position curve is possible within certain limits. Even strongly nonlinear pulling-force characteristics can be achieved by such a modification. Plunger-type magnets are available with additionally magnets and with more coils. In these more complex designs, they provide mono- and bi-stable switching properties. By variation of the wires diameter and the number of turns, they can be adapted easily to any power level.

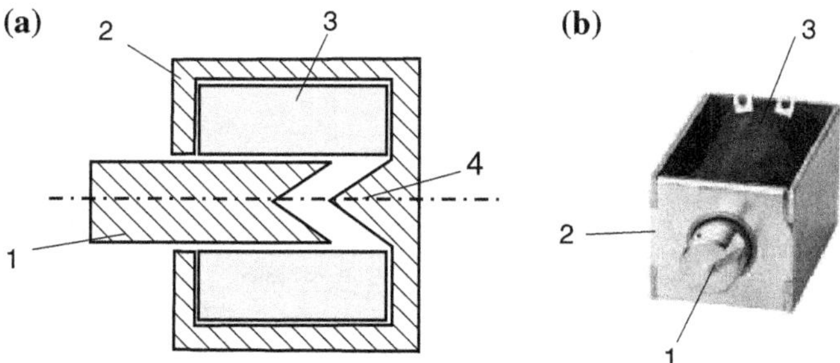

Fig. 9.55 Plunger-type magnet (**a**) with altered force–position curve (*4*), and realization as pulling anchor (**b**) with metal-sheet-made magnetic circuit (*2*)

9.4.4 *Magnetic Actuators in Haptic Devices*

For haptic applications, electromagnetic actuators are mainly used within tactile devices. Nevertheless, admittance-controlled devices can be found providing impressive output forces of high quality even by use of stepper motors. Besides commercial systems such as HAPTICMASTER of *Moog FCS* (Sect. 6.4) especially an idea of LAWRENCE has attracted increased attention in the past few years.

9.4.4.1 Spring-Tendon Actuator

In [76], LAWRENCE describes an inexpensive actuator for kinaesthetic haptic systems (Fig. 9.56) based on an electromagnetic stepper motor coupled via a tendon to a pen and with a spring mechanically connected in parallel. Analogous to other haptic devices, the pen is the interface to the user. Between pen and tendon and spring, there is a bending body with DMS as a force sensor. To additionally minimize the torque ripples of the stepper drive resulting from the latching of the poles, a high-resolution external encoder is attached to the motor and a torque/angle curve measured. A mathematical spline fit of this curve is used for the actuator's control to compensate the torque oscillations. In addition, the closed loop control of the actuator via the force sensor near the pen also includes a compensation of frictional effects. The result of these efforts is a force source, providing a force transmission with little noise and high stiffness up to 75 kN/m with movements of limited dynamics.

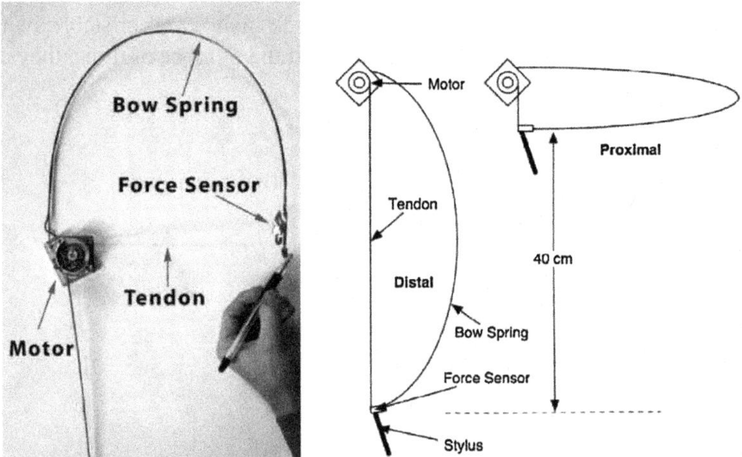

Fig. 9.56 Electromagnetic stepper motor with spring, actuated in closed-loop admittance control mode [76]

9.4.4.2 Electromagnetic Pin Array

The use of electromagnetic actuators for control of single pins in array design is very frequent. The earliest uses for haptic applications go back to printer heads of dot matrix printers that were used in the 1980s and early 1990s of the previous century. Modern designs are more specific to haptics and make use of manufacturing technologies available from microtechnology. In [24], an actuator array (Fig. 9.57) is shown made of coils with 430 windings each and a 0.4 mm-wide iron core. Above it, a magnet is embedded in a polymer layer attracted by the flux induced into the core. With such an actuator and diameter of 2 mm, a maximum force of up to 100 mN is possible. A position measure is realized via inductive feedback. Further realizations of tactile arrays based on electromagnetic actuators and different manufacturing techniques can be found in the work of STREQUE ET AL. [115].

9.4.4.3 Electromagnetic Plunger-Type Magnet for the Tactile Transmission of Speech Information

One fundamental motivation for the design of haptic devices is the partial substitution of lost senses. In particular, methods to communicate information from the sense of sight or hearing by aid of tactile devices have some tradition. BLUME designed and tested an electromagnetic plunger-type magnet based on the reluctance effect 1986 at the University of Technology, Darmstadt. Such actuators were attached to the forearm and stimulated up to eight points by mechanical oscillations encoded from speech signals. The actuator (Fig. 9.58) was made of two symmetrical plunger-type magnets (on the horizontal axis) acting upon a flux-conducting element integrated into the plunger. The whole anchor takes a position symmetrically within the actuator due to the integrated permanent magnet. In this symmetrical position, both magnetic

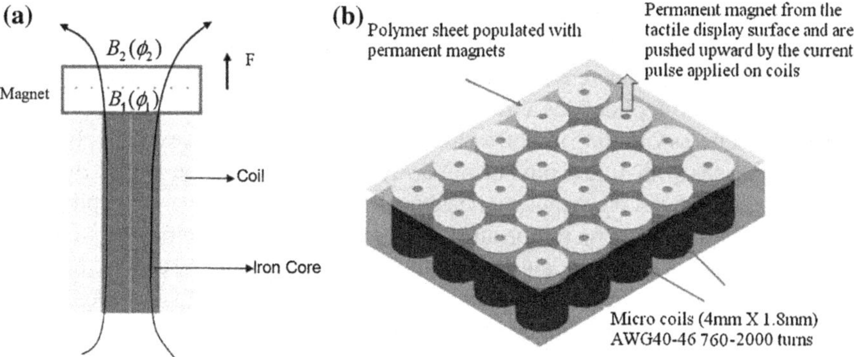

Fig. 9.57 Electromagnetic monostable actuator with permanent magnet: principal sketch (**a**) and actuator design (**b**) [24]

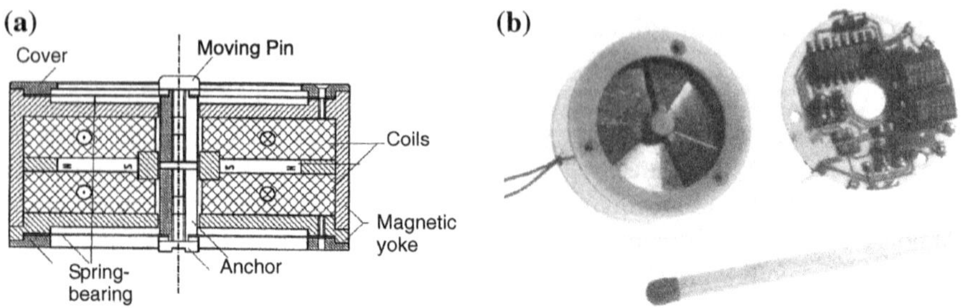

Fig. 9.58 Electromagnetic actuator according to the reluctance principle in a "counteractive plunger-type" design with permanent magnet: cross section (**a**) and design incl. driver electronics (**b**) [13]

circuits conduct a magnetic flux resulting in identical reluctance forces. In active mode, the flux in the upper or lower magnetic circuit is amplified depending on the direction of current flow. The reluctance forces on the amplified side pull the anchor in a current-proportional position working against the magnetic pretension from the permanent magnets, the mechanic pretension from the springs, and the load of the actuator. The plunger is displaced in the direction of the weakened magnetic field. At diameter of 20 mm, this actuator covers a dynamic range of 500 Hz at a efficiency factor of 50 %. The forces lie in the range of about 4 N per ampere.

9.4.5 Conclusion on the Design of Magnetic Actuators

Electromagnetic actuators are—identical to electrodynamic systems —mainly force sources. In rotary drives, especially the reluctance effect is used to generate a continuous movement. With linear drives, mainly plunger-type magnets are used based on the nonlinear transversal effect, although there are exceptions to both showing some surprising properties (Sect. 9.4.4.3). The translational systems are used to actuate as either bistable switches between two discrete states or monostable against a spring (plunger type, break, and valve). There are applications within haptics based on either or both effects. Whereas reluctance-based actuators can be found equally often within kinaesthetic applications as drives and in an admittance-controlled application in tactile systems as vibrating motor, switching actuators are almost exclusively found in tactile devices with single pins or pin arrays. In contrast to the highly dynamic electrodynamic drives, electromagnetic actuators excel in less dynamic applications with higher requirements on torque and self-holding. During switching between two states, however, acceleration and deceleration at the mechanical stop are a highly dynamic but almost uncontrollable action. The dynamic design of switching actions are not the subject of this chapter, but are usually based on modeling a nonlinear force source of the electromagnet and assuming the moving parts as concentrated elements of masses, springs, and dampers. Due to their relatively high masses within the moving parts, the hard to control nonlinearities of fluxes and forces, and the low efficiency factor of the transversal effect in many designs, electromagnetic actuators occupy niches within haptic applications only. However, in those niches, there is no way around for their use. If an appropriate area is found, they excel by an extremely high efficiency factor for the specific design and huge robustness against exterior influences.

9.5 Electrostatic Actuators

Henry Haus, Marc Matysek

Electrostatic transformers are part of the group of electric transformers, such as piezoelectric actuators. *Electric transformers* show direct coupling between the electrical value and the mechanical value. This is contrary to electrodynamic and electromagnetic actuators, which show an intermediate transformation into magnetic values as part of the actuation process. In principle, the transformation may be used in both directions. Hence, all these actuators can be used as sensors as well.

Electrostatic field actuators are used due to their simple design and low power consumption. As a result of the technical progress of micro-engineering, the advantages of integrated designs are fully utilized. For miniaturized systems especially, electrostatic field actuators gain increased importance compared to other actuator principles. This is more surprising as their energy density is significantly lower in macroscopic designs. However, during miniaturization the low efficiency factor and the resulting power loss and heat produced become limiting factors for magnetic actuators [78].

An important subgroup of electrostatic field actuators is given *by solid-state actuators*, with an elastomeric dielectric. It has high breakdown field strength compared to air, builds the substrate of the electrodes, and can simultaneously provide an isolating housing.

In addition to the classic field actuators mentioned above, *electrorheological* fluids are part *of electrostatic actuators* as well. With these actuators, an electric field of arbitrary external source results in a change in the physical (rheological) properties of the fluid.

9.5.1 Definition of Electric Field

The following sections define the electric field and relevant variables for the design of electrostatic actuators.

9.5.1.1 Force on Charge

The magnitude of a force F acting on two charges Q_1 and Q_2 *with* a distance r is given by Coulomb's law [Eq. (9.72)].

$$F = \frac{1}{4\pi\epsilon_0}\frac{Q_1 Q_2}{r^2}. \tag{9.72}$$

9.5.1.2 Electric Field

The electric field E describes the space where these forces are present. The field strength is defined as the relation of the force $\mathbf{F}$ acting on the charge in the field and the charge's magnitude Q.

$$\mathbf{E} = \frac{\mathbf{F}}{Q} \tag{9.73}$$

The charges cause the electric field; the forces on the charges within an electric field are the effect. Cause and effect are proportional. With the electric constant $\epsilon_0 = 8{,}854 \times 10^{-12}$ C/Vm within vacuum and air Eq. (9.74) results:

$$\mathbf{D} = \epsilon_0 \mathbf{E} \tag{9.74}$$

The electric displacement field **D** describes the ratio of the bound charges and the area of the charges. The direction is given by the electric field pointing from positive to negative charges. If the electric field is filled with an insulating material (dielectric), the electric displacement field is bound partly due to the polarizing of the dielectric. Accordingly, the field strength drops from E_0 to E (with still the same electric displacement field). Consequently, the ratio of the weakened field depends on the maximum polarization of the dielectric and is called "permittivity" $\epsilon_r = E_0/E$.

9.5.1.3 Capacity

The electrical capacity is defined as the ratio of charge Q on each conductor to the voltage U between them. A capacitor with two parallel plates charged contrary to a surface of the plates A and a fixed distance d shows a capacity C depending on the dielectric:

$$C = \frac{Q}{U} = \epsilon_0 \epsilon_r \frac{A}{d}. \tag{9.75}$$

9.5.1.4 Energy Storage

Work must be done by an external influence to move charges between the conductors in a capacitor. When the external influence is removed, the charge separation persists and energy is stored in the electric field. If charge is later allowed to return to its equilibrium position, the energy is released. The work done in establishing the electric field, and hence, the amount of energy stored, is given by Eq. (9.76) and for the parallel-plate capacitor by the use of Eq. (9.75) according to Eq. (9.77).

$$W_{\text{el}} = \frac{1}{2} \text{CU}^2 = \frac{1}{2} \frac{Q^2}{C} \tag{9.76}$$

$$W_{\text{el}} = \frac{1}{2}\epsilon_0\epsilon_r\frac{A}{d}U^2 \tag{9.77}$$

This stored electric energy can be used to perform mechanical work according to Eq. (9.78).

$$W_{\text{mech}} = Fx. \tag{9.78}$$

9.5.2 Designs of Capacitive Actuators with Air-Gap

A preferred setup of electrostatic actuators is given by parallel-plate capacitors with air-gap. In these designs, one electrode is fixed to the frame, while the other one is attached to an elastic structure enabling the almost free movement in the intended direction (DoF). All other directions are designed stiff enough to prevent a significant displacement of this electrode. To perform physical work (displacement of the plate), the energy of the electric field according to Eq. (9.77) is used. Considering the principle design of these actuators, two basic variants can be distinguished: The displacement may result in a change in the distance d, or the overlapping area A. Both variants are discussed in the following sections.

9.5.2.1 Movement Along Electric Field

Looking at the parallel-plate capacitor from Fig. 9.59, the capacity C_L can be calculated with

$$C_L = \epsilon_0 \cdot \frac{A}{d} \tag{9.79}$$

As shown before, the stored energy W_{el} can be calculated for an applied voltage U:

$$W_{\text{el}} = \frac{1}{2}CU^2 = \frac{1}{2}\epsilon_0\frac{A}{d}U^2 \tag{9.80}$$

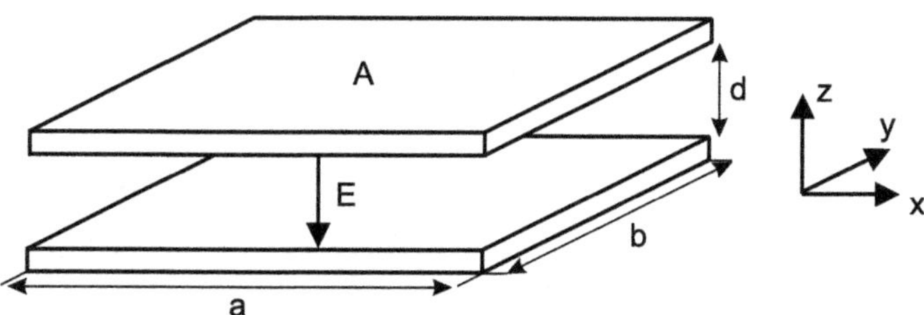

Fig. 9.59 Parallel-plate capacitor with air-gap

The force between both plates in z-direction can be derived by the principle of virtual displacement:

$$F_{z,\mathrm{el}} = \frac{\partial W}{\partial z} = \frac{1}{2}U^2\frac{\partial C}{\partial z} \tag{9.81}$$

$$\mathbf{F}_{z,el} = -\epsilon_0\frac{A}{2d^2}U^2\mathbf{e}_z \tag{9.82}$$

The inhomogeneities of the electric field at the borders of the plates are neglected for this calculation, which is an acceptable approximation for the given geometrical relations of a large plate surface A and a comparably small plate distance d. A spring pulls the moving electrode into its idle position. Consequently, the actuator has to work against this spring. The schematic sketch of this actuator is shown in Fig. 9.60. The plate distance d is limited by the thickness of the insulation layer d_I. Analyzing the balance of forces according to Eq. (9.83), the interdependency of displacement z and electrical voltage U can be calculated as

$$F_z(z) = F_{\mathrm{spring}}(z) + F_{z,\mathrm{el}}(U, z) = 0 \tag{9.83}$$

$$-k \cdot z - \frac{1}{2}\epsilon_0 A\frac{U^2}{(d+z)^2} = 0 \tag{9.84}$$

$$U^2 = -2\frac{k}{\epsilon_0 A}(d+z)^2 \cdot z \tag{9.85}$$

Analyzing the electrical voltage U in dependency of the displacement z, a maximum can be identified as

$$\frac{\mathrm{d}U^2}{\mathrm{d}z} = -2\frac{k}{\epsilon_0 A}(\mathrm{d}^2 + 4\mathrm{d}z + 3z^2) = 0 \tag{9.86}$$

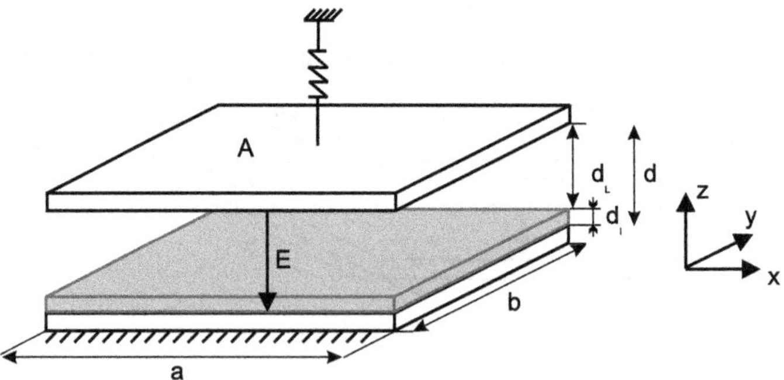

Fig. 9.60 Schematic setup of an actuator with variable air-gap

$$z^2 + \frac{4}{3}dz + \frac{1}{3}d^2 = 0$$

$$z_1 = -\frac{1}{3}d; \ z_2 = -d \tag{9.87}$$

To use the actuator in a stable state, the force of the retaining spring has to be larger than the attracting forces of the charged plates. This condition is fulfilled for distances z

$$0 > z > -\frac{1}{3}d$$

Smaller distances cause attracting forces larger than the retaining force, and the moving plate is strongly pulled onto the fixed plate ("pull-in" effect). As this would immediately result in an electric shortcut, typical designs include an insulating layer on at least one plate. Equations (9.85) and (9.87) are used to calculate the operating voltage for the pull-in:

$$U_{pull-in} = \sqrt{\frac{8}{27}\frac{k}{\epsilon_0 A}d^3} \tag{9.88}$$

The retention force to keep this state is much less than the actual force at the point of pull-in. It should be noted that force increases quadratically with decreasing distance. A boundary value analysis for $d \rightarrow 0$ provides the force $F \rightarrow \infty$. Consequently, the insulation layer fulfills the purpose of a force limitation.

9.5.2.2 Moving Wedge Actuator

A special design of air-gap actuators with varying plate distance is given by the moving wedge actuator. To increase displacement, a bended flexible counter-electrode is placed on a base electrode with a non-conductive layer. The distance between the electrodes increases wedge-like from the fixation to its free end. The resulting electrical field is higher inside the area where the flexible electrode is closest to the counter-electrode and decreases with increasing air-gap. When designing the stiffness of the flexible electrode, it has to be guaranteed that it is able to roll along with the tightest wedge on the isolation. Figure 9.61 shows the underlying principle in idle state and during operation [109].

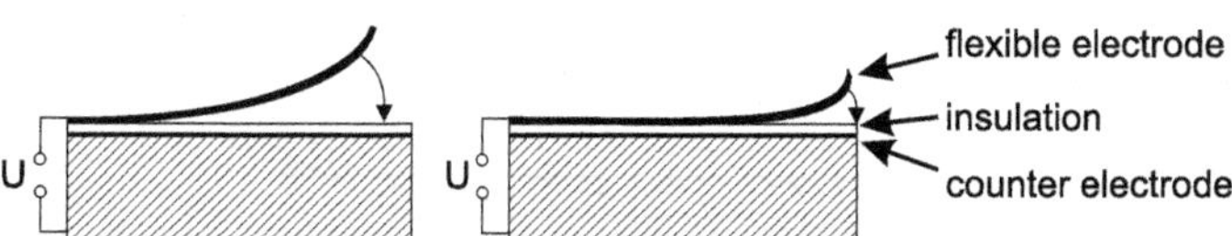

Fig. 9.61 Schematic view of a moving wedge actuator

9.5.2.3 Movement Perpendicular to Electric Field

The major difference compared to the prior design is given by the fact that the plates are moving parallel to each other. The plate distance d is kept constant, whereas the overlapping area varies. Analogous to Eq. (9.80), the forces for the displacement can be calculated in both directions of the plane:

$$F_x = \frac{\partial W}{\partial x} = \frac{1}{2}U^2\frac{\partial C}{\partial x} \tag{9.89}$$

$$\mathbf{F}_x = \frac{1}{2}\epsilon_0\frac{b}{d}U^2\mathbf{e}_x \tag{9.90}$$

$$F_y = \frac{\partial W}{\partial y} = \frac{1}{2}U^2\frac{\partial C}{\partial y} \tag{9.91}$$

$$\mathbf{F}_y = \frac{1}{2}\epsilon_0\frac{a}{d}U^2\mathbf{e}_y \tag{9.92}$$

The forces are independent on the overlapping length only. As a consequence, they are constant for each actuator position. Figure 9.62 shows the moving electrode attached to a retaining spring.

If an electrical voltage is applied on the capacitor, the surface A increases along the border a. Hence, the spring is deflected and generates a counter-force $\mathbf{F}_F$ according to

$$\mathbf{F}_F = -kx\mathbf{E}_x \tag{9.93}$$

The equilibrium of forces acting upon the electrode is given by

$$F_x(x) = F_F(x) + F_{x,el}(U) \tag{9.94}$$

From idle position ($F_x(x) = 0$), the displacement of the electrode in x-direction is given as

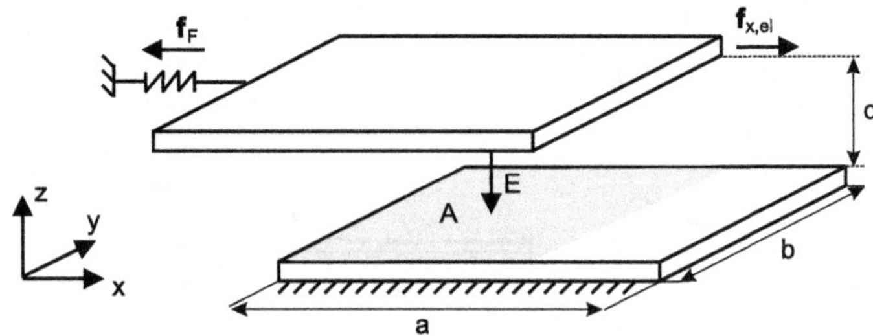

Fig. 9.62 Electrostatic actuator with variable overlapping area

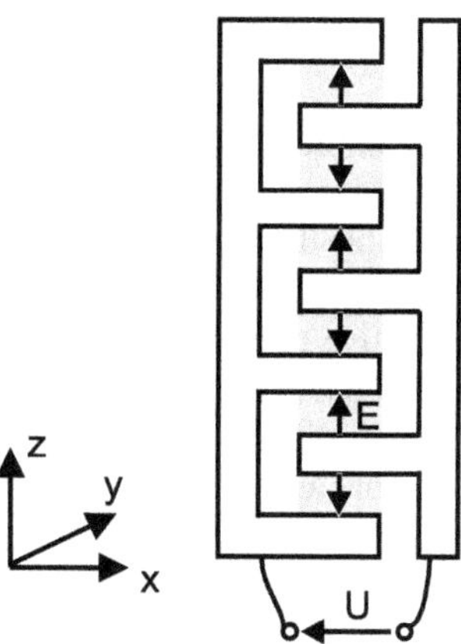

Fig. 9.63 Actuator with comb electrodes and variable overlapping area

$$x = \frac{1}{2}\epsilon_0 U^2 \frac{b}{d}\frac{1}{k} \tag{9.95}$$

Typically, this design is realized in a comb-like structure, with one comb of electrodes engaged in a counter-electrode comb. This equals an electrical parallel circuit of n capacitors, which is identical to a parallel circuit of force sources complementing each other. Figure 9.63 shows such a design. The area of the overlapping electrodes is given by a in x-direction and b in y-direction. With the plate distance d, the capacity based on Eq. (9.96) can be calculated.

$$C_Q = \epsilon_0 \cdot \frac{ab}{d} \cdot n \tag{9.96}$$

By differentiating the energy according to the movement direction, the electromotive force can be calculated as

$$F_x = \frac{\partial W}{\partial x} = \frac{1}{2}U^2 \frac{\partial C}{\partial x} = \frac{1}{2}U^2 \epsilon_0 \frac{b}{d} \cdot n. \tag{9.97}$$

9.5.2.4 Summary and Examples

For all actuators shown, the electrostatic force acts indirectly against the user and is transmitted by a moveable counter-electrode. A simpler design of tactile displays makes use of the user's skin as a counter-electrode, which experiences the whole electrostatic field force. Accordingly, tactile electrostatic applications can be distinguished in direct and indirect principles.

Direct Field Force

The simplest design combines one electrode, respectively, a structured electrode array and an isolating layer. A schematic sketch is given in Fig. 9.64. The user and his finger resemble the counter-electrode. With the attractive force between the conductive skin and the electrodes, a locally distributed increase in friction can be achieved. It hinders a relative movement and can be perceived by the user. Such systems can be easily realized and excellently miniaturized. Their biggest disadvantage is their sensitivity on humidity on the surface, which is brought onto the electrodes during any use in form of sweat. This leads to a blocking of the electrical by the conductive sweat layer above the isolation, preventing the user to feel any relevant force.

Indirect Field Force

With these systems, the field force is used to move an interacting surface. The finger of the user interacts with these surfaces (sliders) and experiences their movements as a perceivable stimulation. A realization with a comb of actuators moving orthogonal to the field direction is given in Fig. 9.65. The structural height is 300 μm, providing 1 mN at operating voltages of up to 100 V. The same design with an actuator made of parallel electrodes can achieve displacements of 60 μm. The comb electrodes shown here displace 100 μm.

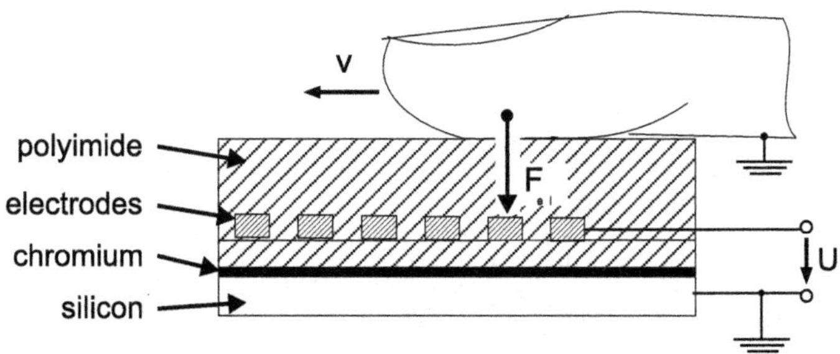

Fig. 9.64 Electrostatic stimulator with human finger as counter-electrode as presented in [119]

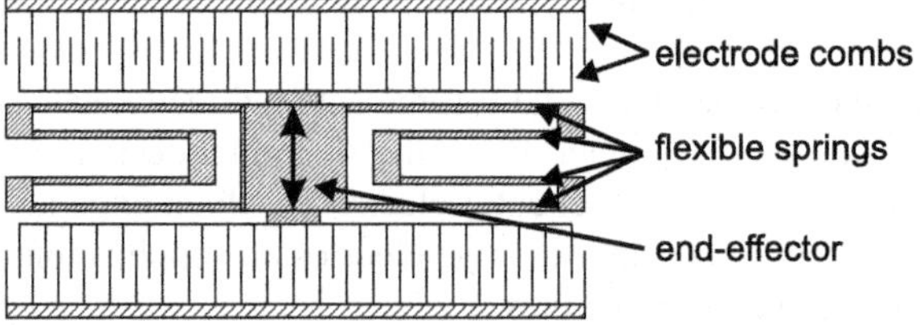

Fig. 9.65 Electrostatic comb-actuator for tangential stimulation [37]

Summary

Electrostatic drives with air-gap achieve force in the range of mN–N. As the actuators are driven by fields, the compromise between plate distance and electrical operation voltage has to be validated for each individual application. The breakdown field strength of air (approx. $3\,\mathrm{V}/\mu\mathrm{m}$) is the upper, limiting factor. The actuators' displacement is limited to several μm. At the same time, the operating voltages reach several hundred volts. Due to the very low achievable displacement, the application of electrostatic actuators is limited to tactile stimulation only.

For the concrete actuator design, it is recommended to deal with the modeling of such actuators, e.g., based on concentrated network parameters (see Sect. 4.3.2). This allows the analysis of the complete electromechanical system starting from the applicable mechanical load situation to the electrical control with a single methodological approach.

9.5.3 Dielectric Elastomer Actuators

As with many other areas, new synthetic materials replace classic materials such as metals in actuator design. Thanks to the enormous progress in material development their mechanical properties can be adjusted to a large spectrum of possible applications. Other advantages are the cheap material costs. Additionally, almost any geometrical shape can be manufactured with relatively small efforts.

Polymers are called "active polymers" if they are able to change their shape and form under the influence of external parameters. The causes for these changes may be manifold: electric and magnetic fields, light, and even pH-value. Being used within actuators, their resulting mechanical properties like elasticity, applicable force, and deformation at simultaneously high toughness and robustness are comparable to biological muscles [4].

To classify the large variety of "active polymers," they are distinguished according to their physical working principle. A classification into "non-conductive polymers" is activated, e.g., by light, pH-value, or temperature, and "electrical polymers" is activated by an arbitrary electrical source. The latter are called "electroactive polymers" (EAP) and are further distinguished into "ionic" and "electronic" EAPs. Generally speaking, electronic EAP are operated at preferably high field strengths near the breakdown field strength. Depending on the thickness layer of the dielectrics, $1-20\,\mathrm{kV}$ are considered typical operation voltages. Consequently, high energy densities at low reaction times (in the area of milliseconds) can be achieved. In contrast, ionic EAP are operated at obviously lower voltages of $1-5\,\mathrm{V}$. However, an electrolyte is necessary for transportation of the ions. It is frequently provided by a liquid solution. Such actuators are typically realized as bending bars, achieving large deformations at their tip with long reaction times (several seconds).

All EAP technologies are the subject of actual research and fundamental development. However, two actuator types are already used in robotics: "Ionic polymer

Table 9.6 Comparison of human muscle and DEA according to PEI [98]

Parameter	Human muscle	DEA
Strain (%)	20–40	10 bis > 100
Stress (MPa)	0, 1–0, 35	0, 1–2
Energy density (kJ/m^3)	8–40	10–150
Density (kg/m^3)	1, 037	≈1,000
Velocity of deformation (%/s)	>50	450 (acrylic) 34,000 (silicone)

metal composite" (IPMC) and "dielectric elastomer actuators" (DEA). A summary and description of all EAP-types is offered by KIM [69]. Their functionality is discussed in the following sections as they affiliate to the group of electrostatic actuators. A comparison between characteristic values of dielectric elastomer actuators and the muscles of the human is shown in Table 9.6. By use of an elastomer actuator with large expansion, additional mechanical components such as gears or bearings are needless. Additionally, the use of these materials may be combined with complex designs similar to and inspired by nature. One application is the locomotion of insects and fish within bionic research [5].

9.5.3.1 Dielectric Elastomer Actuators: Electrostatic Solid-State Actuators

The design of dielectric elastomer actuators is identical to the design of a parallel-plate capacitor, but with an elastic dielectric (a polymer, respectively, elastomer) sandwiched by compliant electrodes. Hence, it is a solid-state actuator. The schematic design of a dielectric elastomer actuator is visualized in Fig. 9.66, left. In an uncharged condition, the capacity and the energy stored is identical to an air-gap actuator [Eqs. (9.75) and (9.76)]. A change in this condition happens by application of a voltage U and is visualized in Fig. 9.66, on the right: The charged capacitor contains more charges ($Q + \Delta Q$), the electrode area increases ($A + \Delta A$), while the distance ($z - \Delta z$) simultaneously decreases. The change of energy after an infinitesimal change dQ, dA, and dz is calculated in Eq. (9.98):

$$\mathrm{d}W = \left(\frac{Q}{C}\right)\mathrm{d}Q + \left(\frac{1}{2}\frac{Q^2}{C}\frac{1}{z}\right)\mathrm{d}z - \left(\frac{1}{2}\frac{Q^2}{C}\frac{1}{A}\right)\mathrm{d}A \qquad (9.98)$$

$$\mathrm{d}W = U\mathrm{d}Q + W\left[\left(\frac{1}{z}\right)\mathrm{d}z - \left(\frac{1}{A}\right)\mathrm{d}A\right] \qquad (9.99)$$

The internal energy change equals the change of the electrical energy by the voltage source and the mechanical energy used. The latter depends on the geometry (parallel (dz) and normal (dA) to the field's direction). In comparison with the air-gap actuator in Sect. 9.5.2, an overlay of decreasing distance and increasing electrodes'

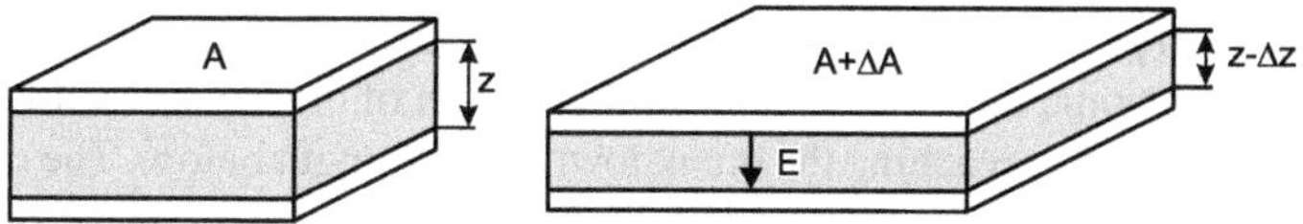

Fig. 9.66 DEA in initial state (*left*) and charged state (*right*)

area occurs. This is caused by a material property that is common to all elastomers and to almost all polymers: the aspect of volume constancy. A body compressed in one direction will extend in the remaining two dimensions if it is incompressible. This gives a direct relation between distance change and the change in electrodes' area. As a consequence, Eq. (9.100) results

$$A\,\mathrm{d}z = -z\mathrm{d}A \tag{9.100}$$

simplifying Eq. (9.99) to

$$\mathrm{d}W = U\mathrm{d}Q + 2W\left(\frac{1}{z}\right)\mathrm{d}z \tag{9.101}$$

The resulting attractive force of the electrodes can be derived from this electrical energy. With respect to the electrode surface A, the electrostatic pressure p_{el} at $\mathrm{d}Q = 0$ is given according to Eq. (9.102)

$$p_{\mathrm{el}} = \frac{1}{A}\frac{\mathrm{d}W}{\mathrm{d}z} = 2W\frac{1}{Az} \tag{9.102}$$

and by application of Eq. (9.76)

$$p_{\mathrm{el}} = 2\left(\frac{1}{2}\epsilon_0\epsilon_r Az\frac{U^2}{z^2}\right)\frac{1}{Az} = \epsilon_0\epsilon_r E^2 \tag{9.103}$$

Comparing this result with Eq. (9.81) as a reference for a pressure of an air-gap actuator with variable plate distance, dielectric elastomer actuators are capable of generating a pressure twice as high with otherwise identical parameters [99]. Additional reasons for the obviously increased performance of the dielectric elastomer actuators are also based on their material. The relative permittivity is given by $\epsilon_r > 1$, depending on the material $\epsilon_r \simeq 3 - 10$. By chemical processing and implementation of fillers, the relative permittivity may be increased. However, it may be noted that other parameters (such as the breakdown field strength and the E-modulus) may become worse, possibly the positive effect of the increased ϵ_r gets lost. The breakdown field strength especially is one of the most limiting factors. With many materials, an increase in breakdown field strength can be observed after planar prestrain. In these cases, breakdown field strengths of 100–400 V/μm are typical [17].

The pull-in effect does not happen at $z = 1/3 \cdot z_0$ (air-gap actuators), but at higher deflections. With some materials mechanical prestrain of the actuator allows to displace the pull-in further, reaching the breakdown field strength before. The reason for this surprising property is the volume constant dielectric layer showing viscoelastic properties. It complies with a return spring with strong nonlinear force–displacement characteristics for large extensions. Its working point is displaced along the stress–strain curve of the material as an effect of the mechanical prestrain.

For application in dielectric elastomer actuators, many materials may be used. The material properties cover an extreme wide spectrum ranging from gel-like polymers up to relatively rigid thermoplastics. Generally speaking, every dielectric material has to have a high breakdown field strength and elasticity besides high relative permittivity. Silicone provides highest deformation velocities and a high temperature resistance. Acrylics have high breakdown field strength and achieve higher energy densities. The following list is a selection of the dielectric materials frequently used today:

- silicone
  - HS 3 (Dow Corning)
  - CF 19-2186 (Nusil)
  - Elastosil P7670 (Wacker)
  - Elastosil RT625 (Wacker)
- acrylics
  - VHB 4910 (3M)

The most frequently used materials for the elastic electrodes are graphite powder, conductive carbon, and carbon grease.

9.5.4 Designs of Dielectric Elastomer Actuators

As mentioned before, dielectric elastomer actuators achieve high deformations (compression in field direction) of 10–30 %. To keep voltages within reasonable ranges, layer thicknesses of 10–100 μm are used depending on the breakdown field strength. The resulting absolute displacement in field direction is too low to be useful. Consequently, there are several concepts to increase it. Two principle movement directions are distinguished for this purpose: the longitudinal effect in parallel to the field (thickness change) and the transversal effect orthogonal to field (surface area change). The importance of this discrimination lies in the volume constancy of the material: uniaxial pressure load equals a two-axial tension load in the remaining spatial directions. Hence, two transversal tensions within the surface result in a surface change. For materials fulfilling the concept of "volume constancy," Eq. (9.104) is valid, providing the following properties for the longitudinal compression S_z and the transversal elongation S_x:

- with a longitudinal compression of 62 %, both extensions are identical;
- for smaller values of the longitudinal compression, the resulting transversal extension is smaller;
- for a longitudinal compression >62 % them, transversal extensions increases faster than the longitudinal compression.

$$S_x = \frac{1}{\sqrt{1 - S_z}} - 1 \tag{9.104}$$

The extension of the surface area S_A depends on the longitudinal compression S_z according to Eq. (9.105):

$$S_A = \frac{dA}{A} = \frac{S_z}{1 - S_z} \tag{9.105}$$

The increase of the area with uniaxial compression is always larger than the change in thickness. Actuators built according to this principle are the most effective ones. Figure 9.67 shows three typical designs. A roll-actuator (left) built as full or tubular cylinder can achieve length changes of more than 10 %. KORNBLUH [71] describes an acrylic roll-actuator achieving a maximum force of 29 N with an own weight of no more than 2.6 g at a extension of 35 mm. The manufacture of electrodes with a large area is very simple. On the other hand, the rolling of the actuators with simultaneous pre-strain (up to 500 %) can be challenging. With a stack-actuator (middle), very thin dielectric layers down to 5 μm with minimized operational voltages can be achieved, depending mainly on the manufacturing technique [81]. As the longitudinal effect is used, extension is limited to approximately 10 %. However, due to their design and fabrication process, actuator arrays at high density can be built [63] and offer typically lifetimes of more than 100 million cycles depending on their electrical interconnection [84]. The most simple and effective designs are based on a restrained foil, whose complete surface change is transformed into an up-arching (diaphragm-actuator, right) [70]. If this actuator experiences a higher external load, such as from a finger, an additional force source, e.g., a pressure, has to be provided to support the actuators own tension.

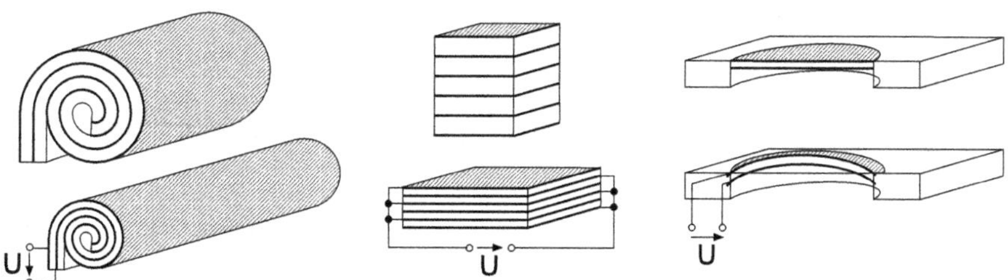

Fig. 9.67 Typical designs of Dielectric Elastomer Actuators: roll-actuator (*left*), stack-actuator (*center*), and diaphragm-actuator (*right*)

9.5.4.1 Summary and Examples

As with air-gap actuators, a dielectric solid-state actuator's major limit is given by the breakdown field strength of the dielectric. However, in contrast to the air-gap actuators, a carefully chosen design can easily avoid any pull-in effect. Consequently, these actuators show larger workspace, and with the high number of different design variants a wide variety of applications can be realized depending on the requirements of displacement, maximum force, and actuator density.

Tactile Displays

The simplest application of a tactile display is a Braille device. Such devices are meant to display Braille letters in patterns of small, embossed dots. In standard Braille, six dots are used, in computer compatible Euro-Braille eight dots. These points are arranged in a 2×3 or 2×4 matrix, respectively (Fig. 9.68). In a display device, 40 to 80 characters are displayed simultaneously. In state-of-the-art designs, each dot is actuated with one piezoelectric bending actuator (Sect. 9.3.5). This technical effort is the reason for the high price of these devices. As a consequence, there are several functional samples existing, which prove the applicability of less expensive drives with simplified mechanical designs, but still sufficient performance. Each of the three variants for dielectric elastomer actuators has already been used for this application.

Figure 9.68 shows the schematic sketch of roll-actuators formed to an actuator column [107]. Each roll-actuator moves one pin, which itself is pushed up above the base plate after applying a voltage. The elastomer film is coiled around a spring of 60 mm length with diameter of 1.37 mm. With an electric field of 100 V/μm applied, the pre-tensioned spring can achieve a displacement of 1 mm at a force of 750 mN. The underlying force source is the spring with spring constant 225 N/m pre-tensioned by a passive film. The maximum necessary displacement of 500 μm is achieved at field strengths of 60 V/μm.

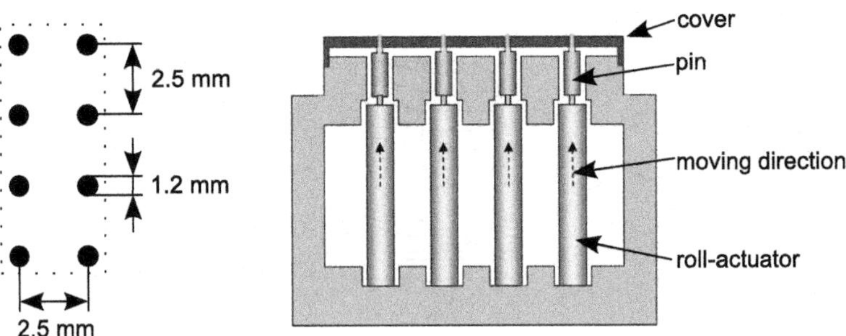

Fig. 9.68 Presenting a Braille sign with roll-actuators, *left* geometry, *right* schematic setup of a Braille row [107]

The application of stack-actuators according to JUNGMANN [63] is schematically sketched in Fig. 9.68, left. The biggest advantage of this variant is given by the extremely high actuator density combined with a simple manufacturing process. Additionally, the closed silicone elements are flexible enough to be mounted on an almost arbitrary formed surface. The surface—by itself made of silicone—shows an adequate roughness and temperature conductivity. It is perceived as "convenient" by many users. With a field strength of 30 V/μm, a stack made of 100 dielectric layers displacements of 500 μm can be achieved. The load of a reading finger on the soft substrate generates a typical contact pressure of 4 kPa, resulting in a displacement of 25 μm. This extension is considerably less than the perception threshold of 10 % of the maximum displacement. For control of the array, it has to be noted that the actuators are displaced in a negative logic. With applied voltage, the individual pin is pulled downwards.

A remote control providing tactile feedback based on the same type of stack-actuators is presented in [85]. The mobile user interface consists of five independent actuating elements. Besides the presentation of tactile feedback, the stack transducers offer to acquire user's input using the transducers intrinsic sensor functionality. The actuators are driven with a voltage of up to 1,100 V generated out of a primary lithium-ion battery cell. A free-form touch pad providing tactile feedback to the human palm is described in [89]. Four actuators are integrated in a PC-mouse to enhance user experience and substitute visual feedback during navigation tasks (see Fig. 9.69). The stacks consist of 40 dielectric layers, each 40 μm in thickness and are supplied by a maximum voltage of 1 kV in a frequency range from 1.5 to 1 kHz. The mouse contains all the required driving electronics and can be customized by a software configuration tool.

The design of a Braille display with diaphragm-actuators according to HEYDT [48] demonstrates the distinct properties of this variant. The increase of elastomer surface results in a notable displacement of a pin from the device's contact area. However, a mechanical prestrain is necessary to provide a force. This can be either generated by a spring or air pressure below the actuator. Figure 9.70 on the right

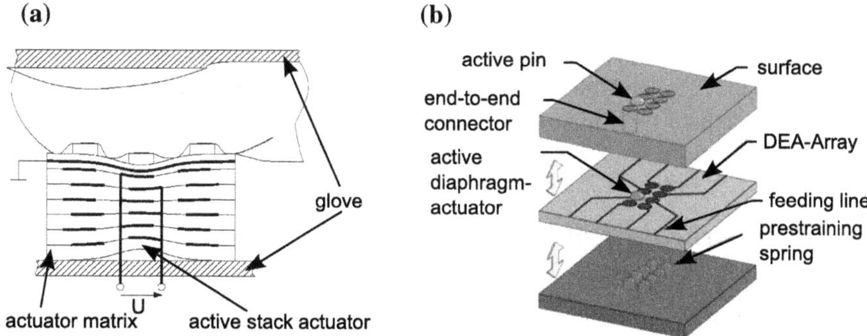

Fig. 9.69 Tactile feedback enhanced PC-mouse: CAD model (**a**), and demonstration device (**b**) [89]

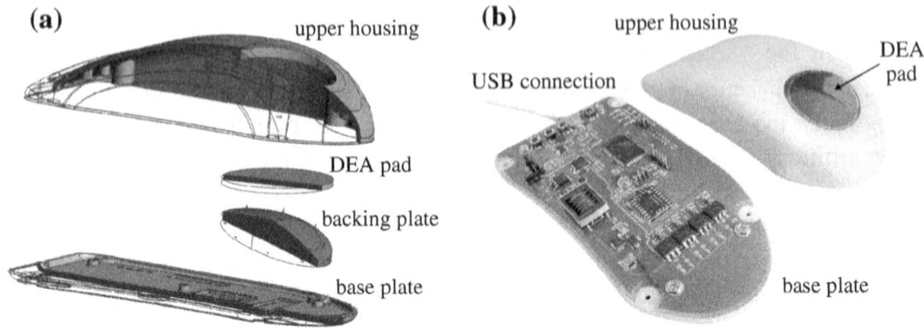

Fig. 9.70 **a** Actuator row with stack-actuators [63]; **b** Use of diaphragm-actuators [48]

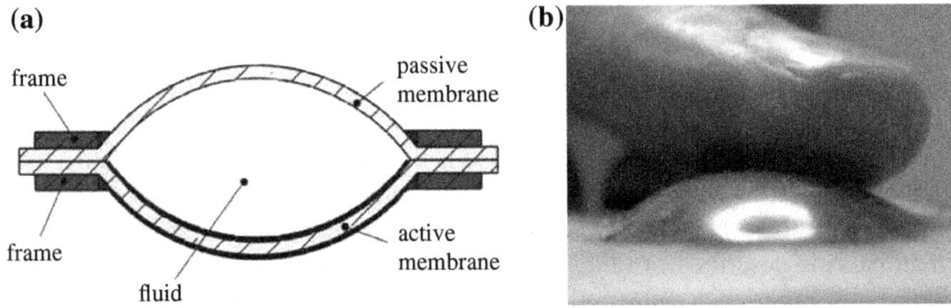

Fig. 9.71 Hydrostatically coupled membrane actuator: **a** schematic drawing, **b** and finger touching active display as presented in [19]

gives a schematic sketch for a single point being pretensioned by a spring with a diameter of 1.6 mm. At an operating voltage of 5.68 kV, the actuator displaces in idle mode 450 μm.

CARPI combines the principle of the diaphragm-actuators with fluid-based hydrostatic transmission [19]. The result is a wearable tactile display intended for providing feedback during electronic navigation in virtual environments. As shown in Fig. 9.71, the actuators are based on an incompressible fluid that hydrostatically couples a dielectric elastomer membrane to a passive membrane interfaced to the user's finger. The transmission of actuation from the active membrane to the finger, without any direct contact, allows suitable electrical safety. However, the actuator is driven with comparatively high voltages up to 4 kV.

9.5.5 Electrorheological Fluids

Fluids influenced in their rheological properties (especially the viscosity) by electrical field varying in direction and strength are called ↪ Electrorheological Fluid (ERF). Consequently, ERF are classified as non-Newton fluids, as they have a variable viscosity at constant temperature. The electrorheological effect has been observed for the first time in 1947 at a suspension of cornstarch and oil by WILLIS WINSLOW.

Electrorheological fluids include dipoles made of polarized particles, which are dispersed in a conductive suspension. These particles are aligned in an applied electrical field. An interaction between particles and free charge carriers happens. Chain-like microstructures are built between the electrodes [23, 51, 111] in this process. However, it seems as if this is not the only effect responsible for the viscosity change, as even when the microstructures [97] were destroyed, a significant viscosity increase remained. The exact analysis of the mechanism responsible for this effect is a subject of actual research.

The viscosity of the fluid changes depending on the strength of the applied electrical field. With an electric field of 1−10 kV/mm, the viscosity may change up to a factor of 1,000 compared to the field-free state. This enormous change equals a viscosity difference between water and honey. An advantage of this method can be found in the dynamics of the viscosity change. It is reversible and can be switched within one millisecond. Therefore, electrorheological fluids are suitable for dynamic applications.

If large field strengths are assumed, the ERF can be modeled as BINGHAM fluid. It has a threshold for linear flow characteristics: starting at a minimum tension $\tau_{F,d}$ (flow threshold) the fluid actually starts to flow. The fluid starts flowing right below this threshold. The shear forces τ are calculated according to Eq. (9.106):

$$\tau = \mu\dot{\gamma} + \tau_{F,d} \tag{9.106}$$

with μ the dynamic viscosity, $\dot{\gamma}$ the shear rate, and $\tau_{F,d}$ the dynamic flow limit. The latter changes quadratically with the electrical field strength [Eq. (9.107)]. The proportional factor C_d is a constant provided with the material's specifications.

$$\tau_{F,d} = C_d E^2 \tag{9.107}$$

For complex calculations modeling the fluid's transition to and from the state of flow, the model is extended to a nonlinear system according to Eq. (9.108) [for $n = 1$ this equals Eq. (9.106)].

$$\tau = \tau_{F,d} + k\dot{\gamma}^n \tag{9.108}$$

This general form describes the shear force for viscoplastic fluids with flow limit according to VITRANI [129]. For analysis of idle state with shear rate $\dot{\gamma} = 0$, the static flow limit $\tau_{F,s}$ with $\tau_{F,s} > \tau_{F,d}$ is introduced. When exceeding the static flow limit, the idle fluid is deformed. With the specific material constants, C_s and E_{ref} Eq. (9.109) can be formulated as

$$\tau_{F,s} = C_s(E - E_{\text{ref}}) \tag{9.109}$$

The materials used for the particles are frequently metal oxides, silicon anhydride, poly urethane, and polymers with metallic ions. The diameter of particles is 1–

100 μm; their proportion on the fluid's volume is 30–50 %. As carrier medium, typically oils (such as silicone oil) or specially treated hydrocarbon are used. To additionally improve the viscosity change, nanoscale particles are added in the electrorheological fluids ("giant electrorheological effect" [40, 137]). In [28] and [105] further mathematical modeling is presented for the dynamic flow behavior of ER-fluids.

The central property of ERF—to reversibly change the viscosity—is used for force-feedback devices, haptic displays, and artificial muscles and joints. As the change in viscosity is mainly a change in counter-forces but not in shape or direct forces, ERF-actuators are counted to the group of "passive actuators." For the characterization of their performance, the ratio between stimulated and idle state is used. They are built in three principle design variants [14] as described in the following sections.

9.5.5.1 Shear Mode

The ER-fluid is located between two parallel plates, one fixed and the other moving relatively to the fixed one. The only constraint is given by a fixed inter-plate distance d. If a force F is applied on the upper plate, it is displaced by a value x at a certain velocity v. For the configuration shown in Fig. 9.72, the mechanical control ratio λ can be calculated according to Eq. (9.112) from the ratio of dissipative forces (field-dependent flow stresses, Eq. (9.115)) and the field-independent viscosity term (Eq. (9.110)) [100]. η gives the basis viscosity of the ER-fluid (in idle state) and τ_y the low-stress depending on the electrostatic field.

$$F_\eta = \frac{\eta v a b}{d} \tag{9.110}$$

$$F_\tau = \tau_y a b \tag{9.111}$$

$$\lambda = \frac{F_\tau}{F_\eta} = \frac{\tau_y d}{\eta v}. \tag{9.112}$$

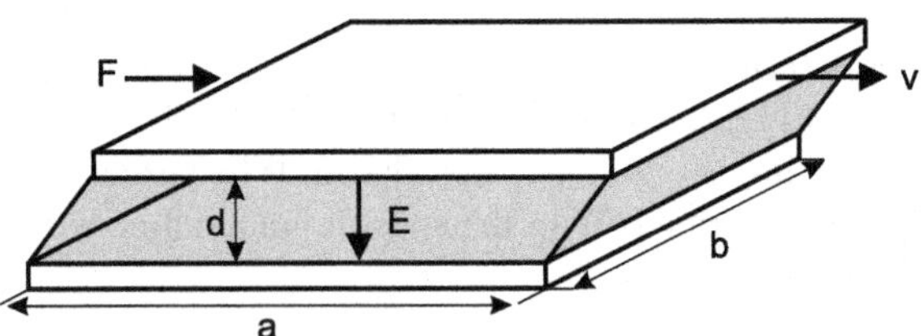

Fig. 9.72 Using ERF to vary the shear force

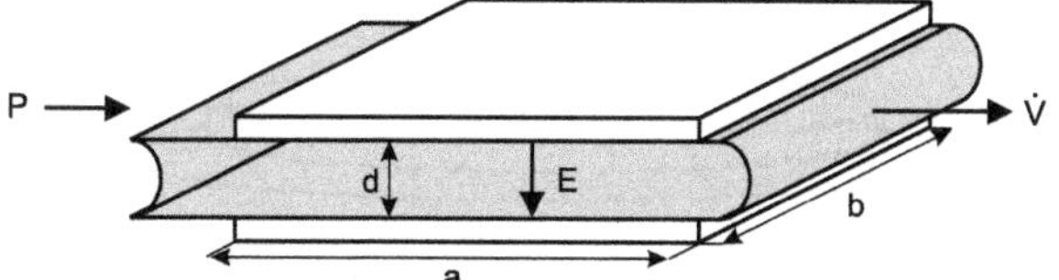

Fig. 9.73 Varying the flow channel's resistivity with ERF-actuators

9.5.5.2 Flow Mode

The schematic sketch of this configuration is shown in Fig. 9.73. Both fixed plates form a channel, with the fluid flowing through it due to an external pressure difference p and a volume flow $\dot{V}$. With an electric field E applied between the plates, the pressure loss increases along the channel and the volume flow is reduced. Analogous to the prior design, a field-independent viscosity-based pressure loss p_η and a field-dependent pressure loss p_τ can be calculated [100]as

$$p_\eta = \frac{12\eta\dot{V}a}{d^3 b} \tag{9.113}$$

$$p_\tau = \frac{c\tau_y a}{d} \tag{9.114}$$

The mechanical control ratio equals

$$\lambda = \frac{p_\tau}{p_\eta} = \frac{c\tau_y d^2 b}{12\eta\dot{V}} \tag{9.115}$$

At an adequate dimensioning of the fluid, the flow resistance can be increased by the electrical field to such a degree that the fluid stops completely when exceeding a specific voltage. This makes the channel a valve without any moving mechanical components.

9.5.5.3 Squeeze Mode

A design to generate pressure is schematically sketched in Fig. 9.74. In contrast to the variants shown before, the distance between both plates is subject to change now. If a force acts on the upper plate, it moves downwards. This results in the fluid being pressed outside. A plate distance d_0 is assumed at the beginning and a relative movement of v of the plate moving downwards. The velocity-dependent viscosity force F_η and the field-dependent tension term F_τ [62] are calculated according to:

$$F_\eta = \frac{3\pi\eta v r^4}{2(d_0 - z)} \tag{9.116}$$

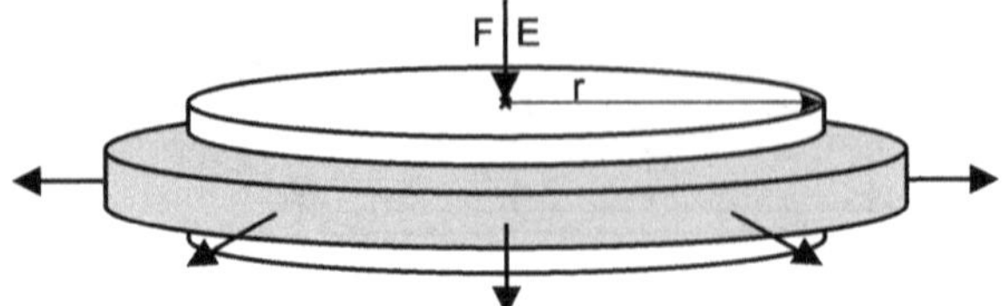

Fig. 9.74 Varying the acoustic impedance with ERF-actuators under external forces

$$F_\tau = \frac{4\pi \tau_y r^3}{3(d_0 - z)}, \tag{9.117}$$

which gives the mechanical control ratio:

$$\lambda = \frac{8\tau_y 3(d_0 - z)^2}{9\eta v r} \tag{9.118}$$

With pressure (force on the upper plate), the fluid is pressed out of the gap. In this configuration, the force–displacement characteristics are strongly influenced by the electrical field strength. An analysis of the dynamic behavior of such an actuator is described in [132].

9.5.5.4 Designing ERF-Actuators

The maximum force F_τ and the necessary mechanical power P_{mech} are the input values for the design of ERF-actuators from the perspective of an application engineer. Equations (9.110) and (9.118) can be combined to calculate the necessary volume for providing a certain power with all three actuator configurations.

$$V = k\frac{\eta}{\tau_y^2}\lambda P_{\text{mech}} \tag{9.119}$$

Consequently, the volume is defined by the mechanical control ratio, the fluid-specific values η and τ_y, such as a constant k dependent on the actual configuration. The electrical energy W_{el} necessary to generate the electrostatic field of the actuator (volume dependent) is calculated according to Eq. (9.120).

$$W_{\text{el}} = V\left(\frac{1}{2}\epsilon_0\epsilon_r E^2\right). \tag{9.120}$$

9.5.5.5 Comparison to Magnetorheological Fluids

↪ Magnetorheological Fluids (MRF) are similar to electrorheological fluids. However, the physical properties of the fluids are influenced by magnetic fields. All calculations shown before are also applicable to MRF. Looking at the volume necessary for an actuator according to Eq. (9.119), considering the viscosities of electrorheological and magnetorheological fluids being comparable, a volume ratio proportional to the reciprocal ratio of the fluid-tensions' square according to Eq. (9.121) results in

$$\frac{V_{\mathrm{ERF}}}{V_{\mathrm{MRF}}} = \frac{\tau_{\mathrm{MRF}}^2}{\tau_{\mathrm{ERF}}^2} \tag{9.121}$$

In a rough but good approximation, the flow stress of a magnetorheological fluid is one magnitude larger than an ERF, resulting in a smaller (approximately factor 100) volume of an MRF-actuator compared to the ERF. However, a comparison between both fluids going beyond the pure volume analysis for similar output power is hard: for an ERF high voltages at relatively small currents are required. The main power leakage is lost by leakage currents through the medium (ERF) itself. With MRF-actuators, smaller electrical voltages at high currents become necessary to generate an adequate magnetic field. The energy for an MRF-actuator is calculated according to Eq. (9.122) with the magnetic flux density B and the magnetic field strength H.

$$W_{\mathrm{el,MRF}} = V_{\mathrm{MRF}} \left(\frac{1}{2} BH\right) \tag{9.122}$$

The ratio between the energies for both fluids is calculated according to Eq. (9.123)

$$\frac{W_{\mathrm{el,ERF}}}{W_{\mathrm{el,MRF}}} = \frac{V_{\mathrm{ERF}}}{V_{\mathrm{MRF}}} \frac{\epsilon_0 \epsilon_r E^2}{BH} \tag{9.123}$$

With typical values for all parameters, the necessary electrical energy for actuator control is comparable for both fluids. An overview of the design of actuators for both types of fluids is given in [18].

9.5.5.6 Summary and Examples

Electrorheological fluids are also called partly active actuators, as they do not transform the electrical values into a direct movement, but change their properties due to the electrical energy provided. The change in their properties covers a wide range. Naturally, their application in haptics ranges from small tactile displays to larger haptic systems.

Tactile Systems

The first application of ERF as tactile sensor in an artificial robot hand was made in 1989 by KENALEY [67]. Starting from this work, several ideas developed to use ERF in tactile arrays for improving systems for virtual reality applications. Several tactile displays, among them a 5 × 5 matrix from TAYLOR [121] and another one from BÖSE [15], were built. Figure 9.75 shows the schematic design of such a tactile element. A piston is pressed in an ERF filled chamber by the user. Varying counter-forces are generated depending on the actuation state of the ERF. Elastic foam is connected to the piston as a spring to move it back to its resting position. With an electric field of 3 V/μm a force of 3.3 N can be achieved at a displacement of 30 mm. Switching the electrical voltages is realized by light emitting diodes and corresponding receivers (GaAs-elements) on the backplane.

Haptic operating controls

Another obvious application for ERF in haptic systems is their use as a "variable brake." This is supported by the fact that typical applications beside haptic systems are variable brakes and bearings (e.g., adaptive dampers). There are several designs with a rotary knob mobbing a spinning disk within an ERF or MRF generating varying counter-torques as shown in [80]. In this case, the measurement of the rotary angle is solved by a potentiometer. In dependency on the rotary angle, the intended counter-force, respectively, torque is generated. The user can perceive a "latching" of the rotary knob with a mature system. The latching depth itself can be varied in a wide range. By the varying friction, hard stops can be simulated, too, such as sticking and of course-free rotation.

An extension of the one-dimensional system is presented in [136]. Two systems based on ERF are coupled to a joystick with two DoF. A counter-force can be generated in each movement direction of the joystick. As ERF are able to generate higher

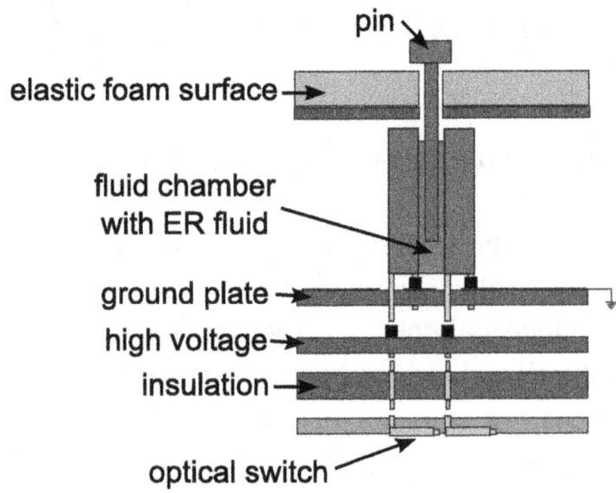

Fig. 9.75 Schematic setup of a tactile actuator based on ER-fluids [15]

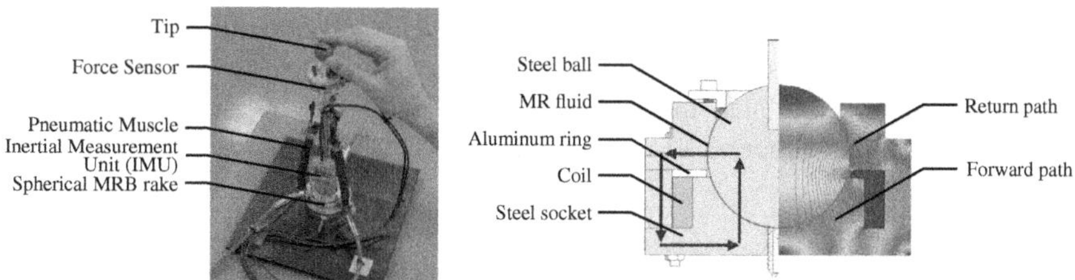

Fig. 9.76 Haptic joystick based on pneumatic actuators and a MRF-brake as presented in [110]

torques with less energy required compared to a normal electrical drive, they are especially suitable to mobile applications like in cars.

SENKAL ET AL. presented a combination of an MRF-brake and pneumatic actuators for a 2D joystick as shown in Fig. 9.76. This hybrid concept uses pneumatic actuators because of the high energy density and the MR brake to increase the fidelity of rigid objects [110]. Further realizations of MRF operating controls can be found in [1].

Force-Feedback Glove

A force-feedback glove was designed as a component for a simulator of surgeries [6]. Surgical interventions shall be trained by the aid of haptic feedback. The system "remote mechanical mirroring using controlled stiffness and actuators" (MEMICO) enables a surgeon to perform treatment with a robot in telemanipulation, whereas the haptic perception is retained. ERF-actuators are used for both ends: on the side of the end-effector, and for the haptic feedback to the user. The adjustable elasticity is based on the same principle as tactile systems. For generating forces, a force source is necessary. A new "electronic controlled force and stiffness" actuator (ECFS) is used for this application. The schematic design is shown in Fig. 9.77. It is an actuator according to the inchworm principle, wherein both brakes are realized by the ER-fluid surrounding it. The driving component for the forward and backward movement is realized by two electromagnets. Both actuators are assembled within a haptic

Fig. 9.77 Schematic setup of an ERF-Inchworm motor [6]

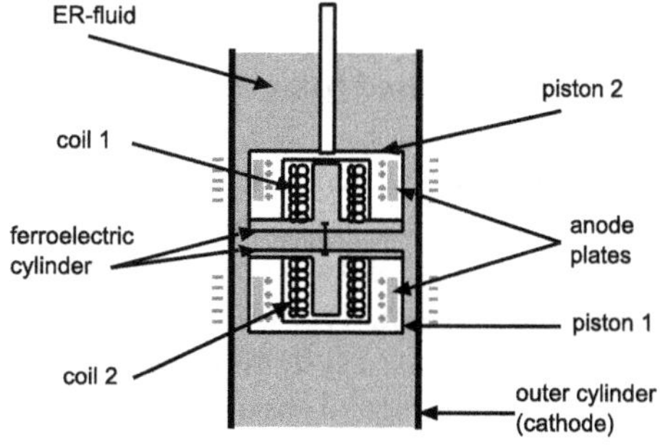

exoskeleton. They are mounted on the rim of a glove to conserve the mobility of the hand. With the actuators in between all finger-joints, arbitrary forces and varying elasticities can be simulated independently. The ECFS actuators are operated at voltages of 2 kV and generate a force of up to 50 N.

9.6 Special Designs of Haptic Actuators

Thorsten A. Kern, Christian Hatzfeld

The actuation principles discussed so far are the most common approaches to the actuation of haptic devices. Besides these principles, there are numerous research projects, singular assemblies, and special classes of devices. The knowledge of these designs is an enrichment for any engineer, yet it is impossible to completely cover all the haptic designs in a single book. This section, nevertheless, intends to give a cross section of alternative, quaint, and unconventional systems for generating kinaesthetic and tactile impressions. This cross section is based on the authors' subjective observations and knowledge and does not claim to be exhaustive. The discussed systems have been selected, as examples suited best to cover one special class of systems and actuators, each. They are neither the first systems of their kind, nor necessarily the best ones. They are thought to be crystallization points of further research, if specific requirements have to be chosen for special solutions. The systems shown here are meant to be an inspiration and an encouragement not to discard creative engineering approaches to the generation of haptic impressions too early during the design process.

9.6.1 Haptic-Kinaesthetic Devices

Haptic-kinaesthetic devices of this category excel primarily due to their extraordinary kinematics and not to very special actuation principles. Nevertheless, every engineer may be encouraged to be aware of the examples of this device class and let this knowledge influence his/her own work.

9.6.1.1 Rope-Based Systems

With rope-based systems, actuators and the point of interaction are connected with ropes, i.e., mechanical elements that can only convey pulling forces. They are especially suited for lightweight systems with large working spaces, as for example simulation of assembly tasks, rehabilitation, and training. von Zitzewitz describes the use of rope-based systems for sport simulation and training (tennis and rowing) as well as an experimental environment to investigate vestibular stimulation in sleeping subjects [131].

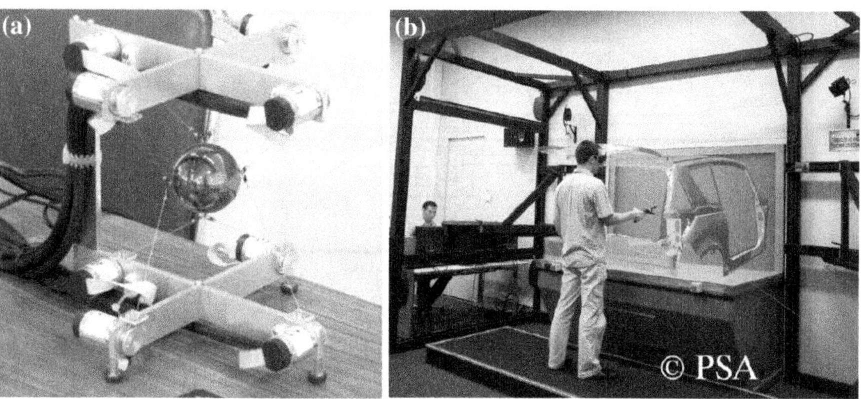

Fig. 9.78 **a** Desktop version of the Spidar with ball-like interaction handle, **b** room-size version INCA 6D with 3D visualization environment by *Haption*

Another system, the Spidar (Fig. 9.78) is based on the work of SATO and has frequently been used in research projects [92, 117] as well as in commercial systems. It is composed of an interaction handle—usually a ball—held by eight strings. Each string is operated by an actuator, which is frequently (but not obligatorily) mounted in the corners of a rectangular volume. The drives are able to generate pulling forces on the strings, enabling the generation of forces and torques in six DoFs on the handle. Typically, the actuators used are based on electrodynamic electronic-commutated units. The Spidar system can be scaled to almost any size, ranging from table-top devices to room-wide installations. It convinces by the small number of mechanical components and the very small friction. As strings are able to provide pull forces only, it is worth noting that just two additional actuators are sufficient to compensate this disadvantage.

9.6.1.2 Screw-and-Cable System

In several projects relating to medical rehabilitation and master slave teleoperation, the CEA-LIST Interactive Robotics Unit use their screw-and-cable system (SCS) shown in a first prototype in 2001 [32, 34–36]. In this prototype six, screw-cable actuators are used to motorize a master arm in a teleoperation system enabling high-fidelity force feedback. In the meantime, the master arm is commercialized by *Haption S.A.* under the name VIRTUOSE 6D 4040 [44]. The patented SCS basic principle can be seen in Fig. 9.79.

A rotative joint is driven by a standard push–pull cable. On one side, the cable is driven by a ball screw that translates directly in its nut (the screw is locked in rotation thanks to rollers moving into slots). The nut is rotating in a fixed bearing and is driven by the motor thanks to a belt transmission [36].

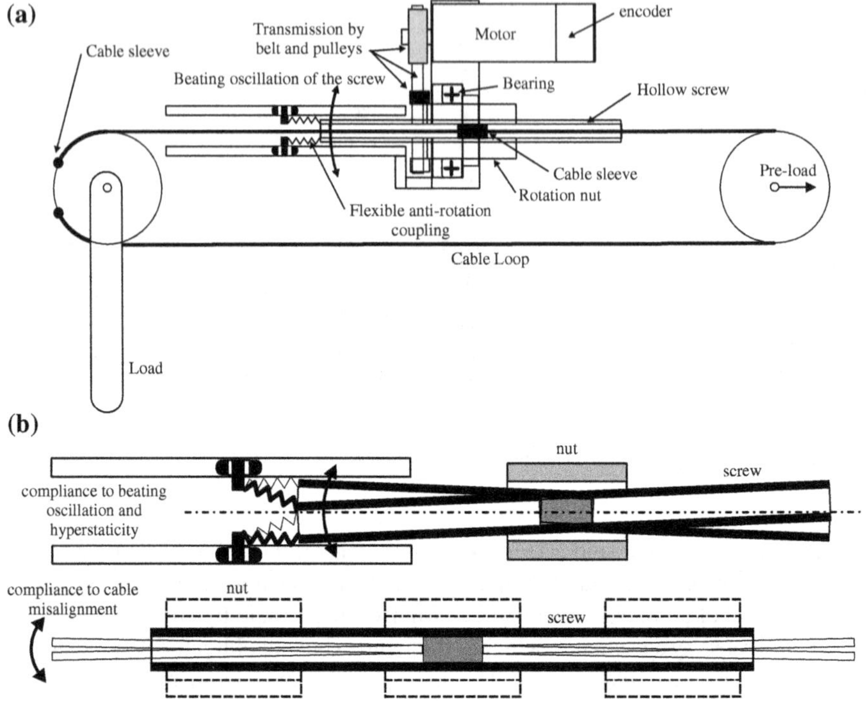

Fig. 9.79 SCS basic principles: **a** driving unit **b** and particular patented mounting [32]. Figure courtesy of *CEA-LIST*

Using SCS allows significantly mass and volume-reduced driving units for joint torque control. The low friction threshold and high backdrivability enable true linear torque control without a sensor, avoiding drift and calibration procedure. Low inertia of the structure leads to high transparency. In the upper-limb exoskeleton ABLE 4D, the SCS is embedded in the moving parts of the arm, resulting in reduced cable length and simplified routing (see Fig. 9.80). The two SCS integrated in the arm module perform alike artificial electrical muscles. Further information about the design of such systems can be found in [33].

9.6.1.3 Magnetorheological Fluids as Three-Dimensional Haptic Display

The wish to generate an artificial haptic impression in a volume for free interaction is one of the major motivations for many developments. The rheological systems shown in Sect. 9.5 provide one option to generate such an effect. For several years, the team of Bicchi has been working on the generation of spatially resolved areas of differing viscosity in a volume (Fig. 9.81) to generate force feedback on an exploring hand. Lately, the results were summarized in [11]. The optimization of such actuators

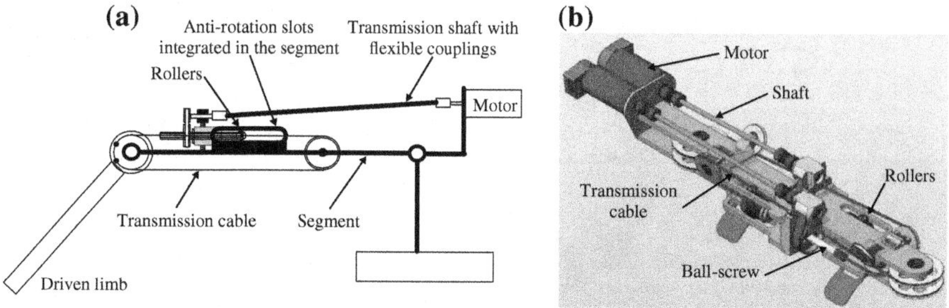

Fig. 9.80 ABLE arm module: **a** optimized architecture to be integrated, **b** structure with its 2 integrated actuators [32]. Figure courtesy of *CEA-LIST*

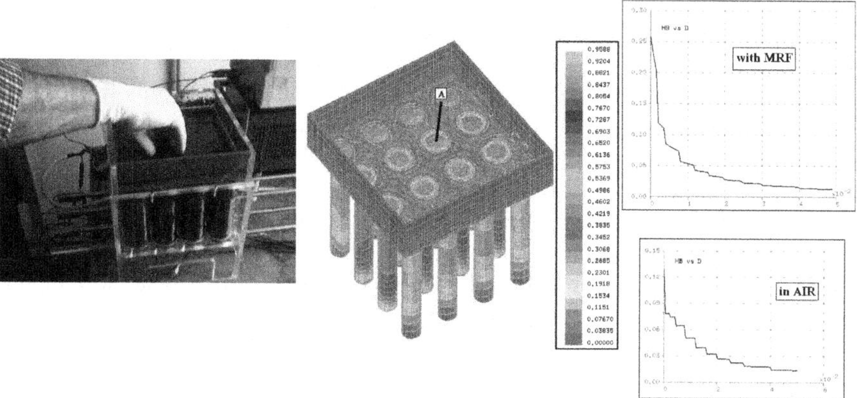

Fig. 9.81 Magnetorheological actuation principle for full-hand interaction based on a 4×4 pattern [11]

is largely dependent on the control of the rheological fluid [10]. The psycho-physical experiments performed until today show that the identification of simple geometrical structures can be achieved on the basis of a 4×4 pattern inside the rheological volume.

9.6.1.4 Self-induction and Eddy Currents as Damping

An active haptic device is designed to generate forces, resp., torques in any direction. By the concept of "active" actuation, the whole spectrum of mechanical interaction objects (e.g., masses, springs, dampers, other force sources like muscles, and moving objects) is covered. Nevertheless, only a slight portion of haptic interaction actually is "active." This has the side effect (of control-engineering approaches) that active systems have continuously to be monitored for passivity. An alternative approach to the design of haptic actuators is given by choosing technical solutions able to dissipate mechanical energy. A frictional brake would be such a device, but its properties are

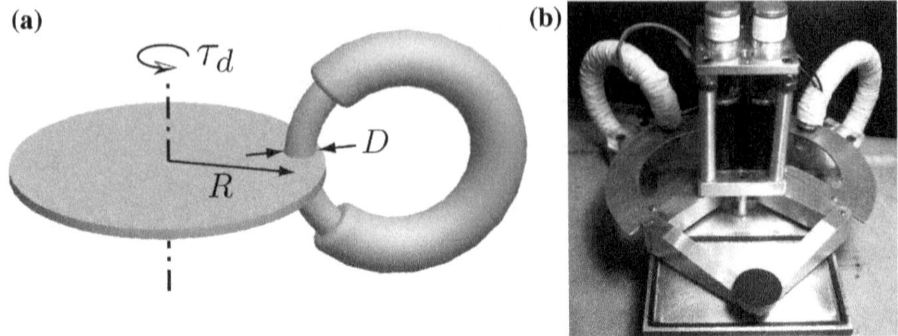

Fig. 9.82 Principle of eddy currents damping a rotating disk (a) and realization as a haptic device (b) by [41]

strongly nonlinear and hard to control. Alternatives are therefore highly interesting. The team of COLGATE showed in [86] how to increase the impedance of an electronic-commutated electrodynamic actuator, whereby two windings were bypassed by a variable resistor. The mutual induction possible by this bypass damped the motor significantly. In [41], the team of HAYWARD went further by implementing an eddy current break into a pantograph-kinematics (Fig. 9.82). This break is a pure damping element with almost linear properties. By this method, a controlled dynamic damping up to 250 Hz was achieved.

9.6.1.5 Serial Coupled Actuators

Serial coupled actuators include an additional mechanical coupling element between actuator and the driven element of the system. In the majority of cases, this is an elastic element that was originally inserted to ease force control of actuators interacting with stiff environments [138]. These so-called serial elastic actuators allow the replacement of direct force control by the position control of both sides of the series elasticity and are used in applications like rehabilitation or man–machine interaction [104]. An example is shown in Fig. 9.83. For haptic applications, this configuration is especially interesting for the display of null-forces and null-torques, i.e., free space movements.

Another application of serial coupled actuators was introduced by MOHAND- OUSAID ET AL. To increase dynamics, lower the impact of inertia and increase transparency, a serial arrangement of two actuators connected with an viscous coupler based on eddy currents was presented in [88] and is shown in Fig. 9.84. By using two motors, the range of displayable forces/torques can be extended, because of the low inertia of the smaller motor higher dynamics can be archived. The viscous clutch couples slip velocity to transmitted torque that is used in the control of the device. With this approach, inertia is decoupled from the delivered torque effectively.

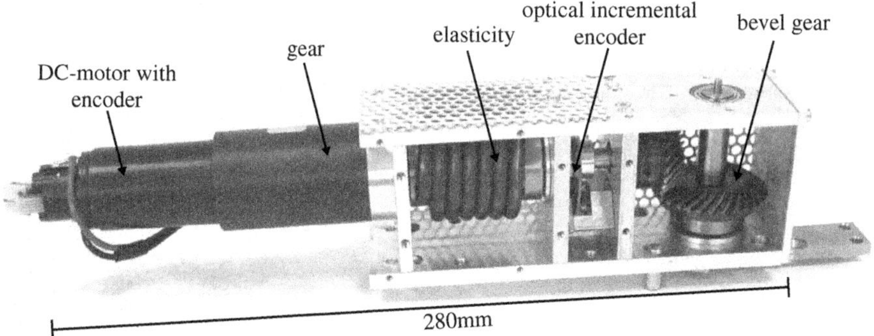

Fig. 9.83 Example for the realization of a serial elastic actuator for use in an active knee orthesis [104]. The bevel gear is needed for better integration of the actuator near the knee of the wearer. Picture courtesy of Roman Müller, Institute of Electromechanical Design, Technische Universität Darmstadt

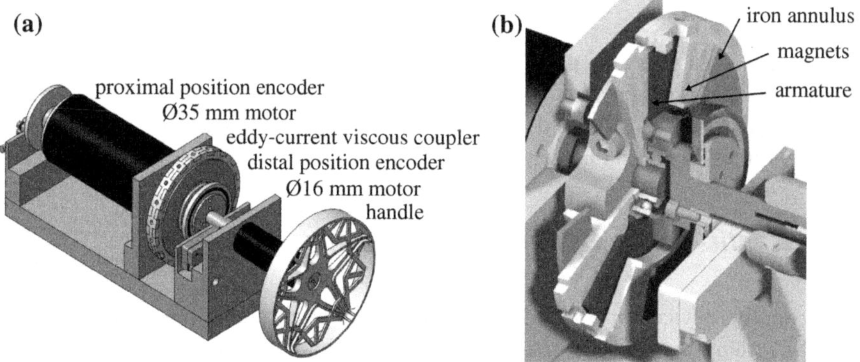

Fig. 9.84 Setup with two serial actuators coupled with an eddy current clutch as presented in [88]

9.6.1.6 MagLev: Butterfly Haptics

In the 1990s, the team of HOLLIS developed a haptic device [9] based on the electrodynamic actuation principle (Fig. 9.85). Since recently, the device has been sold commercially by *Butterfly Haptics*. It is applied to ongoing research projects on psychophysical analysis of texture perception. Six flat coils are mounted in a hemisphere with a magnetic circuit each. The combination of LORENTZ forces of all coils allows an actuation of the hemisphere in three translational and three rotational directions. Via three optical sensors—each of them measuring one translation and one rotation—the total movement of the sphere is acquired. Besides the actuation within its space, the control additionally includes compensation of gravity with the aid of all six actuators. This function realizes a bearing of the hemisphere with Lorentz forces, only. The air-gap of the coils allows a translation of 25 mm and a rotation of $\pm 8°$ in each

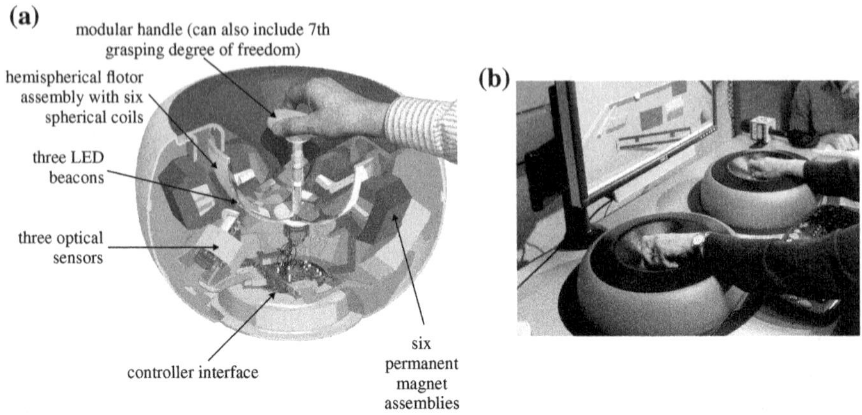

Fig. 9.85 MAGLEV device, **a** inner structure, **b** use of two devices in bimanual interaction. Images courtesy of *Butterfly Haptics, LLC*, Pittsburgh, PA, USA.

direction. Resolutions of $2\,\mu m$ (1σ) and stiffnesses of up to 50 N mm^{-1} can be reached. As a consequence of the small mass of the hemisphere, the electrodynamic actuator principle as a drive and the abandonment of mechanical bearings, forces of a bandwidth of 1 kHz can be generated.

9.6.2 Haptic-Tactile Devices

Haptic-tactile devices of this category are intelligent combinations of well-known actuator principles of haptic systems with either high position resolutions or extraordinary, dynamic properties.

9.6.2.1 Pneumatic

Due to their working principle, pneumatic systems are a smart way to realize flexible high-resolution tactile displays. But these systems suffer acoustic compliance, low dynamics, and the requirement of pressurized air. A one-piece pneumatically actuated tactile 5×5 matrix molded from silicone rubber is described in [90, 91]. The spacing is 2.5 mm with 1-mm-diameter tactile elements. Instead of actuated pins, an array of pressurized chambers without chamber leakage and seal friction is used (see Fig. 9.86). 25 solenoid 3-way valves are used to control the pressure in each chamber resulting in a working frequency of 5 Hz. Instead of closed pressurized chambers, the direct contact between the fingertip and the compressed air is used for tactile stimulation in [2]. The interface to the skin consists of channels each 2 mm in diameter. A similar display is shown in [82]. Using negative air pressure, the tactile stimulus is generated by suction through 19 channels 2.5 mm in diameter with 5 mm intervals.

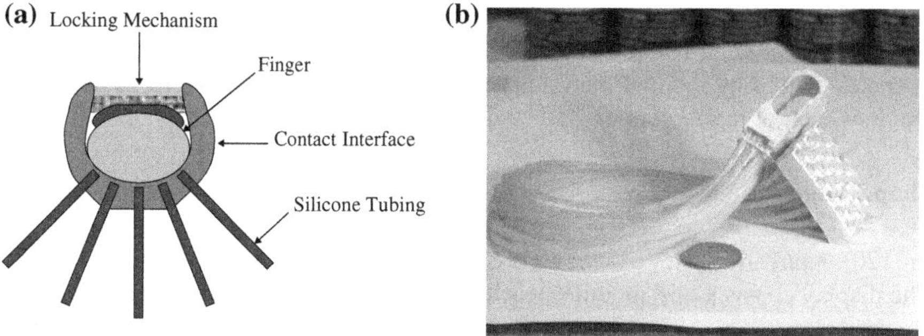

Fig. 9.86 Pneumatic actuated tactile display: sketch (**a**) bidigital teletaction and realization (**b**) [91]

9.6.2.2 Thermopneumatic

A classic problem of tactile pin arrays is given by the high density of stimulator points to be achieved. The space below each pin for control and reconfiguration of the pin's position is notoriously finite. Consequently, a large number of different designs have been tested till date. In [128], a thermopneumatic system is introduced (Fig. 9.87) based on tubes filled with a fluid (methyl chloride) with a low boiling point. The system allows a reconfiguration of the pins within 2 s. However, it has high power requirements, although the individual elements are cheap.

9.6.2.3 Shape-Memory Materials

Materials with shape-memory property are able to remember their initial shape after deformation. When the material is heated up, its internal structure starts to change

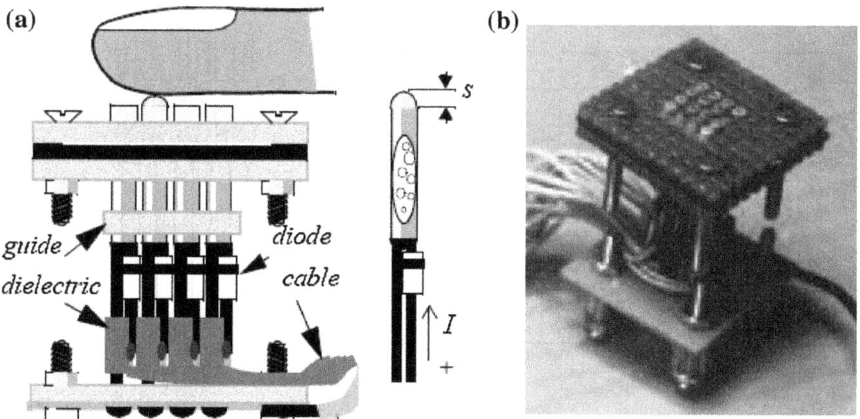

Fig. 9.87 Thermopneumatic actuation principle in a schematic sketch (**a**) and as actual realization (**b**) [128]

and the materials forms back in its pre-deformed state. Due to the material intrinsic actuating effect, high-resolution tactile displays are achievable. The low driving frequencies caused by thermal inertia and needed heating and/or cooling systems are the drawback of this technology.

Shape-Memory Alloys

In [120], a pin array with 64 elements is realized covering an area of $20 \times 40\,\text{mm}^2$. The display consists of 8 modules, each containing eight dots (see Fig. 9.88). Each element comprises a 120 mm length NiTi SMA wire pre-tensioned by a spring. When an electrical current flows through the wire, it heats up and starts to shorten. The result is a contraction of up to 5 mm. Driving frequencies up to a few Hz using a fan to cool the SMA wires down can be reached.

Bistable Electroactive Polymers

Bistable electroactive polymers (BSEP) combine the large-strain actuation of dielectric elastomers with shape-memory properties. The BSEP provide bistable deformation in a rigid structure. These polymers have a glass transition temperature T_g slightly above ambient temperature. Heated above T_g, it is possible to actuate the materials like a conventional dielectric elastomer.

Using a chemically cross-linked poly(tert-butyl acrylate) (PTBA) as BSEP, a tactile display is presented in [94]. The display contains a layer of PTBA diaphragm-actuators and an incorporated heater element array. Figure 9.89 shows the fabricated refreshable Braille display device the size of a smartphone screen.

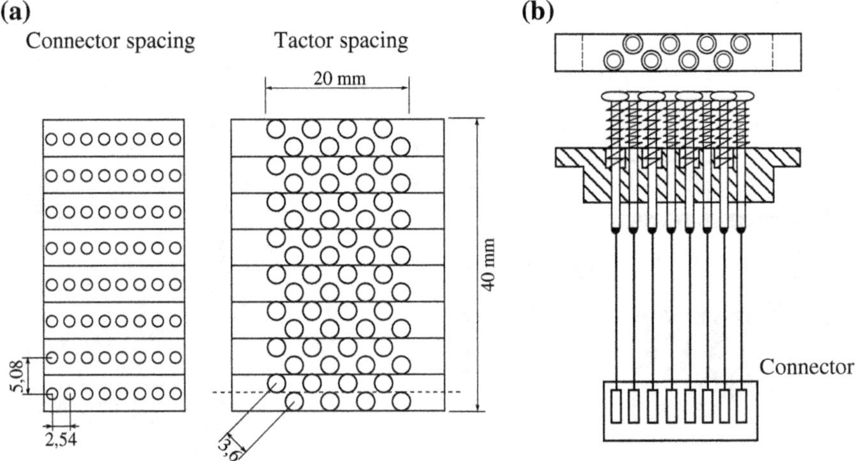

Fig. 9.88 Sketch of tactile display using SMA: Top view (**a**) and side view of one module (**b**) [120]

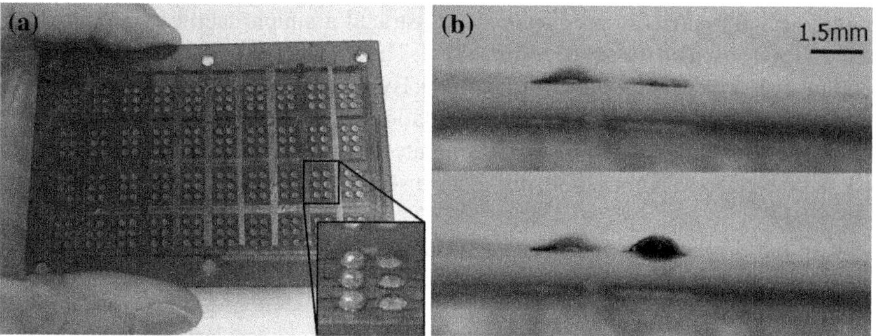

Fig. 9.89 Bistable BSEP Braille display: actuator array (**a**) [94], and zoom of Braille dots in "OFF"? and "ON"? state (**b**) [93]

9.6.2.4 Texture Actuators

Besides an application in Braille-related tasks, the design of tactile displays is also relevant for texture perception. Instead of vibrotactile stimulation on a user's finger, the modification of the friction between a sliding finger and a touchscreen surface is a promising new direction in touchscreen haptics. These kinds of displays are based on two basic technologies. In displays based on the electrovibration effect, a periodic electrical signal is injected into a conductive electrode coated with a thin dielectric layer. The result is an alternating electrostatic force that periodically attracts and releases the finger, producing friction-like rubbery sensations as mentioned in Sect. 9.5.2.4.

In friction displays based on the squeeze film effect, a thin cushion of air under the touching finger is created by a layer placed on top of the screen and is vibrated at an ultrasonic frequency. The modulation of the frequency and intensity of vibrations allows to put the finger touching the surface in different degrees of levitation, thus actually affecting the frictional coefficient between the surface and the sliding finger [68] (Fig. 9.90).

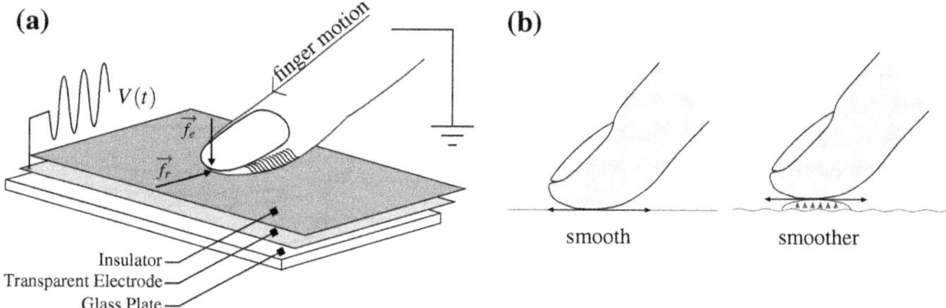

Fig. 9.90 Basic principles of friction tactile displays: **a** Electrovibration effect [7] and **b** squeeze film effect as shown in [20]

In 2007, WINFIELD impressively demonstrated a simple tactile texture display called TPAD based on the squeeze film effect. The actuating element is a piezoelectric bending disk driven in resonance mode [139]. By aid of an optical tracking right above the disk, and with a corresponding modulation of the control signal, perceivable textures with spatial resolutions were generated. The 25-mm-diameter piezoelectric disk is bonded to a 25-mm-diameter glass disk and supported by an annular mount (Fig. 9.91).

A similar display with an increased surface area of $57 \times 76\,\text{mm}^2$ is shown in [83]. The vibrations are created by piezoelectric actuators bonded along one side of a glass plate placed on top of an LCD screen.

9.6.2.5 Flexural Waves

A transparent display providing localized tactile stimuli is presented in [53]. The working principle is based on the concept of computational time reversal and is able to stimulate one or several regions, and hence several fingers, independently. According to the wave propagation equation, the direct solution of a given propagation problem is a diverging wave front and a converging wave front for the time reversed one if the initial condition is an impulse force. Consequently, it is possible to generate peaks of deflection localized in space and in time using constructive interference caused by a multiple of stimulating actuators. Of course, the quality of the focusing process is increasing with the number of transducers and has to be optimized. Depending on the requirements, the noise occurring at the passive areas might be reduced below the tactile perception threshold.

The display shown in Fig. 9.92 consists of a glass sheet with the dimensions of $63 \times 102\,\text{mm}^2$ and 0.2 mm in thickness bonded to a rigid supporting frame at its edges. The used transducers are piezoelectric diaphragms (Murata 7BB-12-9) with a diameter of 12 mm and a thickness of 0.12 mm.

Relating to their patented bending wave technology, the company *Redux Laboratories offers* electromagnetic transducers for medium and large applications as well as piezo-exciters for small form factor applications such as mobile devices (see Fig. 9.93).

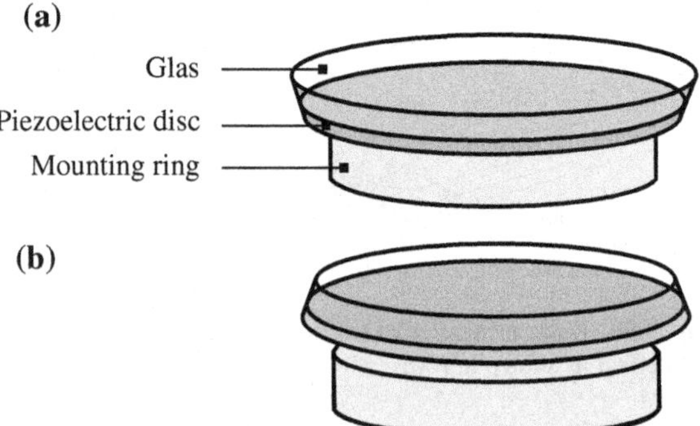

Fig. 9.91 Schematic view of the T-PaD with convex (**a**) and concave (**b**) oscillation states: Piezoelectric disk underneath a glass substrate modulates the friction between finger and display based on the position of the finger and the model of the simulated texture. Figure based on [22]

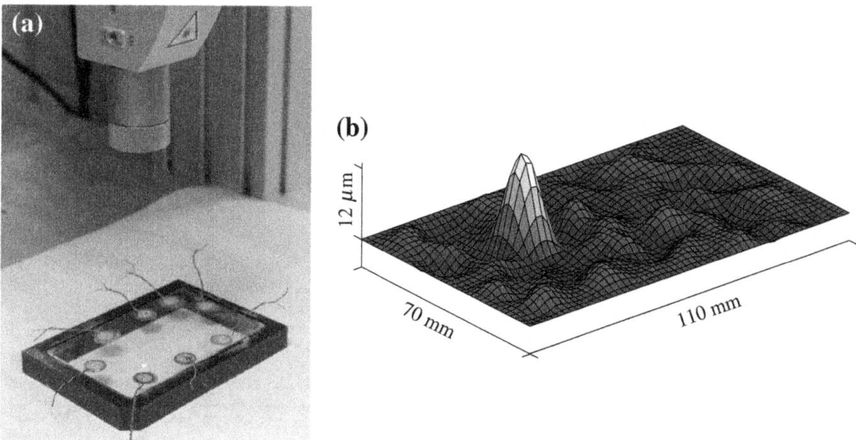

Fig. 9.92 Tactile display using flexural waves: display during characterization with laser vibrometer (**a**) and reconstructed experimental out of plane displacement at focus time without finger load (**b**) [53]

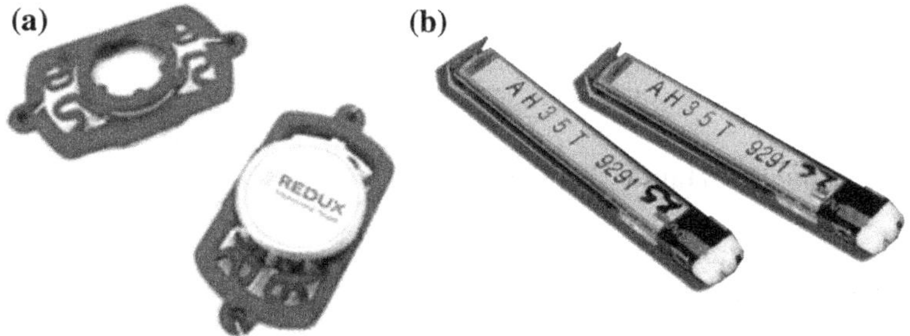

Fig. 9.93 *Redux* transducers for bending wave technology: Moving-coil exciters (**a**), and multi-layered piezos (**b**) [106]

9.6.2.6 Volume-Ultrasonic Actuator

IWAMOTO built tactile displays (Fig. 9.94) that are made of piezoelectric actuators and are actuated in the ultrasonic frequency range. They use sound pressure as a force transmitter. The underlying principle is given by generating a displacement of the skin and a corresponding haptic perception by focused sound pressure. Whereas in the first realization an ultrasonic array had been used to generate tactile dots in a fluid [60], later developments used the air for energy transmission [59]. The pressures generated by the designs (Fig. 9.94) provide a weak tactile impression, only. But

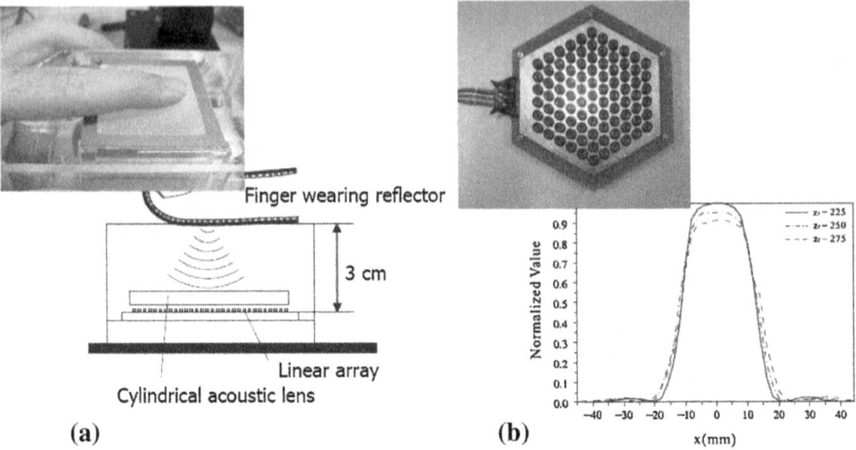

Fig. 9.94 Tactile display based on ultrasonic sound pressure transmitted by a fluid [60] (**a**) or as array of senders for a transmission in the air [59] (**b**)

especially the air-based principle works without any mechanical contact and could therefore become relevant for completely new operation concepts combined with gesture recognition.

9.6.2.7 Ungrounded Haptic Displays

In case of the interaction with large virtual worlds, it is frequently necessary to design devices that are worn on the body, i.e., that do not exhibit a fixed ground connection. An interesting solution has been shown in [123], generating a tactile sensation with belts at the palm and at each finger. The underlying principle is based on two actuators for each belt, generating a shear force to the skin when operated in the same direction, and a normal force when operated in the opposite direction. This enables to provide tactile effects when grasping or touching objects in a virtual world, but without the corresponding kinaesthetic effects.

Other realizations of ungrounded devices include the use of gyro effects and the variation of angular momentum and designs incorporating nonlinear perception properties of the human user as presented in [113]. The device shown in Fig. 9.95 is based on the display of periodical steep inertial forces generated by a spring-mass system and an electrodynamic voice-coil actuator.

9.6.2.8 Electrotactile

As haptic receptors can be stimulated electrically, it is not far-fetched to design haptic devices able to provide low currents to the tactile sense organs. The design

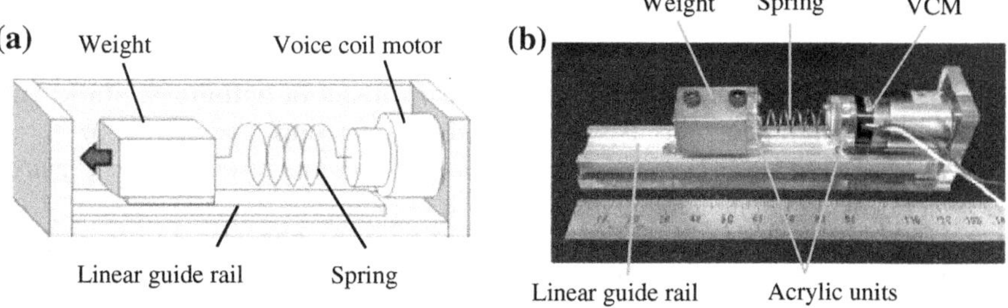

Fig. 9.95 Ungrounded haptic display to convey pulling impressions as shown in [113] **a** scheme , **b** realization

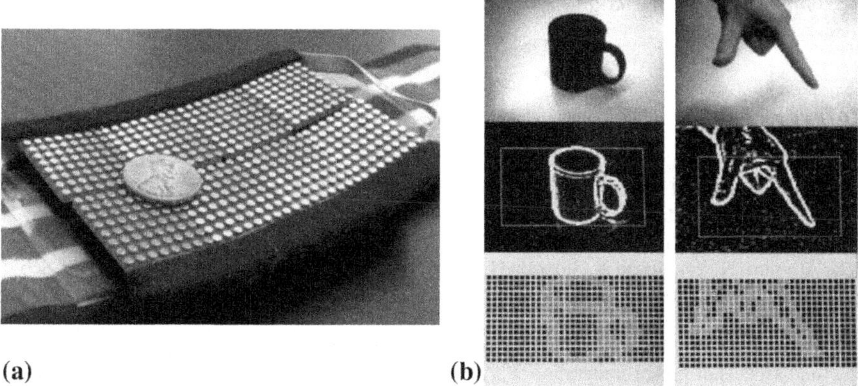

Fig. 9.96 Electrotactile display worn on the forehead: Electrodes (**a**) and edge recognition and signal conditioning principle (**b**) [64]

of such devices can be traced back to the 1970s. One realization is presented in [64] (Fig. 9.96). Electrotactile displays do work—no doubt—however, they have the disadvantage also to stimulate noci-receptors for pain sensation besides the mechanoreceptors. Additionally, the electrical conductivity between display and skin is subject to major variations. These variations are inter-person differences due to variations in skin thickness, but they are also a time-dependent result of electrochemical processes between sweat and electrodes. The achievable tactile patterns and the abilities to distinguish tactile patterns are subject to current research.

Recommended Background Reading

[3] Ballas, R.: **Piezoelectric multilayer beam bending actuators: static and dynamic behavior and aspects of sensor integration**. Springer, 2007. *Dynamic modeling of piezoelectric bimorph actuators based on concentrated and distributed network elements.*

[52] Huber, J.; Fleck, N. & Ashby, M.: **The Selection of Mechanical Actuators Based on Performance Indices**. In: Proceedings of the Royal Society of London. Series A: Mathematical, Physical and Engineering Sciences, 1997. *Analysis of several performance indices for different actuation principles.*

[65] Kallenbach, E.; Stölting, H.-D. & Amrhein, W. (Eds.): **Handbook of Fractional-Horsepower Drives**. Springer, Heidelberg, 2008. *Overview about the most common fractional horsepower actuators and transmissions with focus on the design parameters.*

References

1. Ahmadkhanlou F (2008) Design, modeling and control of magnetorheological fluid-based force feedback dampers for telerobotic systems. In: Proceedings of the Edward F. Hayes graduate research forum. http://hdl.handle.net/1811/32023
2. Asamura N, Yokoyama N, Shinoda H (1998) Selectively stimulating skin receptors for tactile display. IEEE Comput Graph Appl 18(6):32–37. ISSN: 02721716. doi:10.1109/38.734977. (Visited on 02/09/2014)
3. Ballas R (2007) Piezoelectric multilayer beam bending actuators: static and dynamic behavior and aspects of sensor integration. Springer, Berlin, pp XIV, 358. ISBN: 978-3-540-32641-0
4. Bar-Cohen Y (2001) Electroactive polymer (EAP) actuators as artificial muscles–reality, potential, and challenges. SPIE Press monograph, vol 98. SPIE Press, Bellingham, pp XIV, 671. ISBN: 0-8194-4054-x
5. Bar-Cohen Y (2006) Biomimetics–biologically inspired technologies. CRC, Taylor und Francis, Boca Raton, pp XVIII, 527. ISBN: 0-8493-3163-3
6. Bar-Cohen Y et al (2001) Virtual reality robotic telesurgery simulations using MEMICA haptic system. In: 8th annual international symposium on smart structures and material—EAPAD. doi:10.1117/12.432667
7. Bau O et al (2010) TeslaTouch: electrovibration for touch surfaces. In: Proceedings of the 23nd annual ACM symposium on user interface software and technology (UIST '10). ACM, New York, pp 283–292. doi:10.1145/1866029.1866074
8. Belford JD The stepped horn–Technical Publication TP-214. Technical report http://www.morganelectroceramics.com/resources/technical-publications/
9. Berkelman PJ, Butler ZJ, Hollis R L (1996) Design of a hemispherical magnetic levitation haptic interface device. In: ASME (ed) ASME international mechanical engineering congress and exposition., pp 483–488. http://www.ri.cmu.edu/pub_files/pub1/berkelman_peter_1996_1/berkelman_peter_1996_1.pdf
10. Bicchi A et al. (2005) Analysis and design of an electromagnetic system for the characterization of magnetorheological fluids for haptic interfaces. IEEE Trans Magn 41(5). Interdepartmental Res. Centre, Pisa University, Italy, pp 1876–1879. doi:10.1109/TMAG.2005.846280
11. Bicchi A (2008) The sense of touch and its rendering—progress in haptics research, vol 45. Springer tracts in advanced robotics. Springer, Berlin, pp XV, 280. ISBN: 9783540790341

12. Biet et al M (2006) A piezoelectric tactile display using travelling lamb wave. In: Proceedings of eurohaptics. Paris, pp 989–992
13. Blume H-J, Boelcke R (1990) Mechanokutane Sprachvermittlung, vol 137. Reihe 10 137. Düsseldorf: VDI-Verl. ISBN: 3-18-143710-7
14. Böse H, Trendler A (2003) Smart fluids–properties and benefit for new electromechanical devices. In: AMAS workshop SMART'03. pp 329–336. http://publica.fraunhofer.de/documents/N-51527.html
15. Böse H et al. (2004) A new haptic sensor-actuator system based on electrorheological fluids. In: Actuator 2004: 9th international conference on new actuators. HVG Hanseatische Veranst. GmbH, Div. Messe Bremen, pp 300–303. doi:10.1016/S0531-5131(03)00443--6
16. Brissaud M (1991) Characterization of piezoceramics. IEEE Trans Ultrason Ferroelectr Freq Control 38(6):603–617. doi:10.1109/58.108859
17. Brochu P, Pei Q (2010) Advances in dielectric elastomers for actuators and artificial muscles. Macromol Rapid Commun 31(1):10–36. doi:10.1002/marc.200900425. (Visited on 04/03/2012)
18. Carlson JD, Stanway R, Johnson AR (2004) Electro-rheological and magneto-rheological fluids: a state of the art report. In: Actuator 2004: 9th international conference on new actuators. HVG Hanseatische Veranst.-GmbH, Div. Messe Bremen, Bremen, pp 283–288
19. Carpi F, Frediani G, De Rossi D (2012) Electroactive elastomeric actuators for biomedical and bioinspired systems. In: IEEE, June 2012, pp 623–627. doi:10.1109/BioRob.6290761. (Visited on 02/26/2014)
20. Casiez G et al. (2011) Surfpad: riding towards targets on a squeeze film effect. In: Proceedings of the SIGCHI conference on human factors in computing systems. ACM Press, New York, p 2491. doi:10.1145/1978942.1979307
21. Chapuis D, Michel X (2007) A haptic knob with a hybrid ultrasonic motor and powder clutch actuator. In: Michel X, Gassert R (eds) Euro haptics conference, 2007 and symposium on haptic interfaces for virtual environment and teleoperator systems. World Haptics 2007. Second Joint. pp 200–205. doi:10.1109/WHC.2007.5
22. Colgate JE, Peshkin M (2009) Haptic device with controlled traction forces. 8525778:B2
23. Conrad H, Fisher M, Sprecher AF (1990) Characterization of the structure of a model electrorheological fluid employing stereology. In: Proceedings of the 2nd international conference on electrorheological fluids
24. Deng K, Enikov E, Zhang H, (2007) Development of a pulsed electromagnetic micro-actuator for 3D tactile displays. In:(2007) IEEE/ASME international conference on advanced intelligent mechatronics. University of Arizona, Tucson, pp 1–5. doi:10.1109/AIM.2007.4412457
25. Doerrer C (2004) Entwurf eines elektromechanischen Systems für flexibel konfigurierbare Eingabefelder mit haptischer Rückmeldung". PhD thesis. Technische Universität Darmstadt, Institut für Elektromechanische Konstruktionen. http://tuprints.ulb.tu-darmstadt.de/435/
26. Dr. Fritz Faulhaber GmbH & Co. KG. Technical Informations. Tech. rep. 2013. www.faulhaber.com
27. Eisner E (1966) Complete solutions of the 'Webster' horn equation. J Acoust Soc Am 41(4B):1126–1146. doi:10.1121/1.1910444
28. El Wahed AK et al (2003) An improved model of ER fluids in squeeze-flow through model updating of the estimated yield stress. J Sound Vibr 268(3):581–599. doi:10.1016/S0022-460X(03)00374-2
29. Fernandez JM, Perriard Y (2003) Comparative analysis and modeling of both standing and travelling wave ultrasonic linear motor. In: IEEE ultrasonic symposium, pp 1770–1773. doi:10.1109/ULTSYM.2003.1293255
30. Fleischer M, Stein D, Meixner H (1989) Ultrasonic piezomotor with longitudinally oscillating amplitude-transforming resonator. IEEE Trans Ultrason Ferroelectr Freq Control 36(6):607–613. doi:10.1109/58.39110
31. Flueckiger M et al. (2005) fMRI compatible haptic interface actuated with traveling wave ultrasonic motor. In: Industry applications conference, 2005. Fortieth IAS annual meeting. Conference record of the 2005, vol 3. pp 2075–2082. ISBN: 0197-2618, doi:10.1109/IAS.2005.1518734

32. Garrec P (2010) Design of an anthropomorphic upper limb exoskeleton actuated by ball-screws and cables. In: University politecnicae of bucharest. Sci Bull D Mech Eng 72(2):23–34. ISSN 1454–2358. http://www.scientificbulletin.upb.ro/rev_docs_arhiva/full9662.pdf
33. Garrec P (2010) Screw and cable actuators SCS and their applications to force feedback teleoperation, exoskeleton and anthropomorphic robotics. In: Abdellatif H (ed) Robotics 2010 current and future challenges. InTech. ISBN: 978-953-7619-78-7. http://www.intechopen.com/download/get/type/pdfs/id/9370 (Visited on 02/09/2014)
34. Garrec P et al. (2006) A new force-feedback, morphologically inspired portable exoskeleton. In: IEEE, pp 674–679. doi:10.1109/ROMAN.2006.314478. (Visited on 02/09/2014)
35. Garrec P et al (2007) Evaluation tests of the telerobotic system MT200-TAO in AREVA NC La Hague hot cells. In: Proceedings of ENC, Brussels. http://www-ist.cea.fr/publicea/exldoc/200800002158.pdf
36. Garrec P et al. (2008) ABLE, an innovative transparent exoskeleton for the upper-limb. In: IEEE, pp 1483–1488. doi:10.1109/IROS.2008.4651012. (Visited on 02/09/2014)
37. Ghoddsi R et al (1996) Development of a tangential tactor using a LIGA/MEMS linear microactuator technology microelectromechanical system (MEMS). In: Microelectromechanical system (MEMS), DSC, vol 59. Atlanta, pp 379–386. ISBN: 978-0791815410
38. Giraud F, Giraud F, Semail B (2004) Analysis and phase control of a piezoelectric traveling-wave ultrasonic motor for haptic stick application. In: Semail B, Audren J-T (eds) IEEE Transactions on industry applications, vol 40(6), pp 1541–1549. doi:10.1109/TIA.2004.836317
39. Giraud F, Lemaire-Semail B, Martinot F (2006) A force feedback device actuated by piezoelectric travelling wave ultrasonic motors. In: ACTUATOR 2006, 10th international conference on new actuators. pp 600–603. ISBN: 9783933339089
40. Gong X et al (2008) Influence of liquid phase on nanoparticle-based giant electrorheological fluid. Nanotechnol 19(16):165602. doi:10.1088/0957-4484/19/16/165602
41. Gosline AH, Campion G, Hayward V (2006) On the use of eddy current brakes as tunable, fast turn-on viscous dampers for haptic rendering. In: Eurohaptics conference. Paris, pp 229–234. http://www.cirmmt.org/research/bibliography/GoslineEtAl2006
42. Hagedorn P et al (1998) The importance of rotor flexibility in ultrasonic traveling wave motors. Smart Mater Struct 7:352–368. doi:10.1088/0964-1726/7/3/010
43. Hagood NW, McFarland AJ (1995) Modeling of a piezoelectric rotary ultrasonic motor. IEEE Trans Ultrason Ferroelectr Freq Control 42(2):210–224. doi:10.1109/58.365235
44. Haption SA (2014) http://www.haption.com/ (visited on 02/19/2014)
45. Hayward V, Cruz-Hernandez M (2000) Tactile display device using distributed lateral skin stretch. In: Wikander J (ed) Symposium on haptic interfaces for virtual environment and teleoperator systems, IMECE 2000 conference. http://www.cim.mcgill.ca/jay/index_files/research_files/VH-MC-HAPSYMP-00.pdf
46. He S et al (1998) Standing wave bi-directional linearly moving ultrasonic motor. IEEE Trans Ultrason Ferroelectr Freq Control 45(5):1133–1139. doi:10.1109/58.726435
47. Helin P et al (1997) Linear ultrasonic motors using surface acoustic waves mechanical model for energy transfer. In: Solid state sensors and actuators, TRANSDUCERS '97 Chicago., 1997 international conference on, vol 2. Chicago, IL, pp 1047–1050, 16–19 June 1997. doi:10.1109/SENSOR.1997.635369
48. Heydt R, Chhokar S (2003) Refreshable braille display based on electroactive polymers. In: 23rd international display research conference, pp 111–114
49. Hirata H, Ueha S (1995) Design of a traveling wave type ultrasonic motor. IEEE Trans Ultrason Ferroelectr Freq Control 42(2):225–231. doi:10.1109/58.365236
50. Hu M et al (2005) Performance simulation of traveling wave type ultrasonic motor. In: Proceedings of the eighth international conference on electrical machines and systems, 2005. ICEMS 2005, vol 3. pp 2052–2055, 27–29 Sept 2005. ISBN: 7-5062-7407-8. doi:10.1109/ICEMS.2005.202923
51. Huang X et al (2007) Formation of polarized contact layers and the giant electrorheological effect. Int J Mod Phy B (IJMPB) 21(28/29):4907–4913. doi:10.1142/S0217979207045827

52. Huber J, Fleck N, Ashby M (1997) The selection of mechanical actuators based on performance indices. In: Proceedings of the Royal Society of London. Series A: Mathematical, physical and engineering sciences, vol 453. 1965, pp 2185–2205. doi:10.1098/rspa.1997.0117
53. Hudin C, Lozada J, Hayward V (2013) Localized tactile stimulation by time-reversal of flexural waves: case study with a thin sheet of glass. In: IEEE, pp 67–72. doi:10.1109/WHC.2013.6548386
54. HyperBraille. http://hyperbraille.de/ (visited on 02/09/2014)
55. IEEE (1988) Standard on piezoelectrity. doi:10.1109/IEEESTD.1988.79638
56. Ikeda T (1990) Fundamentals of piezoelectricity. Oxford University Press, Oxford, pp XI, 263. ISBN: 0-19-856339-6
57. Ikei Y, Ikei Y, Shiratori M (2002) TextureExplorer: a tactile and force display for virtual textures. In: Haptic interfaces for virtual environment and teleoperator systems. In: Shiratori M (ed) 10th symposium on HAP TICS 2002. Proceedings, pp 327–334. doi:10.1109/HAPTIC.2002.998976
58. Ikei Y, Yamada M, Fukuda S (1999) Tactile texture presentation by vibratory pin arrays based on surface height maps. In: Olgac N (ed) ASME dynamic systems and control division. 67:97–102. ISBN: 0791816346
59. Iwamoto T, Tatezono M, Shinoda H (2008) Non-contact method for producing tactile sensation using airborne ultrasound. In: Haptics: perception, devices and scenarios. Springer, Berlin, pp 504–513. doi:10.1007/978-3-540-69057-3_64
60. Iwamoto T, Shinoda H (2005) Ultrasound tactile display for stress field reproduction–examination of non-vibratory tactile apparent movement. In: First joint eurohaptics conference and symposium on haptic interfaces for virtual environment and teleoperator systems. WHC 2005, pp 220–228. doi:10.1109/WHC.2005.140
61. Jendritza DJ (1998) Technischer einsatz neuer aktoren: grundlagen, werkstoffe, designregeln und anwendungsbeispiele. 2nd ed. vol 484. Kontakt Studium 484. Renningen-Malmsheim: expert-Verl., p 493. ISBN: 3-8169-1589-2
62. Jolly MR, Carlson JD (1996) Controllable squeeze film damping using magnetorheological fluid. In: Actuator 96, 5th international conference on New Actuators, Bremen
63. Jungmann M (2004) Entwicklung elektrostatischer festkörperaktoren mit elastischen dielektrika für den einsatz in taktilen anzeigefeldern. Technische Universität Darmstadt, Dissertation, p 138. http://tuprints.ulb.tu-darmstadt.de/500/
64. Kajimoto H, Kanno Y, Tachi S (2006) Forehead electro-tactile display for vision substitution. In: Eurohaptics conference. Paris. http://lsc.univ-evry.fr/eurohaptics/upload/cd/papers/f62.pdf
65. Kallenbach E, Stölting H-D, Amrhein W (eds) (2008) Handbook of fractional-horsepower drives. Springer, Berlin. ISBN: 978-3-540-73128-3
66. Kallenbach E et al (2008) Elektromagnete-Grundlagen, Berechnung, Entwurf und Anwendung. Wiesbaden: Vieweg+Teubner, pp XII, 402. ISBN: 978-3-8351-0138-8
67. Kenaley GL, Cutkosky MR (1989) Electrorheological fluid-based robotic fingers with tactile sensing. In: IEEE international conference on robotics and automation. Proceedings, vol 1. pp 132–136. doi:10.1109/ROBOT.1989.99979
68. Kim S-C, Israr A, Poupyrev I (2013) Tactile rendering of 3D features on touch surfaces. In: ACM Press, pp 531–538. doi:10.1145/2501988.2502020. (Visited on 02/17/2014)
69. Kim KJ, Tadokoro S (2007) Electroactive polymers for robotic applications—artificial muscles and sensors. Springer, London
70. Kornbluh R et al. (1999) High-Field electrostriction of elastomeric polymer dielectrics for actuation. In: 6th annual international symposium on smart structures and material—EAPAD, pp 149–161. doi:10.1117/12.349672
71. Kornbluh R, Pelrine R (1998) Electrostrictive polymer artificial muscle actuators. IEEE Int Conf Robotics Autom 3:2147–2154. doi:10.1109/ROBOT.1998.680638
72. Koyama T, Takemura K (2003) Development of an ultrasonic clutch for multi-fingered exoskeleton haptic device using passive force feedback for dexterous teleoperation. In:

Takemura K, Maeno T (eds) Intelligent robots and systems. (IROS 2003). Proceedings. 2003 IEEE/RSJ international conference on, vol 3, pp 2229–2234. doi:10.1109/IROS.2003.1249202

73. Kyung K-U, Lee JY (2007) Haptic stylus and empirical studies on braille, button, and texture display. J Biomed Biotechnol 2008(369651):11. doi:10.1155/2008/369651
74. Kyung K-U, Park J-S (2007) Ubi-pen: development of a compact tactile display module and its application to a haptic stylus. In: Park J-S (ed) EuroHaptics conference, 2007 and symposium on haptic interfaces for virtual environment and teleoperator systems. World Haptics 2007. Second Joint. pp 109–114. doi:10.1109/WHC.2007.121
75. Kyung K-U, Ahn M (2005) A compact broadband tactile display and its effectiveness in the display of tactile form. In: Ahn M, Kwon D-S (eds) Eurohaptics conference, 2005 and symposium on haptic interfaces for virtual environment and teleoperator systems. World Haptics 2005. First Joint, pp 600–601. doi:10.1109/WHC.2005.4
76. Lawrence D, Pao L, Aphanuphong S (2005) Bow spring/tendon actuation for low cost haptic interfaces. In: Haptic interfaces for virtual environment and teleoperator systems. WHC 2005. First joint eurohaptics conference and symposium on (2005). Aerospace Engineering. Colorado University, Boulder, pp 157–166. doi:10.1109/WHC.2005.26
77. Lenk A et al (eds) (2011) Electromechanical systems in microtechnology and mechatronics: electrical, mechanical and acoustic networks, their interactions and applications. Springer, Heidelberg. ISBN: 978-3-642-10806-8
78. Leondes CT (2006) MEMS/NEMS–handbook techniques and applications. Springer, New York. ISBN 0-387-24520-0
79. Levesque V, Levesque V, Pasquero J (2007) Braille display by lateral skin deformation with the STReSS2 tactile transducer. In: Pasquero J, Hayward V (eds) EuroHaptics conference, 2007 and symposium on haptic interfaces for virtual environment and teleoperator systems. World Haptics 2007. Second Joint, pp 115–120. doi:10.1109/WHC.2007.25
80. Li WH et al (2004) Magnetorheological fluids based haptic device. Sensor Rev 24(1):68–73. doi:10.1108/02602280410515842
81. Lotz P, Matysek M, Schlaak HF (2011) Fabrication and application of miniaturized dielectric elastomer stack actuators. IEEE/ASME Trans Mech 16(1):58–66. doi:10.1109/TMECH.2010.2090164. (Visited on 04/11/2012)
82. Makino Y, Asamura N, Shinoda H (2003) A cutaneous feeling display using suction pressure. In: SICE 2003 annual conference (IEEE Cat. No.03TH8734). SICE 2003 annual conference, vol 3. Society of Instrument and Control Engineers, Fukui, Japan. Tokyo, Japan, pp 2931–2934. 4–6 Aug 2003. ISBN: 0-7803-8352-4
83. Marchuk N, Colgate J, Peshkin M (2010) Friction measurements on a large area TPaD. In: Haptics symposium, 2010 IEEE, pp 317–320. doi:10.1109/HAPTIC.2010.5444636
84. Matysek M, Lotz P, Schlaak H (2011) Lifetime investigation of dielectric elastomer stack actuators. IEEE Trans Dielectrics Electr Insul 18(1):89–96. doi:10.1109/TDEI.2011.5704497 (Visited on 04/10/2012)
85. Matysek M et al (2011) Combined driving and sensing circuitry for dielectric elastomer actuators in mobile applications. In: Proceedings of SPIE. vol 7976. San Diego. doi:10.1117/12.879438. (Visited on 03/29/2012)
86. Mehling J, Colgate J, Peshkin M (2005) Increasing the impedance range of a haptic display by adding electrical damping. In: Eurohaptics conference, 2005 and symposium on haptic interfaces for virtual environment and teleoperator systems. World haptics 2005. First Joint. NASA Johnson Space Center, Houston, pp 257–262. doi:10.1109/WHC.2005.79
87. Metec AG. http://web.metec-ag.de/ (Visited on 02/09/2014)
88. Mohand-Ousaid A et al (2012) Haptic interface transparency achieved through viscous coupling. Int J Rob Res 1(3):319–329
89. Mößinger H et al (2014) Tactile feedback to the palm using arbitrarily shaped DEA. In: Proceedings of SPIE, vol 9056. SPIE, San Diego. doi:10.1117/12.2045302
90. Moy G, Wagner C, Fearing R (2014) A compliant tactile display for teletaction. IEEE 4:3409–3415. doi:10.1109/ROBOT.2000.845247. (Visited on 02/09/2014)

91. Moy G (2002) Bidigital teletaction system design and performance. PhD thesis. University of California at Berkeley. http://robotics.eecs.berkeley.edu/ronf/PAPERS/Theses/gmoy-thesis02.pdf
92. Murayama J et al (2004) SPIDAR G&G: two-handed haptic interface for bimanual VR interaction. In: Eurohaptics, vol 1. 1. Universität München, München, pp 138–146
93. Niu X et al (2011) Refreshable tactile displays based on bistable electroactive polymer. In: Proceedings of SPIE, vol 7976. doi:10.1117/12.880185
94. Niu X et al (2012) Bistable electroactive polymer for refreshable Braille display with improved actuation stability. In: Proceedings of SPIE, vol 8340. doi:10.1117/12.915069
95. Nührmann D (1998) Das große Werkbuch Elektronik. 7. Poing: Franzis' Verlag. ISBN: 3772365477
96. Olsson P et al (2012) Rendering stiffness with a prototype haptic glove actuated by an integrated piezoelectric motor. In: Haptics: perception, devices, mobility, and communication. Springer, Berlin, pp 361–372. doi:10.1007/978-3-642-31401-8_33
97. Parthasarathy M, Klingenberg DJ (1996) Electrorheology: mechanisms and models. Mater Sci Eng R Rep 17(2):57–103. doi:10.1016/0927-796X(96)00191-X
98. Pei Q et al (2003) Multifunctional electroelastomer roll actuators and their application for biomimetic walking robots. In: Proceedings of SPIE, vol 5051. [31]. pp 281–290. doi:10.1117/12.484392
99. Pelrine R, Kornbluh R (2008) Electromechanical transduction effects in dielectric elastomers: actuation, sensing, stiffness modulation and electric energy generation. In: Carpi F et al (eds) Dielectric elastomers as electromechanical transducers. 1st edn. Elsevier, Amsterdam, p 344. doi:10.1016/B978-0-08-047488-5.00001-0
100. Phillips RW (1969) Engineering applications of fluids with a variable yield stress. PhD thesis. University of California, Berkeley
101. Physik Instrumete PI GmbH & Co KG (2003) Piezo linear driving mechanism for converting electrical energy into motion, has a group of stacking actuators for driving a rotor in a guiding mechanism
102. Polzin J (2013) Entwurf einer Deltakinematik als angepasste Struktur für die haptische Bedieneinheit eines Chirurgieroboters. Bachelor Thesis. Technische Universität Darmstadt, Institut für Elektromechanische Konstruktionen
103. Popov D, Gaponov I, Ryu J-H (2013) A preliminary study on a twisted strings-based elbow exoskeleton. In: World haptics conference (WHC), 2013. IEEE, pp 479–484. doi:10.1109/WHC.2013.6548455
104. Pott PP et al (2013) Seriell-Elastische Aktoren als Antrieb für aktive Orthesen. In: at-Automatisierungstechnik, vol 61, pp 638–644. doi:10.1524/auto.2013.0053
105. Rajagopal KR, Ruzicka M (2001) Mathematical modeling of electrorheological materials. Continuum Mech Thermodyn 13(1):59–78. doi:10.1007/s001610100034
106. Redux Laboratories. http://www.reduxst.com/ (visited on 02/19/2014)
107. Ren K et al (2008) A compact electroactive polymer actuator suitable for refreshable Braille display. Sens Actuators A Phys 143(2):335–342. doi:10.1016/j.sna.2007.10.083
108. Sattel T (2003) Dynamics of ultrasonic motors. Dissertation. Technische Universität Darmstadt, pp IV, 167. http://tuprints.ulb.tu-darmstadt.de/305/1/d.pdf
109. Schimkat J et al. (1994) Moving wedge actuator: an electrostatic actuator for use in a microrelay. In: MICRO SYSTEM technologies '94, 4th international conference and exhibition on micro, electro, opto, mechanical systems and components. VDE-Verlag, pp 989–996.
110. Senkal D, Gurocak H (2011) Haptic joystick with hybrid actuator using air muscles and spherical MR-brake. Mechatronics 21(6):951–960. doi:10.1016/j.mechatronics.2011.03.001
111. Sheng P (2005) Mechanism of the giant electrorheological effect. Int J Modern Phys B (IJMPB) 19(7/9):1157–1162. doi:10.1016/j.ssc.2006.04.042
112. Sherrit S et al (1999) Modeling of horns for sonic/ultrasonic applications. In: Ultrasonics symposium. Proceedings. IEEE 1:647–651. doi:10.1109/ULTSYM.1999.849482
113. Shima T, Takemura K (2012) An ungrounded pulling force feedback device using periodical vibrationimpact. In: Haptics: perception, devices, mobility, and communication. Springer, Berlin, pp 481–492. doi:10.1007/978-3-642-31401-8_43

114. Sindlinger S (2011) Haptische Darstellung von Interaktionskräften in einem Assistenzsystem für Herzkatheterisierungen. Dissertation. Technische Universität Darmstadt. http://tuprints.ulb.tu-darmstadt.de/2909/
115. Streque J et al (2012) Magnetostatic micro-actuator based on ultrasoft elastomeric membrane and copper—permalloy electrodeposited Structures. In: IEEE international conference on micro electro mechanical systems (MEMS). Paris, pp 1157–1160. doi:10.1109/MEMSYS.2012.6170368
116. Summers IR, Chanter CM (2002) A broadband tactile array on the fingertip. J Acoust Soc Am 112(5):2118–2126. doi:10.1121/1.1510140
117. Takamatsu R, Taniguchi T, Sato M (1998) Space browsing interface based on head posture information. In: Computer human interaction, 1998. Proceedings. 3rd Asia Pacific, pp 298–303. ISBN: 0-8186-8347-3
118. Takasaki M, Kuribayashi Kurosawa M, Higuchi T (2000) Optimum contact conditions for miniaturized surface acoustic wave linear motor. In: Ultrasonics 38:1–8, pp 51–53. doi:10.1016/S0041-624X(99)00093-1
119. Tang H, Beebe DJ (1998) A microfabricated electrostatic haptic display for persons with visual impairments. Rehabilitation Engineering. IEEE Trans Neural Syst Rehabil 6(3):241–248. doi:10.1109/86.712216
120. Taylor P (1997) The design and control of a tactile display based on shape memory alloys. In: Developments in tactile displays (Digest No. 1997/012), IEEE Colloquium on, vol 1997. doi:10.1049/ic:19970080
121. Taylor PM et al (1998) Advances in an electrorheological fluid based tactile array. Displays 18(3):135–141. doi:10.1016/S0141-9382(98)00014-6
122. Tietze U, Schenk C (2002) Halbleiter-Schaltungstechnik. 12th edn. Springer, Berlin, pp. XXV, 1606. ISBN: 3-540-42849-6
123. Tsagarakis N, Horne T, Caldwell D (2005) SLIP AESTHEASIS: a portable 2D slip/skin stretch display for the fingertip. In: First joint eurohaptics conference and symposium on haptic interfaces for virtual environment and teleoperator systems. WHC 2005. pp 214–219. doi:10.1109/WHC.2005.117
124. Uchino K (1997) Piezoelectric actuators and ultrasonic motors. Kluwer, Boston, pp VIII, 349. ISBN: 9780792398110
125. Uchino K, Giniewicz JR (2003) Micromechatronics. Materials engineering, vol 22. Dekker, New York, pp XIV, 489. ISBN: 0-8247-4109-9
126. Uhea S et al (1996) Ultrasonic motors—theory and application. Oxford University Press, Oxford
127. Uhea S (2003) Recent development of ultrasonic actuators, vol 1. Ultrasonics symposium, 2001 IEEE. Atlanta, 7–10 Oct 2001, vol 1, pp 513–520. doi:10.1109/ULTSYM.2001.991675
128. Vidal-Verdú F, Navas-González R (2004) Thermopneumatic approach for tactile displays. In: Mechatronics & robotics. Aachen, Germany, pp 394–399. ISBN: 3-938153-30-X. doi:10.1117/12.607603
129. Vitrani MA et al (2006) Torque control of electrorheological fluidic resistive actuators for haptic vehicular instrument controls. J Dyn Syst Measur Control 128(2):216–226. doi:10.1115/1.2192822. http://link.aip.org/link/?JDS/128/216/1
130. Völkel T, Weber G, Baumann U (2008) Tactile graphics revised: the novel brailledis 9000 pin-matrix device with multitouch input. In: Computers Helping People with Special Needs. Springer, Berlin, pp 835–842. doi:10.1007/978-3-540-70540-6_124
131. von Zitzewitz J (2011) R3—A reconfigurable rope robot as a versatile haptic interface for a cave automatic virtual environment. Dissertation. ETH Zürich
132. Wahed AK (2004) The characteristics of a homogeneous electrorheological fluid in dynamic squeeze. In: Actuator 2004: 9th international conference on new actuators. Bremen, pp 605–608
133. Wallaschek J (1998) Contact mechanics of piezoelectric ultrasonic motors. Smart Mater Struct 3:369–381. doi:10.1088/0964-1726/7/3/011

134. Wang Q, Hayward V (2010) Biomechanically optimized distributed tactile transducer based on lateral skin deformation. Int J Robot Res 29(4):323–335. doi:10.1177/0278364909345289
135. Weaver W, Timosenko SP, Young DH (1990) Vibration problems in engineering, 5th edn. Wiley, New York, pp XIII, 610. ISBN: 0-471-63228-7
136. Weinberg B et al (2005) Development of electro-rheological fluidic resistive actuators for haptic vehicular instrument controls. Smart Mater Struct 14(6):1107–1119. doi:10.1088/0964-1726/14/6/003
137. Wen W, Huang X, Sheng P (2004) Particle size scaling of the giant electrorheological effect. Appl Phys Lett 85(2):299–301. doi:10.1063/1.1772859
138. Williamson MM (1995) Series elastic actuators. Master Thesis. Massachusetts Institute of Technology. http://dspace.mit.edu/handle/1721.1/6776
139. Winfield L et al (2007) T-PaD: tactile pattern display through variable friction reduction. In: IEEE, pp 421–426. doi:10.1109/WHC.2007.105. (Visited on 01/17/2014)
140. Würtz T et al (2010) The twisted string actuation system: modeling and control. In: IEEE/ASME international conference on IEEE/ASME international conference on advanced intelligent mechatronics (AIM). IEEE, pp 1215–1220. doi:10.1109/AIM.2010.5695720
141. Yu O, Zharri (1994) An exact mathematical model of a travelling wave ultrasonic motor, vol 1. Ultrasonics symposium. Proceedings, 1994 IEEE. Cannes, 1–4 Nov 1994, pp 545–548. vol 1. doi:10.1109/ULTSYM.1994.401647
142. Zhu M (2004) Contact analysis and mathematical modeling of traveling wave ultrasonic motors. In: IEEE transactions on ultrasonics, ferroelectrics and frequency control, vol 51, pp 668–679. ISBN: 0885-3010. 6 June 2004

Chapter 10
Sensor Design

Jacqueline Rausch, Thorsten A. Kern and Christian Hatzfeld

Abstract Multiple sensors are applied in haptic devices designs. Even if they are not closed-loop controlled in a narrow sense of force or torque generation, they are used to detect movement ranges and limits or the detection of the presence of a user and its type of interaction with an object or human–machine interface (HMI). Almost any type of technical sensor had been applied in the context of haptic devices. Especially, the emerging market of gesture-based user interaction and integration of haptics due to ergonomic reasons extends the range of sensors potentially relevant for haptic devices. This chapter gives an introduction in technologies and design principles for force/torque sensors and addresses common types of positioning, velocity, and acceleration sensors. Further, sensors for touch and imaging sensors are addressed briefly in this section.

10.1 Force Sensors

Jacqueline Rausch

This section deals with selection and design of force sensors, which are implemented in haptic systems. Approaches like measuring current in actuators to derive occurring force are not part of the chapter. In Sect. 10.1.1, fundamental problems are discussed, which are the basis of every sensor design process. A selection of factors

J. Rausch (✉)
Roche Diagnostics GmbH, Sandhofer Straße 116, 68305 Mannheim, Germany
e-mail: j.rausch@hapticdevices.eu

T.A. Kern
Continental Automotive GmbH, VDO-Straße 1, 64832 Babenhausen, Germany
e-mail: t.kern@hapticdevices.eu

C. Hatzfeld
Institute of Electromechanical Design, Technische Universität Darmstadt,
Merckstr. 25, 64283 Darmstadt, Germany
e-mail: c.hatzfeld@hapticdevices.eu

C. Hatzfeld and T.A. Kern (eds.), *Engineering Haptic Devices*,
Springer Series on Touch and Haptic Systems, DOI 10.1007/978-1-4471-6518-7_10

to be taken in account is made in Sect. 10.1.1.5. After a short introduction in basic transfer properties, sensor characteristics are analyzed according to haptic aspects and complemented by application examples.

10.1.1 Constraints

The topology of haptic systems significantly influences force sensor design. The application of the haptic device itself has an extraordinary relevance. All systems have in common that an user mechanically contacts objects. It has to be clarified, which use of the device is intended, e.g., if it is going to be a telemanipulator for medical purposes, or a CAD tool with force feedback. The mechanical properties of the user itself, and in case of telemanipulation systems the mechanical properties of manipulating objects, have to be analyzed for the sensor development. All these factors will be discussed within this section.

10.1.1.1 Topology of the Device

The application itself appoints the topology of the haptic device. Taking control engineering aspects into account haptic systems can be classified into four types, which are discussed in Chap. 6. In the following, these topologies are analyzed referring to the measured values:

- Open-loop control of impedance: Measurement of user movements (velocity or displacement)and feedback of a force
- Closed-loop control of impedance: Measurement of both user movements and interaction and feedback of a force
- Open-loop control of admittance: Optional measurement of user force and feedback of a position
- Closed-loop control of admittance: Measurement of both user force and movements and feedback of a position

In case of open-loop control only the mechanical properties of objects have to be taken into account for force sensor design, independent if objects are physical or virtual ones. In case of haptic simulators like flight simulators, virtual objects are acting. Mechanical properties are often stored in look-up tables and force sensors are dispensable. In case of telemanipulation systems, the end effector of the haptic system interacts with physical objects. Their mechanical properties have to be detected with capable force sensors.

Most telemanipulation systems are impedance controlled. In case of closed-loop control, the mechanical impedance of both user and manipulating object are considered. In designing closed-loop impedance-controlled systems force sensors have to be integrated into the device detecting the user force. In designing closed-loop admittance-controlled systems the output movements of the haptic interface have to be measured using a velocity sensor (see Sect. 6.2)

Consequently, the measuring object can be both the user itself and a real, physical object. Beside its mechanical properties, the modality of the interaction with haptic systems has to be analyzed to identify fundamental sensor requirements such as dynamic bandwidth, nominal load, and resolution. The main factors influencing the sensor design are both contact situation and objects' mechanical properties. In the following, they are analyzed by examining the mechanical properties and the texture of the objects' surface separately.

10.1.1.2 Contact Situation

It is necessary to distinguish between the user of the haptic system and the physical object due to different interaction modalities identifying mechanical properties. If the user is the "measuring object," interaction forces have to be measured. Universally valid conclusions concerning amplitude, direction, and frequency of the acting force cannot be done. Mechanical impedance depends on the manner of grasping the device, age, and gender of the user itself (Chap. 3). In Sect. 3.1.3, the manners of grasping are classified: power-grasps, precision-grasps, and touch-grasps. In case of power- and precision-grasps, finger or palm are used as counter bearing, which results in a high absolute value of force up to 100 N [12, 31] and a stiffer contact.

Additionally, the direction of the force vector has to be taken into account. Depending on application of the haptic device and the manner of grasping up to six degrees of freedom results three force components and sometimes three torques. Neglecting torques between user and device three components of force have to be measured. If the user is in static contact with the handheld device, measuring normal force components with respect to orientation of the contact plane is sufficient. If the user is exerting relative movements to the device, shear forces also occur and three components have to be measured.

Considering the frequency dependence of humans' haptic perception, both static and dynamic signal components have to be considered equally. The lower cutoff frequency of haptic devices tends to quasistatic action at almost 0 Hz, which may happen when a device is held without movement in free space. If the force signal is subject to noise or even the slightest drift, the haptic impression will be disturbed soon (compare perception thresholds in Sect. 2.1). Manner and preload of grasping affect the upper cutoff frequency of the sensor. In case of power- and precision-grasps, the absolute value of force achieves higher values which results in an upper cutoff frequency being $\ll$ 10,000 Hz. Values of about 300 Hz are sufficient (Sect. 2.1). Within contact grasps preload is much lower than before enabling high-frequency components to be transmitted directly to the skin up to a range of approximately 1,000 Hz.

In case of telemanipulation systems, the end effector interacts with a real, physical object. Assumptions made for the measuring object "user" can partially be transferred to this situation. Following NEWTON's law *actio et reactio*, the absolute value of force depends on intensity and way of interaction. Possible examples are compression and lift of objects with a gripper, or exploration with a stick. For telemanipulation systems in minimally invasive surgery, the absolute value of force ranges from 1 to 60 N (comp. e.g., [76]). The most promising approach is given by analyzing the intended application within preliminary tests and derivation of a model. The dynamics of the interaction, especially of the upper cutoff frequency, is dominated by the mechanical impedance of the object itself, which will be described within the following section.

10.1.1.3 Mechanical Properties of Measuring Objects

As stated for the user in Chap. 3, the mechanical impedance of objects can be subdivided into three physical actions: elastic compliance n, damping d, and mass m. In case of rigid objects made of, e.g., metal or ceramics, the property of elasticity is dominant. Interaction between haptic systems and objects can be considered as a rigid contact. Consequently, the force signal includes high-frequency components. The upper cutoff frequency should take a value of minimum 1,000 Hz to make sure to cover all dynamics responsible for haptic perception. Soft objects, such as silicone or viscera have a viscoelastic material performance. Following KELVIN viscoelastic behavior can be simulated by a network made of elastic compliances n_i and damping elements d_i, such as masses m_i. Using such an equivalent network, dynamic effects like relaxation and creeping can be modeled (Figs. 10.1 and 10.2).

First of all the elasticity of measuring objects has to be investigated for designing a haptic sensor. An arithmetic example in Sect. 2.4.2 compares the different cutoff frequencies of materials. For soft materials such as rubber, upper cutoff frequency takes values below 10 Hz. During interaction with soft materials mainly low-frequency components appear. The upper cutoff frequency is defined by the interaction frequency of 10 Hz at maximum [36, 73, 86]. If the measuring object is a soft one with embedded rigid objects, like for example tumors in soft body tissue, an upper cutoff frequency of about 1,000 Hz should be realized. To get more precise information about frequency requirements, it can hardly be done without an analysis of the interaction object. For a first rule of thumb calculated cutoff frequency as derived in Sect. 2.4.2 are sufficient. In case of doubt, the frequency range of the sensor should

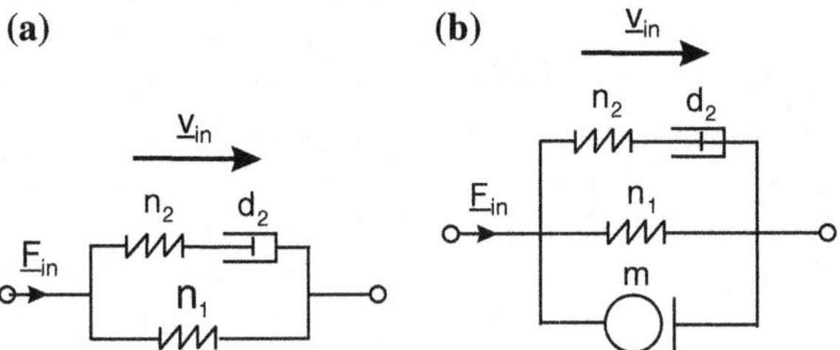

Fig. 10.1 Kelvin model (standard linear solid) modeling viscoelastic behavior of objects. For calculating the resonance frequency, a mass element has to be added. By adding further damping and spring elements, dynamic behavior of every object material can be modeled. **a** Kelvin model modeling dynamic effects. **b** Kelvin model extended by objects weight for calculating the resonance frequency

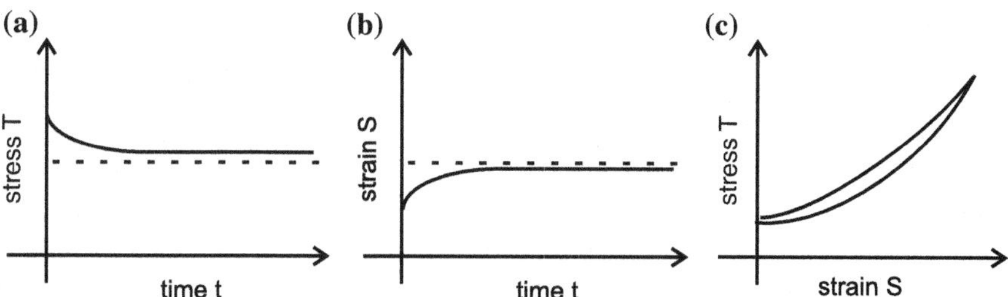

Fig. 10.2 Visualization of viscoelastic phenomena **a** relaxation, **b** creep, and **c** hysteris

always be oversized, not to already loose relevant haptic information at this very first point in the processing chain.

Beside dynamics, the required force resolution depends on a physiological value, too. The ↪JND lies in the range between 5 and 10 % of the absolute force value (Sect. 2.1). From the JND, the sensor characteristics of measurement uncertainty can be derived. If realized as a constant value—which is common to many technical sensor solutions—5 % of the lowest appearing value of force should be chosen to prevent distortion of the haptic impression of the object. Nevertheless, there is no actual requirement for haptic applications to have a constant or even linear sensor resolution. With telemanipulation systems, the interaction of the haptic system and real, physical objects is the main application. Depending on the type of interaction, frequently the surface structure of objects, the so-called texture becomes equally or even more important than the object's elastic compliance. Helpful literature for modeling dynamics of mechanical or electromechanical systems are [24, 47]. The resulting challenges for sensor development are discussed within the following section.

10.1.1.4 Texture of Measuring Objects

Properties, which are relevant for the human perception of texture, are geometrical surface structure on the one hand (e.g., the wood grain), on the other hand some kind of "frequency image" generated by the geometrical structure in the (vibro-)tactile receptors when being touched by skin. To detect the surface structure of an object, variation of force against the contact area can be derived. For **static measurement** sensor arrays of single-component force or pressure sensors are a common technical solution. These arrays are placed onto the object. The objects structure generates different values of contact forces, providing a force distribution on the sensor surface. Size of array and individual array elements cannot be defined in general, but depends on the smallest detectable structure on the measurement object itself. In case of static measurement sketched above, number and size of the sensor array elements should be dimensioned slightly smaller than the minimum structure of the measuring object. The size of each element may not be larger than half of the size of the smallest structure to be measured. However, even fulfilling this requirement aberration will appear. Figure 10.3 shows that in case of the width of the sensor element being larger or identical to the smallest structure, the distance between the elements is detected smaller than in reality. With n sensor elements, the width of the structure element is replayed to $\frac{n+1}{n}$ and the distance to $\frac{n-1}{n}$. If the number of sensor elements per surface area increase, the aberration is diminishing and the structure is approximated more realistic (Fig. 10.4). However, with the number of elements, the effort of signal conditioning and analysis is increasing.

Beside the described aberration, an additional disadvantage of static measurements is given by the fact that the knowledge of the texture is not sufficient to get information about the object's material. The complete haptic impression needs frequency information depending on the elastic properties of texture and surface friction too. To gain these data, a relative movement between object and haptic system should be performed, to measure the texture **dynamically and spatially**. Depending on velocity of the relative movement and speed of the signal detection algorithms, the spatial resolution can be multiplied using the same number of sensor elements as in the example shown before. Even the use of sensor array with a simultaneous detection of multiple points becomes unnecessary. With knowledge about the exploration velocity and its direction, the information can be put into relation to each other. For texture analysis, multicomponent force sensors should be used, as especially the combined forces in the direction of movement and normal to the surface contribute to haptic perception [60]. This dynamic measurement principle is comparable with the intuitive exploration made by humans: To gain the texture of an object, humans gently touch and stroke over its surface. The surface structure excites the fingerprint to oscillate and the vibrotactile sensors acquire the frequency image. The absolute value of normal forces reached during such explorations are in a range of 0.3–4.5 N [13]. As stated earlier, force resolution is defined by the ↪JND. Haptic information about texture is included into the high-frequency components of the signal. For haptic

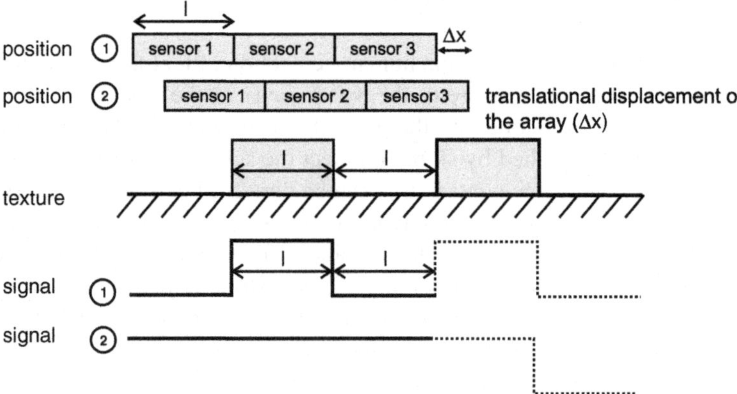

Fig. 10.3 Illustration of static and spatially resolved force measurement using as $3 \times n$ array. One sensing element has the same dimension like a texture element. At position 1 the array is optimally placed. If the array is shifted about Δx to position 2, the texture is incorrectly detected

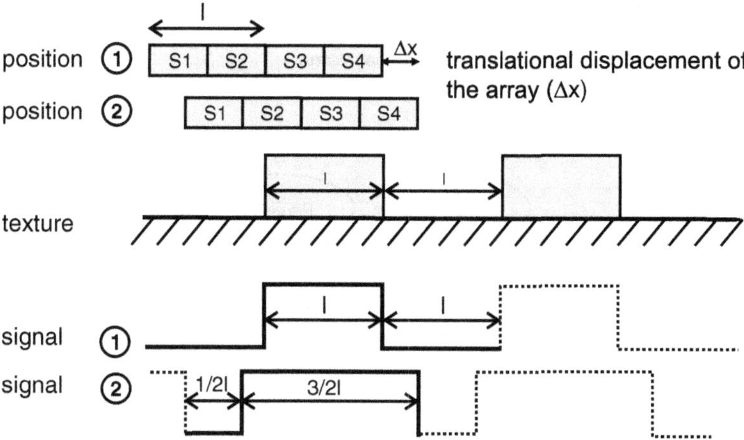

Fig. 10.4 Illustration of static and spatially resolved force measurement using as $6 \times n$ array. Size of one sensing element is half of a texture element. At position 1 the array is optimally placed. In case of any other position an aberration occurs. Aberration decreases with increasing number of sensing elements in an appropriate array

applications, the maximum frequency should be located at 1,000 Hz. The absolute value of nominal force should be chosen depending on the elastic compliance of the object. If the object is softer, the absolute value can be chosen lower as surface structures will deform and cannot be detected anymore. To be able to measure equally good at soft and rigid objects, the nominal force should take values ≤ 4.5 N. CALDWELL [13] for example decided to use $F = 0.3$ N.

10.1.1.5 Selection of Design Criteria

Following the description of the most relevant constraints, limiting factors for sensor design in haptic applications can be found in physiological values. Nominal force, force resolution, covered frequency range, and measurement uncertainty can be derived from humans' haptic perception. For a quantitative analysis of these requirements, the contact between measuring object and force sensor is to bring into focus. Measurement range and number of detectable force components are defined by the application and the structure of the device.

The geometrical dimensions and other mechanical requirements are given depending on the point of integration into the haptic system. The diagram displayed in Fig. 10.5 visualizes the procedure of how to identify most important requirements for sensor design.

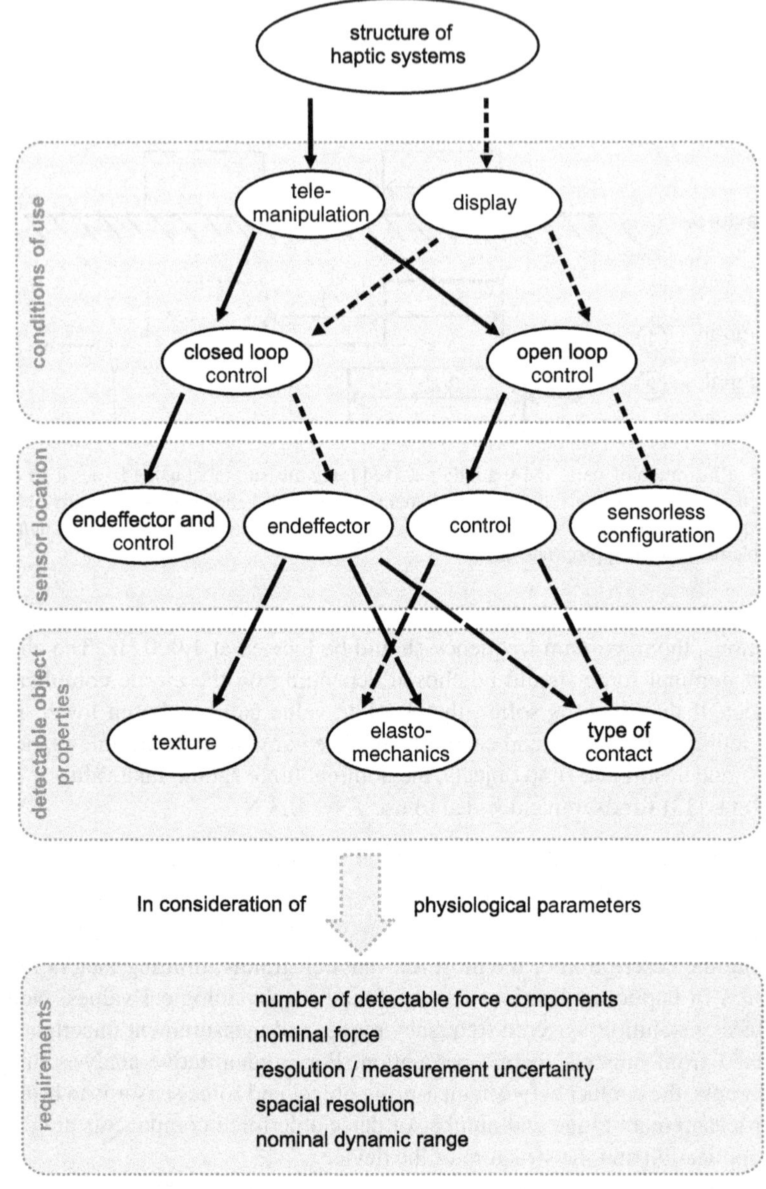

Fig. 10.5 Tree diagram to identify the principle requirements on haptic force sensors. Beside mechanical characteristics of the object physiological parameters of human haptic perception also have to be considered

10.1.2 Sensing Principles

Within the previous section, the most important criteria for the design and development of a haptic sensor were named and introduced. Section 10.1.3 summarizes major requirements once again in tabular form. In order to help choosing a suitable sensor principle, variants according to Fig. 10.6 are presented in this section. Beside established measurement elements, such as resistive, capacitive, optic, or piezoelectric ones, other less common sensor designs based on electroluminescence or active moving coils are discussed, too.

Most sensor principles are active transformers using the displacement principle for force measurement, which means that elastomechanic values such as stress or strain are detected and the corresponding force is calculated. Sensors belonging to the group of active transducers are resistive, capacitive, optic, and magnetic ones, working according to the displacement principle, too. Piezoelectric, electrodynamic, or electrostatic sensors are part of the group passive transducers. After a short introduction in elastomechanics, each sensing principle will be discussed according to its operating mode and several applications will be presented. All sensor principles will be estimated concerning their applicability for kinaesthetic and tactile force measurement, and put into relation to requirements known from Chap. 5. At the end of this chapter, a ranking method for the selection of suitable sensor principles is being given.

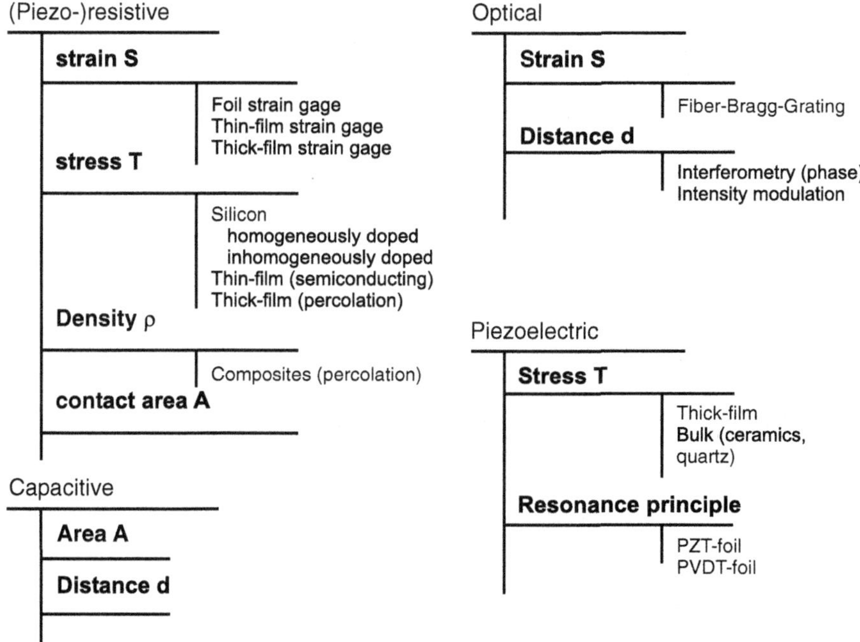

Fig. 10.6 Overview of established measurement principles for detecting forces in haptic systems. Furthermore, active sensor systems are also discussed in the following section

10.1.2.1 Basics of Elastomechanics

As mentioned before, large number of sensor principles are based upon elastomechanics. This section will summarize fundamental knowledge which is necessary for sensor design. If force is exerted to an elastic body, it deforms elastically depending on the amount of force. Internal stress T occurs resulting in a shape change—the strain S. Stress and strain are correlated with specific material parameters, the so-called elastic moduli s_{ij}.

For a better comprehension, a short *gedanken experiment* will be performed [28]. If a volume element ΔV is cut from an object under load (Fig. 10.7), substitute forces ΔF will act upon the surfaces of the cuboid to keep the state of deformation. Due to the required state of equilibrium, the sum of all forces and torques acting upon ΔV must equal zero.

Subdividing the force ΔF in its three components ΔF_1, ΔF_2, and ΔF_3, just those components remain orthogonal to the surface elements ΔA_j. The quotient of the acting force component ΔF_i and the corresponding surface element ΔA_j results in a mechanical stress T_{ij}. Following the equilibrium condition $T_{ij} = T_{ji}$, six independent tension components remain, resulting in the stress tensor. Tensor elements can be factorized into normal (stress parallel to surface normal) and shear stress components (stress orthogonal to surface normal). Analyzing the volume element ΔV before and after load, a displacement of the element ΔV with relation to the coordinate system $\langle 123 \rangle$ such as a deformation happens. The sides of the cube change their lengths and are not orthogonal to each other anymore (Fig. 10.8).

To describe that shape change, strain S_{ij} is introduced. The quantity strain is a tensor too, consisting of nine elements (Eq. 10.1)

$$\begin{pmatrix} d\xi_1 \\ d\xi_2 \\ d\xi_3 \end{pmatrix} = \begin{pmatrix} S_{11} & S_{12} & S_{13} \\ S_{21} & S_{22} & S_{23} \\ S_{31} & S_{32} & S_{33} \end{pmatrix} \cdot \begin{pmatrix} \Delta x_1 \\ \Delta x_2 \\ \Delta x_3 \end{pmatrix} \tag{10.1}$$

Due to volume constancy, the following correlation can be defined as

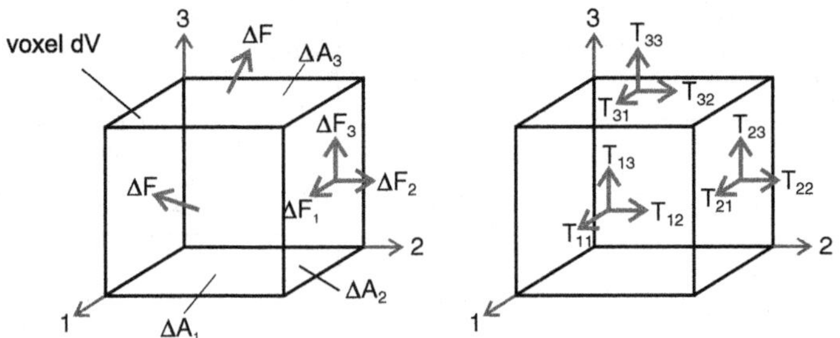

Fig. 10.7 Voxel dV of an elastic object. Due to external deformation, internal stress occurs which can be described by the component T_{ij} of the stress tensor [28]

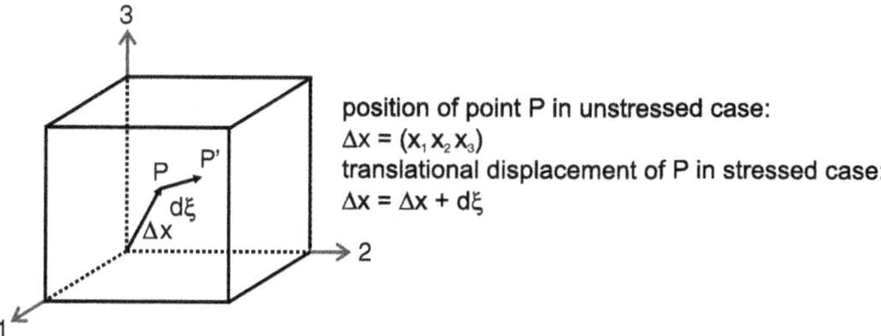

Fig. 10.8 Displacement of point P to P' due to application of force visualizes the state of strain [28]

$$S_{ij} = S_{ji} = \frac{1}{2} \cdot \left(\frac{\delta \xi_i}{\delta x_j} + \frac{\delta \xi_j}{\delta x_i} \right) \tag{10.2}$$

and thus, the matrix can be reduced to six linear independent elements. Normal strain components acting parallel to the corresponding normal surface result in volume change. Shear components, acting normal to the surface, describe the change of the angle between the borders of the volume element. In case of isotropic materials, such as metals or Al_2O_3 ceramics, the correlation between shape change mentioned before and after mechanical strains can be formulated as follows:

$$\begin{pmatrix} S_1 \\ S_2 \\ S_3 \\ S_4 \\ S_5 \\ S_6 \end{pmatrix} = \begin{pmatrix} s_{11} & s_{12} & s_{12} & 0 & 0 & 0 \\ s_{12} & s_{11} & s_{12} & 0 & 0 & 0 \\ s_{12} & s_{12} & s_{11} & 0 & 0 & 0 \\ 0 & 0 & 0 & 2(s_{11} - s_{12}) & 0 & 0 \\ 0 & 0 & 0 & 0 & 2(s_{11} - s_{12}) & 0 \\ 0 & 0 & 0 & 0 & 0 & 2(s_{11} - s_{12}) \end{pmatrix} \cdot \begin{pmatrix} T_1 \\ T_2 \\ T_3 \\ T_4 \\ T_5 \\ T_6 \end{pmatrix} \tag{10.3}$$

For simplification, six independent strains, respectively, stress components are summarized in a vector. Components with index 1, 2, and 3 mark normal components, those with indices 4, 5, and 6 mark shear components [28]. Parameters s_{ij} are regardless of direction. Taking YOUNGs modulus E and shear modulus G into account, parameters can be derived:

$$s_{11} = \frac{1}{E}, s_{12} = \frac{\nu}{E}, \frac{1}{G} = 2(s_{11} - s_{12}) = \frac{2}{E}(1 + 2\nu) \tag{10.4}$$

ν marks the so-called POISSON ratio, which is material dependent. Using metal ν values between 0.25 and 0.35 can be achieved. In case of homogeneous materials, Eq. (10.3) can be reduced to a linear correlation $T = E \cdot S$. For anisotropic materials such as silicon or quartz, elastomechanic properties depending on the orientation of the coordinate system ((comp. the following Sect. "Piezoresistive Silicon Sensors"), result in a matrix of elastic coefficients with up to 21 elements. For further reading on elastomechanics, e.g., [28, 105] are recommended.

Example "Beam Bending"

If a force vector is exerted to the tip of a beam bender made of isotropic materials and clamped on one side (Fig. 10.9), a bending moment M_B occurs.

Mechanical stress components $T(y)$ are linear distributed on the cross section and take values of $T(y) = c \cdot y$, whereas c is a proportional factor. Bending moment equals the integral of the stress $T_3(y)$ distributed on the cross section.

$$M_B = \int_A y \cdot T_3(y)\mathrm{d}A = c \cdot \int_A y^2\mathrm{d}A \tag{10.5}$$

As the integral $\int_A y^2\mathrm{d}A$ equals the axial moment of inertia I, c is calculated as

$$c = \frac{M_B}{I}. \tag{10.6}$$

The resulting strain components S_1 and S_2 act transversal to the beam's surface. For elastic deformation, strain component S_1 and stress component T_2 are correlated via the YOUNGs modulus E

$$S_2 = \frac{T_2}{E} = \frac{M_B}{I \cdot E} = \frac{F \cdot (l - z)}{I \cdot E} \tag{10.7}$$

and therefore depending on the geometry of the cross section A of the beam, position z at the beam's surface and acting force F. For calculations of strain component S_1 transversal contraction has to be considered as follows:

$$S_1 = -\nu \cdot S_2. \tag{10.8}$$

Further readings of elastomechanics, for example the calculations of deformation of fiber-reinforced composites, the works of GROSS [28], WERTHSCHÜTZKY [105], and BALLAS [4] are recommended.

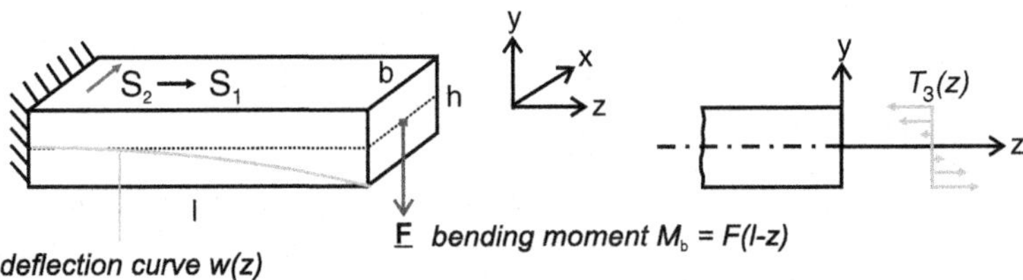

Fig. 10.9 Behavior of a bending beam, the *right-hand detail* shows stress distribution along the profile

10.1.2.2 Detection of Force

According to Fig. 10.9, acting forces can be measured evaluating both resulting strain distribution on the surface and displacement of beam. According to the example above, detection of strain S_2 can be derived using BERNOULLIs theory. Thus, strain components acting transversal to the surface can be neglected for slender and long-beam geometries. Stress- or strain-sensitive elements should be placed in such a way, that a maximum surface strain change can be detected.

Correlations described above are examples for a cantilever beam. Being able to measure more than just one force component, a suitable deformation element has to be designed considering the elastomechanic correlations. For example, works of BRAY [11] and RAUSCH [72] can help designing such an element. Primary objective is to generate a strain distribution in loading case, which enables to deduce the force components.

The correlation of force F_i and electric signal v_i of the sensor element usually is given by a linear system of equations (e.g., [104]). Equation (10.9) shows an example for a three-axial sensor:

$$\begin{pmatrix} v_1 \\ v_2 \\ v_3 \end{pmatrix} = \begin{pmatrix} a_{11} & a_{12} & a_{13} \\ a_{21} & a_{22} & a_{23} \\ a_{31} & a_{32} & a_{33} \end{pmatrix} \cdot \begin{pmatrix} F_1 \\ F_2 \\ F_3 \end{pmatrix} \tag{10.9}$$

It can be assumed that all force components contribute to each individual voltage signal v_i. The elements a_i of the matrix can be found by calibrating the sensor. During the calibration process, only one independent force component for each direction is applied to the sensor and the resulting voltage components are measured. After inverting the matrix $\mathbf{A}$ to $\mathbf{A}^{-1}$, the force vector can be calculated easily.

A lot of research is done in reducing the number of measuring cycles for calibration of multiaxial force sensors. Most common methods are:

- Least squares method [7]: Most accurate method. Execution of load cycle with n load steps for each direction. For a tri-axial force sensor $6n$ measuring cycles are necessary.
- Shape from motion method [39, 104]: Application of a force vector with known absolute value, which is randomly rotating in space. Accuracy comparable to the first method, but less time consuming. Only valid, if all components have the same amount of nominal load. If not so, then the third method is advisable.
- Hyperplane calibration method [62]: Accuracy and time consumption comparable to the second method. Three quasi-orthogonal load vectors must be applied to the sensor.

For further information on calibration check the above-mentioned literature.

10.1.2.3 Resistive Strain Measurement

One of the most commonly used sensing principles for force sensing is based on resistive detection of strain or rather stress components occurring in a (measuring) object. With resistive strain measurement, a resistor pattern is applied on the bending elements surface. Resistors must be located in areas of maximum strain. As a quick reminder: Electrical resistance is defined via

$$R_0 = \rho \cdot \frac{l}{A} = \rho \cdot \frac{l}{b \cdot h}, \tag{10.10}$$

ρ marks specific resistance, l, b, h (length, width, height) define volume of the resistor itself. The total differential shown in Eq. (10.11) gives relative resistivity change resulting from the deformation:

$$\frac{\mathrm{d}R}{R_0} = \underbrace{\frac{\mathrm{d}l}{l} - \frac{\mathrm{d}b}{b} - \frac{\mathrm{d}h}{h}}_{\text{rel. volume changing}} + \underbrace{\frac{\mathrm{d}\rho}{\rho}}_{\text{piezoresistive part}}. \tag{10.11}$$

Deformation causes on the one hand change the geometrical part $\frac{l}{A}$. Taking YOUNGs modulus E and POISSONs ratio ν account, plain stress for isotropic material can be derived [28]:

$$\frac{\mathrm{d}l}{l} = S_1 = \frac{1}{E} \cdot T_1 - \frac{\nu}{E} \cdot T_2, \tag{10.12}$$

$$\frac{\mathrm{d}b}{b} = S_2 = -\frac{\nu}{E} \cdot T_1 + \frac{1}{E} \cdot T_2, \tag{10.13}$$

$$\frac{\mathrm{d}h}{h} = S_3 = -\frac{\nu}{E} \cdot T_1 - \frac{\nu}{E} \cdot T_2. \tag{10.14}$$

Indices 1, 2, and 3 mark the direction components. Concerning the geometrical change, the resulting gage factor k describing the sensitivity of the material takes a value of about two (see Eq. 10.15). On the other hand, plane stress provokes a chance of specific resistivity ρ.

Material-specific changes will be discussed within Sect. "Piezoresistive Silicon Sensors." Using Eq. (10.15), the correlation between strain and relative resistivity change is formulated:

$$\frac{\mathrm{d}R}{R_0} = \underbrace{\left(2 - \frac{\mathrm{d}\,(N \cdot \mu)}{S \cdot N \cdot \mu}\right)}_{:=k,\ \text{gage factor}} \cdot S \tag{10.15}$$

whereas μ represents the electron mobility and N the number density of molecules. The change of the resistivity can be measured using a so-called WHEATSTONE bridge circuit. This circuit is built of one up to four active resistors connected in a bridge circuit and fed by a constant voltage or constant current (Fig. 10.10). Equation 10.16

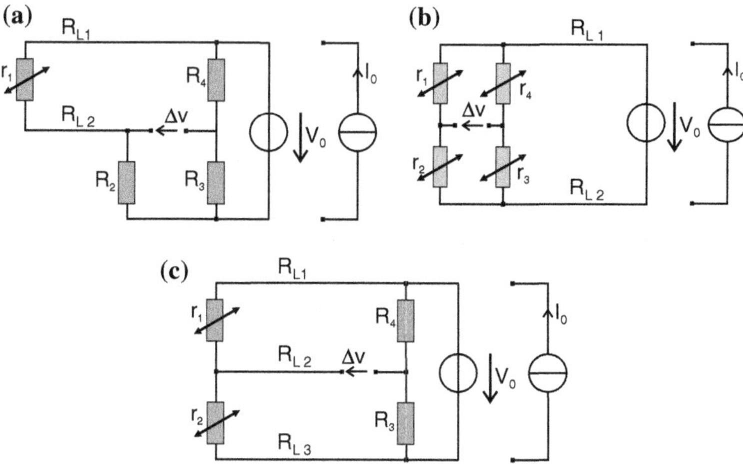

Fig. 10.10 WHEATSTONE bridge configurations for evaluating one up to four resistors. **a** Quarter bridge (1 active resistor). **b** Full bridge (4 active resistors). **c** Half bridge (2 active resistors)

calculates bridge Fig. 10.10c with the assumption that the basic resistances R_{0i} equal the resistance R_0. The values of R_0 such as gage factors are specific to material and listed in Table 10.1 (further informations e.g., [72, 81]).

$$\Delta v = \frac{V_{cc}}{R_0 \cdot I_0} = \frac{1}{4} \cdot \left\{ \frac{r_1}{R_{01}} - \frac{r_2}{R_{02}} + \frac{r_3}{R_{03}} - \frac{r_4}{R_{04}} \right\} \tag{10.16}$$

The supply with constant current I_0 has the great advantage that a temperature-dependent drift of the measurement signal will be compensated. More advanced information can be found in [20, 106].

In case of metallic resistors, a gage factor of approximately two occurs. The material-specific component of metals is less important and affects the first decimal place only. In case of semiconductors and ceramic materials, the material-specific component is dominant. In case of semiconductor-strain gages, the gage factor takes values up to 150. Using resistor pastes, applied in thick-film technology on substrates,[1] and polysilicon layers, sputtered in thin-film technology the material-specific component is dominant. On this, gage factors achieve values of up to 18 in case of thick-film resistors and up to approximately 30 for thin-film resistors. Table 10.1 lists the gage factor for several materials usually used in strain measurement. As mentioned earlier, strain gages are manufactured in different technologies. The most commonly used types are foil-strain gages; thick- and thin-film manufactured measurement elements are found mainly in OEM-sensors and for specific solutions in automation industry due to the necessary periphery and the manufactur-

[1] For substrate material mainly (layer-) ceramics are used. Less frequent is the use of metals, as isolating layers have to be provided then.

Table 10.1 Gage factor, strain resolutio,n and nominal strain of important resistive materials according to [72]

Technology	Material	Gage factor	R_0 in Ω	S_{min}	S_N (%)	References
Foil-strain gage	CuNi	About 2	120, 350, 700	$\pm 10^{-7}$	±0.1	[32, 35, 91]
Thick-film	$Bi_2Ru_2O_7$	12.1–18.3	1,000	$\pm 10^{-6}$	±0.1	[3, 65]
	PEDOT:PSS	0.48–17.8	–	≥ 10	–	[44, 46]
Thin-film	TiON	About 1k	4–5	$\pm 10^{-7}$	±0.1	[3, 42, 105]
	Poly-Si	20–30	About 1k	$\pm 10^{-7}$	±0.1	[3, 105]
Si-technology	Homogeneous	100–255	120–1k	$\pm 10^{-6}$	±0.2	[58]
	Inhomogeneous	80–255	1k–5 k	$\pm 10^{-7}$	±0.05	[5, 23, 105]
Fiber-sensors	Carbon	1.3–31	About 10 k	–	0.2–15	[16, 43]

ing process. Relevant literature can be found in the publications of PARTSCH [65] and CRANNY [18].

To deposit thin-film sensing layers, other technologies like inkjet or aerosoljet printing can be used. The inks are suspensions containing electrically conducting particles made of carbon, copper, gold, silver, or even conducting polymers like PEDOT:PSS. One advantage is that compared to conventional thick-film pastes, the finishing temperature is below 300 °C and thus various substrates can be functionalized. Further information can be found in [44, 49, 72].

Foil-strain gages are multilayer systems made of metallic measurement grids and organic substrates. It is applied (Fig. 10.11) and fixated on bending elements via cold hardening cyanoacrylate adhesive (strain analysis) or via hot hardening adhesives such as epoxy resin (transducer manufacture). These gages are long-term stable, robust, and especially used for high-precision tasks in wind-tunnel-scales and balance sensors. Achievable dynamics, resolution, and measurement range are solely depending on the deformation element. The minimum size of the individual strain gages taken of the shelf is in the area of 3 mm width and 6 mm length. The measurement pattern itself is smaller in its dimensions. On this, it is possible to shorten the organic substrate to finally achieve 1.5 mm width and 5 mm length as a typical min-

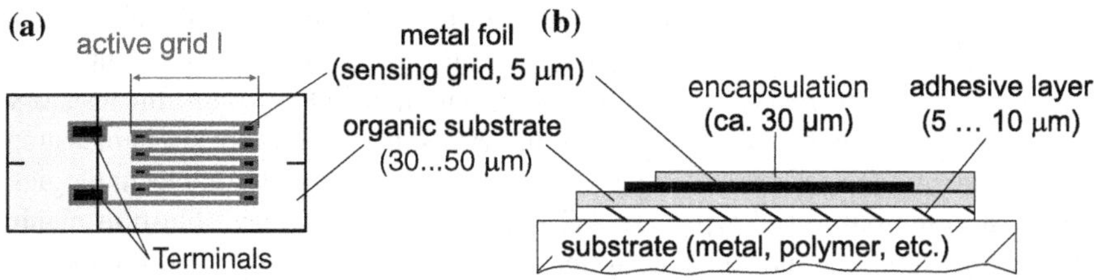

Fig. 10.11 Assembly of conventional strain gages: measuring grid is usually made of a patterned metal foil. In case of special applications, metal wires are applied. **a** Top view of strain gage. **b** Cross section of integrated gage

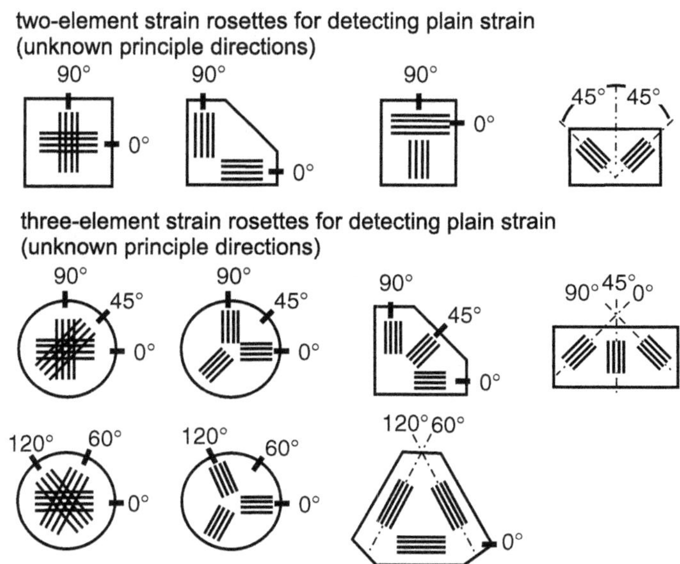

Fig. 10.12 Compilation of possible grid configurations of strain gages. See also [105]

imum size. If foil-strain gages are considered, the surface strains resulting from the nominal load should be 1,000 μm/m for an optimum usage of the strain gage. Many measurement pattern are applied for force and torque sensors. Figure 10.12 shows a selection of commercialized measuring grids ready for application on deformation elements.

Beside resistive foil-strain gages, semiconductor-strain gages are available. Their general design is comparable to conventional strain gages, as the semiconducting elements are assembled with organic substrates.[2] Measurement elements are used identical to foil-strain gages and are available in different geometrical configurations such as T-rosettes.

Using measuring elements with a higher gage factor (Table 10.1), deformation elements can be designed stiffer, allowing smaller nominal strains. Such elements are especially relevant for the design of miniaturized sensors for haptic systems, as small dimensions and high cutoff frequencies have to be achieved. A commercially available example is the OEM-sensor *nano 17* from ATI (Fig. 10.13). Strain elements are piezoresistive ones and their gage factor takes values of approximately 150. Due to high potential for miniaturization and manifold application in haptic systems, piezoresistive sensors—especially silicon sensors—will be discussed in an independent subsection.

---

[2] Also single semiconducting elements without organic substrate are available. They are highly miniaturized (width of about 230 μm, length of about 400 μm), but has to be insulated from the deformation element.

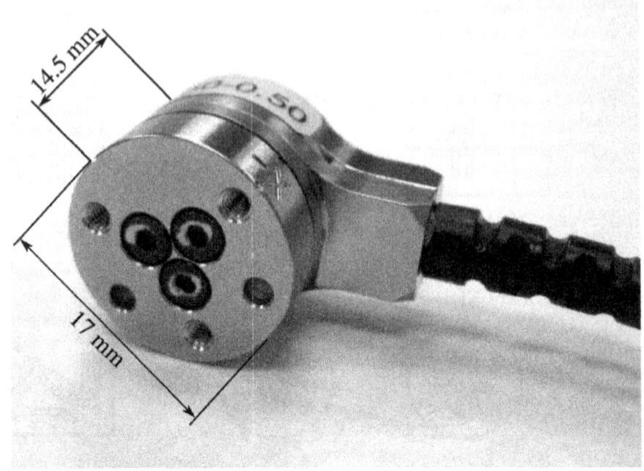

Fig. 10.13 Miniaturized force/torque sensor nano17 (*ATI Industrial Automation, Inc.*, Apex, NC, USA). Resonance frequency of the sensor takes a value of about 7.2 kHz

Piezoresistive Silicon Sensors

Published by SMITH in 1954 for the first time [89], semiconducting materials with a symmetric crystal structure such as silicon or germanium possess a change in their conductivity σ due to an applied force or pressure. In the following paragraphs, this effect is discussed more deeply for monocrystalline silicon.

The Piezoresistive Effect

If a semiconducting material is deformed due to a load, stress components T_i are generated inside the material. For your information: Due to the anisotropic properties of the material, the elastomechanic properties are depending on the position of the coordinate system, and consequently on the orientation of the crystal lattice. These stress components affect the electron mobility μ and—as a consequence—the specific resistivity ρ. ρ is a material-specific value, characterized via the parameters electron mobility μ and number of charge carriers N (comp. Sect. 10.1.2.1). Considering these parameters, correlation between relative resistivity change and the resulting strain tensor can be expressed to:

$$\frac{\mathrm{d}\rho}{\rho} = \frac{\mathrm{d}V}{V} - \frac{\mathrm{d}(N \cdot \mu)}{N \cdot \mu}, \quad \text{with } \rho = \frac{V}{N \cdot \mu \cdot |q|}, \tag{10.17}$$

whereas V is the volume of the resistive area, and $|q|$ is the charge of the particles.

Following the OHM's law the specific resistance ρ is connected by the vector $\mathbf{E} = (E_1; E_2; E_3)^T$ of the electrical field and the current density $\mathbf{J} = (J_1; J_2; J_3)^T$:

$$\begin{pmatrix} E_1 \\ E_2 \\ E_3 \end{pmatrix} = \begin{pmatrix} \rho_{11} & \rho_{12} & \rho_{13} \\ \rho_{21} & \rho_{22} & \rho_{23} \\ \rho_{31} & \rho_{32} & \rho_{33} \end{pmatrix} \cdot \begin{pmatrix} J_1 \\ J_2 \\ J_3 \end{pmatrix} = \begin{pmatrix} \rho_1 & \rho_6 & \rho_5 \\ \rho_6 & \rho_2 & \rho_4 \\ \rho_5 & \rho_4 & \rho_3 \end{pmatrix} \cdot \begin{pmatrix} J_1 \\ J_2 \\ J_3 \end{pmatrix} \tag{10.18}$$

Due to the symmetric crystalline structure of silicon,[3] six independent resistive components ρ_i result, which are symmetrical to the diagonal of tensor ρ. Taking the matrix of piezoresistive coefficients π into account, the influence of the six acting stress components T_i can be formulated. The cubic symmetry results in reduction of the number of piezoresistive and direction-dependent coefficients to three. By doping silicon with impurity atoms such as boron or phosphor areas of higher resistivity are generated. By influencing the type and the concentration of dopant, the three π-coefficients can be influenced. Further information on doping can be found, e.g., in [5, 6].

$$\begin{pmatrix} \rho_1 \\ \rho_2 \\ \rho_3 \\ \rho_4 \\ \rho_5 \\ \rho_6 \end{pmatrix} = \begin{pmatrix} \rho_0 \\ \rho_0 \\ \rho_0 \\ 0 \\ 0 \\ 0 \end{pmatrix} + \begin{pmatrix} \pi_{11} & \pi_{12} & \pi_{12} & 0 & 0 & 0 \\ \pi_{12} & \pi_{11} & \pi_{12} & 0 & 0 & 0 \\ \pi_{12} & \pi_{12} & \pi_{11} & 0 & 0 & 0 \\ 0 & 0 & 0 & \pi_{44} & 0 & 0 \\ 0 & 0 & 0 & 0 & \pi_{44} & 0 \\ 0 & 0 & 0 & 0 & 0 & \pi_{44} \end{pmatrix} \cdot \begin{pmatrix} T_1 \\ T_2 \\ T_3 \\ T_4 \\ T_5 \\ T_6 \end{pmatrix} \cdot \rho_0 \tag{10.19}$$

For homogenous silicon with a small concentration of dopants, the values in Table 10.2 can be used.

Depending on angle between current density vector $\mathbf{J}$ and stress component T_i, three effects can be distinguished. Within the so-called longitudinal effect current i is guided parallel to the normal component of stress, within transversal effect i is guided normal to the normal component of stress, and thus shear effect i is guided parallel or normal to the shear component of stress. Figure 10.14 visualizes the mentioned correlations.

For the resistivity change, depending on the orientation of the resistive area from Fig. 10.14, the following equation becomes valid:

$$\frac{\mathrm{d}R}{R} \approx \frac{\mathrm{d}\rho}{\rho} = \pi_\mathrm{L} \cdot T_\mathrm{L} + \pi_\mathrm{Q} \cdot T_\mathrm{Q} \tag{10.20}$$

Table 10.2 Piezoresistive coefficients of homogeneously doped silicon [5]

Doping	N in $\frac{1}{\mathrm{cm}^{-3}}$	ρ in Ω cm	π_{11} in $\frac{\mathrm{mm}^2}{N}$	π_{12} in $\frac{\mathrm{mm}^2}{N}$	π_{44} in $\frac{\mathrm{mm}^2}{N}$
n-Si	6×10^{14}	11.7	-102.2×10^{-5}	$+53.4 \times 10^{-5}$	-13.6×10^{-5}
p-Si	1.8×10^{14}	7.8	$+6.6 \times 10^{-5}$	-1.1×10^{-5}	$+138.1 \times 10^{-5}$

[3] Face centered cubic.

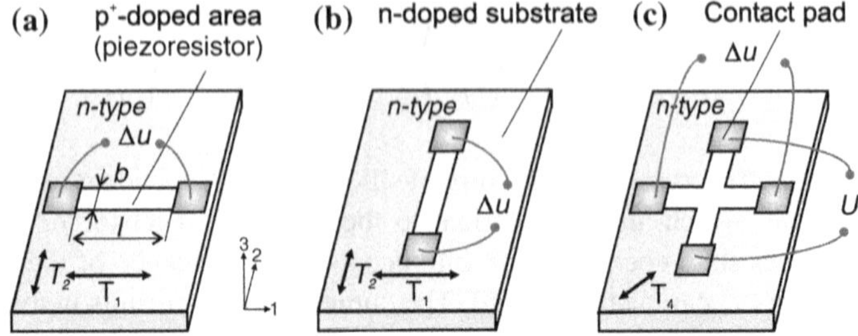

Fig. 10.14 Visualization of the piezoresistive effects: longitudinal, transversal, and shear effect in silicon [72]. Transversal and longitudinal effect is normally used for commercial silicon sensors **a** longitudinal effect ($T_1 \| J \rightarrow T_1 = T_L$), **b** transversal effect ($T_1 \perp J \rightarrow T_1 = T_Q$), **c** shear effect (piezores. Hall effect)

Table 10.3 Compilation of π_l- und π_q-coefficients for selected resistor assemblies dependent on the crystallographic orientation [90]

Surface orientation	Longitudinal	π_l	Transversal	π_q
(100)	[100]	π_{11}	[010]	π_{12}
	[110]	$\frac{\pi_{11}+\pi_{12}+\pi_{44}}{2}$	$[\bar{1}10]$	$\frac{\pi_{11}+\pi_{12}-\pi_{44}}{2}$
(110)	[111]	$\frac{\pi_{11}+2\pi_{12}+2\pi_{44}}{3}$	$[\bar{1}12]$	$\frac{\pi_{11}+2\pi_{12}-\pi_{44}}{3}$
	[110]	$\frac{\pi_{11}+\pi_{12}+\pi_{44}}{2}$	[001]	$\frac{\pi_{11}+5\pi_{12}-\pi_{44}}{6}$

As a consequence, longitudinal and transversal stress components are influencing the calculation of the resistivity change. Depending on the crystallographic orientation of the resistive areas the π-coefficient is formed by longitudinal and transversal coefficient (Table 10.3). For homogeneous Boron concentration of $N_R \approx 3 \times 10^{18}\,\mathrm{cm}^{-3}$ the following values are achieved [5]:

$$\pi_L = 71.8 \times 10^{-5}\,\mathrm{MPa}^{-1},$$

$$\pi_Q = -65.1 \times 10^{-5}\,\mathrm{MPa}^{-1}.$$

More advanced information for the design of piezoresistive silicon sensors can be found in the publications of Bao [5], Barlian [6], Meiss [53], Rausch [72] and Werthschützky [68].

Examples of Piezoresistive Silicon Sensors

Piezoresistive silicon sensors for physical quantities like pressure and force are commonly integrated in silicon deformation elements. In case of pressure transducers, this kind of manufacture is state-of-the-art. For all pressure ranges sensor elements

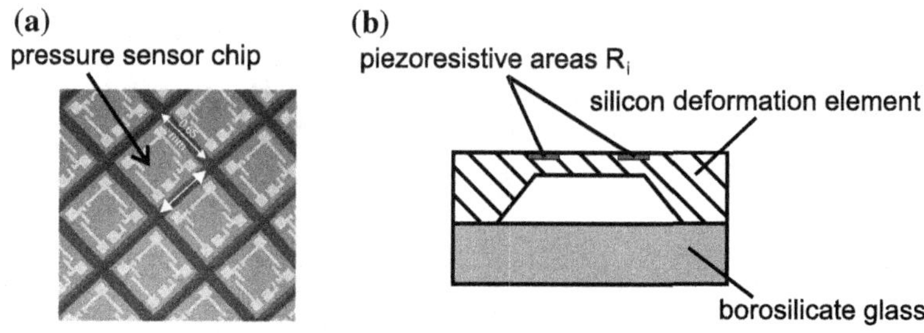

Fig. 10.15 Example of piezoresistive silicon-pressure sensors [88], **a** unseperated chips, edge length of about 650 μm, **b** sectional drawing of the sensor

can be purchased. For example, the company *Silicon Microstructures Inc. (SMI)* sells chips with glass counterbody for absolute pressure measurement with an edge length of 650 μm (Fig. 10.15a). In case of suitable packaging, these sensors could be arranged in an array to measure the uniaxial force- or pressure-distribution on a surface.

In case of force sensors, realization of miniaturized multicomponent force sensors is the current issue in research. Dimensions of single sensor elements range from 200 μm to 2 mm. Nominal force covers a range of 300 mN to 2 N. Due to batch-manufacture of measurement elements, realization of both single sensor elements and array-design[4] is possible. Sensitivity of sensors takes values of 2 % relative resistivity change in loading case. Figure 10.16 shows four examples of current topics in research. Variants (a) [100], (b) [103], and (d) [52] were designed for force measurement in haptic systems. Variant (c) [10] was built for tactile, dimensional measurement technology. Force transmission is always realized by beam- or rod-like structures.

Since 2007, a Hungarian manufacturer is selling the *Tactologic* system. Up to 64 miniaturized sensor elements are connected in an array of $3 \times 3\,\text{mm}^2$. Sensor elements have a size of $0.3 \times 0.3\,\text{mm}^2$ and are able to measure shear forces up to 1 N and normal force up to 2.5 N at nominal load. The force transmission is realized by soft silicone dots, applied to every individual sensor element (Fig. 10.17a, b). Using this array, static and dynamic loads in the range of kilohertz are measurable. But the viscoelastic material properties of the force transmission influence the dynamics due to creeping, especially the measurement of the normal forces [101, 102]. Another approach is to use piezoresistivity of silicon micromachined transistors using the above-mentioned shear effect (especially MOSFET, see [21, 40, 99]). As well as strain measurement these sensors are used to monitor the state of stress occurring in packaging process [21, 99]. Polyimide foil containing sensor elements (strain-sensitive transistors) with a thickness of about 10 μm are available since at least around year 2000 [40].

[4] By isolating arrays instead of single sensors in the last processing step.

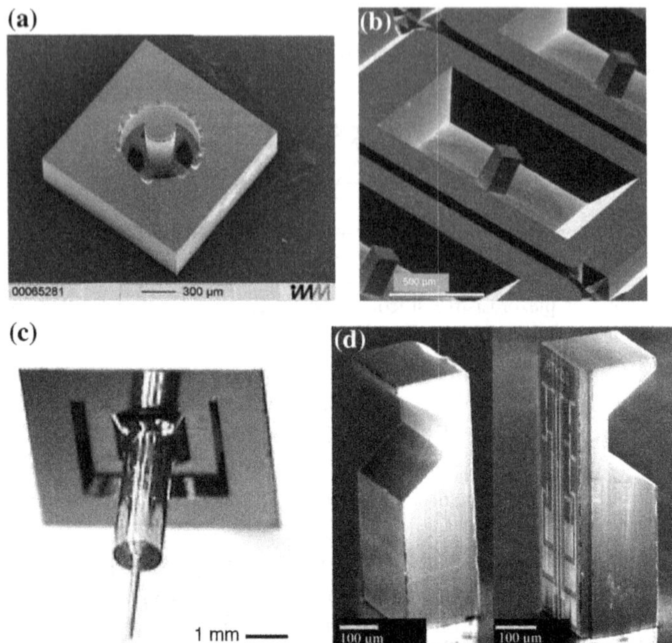

Fig. 10.16 Different realizations of piezoresistive silicon force sensors: **a** [100], **b** [103], **c** [10], and **d** [52]; **a** Single tri-axial sensor. **b** Array of tri-axial sensors. **c** Single tri-axial sensor. **d** Single tri-axial sensor [52]

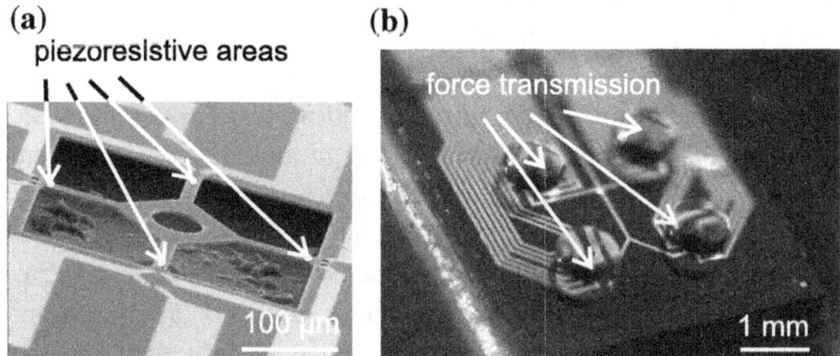

Fig. 10.17 Tactile multicomponent force sensor [101]. **a** One taxel of the sensing array (Tactologic), **b** 2 × 2 array for triaxial-force measurement (Tactologic)

Further Resistive Sensors

Besides resistive transducers presented until now, other more "exotic" realizations exist, which will be introduced within three examples. All sensors are suitable for array assembly to measure position-dependent pressure and a single-force component. The used measurement principles are based on the change of geometrical parameters of the force elements. The examples shown in Fig. 10.18a and b [41, 78]

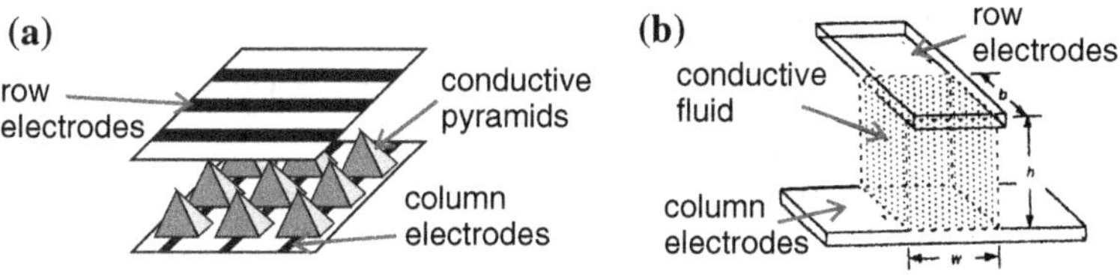

Fig. 10.18 Selected examples of foil sensors using the effect of a load-dependent constriction resistance. **a** Micromachined tactile array developed by Fraunhofer Institute IBMT. **b** Variation of the electrodes distance [41, 78]

use the load-dependency of the constriction resistance. With increased pressure,[5] the electrical contact area A increases and the resistance decreases.

The companies *Interlink Electronic* and *TekScan* use this effect for their sensor arrays (also called *Force Sensing Resistors—FSR*) (Fig.10.19). Interlink distributes polymer foils printed with resistor pastes in thick-film technology. Their basic resistance takes values in the region of $M\Omega$. The sensor foils have a height of 0.25 mm and a working range of zero to one Newton, respectively, 100 N. Beside the sensitivity to force or pressure, the sensors show a temperature dependency of 0.5 % K.

The sensor foils from *TekScan* are located in the range of 4.4 up to 440 N, the spatial distribution reaches up to 27.6 elements per centimeter. The available array size reach from approximately $13 \times 13\,\text{mm}^2$ up to $0.5 \times 0.5\,\text{m}^2$. The height of the foils is around 0.1 mm. The measurement inaccuracy takes a value of 10 %. The frequency range reaches from static up to 100 Hz. Beside the application in data gloves, as described by Burdea [12], the foil sensors are used in orthopedics to detect the pressure-distribution in shoes and prosthesis and within automotive industry for ergonomic studies.

Another approach is the variation of the distance between two electrodes (see Fig. 10.18b). The sensing element is made of flexible substrates. The electrodes are arranged in rows and columns. The gaps in between are filled with an electrical

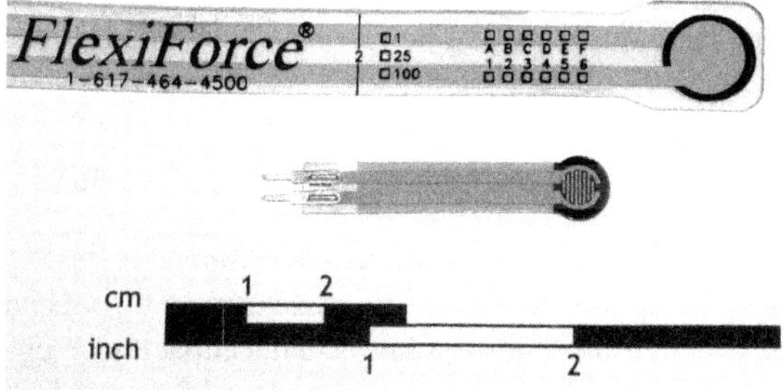

Fig. 10.19 Foil sensors for compressive force detection, *top* FlexiForce by *TekScan Inc.*, South Boston, MA, USA, *bottom* FSR by *Interlink Electronics*, Camarillo, CA, USA. These sensors are often used to detect grasp forces, for example

[5] The force can be calculated taking the contact area into account.

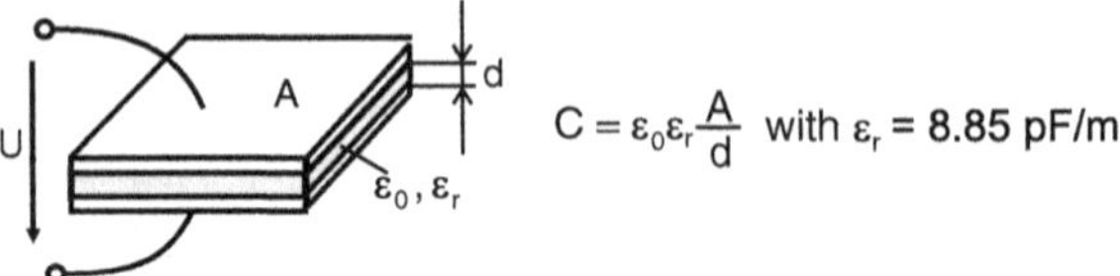

Fig. 10.20 Assembly of a single capacitance

conductive fluid. In loading case, the fluid is squeezed out and the distance of the electrodes varies. A disadvantage of this principle is given by the necessity for very large distance variations up to 10 mm to achieve usable output signals. Until today, this principle is still a topic of research.

10.1.2.4 Capacitive Sensors

Within every capacitive sensor at least two electrodes are located parallel to each other. Figure 10.20 shows a design based on a single measurement capacity. In contrast to the resistive principle—measuring the mechanical variables stress and strain—the capacitive principle measures the integral values displacement (or elongation) directly.

Concerning the working principle, three classes can be identified, which shows some similarities to electrostatic actuators discussed in Sect. 9.5. The first class uses the displacement principle. On this, the mechanical load changes the electrode distance d or the active electrode area A. In the third class, the relative dielectric ε_r is influenced. The change of electrode distance is usually used for measuring force, pressure, displacement, and acceleration. In these cases, the mechanical load is directly applied to the electrode and displaces it relatively to the other one. The resulting capacitance change can be calculated:

$$\frac{\Delta C}{C_0} = \frac{1}{1 \pm \xi/d} \approx \pm\frac{\xi}{d}. \tag{10.21}$$

ξ marks the change of distance. Additionally, the electrode distance can be kept constant, and only one electrode can be parallel displaced (Fig. 10.21). The active electrode area varies accordingly and the resulting capacitance change can be used to measure angle, filling level, or displacement. It is calculated according to:

$$\frac{\Delta C}{C_0} = 1 \pm \frac{\Delta A}{A_0}. \tag{10.22}$$

The third option for a capacitance change is the variation of the relative dielectric. This principal is often used for measuring a filling level, e.g., of liquids, or as a proximity switch for layer thickness. This capacitance change is calculated according to

$$\frac{\Delta C}{C_0} = 1 \pm \frac{\Delta \varepsilon_r}{\varepsilon_{r0}}. \tag{10.23}$$

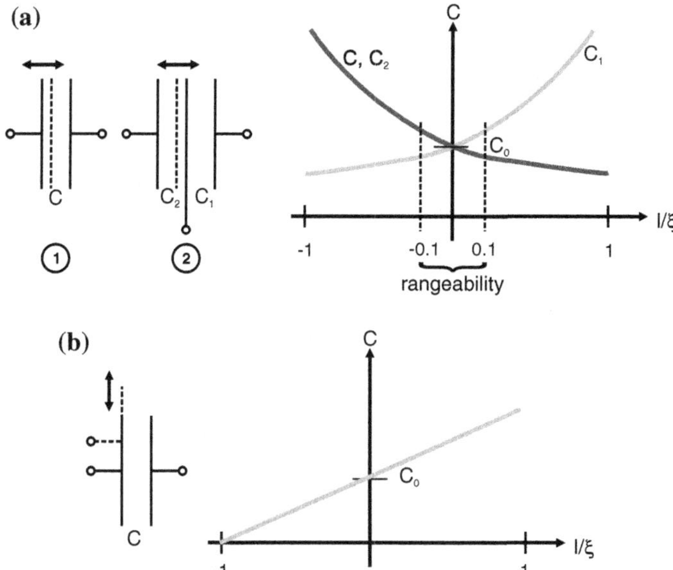

Fig. 10.21 Schematic view of capacitive sensing principle and characteristic curve of capacitance. **a** Variation of electrode distance in case of single (*1*) and differential setup (*2*). **b** Variation of effective electrode area

Characteristics of Capacitive Pressure and Force Sensors

The main principle used for capacitive force, respectively, pressure transducers is measuring displacements. Consequently, the following paragraph will concentrate on this principle. As stated within Eq. (10.21) for the change of distance, the interconnection between capacitance change and mechanical load is nonlinear for single capacities. The displacement ξ lies in the range of 10 nm to 50 μm [105]. For linearization of the characteristic curves, an operating point has to be found, e.g., by arranging three electrodes as a differential capacitor. The displacements ξ typical for the working range are $\leq 10\,\%$ than the absolute electrode distance d. In this range, the characteristic curve can be approximated as linear (Fig. 10.21a). With the principle varying the electrode's surface the capacitance changes proportional to it, resulting in a linear capacitance change (Fig. 10.21b).

The evaluation of the capacitance change can be made by an open- or closed-loop measuring method. Concerning open-loop method, the sensor is either integrated in a capacitive bridge circuit, or it is put into a passive oscillating circuit with coil and resistor. Alternatively, the impedance can be measured at a constant measurement frequency. An alternative could be the application of a pulse-width-modulation (also called: recharging method). A closed-loop approach is characterized by the compensation of the displacement by an additional energy. The main advantage of the closed-loop signal conditioning is the high linearity achieved by very small displacements. Additional information can be found in [105, 106].

The advantage of capacitive sensor in contrast to resistive sensors lies in the effect of little power consumption and high sensitivity. Additionally, the simple structure enables a low-cost realization of miniaturized structures in surface micromachining (Fig. 10.22). In contrast to the resistive sensors—where positions and dimensions of the resistive areas have a direct influence on transfer characteristics—the manufacture tolerances for capacitive sensors are quite high. Mechanically induced stress due to packaging and temperature influence has almost no influence on their performance. Even mispositioning of electrodes next to each other do not change the transfer characteristics, only the basic capacitance. The manufacturing steps with silicon capacitive sensors are compatible to CMOS-technology. This allows a direct integration of the sensor electronics on the chip, to minimize parasitic capacities. Especially with miniaturized sensors,[6] a good signal-to-noise ratio can be achieved [71]. The problem of parasitic capacities or leakage fields is one of the major challenges for the design of capacitive actuators, as it achieves easily a level comparable to the capacitance used for measurement. An additional challenge can be found in the constancy of the dielectric value, which is subject to changes in open-air gap actuators due to humidity or other external influence factors.

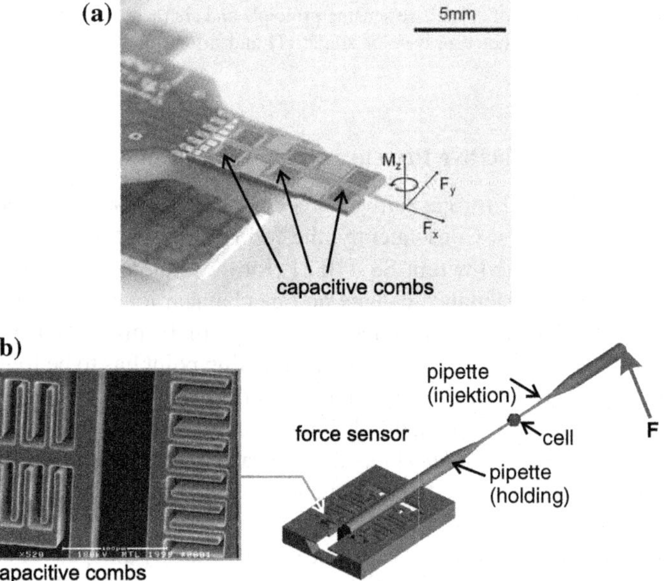

Fig. 10.22 Examples of capacitive silicon multicomponent force sensors. **a** 2-component force sensor, nominal load 1.5 mN [8], **b** 6-component force/torque sensor, nominal load 500 μN [93]

[6] Due to the small electrodes a small basic capacitance is achieved, comp. equation in Fig. 10.20.

Examples of Capacitive Sensors

Concerning the manufacturing technology, capacitive sensors integrated in haptic systems can be distinguished in three classes. The first class is represented by miniaturized pressure sensors, being realized using silicon microtechnology. Due to their small size of few millimeters, the moving masses of the sensor are low and thus cover a wide dynamic range (frequencies from static to several kilohertz). As shown before, the micromachined capacitive sensors may be combined to arrays for measuring spatially distributed load. As an example, SERGIO [85] reports the realization of a capacitive array in CMOS-technology. A challenge is given by the capacity changes in the range of femto-Farad, which is similar to the capacity of the wiring. A relief is given by a parallel circuit of several capacities to a sensor element [105]. The frequency range of the shown examples range from static measurement up to several MHz upper cutoff frequency. Consequently, it is suitable for haptic-related measurements of tactile information. Another example is given by an array made of polysilicon. It has an upper cutoff frequency of 1,000 Hz and a spatial resolution of 0.01 mm^2 suitable for tactile measurements. It was originally designed for acquisitions of fingerprints. REY [75] reports the use of such an array for intracorporal pressure measurement at the tip of a gripper. Once again the leakage capacities are a problem, as they are within the range of the measured capacity changes.

Two examples of multicomponent force sensors built in surface micromachining are shown in Fig. 10.22a, b [8, 93]. The two-axial sensor[7] is designed for atomic force microscopy. The nominal load of this application lies in a range of μN. The three-axial sensor was designed for micromanipulation, e.g., in molecular biology, with similar nominal values of several μN. Both sensors are using the displacement change for measurement.

The second class is represented by ceramic pressure load cells. They are widely used in automotive industry and industrial process measurement technology. Substrate and measurement diaphragm are typically made of Al_2O_3 ceramics. The electrodes are sputtered on ceramic substrates. Substrate and measurement diaphragm are connected via solder applied in thick-film technology. In contrast to silicon sensors, ceramic sensors are macroscopic and have dimensions in the range of several centimeters. Based on the technology sensors in differential-, relative-, and absolute-designs with nominal pressures in the range of zero to 200 mbar such as in zero to 60 bar are available (e.g., Fig. 10.23, *Endress und Hauser*). The frequency range of these sensors is low, upper cutoff frequencies of approximately 10 Hz are achieved.

The third class is built from foil sensors, distributed, e.g., by the company *Althen GmbH*, Kelkheim, Germany. These capacitive sensor elements are arranged in a matrix with a spacial resolution of $\leq 2 \times 2\,mm^2$. As substrate a flexible polymer foil is used. The height of such an array is 1 mm. The frequency range ranges from static to approx. 1,000 Hz. Nominal loads up to 200 kPa can be acquired with a resolution of 0.07 kPa. Due to creeping (comp. Sect. 10.1.1.3) of the substrate and parasitic capacities, a high measurement inaccuracy exists.

---

[7] With respect to "force" component.

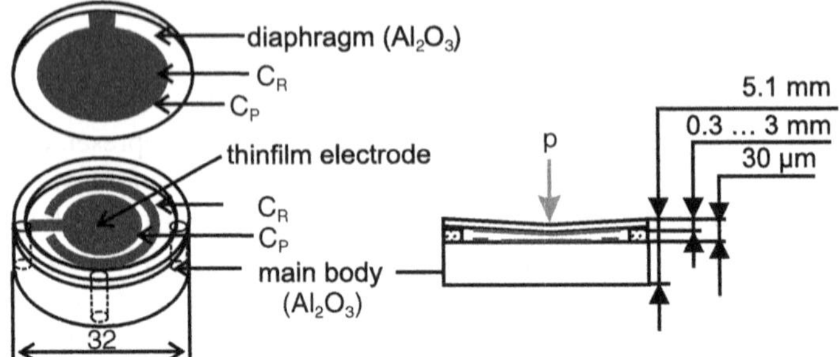

Fig. 10.23 Schematic view of a ceramic pressure sensor fabricated by *Endress und Hauser*, Weil am Rhein, Germany [105]

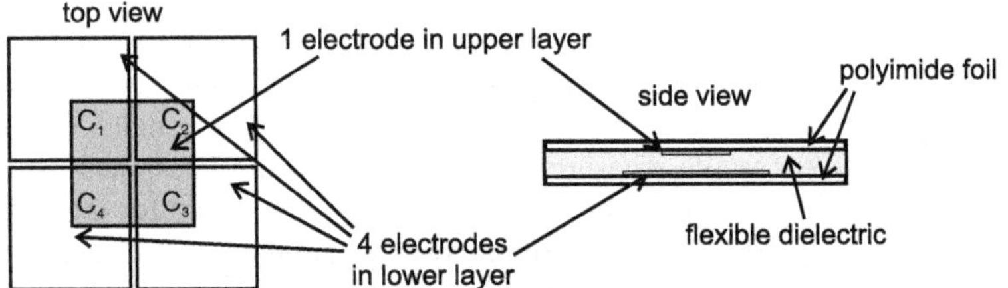

Fig. 10.24 Schematic view of capacitive shear force sensors as presented in [15]

Another polymeric foil sensor in the field of investigation is that one shown in Fig. 10.24 [15]. In contrast to prior examples, this array is used for direct force measurement. Normal forces are detected measuring the change of electrode distance, shear forces by detecting the change of active electrode surface. Similar to the sensors of the company *Althen* static and dynamic changes up to 1,000 Hz can be measured. The spatial resolution is given with $1 \times 1\ \text{mm}^2$. A disadvantage of the design is the high measurement inaccuracy through creeping of the polymer and leakage capacities.

10.1.2.5 Optical Sensors

In the area of optical measurement technology, sensors based on freely propagating beams and fiber optics are available. For force and pressure sensing mainly fiber optic sensors are used, which will be introduced further within this subsection. All fiber optic sensors have in common that mechanical load influences transmission characteristics of the optical transmission network, resulting in an influence of parameters of a reflected or simply transmitted electromagnetic wave. The electromagnetic wave is defined by its wave equation [54].

$$\nabla^2 \Psi = \frac{\delta^2 \Psi}{\delta x^2} + \frac{\delta^2 \Psi}{\delta y^2} + \frac{\delta^2 \Psi}{\delta z^2} \qquad (10.24)$$

Ψ represents an arbitrary wave. A possible solution for this differential equation is the propagation of a plane wave in open space. In this case, electrical field E and magnetic field B oscillate orthogonal to each other. Electrical field propagating in z-direction is described by Eq. (10.25).

$$E(z, t) = \frac{1}{2} A(z, t) \cdot e^{j\omega_0 t - \beta_0 z} \tag{10.25}$$

A marks the amplitude of the envelope, ω_0 is the optical carrier frequency, and the propagation constant β_0. With the propagation group velocity $v_g(\lambda)$,[8] the E- and B-field are connected. Depending on the transmitting medium group velocity can be calculated via refraction index n_g [55].

$$v_g(\lambda) = \frac{c_0}{n(\lambda)} \tag{10.26}$$

According to wave length λ, n different values result. Waves are propagating differently depending on their frequency and wavelength. A pulse "spread out". For further information, sources [54, 55, 63, 108] and [56] are recommended. If only mechanical loads such as force or pressure influences the transmission network, the resulting deformation can influence thc transmission in two different ways:

1. Material specific: Change of the refraction index n (photoelastic effect)
2. Geometric: Change of beam guidance

The photoelastic effect describes the anisotropy of the refraction index influenced by mechanical stress. Figure 10.25 visualizes this effect. Resulting refraction index change is dependent on applied stress T and is given by the following equation [33]:

$$\Delta n = (n_1 - n_2) = C_0 \cdot (T_1 - T_2) \tag{10.27}$$

C_0 is a material-specific, so-called photoelastic coefficient. T_i marks the resulting internal stress. Depending on refraction index polarization, wave length and phase of beam are changing. In the geometric case, mechanical load changes the conditions of the beam guidance. Using geometrical optics influences of mechanical loads on intensity and phase of radiation can be characterized.

A disturbing source for all fiber optical sensors cannot be neglected: the temperature. Refraction index is depending on temperature changes, and consequently influences the properties of the guided wave. Beside thermal-elastic coefficients describing the strain resulting from temperature changes within any material, temperature directly influences the refraction index itself (Sect. 10.1.2.5). For temperature compensation, a reference fiber has to be used, unloaded, and only influenced by temperature change. An advantage of all fiber optical sensors is given by their immunity to electromagnetic radiation. The following paragraphs introduce the most important principles for optical force and pressure measurement.

---

[8] In vacuum it is equal to speed of light $c_0 = 2.99792458 \times 10^8$m/s.

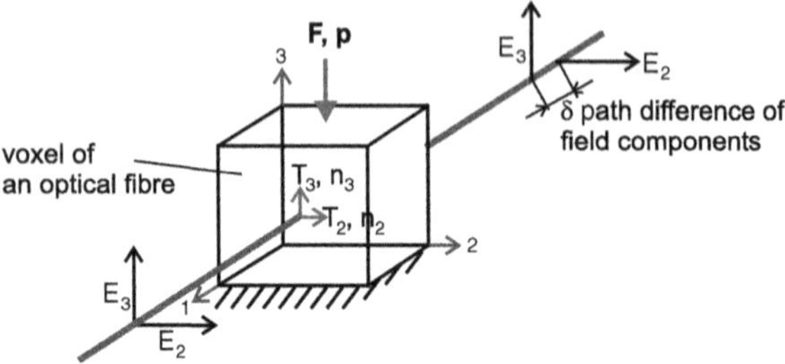

Fig. 10.25 Visualization of photoelastic effect [33]. Due to different refraction indices perpendicularly to the propagation direction propagation velocity of each field component is different and an optical path difference δ occurs. Polarization is changing

Change of Intensity

In principle, two transducer types varying the intensity can be distinguished. Both have in common that mechanical load varies the condition of total reflection (Fig. 10.26). The angle α_c is defined as the critical angle for total reflexion and defined by SNELLIUS' law:

$$\sin(\alpha_c) = \frac{n_2}{n_1} \tag{10.28}$$

The numerical aperture *NA* gives the appropriate critical angle θ_c for coupling radiation into a multimode fiber:

$$\sin(\theta_c) = \sqrt{n_1^2 - n_2^2} \tag{10.29}$$

If the angle varies due to mechanical load and takes values larger than θ_c, respectively, smaller values than α_c, conditions for total reflections are violated. The beam will not be guided within the core of the fiber. Total intensity of the transmitted radiation will become less. Figure 10.27 shows a schematic sketch of the design of the very first variant. The sensor element is attached to the end of a multimode fiber.

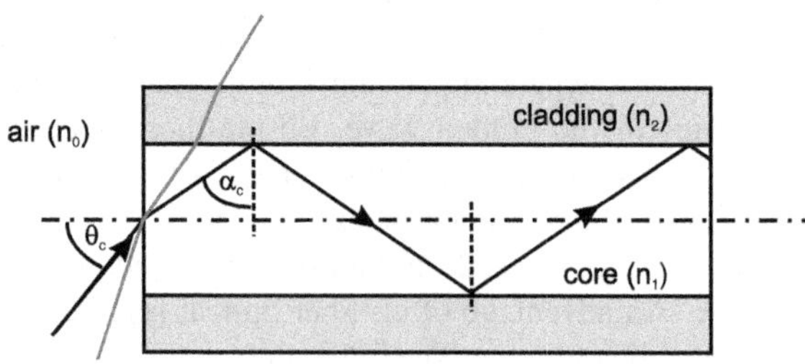

Fig. 10.26 Guidance of multimode fibers. Beams injected with angles above θ_c are not guided in the core

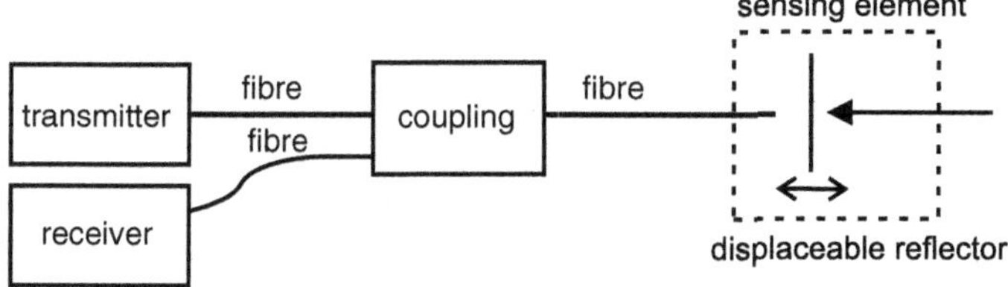

Fig. 10.27 Schematic view of a fiber optic sensor with intensity modulation

In a first variant, the light (e.g., emitted by a laser-diode $\lambda = 1{,}550$ nm) is coupled into a multimode fiber. A reflective element is attached to the end of the transmission line. The element can be designed as a deformable object or a rigid one mounted on a deformable substrate. Mechanical load acts on this object. Due to the load, the reflective element will be deformed (in case of a flexible surface) or displaced (in case of a rigid surface). Varying the displacement, the mode of operation is comparable to a displacement sensor. The intensity is directly proportional to the displacement (Fig. 10.28). The load itself is a function of displacement and directly proportional to the elastic compliance n of the sensor element:

$$F(z) = n \cdot z. \tag{10.30}$$

If the geometry of the area changes, a part of the beam—according to the laws of geometrical optics—is decoupled into the cladding (dispersion) and an intensity loss can be measured at the detector (Fig. 10.28).

In academic publications from PEIRS [67] and KERN [37] such a mode of operation is suggested for multicomponent force measurement. In this case, the measurement

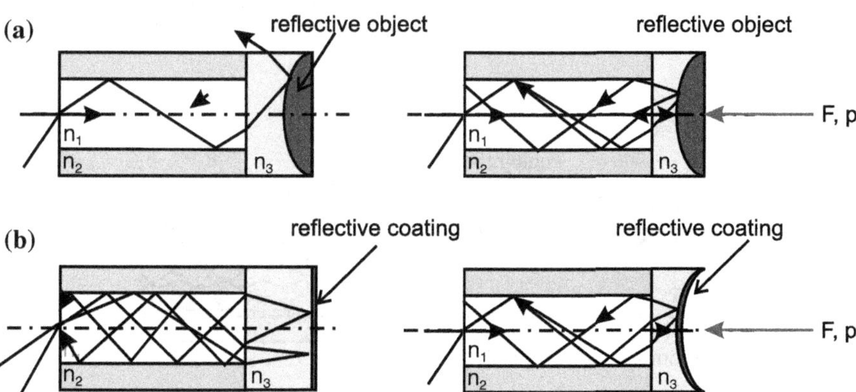

Fig. 10.28 Variation of intensity due to displacement of rigid and flexible elements. **a** Stiff element: intensity is depending on the displacement. **b** Flexible element: intensity is depending on the deformation

range is directly proportional to the mechanical properties of fixation of the reflective body. Using the calculation method known from Sect. 10.1.2.1, this fixation can be designed. A disadvantage of this principal is the use of polymers for the coupling of the reflective object. This leads to creeping of the sensor signal. The measurement inaccuracy of these sensors lies in a range of 10 % [37]. Their diameter takes a value of few millimeters. The length depends on the application. Another source of noise is the temperature. A temperature change leads to a dilatation (or shrinkage) of the polymer itself and displaces the reflective element. The displacement change results in a defective measurement signal. Due to the small size, an array assembly is possible.

The second variant is a so-called microbending sensor. Its fundamental design is schematically given in Fig. 10.29. Like before, a beam is coupled into a multimode fiber. Force, pressure, or strain applied by a comb-like structure results in microbending of the fiber (Fig. 10.29b).

In case of deformation—similar to the first variant—a part of the light is decoupled into the cladding. The intensity of the measured light diminishes.[9] The gaps between the comb-like structures for microbending sensors are in the range of one millimeter. The height of the structure is in the same dimension [92]. To apply mechanical loads, an area of $\cong$1 cm length and a width of $\geq$5 mm is used. Measurement range depends on displacement of the bending structure and diameter of the fiber itself. Pandey [64] describes the realization of a pressure sensor for loads up to 30 bar. If the bending diameter becomes smaller, lower nominal pressures and forces are possible. Concerning the detection of force components, only one-component sensors can be realized using this principle.

If spatially distributed mechanical load has to be measured, multiple microbending structures can be located along one fiber. To evaluate the several measuring points, for instance optical time domain reflectometry (ODTR) can be used. This device sends a pulsed signal (light pulses of around 10 μs length) guided in a fiber, and measures the reflexion depending on time. Based on the propagation velocity of the

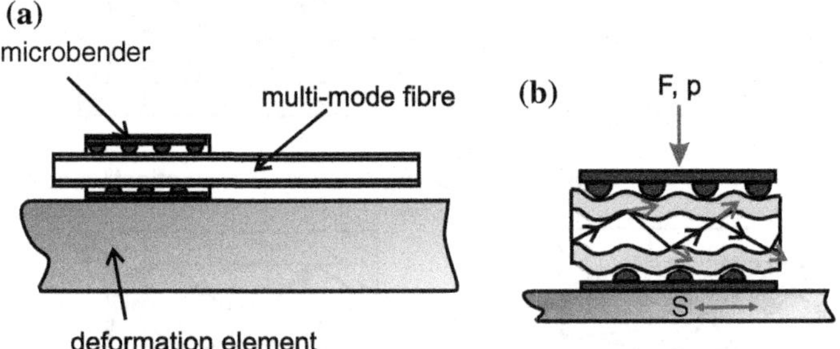

Fig. 10.29 Variation of beam guidance in case of microbending. **a** Assembly of microbending elements. **b** Beam guiding in loading case

[9] Both versions are possible: Measuring the transmitted and the reflected radiation.

beam inside the fiber v, the time delay for each measuring point can be calculated by relating them. Additional information can be found in [48, 92] or [64]. The dynamics of these sensors is only limited by the sensor electronics and could theoretically be applied to the whole range of haptic applications.

Change of Phase

The variation of the phase of light by mechanical load is used for interferometric sensors. The most commonly used type is based on the Fabry-Pérot interferometer, discussed in the following paragraph. Other variants are Michelson- and Mach-Zehnder-interferometers. The assembly is made of two plane-parallel, reflective, and semitransparent objects, e.g., at the end of a fiber, building an optical resonator (Fig. 10.30). The beam is reflected several times within the resonator and interferes with each reflection. The resonance condition of this assembly is given by the distance d of the reflective elements and the refraction index n within the resonator. The so-called free spectral range marks the phase difference δ, generating a constructive superposition of beams:

$$\delta = \frac{2\pi}{\lambda} \cdot 2 \cdot n \cdot d \cdot \cos(\alpha) \tag{10.31}$$

Figure 10.30b shows the typical characteristics of the transmission spectrum of a Fabry-Pérot interferometer. According to the formula shown above the corresponding wavelength yields a transmission peak; all other wavelengths are damped and annihilated. Due to the mechanical load the distance d of the surfaces is varied, changing the conditions for constructive interference. Sensors using this principle are used by the company *LaserComponents Gmbh*, Olching, Germany, for uniaxial

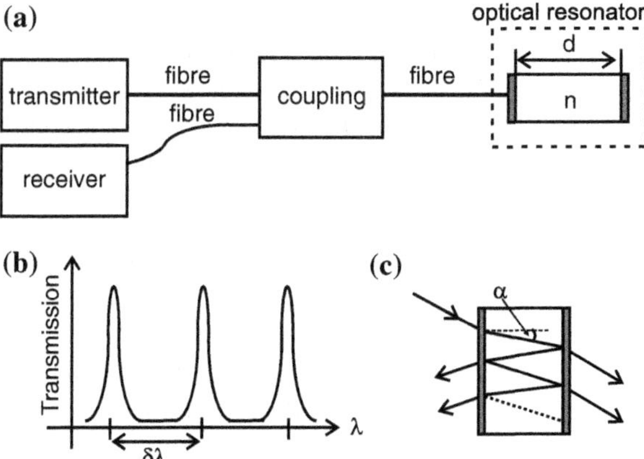

Fig. 10.30 Assembly and operating mode of a Fabry-Pérot interferometer. **a** Schematic assembly of a Fabry-Pérot interferometer. **b** Transmission spectrum. **c** Interferences in a resonator

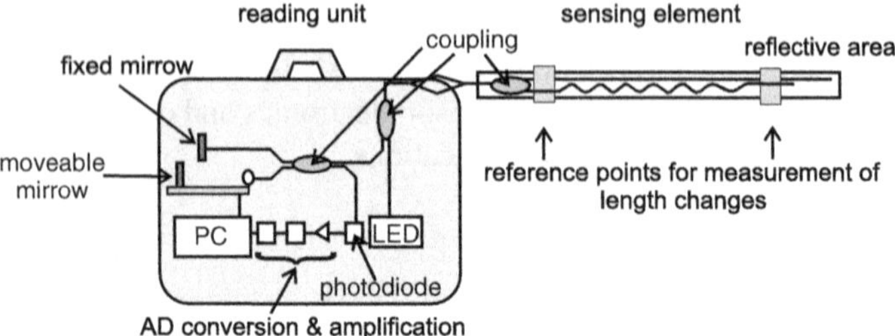

Fig. 10.31 Temperature compensation in interferometric strain-sensing elements [34]

force or pressure measurement, and can be bought for nominal pressures up to 69 bar [45]. The influence of temperature would also appear to be problematic too and has to be compensated by a reference configuration parallel working.

Beside pressure transducers, single-component forces and strains can be measured (Fig. 10.31). The design equals a Michelson-Interferometer. The sensor element is made of two multimode fibers, whereas the strain acts upon only one fiber. Identical to the Fabry-Pérot-configuration, the sensor element is made of two plane-parallel reflective surfaces, whose distance varies according to varying strain. Inside the measuring electronics, a reference design is included. To measure the mechanical load, the phase of reference and measuring assembly is compared. This measurement principle enables to measure frequencies in the range of several kilohertz. The geometrical dimension is given by the diameter of the fiber including some protective coating ≤ 1 mm, and the length of 2–20 mm depending on the application itself. For pressure sensors, the measuring error with respect to nominal load takes a value of about 0.5 %, with strain gages at a factor of 15×10^{-6}.

Change of Wavelength

For optical detection of strain, the so-called fiber BRAGG grating sensors (FBG sensor) are widely used. To realize the sensing element, the refractive index of the core in a single mode fiber is varied due to the position (Fig. 10.32) and a grating arise [83]. The refractive index modulation can be described by

$$n(z) = n_0 + \delta n_{\text{effective}}(z) = n_0 + \delta \overline{n}_{\text{effective}} \cdot \left(1 + s \cdot \cos\left(\frac{2\pi}{\Lambda} z + \phi(z)\right)\right) \quad (10.32)$$

whereas n_0 is the refractive index within the core, $\delta \overline{n}_{\text{effective}}$ is the average of the index's modulation, and s a measure of the intensity of the index's modulation. Λ marks the grating period and the phase shift $\phi(z)$ resulting from the measured value. In idle situation results $\phi(z) = 0$. Figure 10.32 gives a schematic drawing of the assembly.

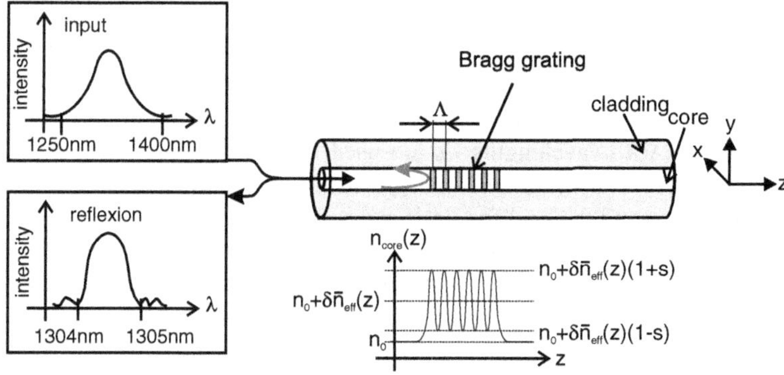

Fig. 10.32 Operational mode of FBG sensors [83]

If light is coupled into the fiber, only parts of it are reflected according to the law of BRAGG. The reflective spectrum shows a peak at the so-called BRAGG-wavelength λ_b. This wavelength depends on the refractive index $n(z)$ and the grating period Λ:

$$\lambda_b = 2n\Lambda. \tag{10.33}$$

In loading case both grating distance and refractive index varies. The maximum of the spectrum is shifting from λ_0 to another wavelength. According to the wavelength shift, mechanical load can be determined. The following condition is achieved:

$$\frac{\Delta\lambda}{\lambda_0} = \underbrace{(1 - C_0)}_{\text{gage factor}} \cdot (S + \alpha_{VK} \cdot \Delta\vartheta) + \frac{\delta n/n}{\delta\vartheta} \cdot \Delta\vartheta, \tag{10.34}$$

whereas α_{VK} is the coefficient of thermal expansion of the deformation body, and C_0 the photoelastic coefficient. Beside the change induced by mechanical strain S, the change of temperature ϑ influences the wavelength shift in the same dimension. Compensating the influence of temperature, another FBG sensor has to be installed as reference at an unloaded area. The temperature compensation is afterward achieved by comparison between both signals. Analogous to resistive strain sensors, a gage factor of $k \approx 0.78$ can be achieved with constant measurement temperature. Extensions up to 10.000 μm/m can be achieved.

The width of the sensor lies in the area of single mode fibers. The sensors length is defined by the grating, which has to be three millimeters at least to provide a usable reflective spectrum [9, 22, 38]. Resolution takes a value of 0.1 μm/m and is—such as its dynamics—defined by the sensor electronics. Similar to strain gages these sensors can be mechanically applied on deformation elements, whose dimensions and shapes define the measurement range. A challenge with the application of fiber sensors in this context is the differing coefficients of thermal expansion between deformable element, adhesive, and fiber. Additionally, reproducibility of the adhesive-process for fibers is not as high as typically required. Especially creeping of glue results in large measurement errors. Comparable to the microbending principle, FBG sensors are applicable to several spatially distributed measurement points. To distinguish the several positions, gratings with different periods Λ_i and thus different BRAGG-wavelengths λ_b are used. The company *Hottinger Baldwin Messtechnik GmbH*, Darmstadt, Germany, distributes several designs containing an application area around the grid for strain measurement.

Beside monitoring of structures or strain analysis, FBG sensors can be used for realizing force sensors, too. MUELLER describes the use of FBG sensors in a triaxial-force sensor for medical application [59]. Further information on the application of FBG can be found in [9, 22, 83] and [38].

10.1.2.6 Piezoelectric Sensors

Piezoelectric sensors are widely used, especially for measurement of highly dynamic activities. The measurement principle is based on a measure-induced charge displacement within the piezoelectric material, the so-called reciprocal piezoelectric effect (Sect. 9.3). Charge displacement leads to an additional polarization of the material resulting in a change of charge on the materials surface. This can be detected using electrodes (Fig. 10.33). Beside the measurement of force, it is for pressure and acceleration measurement especially. For force measurement, the longitudinal effect is primarily used. Detailed information about the piezoelectric effect and possible designs are found in Sect. 9.3. Materials used for sensing elements will be introduced in the following paragraph.

The general equation of state states for operation in sensor mode:

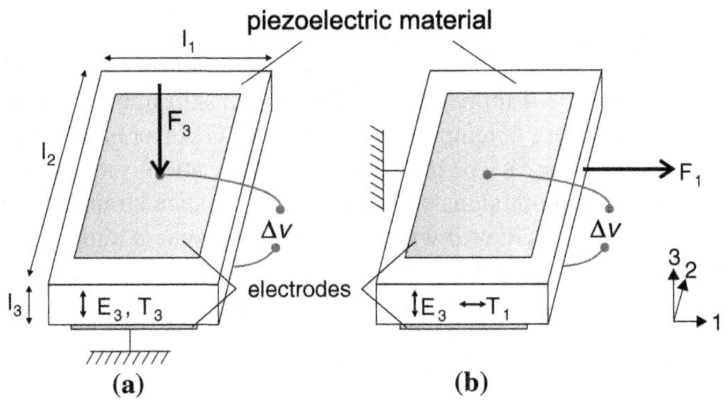

Fig. 10.33 Visualization of piezoelectricity. **a** Longitudinal effect ($E_3 \| T_3$). **b** Transversal effect ($E_3 \| T_1$)

$$D_i = \underbrace{\varepsilon_{ij}^T \cdot E_j}_{\to 0} + d_{im} \cdot T_m, \tag{10.35}$$

$$D_3 = d_{31} \cdot T_1. \tag{10.36}$$

A stress contribution in the sensing material leads to a change of charge density D_i, whereas ε_{ij}^T marks relative permittivity and d_{im} piezoelectric charge constant. Taking geometric parameters of the sensor, electrode area $A = l_1 \cdot l_2$ and thickness l_3 of dielectric, the resulting charge q can be derived. Taking electric parameters into account, sensor output voltage Δu can be calculated [47, 87, 105]:

$$q = D_3 \cdot A_3, \tag{10.37}$$

$$\Delta v = q \cdot \frac{1}{C_\mathrm{p}}, \quad \text{with } C_\mathrm{p} = \frac{e_{33}^T \cdot l_2 \cdot l_1}{l_3}, \tag{10.38}$$

C_p marks capacitance and e_{33}^T piezoelectric force constant.

Technically relevant materials can be distinguished into three groups. The first group is built of monocrystals such as quartz, gallium, and orthophosphate.[10] Polarization change in case of mechanical load is direct proportional to the stress. Its transfer characteristic is very linear and does not have any relevant hysteresis. Piezoelectric coefficients are long-term stable. One disadvantage is the small coupling factor k of about 0.1. For remembrance: k is defined as the quotient transformed to the absorbed energy.

The second group is formed by polycrystalline piezoceramics, such as barium titanate ($BaTiO_3$) or lead zirconate titanate (PZT, $Pb(ZiTi)O_3$), being manufactured in a sintering process. The polarization is artificially generated during the manufacturing process (Sect. 9.3). An advantage of this material is the coupling factor, which is seven times higher than that of quartz. A disadvantage is the nonlinear transfer characteristics with a noticeable hysteresis, and a reduced long-term stability. The materials tend to depolarize.

The last group is build from partial crystalline plastic foils made of polyvinylidene fluoride (PVDF). Its coupling factor lies with 0.1–0.2 in the area of quartz. Advantageous are the limit size (foil thickness of a few µm) and the high elasticity of the material.

The first two sensor materials are used in conventional force sensors, as e.g., distributed by the company *Kistler*. Nominal forces take values of 50 N to 1.2 MN. The sensor typically has a diameter of 16 mm and a height of 8 mm. Alternations of load up to 100 kHz are measurable. Single as well as multiple-component sensors are state-of-the-art. Figure 10.34 shows the general design of a three-component force sensor from *Kistler*.

Piezoelectric force sensors are typically used for the analysis of the dynamic forces occurring during drilling and milling or for stress analysis in automotive industry. In haptic system, these sensor variants can hardly be found. Not exclusively but mostly

[10] This crystal is especially applicable for high temperature requirements.

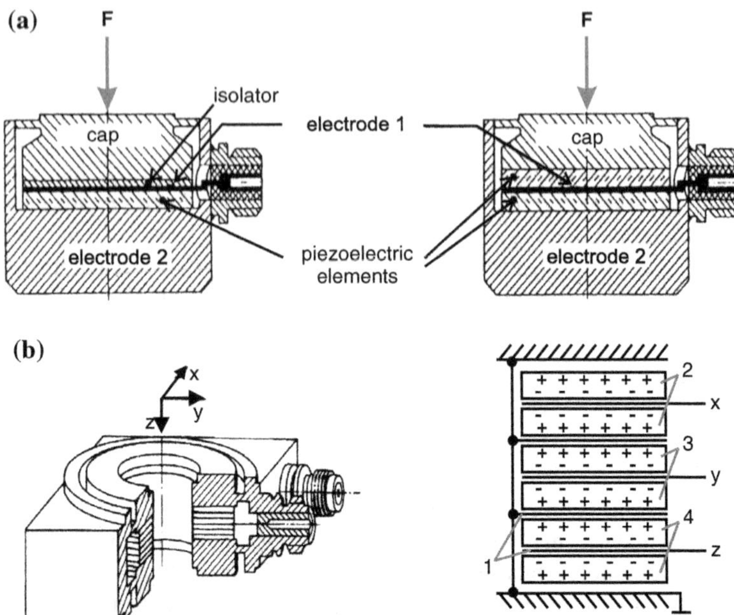

Fig. 10.34 **a** Possible assemblies für piezoelectric force sensors. **b** Assembly of a three-component sensor: *1* electrodes, *2* quartz plates shear effect—F_x, *3* quartz plates longitudinal effect—F_z, and *4* quartz plates shear effect—F_y

because they are not suitable to measure static loads. Sensors based on PVDF-foils as piezoelectric material are increasingly used for the measurement of tactile actions. The piezoelectric effect, however, is used for the generation of a displacement and not for its measurement, making this variant being described in Sect. 10.1.2.7.

10.1.2.7 Less Common Sensing Principles

Sensor designs shown in this subsection are not force or pressure sensors for conventional purposes. All of them have been designed for different research projects in the context of haptic systems. Focus of these developments lies in the spatially distributed measurement of tactile information.

Resonance Sensors

For measurement of vibrotactile information, e.g., the so-called resonance principle could be used. Figure 10.35a shows the principal design of such a sensor. A piezoelectric foil (PZT or even PVDF) is used as an actuator. Electrodes on both sides of the foil apply an electrical oscillating signal, resulting in mechanical oscillations of the material due to the direct piezoelectric effect. The structure oscillates at its resonance frequency f_0 calculated by the following formula

$$f_0 = \frac{1}{2d} \cdot \sqrt{\frac{n}{\rho}} \tag{10.39}$$

whereas d is the thickness, n the elasticity, and ρ the density of the used material. The load, responsible for the deformation, is proportional to the frequency change [70]. For spatially distributed measurement, the sensors are connected as arrays of elements with 3×3 and 15×15 sensors. The dimensions of the sensing arrays takes values of 8×8 mm^2, respectively, 14×14 mm^2. The thickness of the foil is $\ll$1 mm. A huge disadvantage of this principle is the high temperature dependency of the resonance frequency from the piezoelectric material used. The coefficient lies at 11.5 Hz per 1 °C within a temperature range between 20 and 30 °C [26, 41].

The so-called surface acoustic wave resonators, SAW sensors, make use of the change of their resonance frequency too. The excitation occurs via an emitter called "Inter-digital structure" (Fig. 10.35b). The mechanical oscillations with frequencies in the range of MHz distribute along the surface of the material. They are reflected on parallel metal structures and detected by the receiving structure. Due to mechanical values applied the material is deformed, the runtime of the mechanical wave changes, and consequently the sensor's resonance frequency. With this design, the temperature is one of the major disturbing values. SAW sensors are used for measurement of force, torques, pressure, and strain. The dynamic range reaches from static to highly dynamic loads.

Electrodynamic Sensor Systems

Within the research project *TAMIC* an active sensor system for the analysis of organic tissue in minimally invasive surgery was developed [29]. The underlying principle is

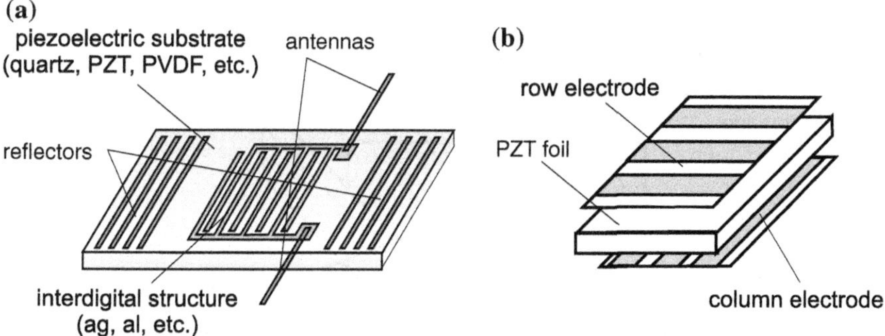

Fig. 10.35 Schematic view of resonance sensors. **a** Principle of a surface acoustic wave resonator [72]. **b** Principle of a resonant sensor array [70]

based on an electrodynamic actuated plunger excited to oscillations (see Sect. 9.1). The plunger is magnetized in axial direction. The movements of the plunger induce voltages within an additional coil included in the system. The material to be measured is damping the movement, which can be detected and quantified by the induced voltage. The maximum displacement of the plunger is set to one millimeter. The system is able to measure dynamically from 10 to 60 Hz. The nominal force lies in the range of 200 mN. The geometrical dimensions of the system are a diameter of $\leq$15 mm, and a length of $\leq$400 mm, which is near to typical minimally invasive instruments. Detailed information can be found in [84].

Another example for a miniaturized sensor for the measurement of spatially distributed tactile information is presented by HASEGAWA in [30]. Figure 10.36 shows the schematic design of one element.

The elements are arranged in an array structure. In quasi-static operation mode the system is able to measure contact force and the measurement object's elasticity. The upper surface is made of a silicon-diaphragm with a small cubical for force-application to the center of the plane. The displacement of the plate is measured identical to a silicon-pressure or -force sensor with piezoresistive areas on the substrate. By the displacement, the applied contact force can be derived. For measuring the elastic compliance of the object, a current is applied to the flat coil (Fig. 10.36). In the center of the diaphragm's lower side a permanent magnet is mounted. The electrically generated magnetic field is oriented in the opposite direction of the permanent magnet. The plate is displaced by this electromagnetic actuator and the cube is pressed back into the object. The force necessary to deform the object is used in combination with the piezoresistive sensors signal for calculation of the object's elastic compliance. In the dynamic operation mode, the coil is supplied with an oscillating signal, operating the diaphragm in resonance. Due to interaction with the measured object, the resonance condition changes. By the changing parameters, such as phase rotation, resonance frequency, and amplitude, elastic coefficients such as damping coefficients of the material can be identified. Due to the high degree of miniaturization, highly dynamic actions up to several kilohertz are possible to be

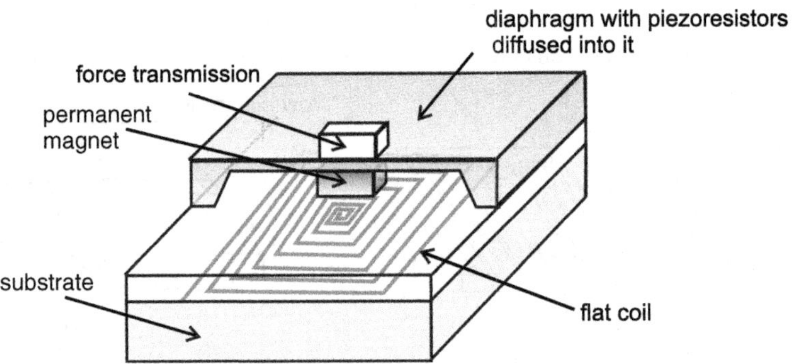

Fig. 10.36 Schematic view of an active element [30]. The dimensions are $6 \times 6 \times 1$ mm^3

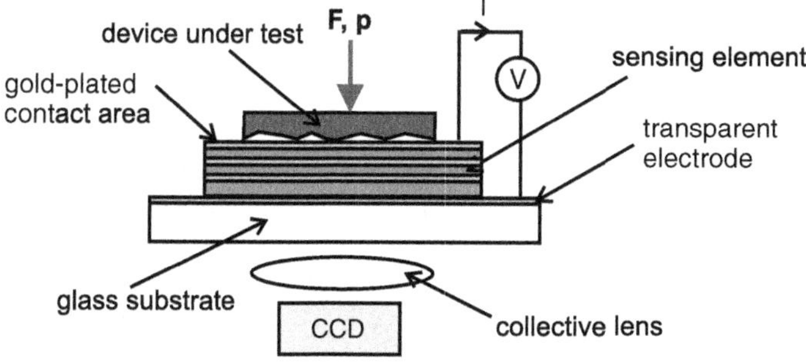

Fig. 10.37 Schematic view of an electroluminiscence sensor [79]

measured. The nominal force lies in the area of 2 N, the resolution of the system is unknown.

Electroluminescence Sensors

A high-resolution touch sensor is presented by SARAF [79]. It is thought to be used for the analysis of texture on organ surfaces. Figure 10.37 shows a schematic sketch. On a transparent glass substrate, a layer-compound of 10 μm height made of gold- and cadmium sulfite particles[11] is applied. The single layers are separated by dielectric barriers. The mechanical load is applied on the upper gold layer, resulting in a breakthrough of the dielectric layer and a current flow. Additionally, energy is released in form of small flashes. This optical signal is detected using a CCD-camera. The signal is directly proportional to the strain distribution generated by the load. The resulting current density is measured and interpreted.

The spatial resolution of the design is given with 50 μm. Nominal pressures of around 0.8 bar can be detected. The sensor area has a size of 2.5×2.5 mm^2, the thickness of the sensor is ≤ 1 mm and thus very thin. Additional information can be found in [79].

10.1.3 Selection of a Suitable Sensor

In earlier sections, sources for the requirements identification have been presented. Afterward, presentation and discussion of the most relevant sensor principles to measure forces were made. This section is intended to help engineers to select or even develop an appropriate force sensor. Depending on the identified requirements found using Sect. 10.1.1, a suitable sensor principle can be chosen.

[11] A semiconducting material.

Table 10.4 Compilation of main requirements on haptic sensors

Type of information	Requirements	Values
Kinaesthetic	Nominal load F_N	(5–100) N
	Resolution ΔF	5 % F_N
	Frequency range	(0–10) Hz
Tactile	Nominal load F_N	$\leq$0.3 N or $\leq$4.5 N
	Resolution ΔF	5 % F_N
	Frequency range	(0–1,000) Hz
	Spatial resolution Δx	Structural dependent, $\geq$0.5 mm

Depending on system topology and measurement task further requirements have to be considered

To get a better overview, the basic requirements described in Sect. 10.1.1.5 are collected in Table 10.4. The requirements are distinguished concerning human perception in kinaesthetic and tactile information. More detailed information concerning force and spatial resolution can of course be found in Sect. 10.1.1. The properties of active and passive transformers—force measurement is done via a mechanical variable such as strain or stress detected via elastomechanics—are strongly dependent on the design of the deformation element. Especially the nominal force, number of components to be measured, and the dynamics are directly influenced by the deformation element's design.

A comparison of all sensor principles can hardly be done. Consequently, the methods will be compared separately from each other. As evaluation criteria, transfer characteristics and geometrical dimensions are chosen. Figure 10.38 classifies the principles according to gage factor and geometry. According to the increasing size of the strain-sensing element, the whole-force sensor can be designed at a higher level

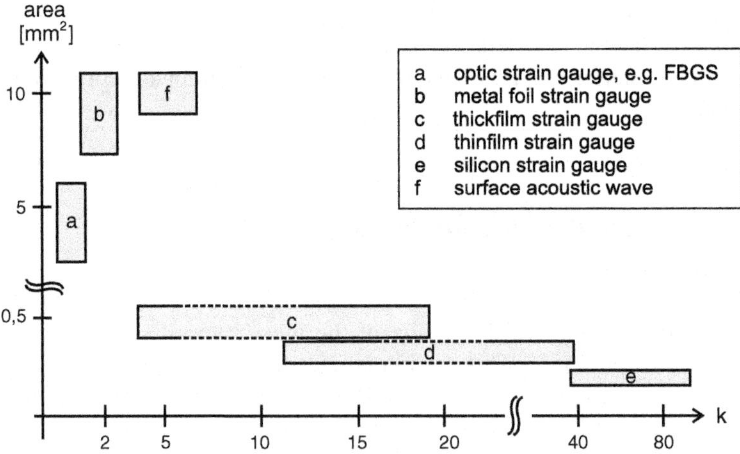

Fig. 10.38 Comparison of different strain measurement technologies due to dimensions and gage factor

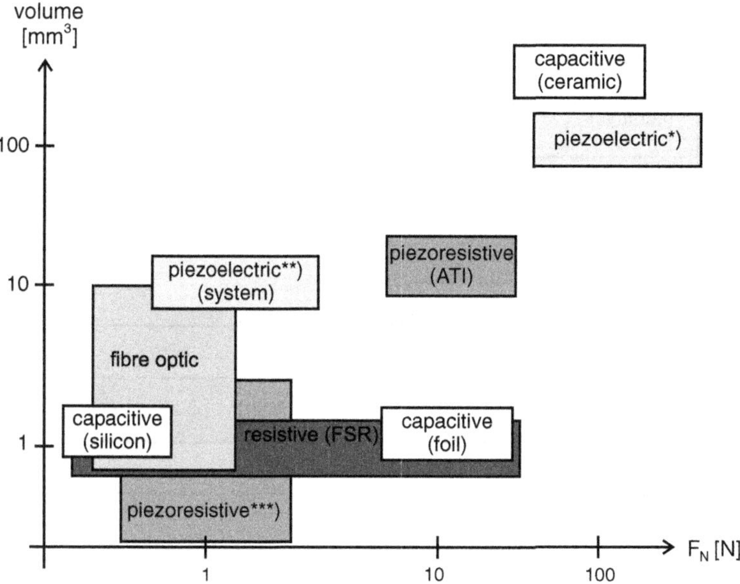

Fig. 10.39 Comparison of different measurement technologies due to dimensions and nominal load. * Commercial force/torque sensors. ** Active sensor systems classified as exotic sensors. *** Tactologic system

of miniaturization. A direct result of smaller size is the minimized mass, providing an increased upper cutoff frequency. If the gage factor of the sensing element is higher, lower absolute value of strain is necessary to get a high output signal. Additionally, the overall design can be designed stiffer. This enables to detect smaller nominal forces and thus higher cutoff frequencies. Concerning the lower cutoff frequency, strain-sensing elements are suitable for measuring static loads. Using piezoresistive and capacitive silicon sensors, an upper cutoff frequency of 10 kHz or more can be measured with high resolution.

The other sensor principles can be compared contingent on nominal load and dimensions. Figure 10.39 classifies the presented principals according to their nominal load and corresponding construction space.

Except the piezoelectric sensors, all sensor principles can be used for measuring static and dynamic loads. The upper cutoff frequency mainly depends on the mass of the sensor which has to be moved. Consequently, the more miniaturized the sensor, the higher the upper cutoff frequency becomes. Figure 10.40 compares the presented sensor principles according to the detectable nominal load and the corresponding dynamic range.

By means of the shown diagrams a preselection of suitable sensor principles for the intended application can be done. Additional sensor properties such as resolution, energy consumption, costs, or impact of noise are strongly depending on the individual realization and will not be taken into account here. Advanced descriptions of

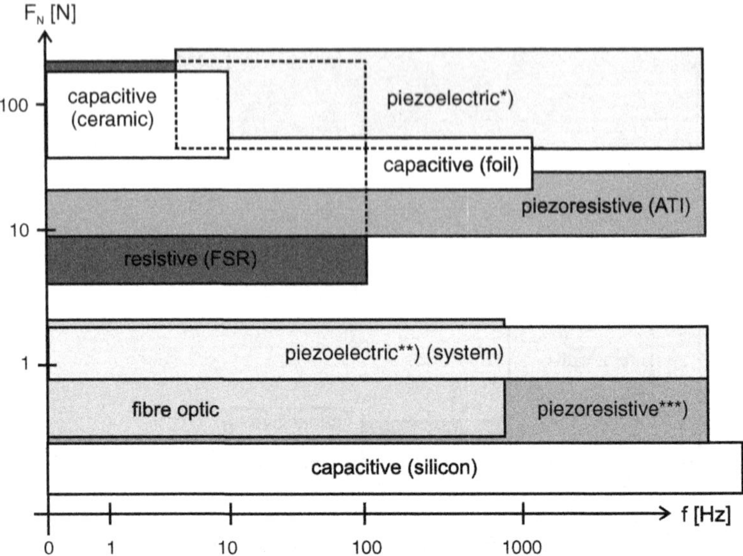

Fig. 10.40 Comparison of different measurement technologies due to nominal load and frequency range. ∗ Commercial force/torque sensors. ∗∗ Active sensor systems classified as exotic sensors. ∗∗∗ Tactologic system

sensor properties have to be taken from the literature highlighted in the corresponding subsections for the individual principle.

To give an example of how to select a suitable force sensor, the task *laparoscopic palpation of tissue* is chosen. Figure 10.41 shows the tree diagram which can be used for analyzing the task and deriving requirements. Laparoscopic palpation is a telemanipulation task for characterizing texture. It is done via closed-loop control. To avoid undesired influences of the laparoscopic instrument itself onto the sensing signal (e.g., friction between instrument and abdominal wall), the sensor should be integrated into the tip. The laparoscope is used to scan the tissues surface. Detecting three directions of contact force, texture and even compliance of tissue can be analyzed.

Taking contact information into account (see Table 10.4) cutoff frequency, resolution, and nominal force can be derived. The dimensions of the laparoscope limit the construction space. Also static information has to be measured, thus an active sensing principle-like (piezo-)resistive, capacitive, inductive, or optic should be considered. Due to limited space, piezoresistive sensing is recommendable.

If no force sensor with the determined requirements is available, a deformation element has to be designed separately taking load condition and elastomechanics into account. For example, [11, 72] are helpful references for designing deformation elements. The strain-sensing element can be chosen depending on the aimed resolution and construction space. Table 10.5 gives an overview of common strain-sensing technologies.

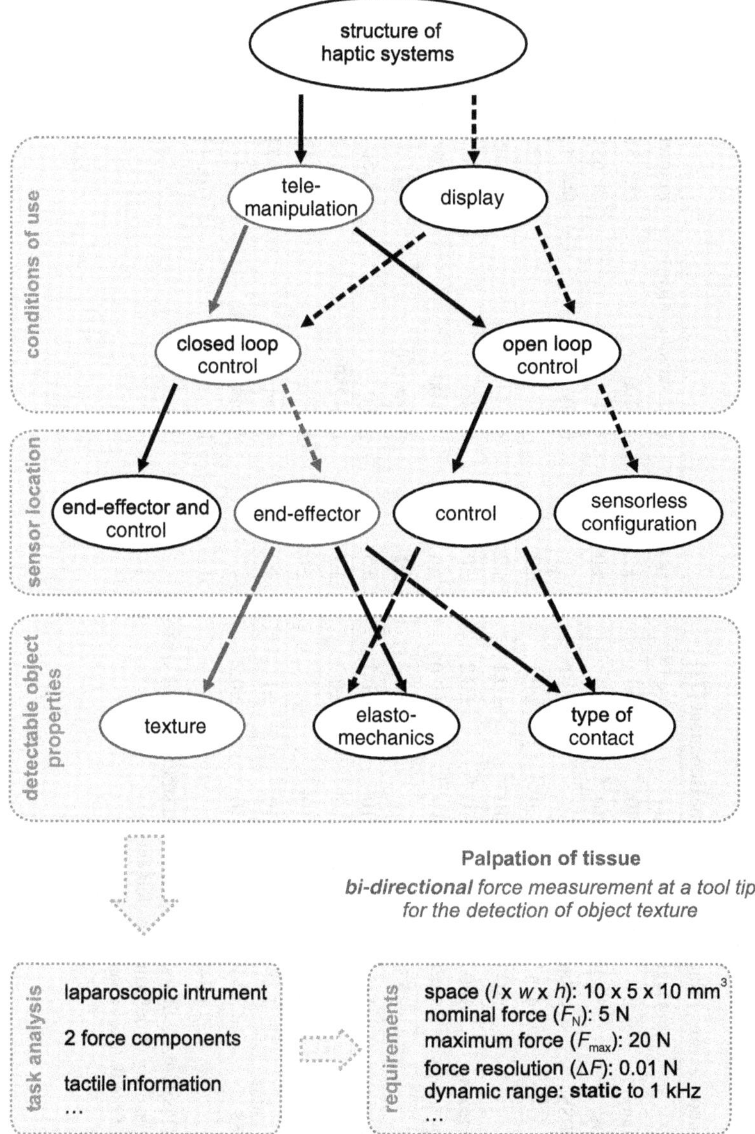

Fig. 10.41 Tree diagram for selecting a force sensor. Exemplarily, the task *laparoscopic palpation of tissue* is chosen

Table 10.5 Comparison of common sensing principles for strain measurement [72]

Principle	Material	Gage factor	S_{min}	S_N (%)	S_{max} (%)	Thickness h_M (μm)	References
Foil-strain gage	Constantan	2.0	$\pm 10^{-7}$	±0.1	±2	80–100	[32, 35, 91]
Thick-film	$Bi_2Ru_2O_7$	12.1–18.3	$\pm 10^{-6}$	±0.1	±0.2	≤50	[3, 65]
	PEDOT:PSS	0.48–17.8	–	≥10[a]	–	1–10	[44, 46]
Thin-film[b]	TiON	4–5	$\pm 10^{-7}$	±0.1	±2	≤1	[3, 42, 105]
	Poly-Si	20–30	$\pm 10^{-7}$	±0,1	±2	≤1	[3, 105]
Si-Technology	Homogeneous	100–255	$\pm 10^{-6}$	±0.2	±0.3	10–15	[58]
	Inhomogeneous	80–255	$\pm 10^{-7}$	±0.05	±0.1[c]	17–100	[5, 23]
Capacitive	PVDF[d]	≤83	–	±0.3	±1	About 150	[2]
	Inter-digital	1–5	$\pm 10^{-7}$	±0.1	±5	≤500	[3, 50, 94]
FBGS	–	0.78	$\pm 10^{-7}$	±0.2	±1	150–250	[77, 82]
Piezoelectric	PZT[e]	$\leq 2 \times 10^6$	$\pm 10^{-10}$	–	±0.1	200	[25, 69]
SAW	Quartz	1.28–20	$\pm 10^{-7}$	±1	±2	About 600	[57, 74, 97, 105, 109]
Magneto-elastic	NiFe45/55	About 1,500	$\pm 10^{-8}$	±0.2	–	About ≤80 μm	[1, 47, 66]
Fibers	Carbon	1.3–31	–	–	0.2–15	12	[16, 43]

[a] Depending on substrate, in [14] polymeric fibers with a maximum strain of 10 % are used
[b] Maximum strain depends on elasticity of deformation element
[c] According to [51] elongation of break takes a value of ±0.2 %
[d] $d\varepsilon/\varepsilon$, sandwich-topology
[e] Patch transducer, PI

10.2 Positioning Sensors

Thorsten A. Kern

To acquire the user's reaction in haptic systems, a measurement of positions, respectively and their time derivatives (velocities and accelerations) is necessary. Several measurement principles are available to achieve this. A mechanical influence of the sensor on the system has to be avoided for haptic applications, especially kinaesthetic ones. Consequently, this discussion focuses on principles which do not affect the mechanical properties significantly. Beside the common optical measurement principles, the use of inductive or capacitive sensors is promising especially in combination with actuator design. This chapter gives an overview about the most frequently used principles, amended by hints for their advantages and disadvantages when applied to haptic systems.

10.2.1 Basic Principles of Position Measurement

For position measurement, two principle approaches can be distinguished: differential and absolute measuring systems.

10.2.1.1 Incremental Principle

Differential systems acquire the change in discrete steps together with the direction of change, and protocol (typically: count) these events. This protocol has to be set back to a reference position by an external signal. If no step loss happens during movement, a prior initialized differential system is able to provide the absolute position as output. If this initializing reference position is set in point which is passed often, a differential system will be referenced frequently during normal operation. Potential step losses would then affect the time till the next initializing event only.

Measurement of the steps is done via a discrete periodic event, typically encoded in a code disc with grooves or a magnetic rotor. This event is transformed by the sensor in a digital signal, whose frequency is proportional to the velocity of the movement (Fig. 10.42a). Some additional directional information is required to be able to measure the absolute position. A typical solution for this purpose is the use of two identical event types with a phase shift (between 1 and 179°, typically 90°. By looking at the status (*high/low*) of these incremental signals (Fig. 10.42b) at, e.g., the rising edges of the first incremental signal (A), a *low* encodes one movement direction, and a *high* encodes the opposite movement direction. Accordingly, the count process either adds or subtracts the pulses generated—in this case—by the second signal (B). State-of-the-art microcontrollers are equipped with counters for incremental measurement already. They provide input pins for count-signal and count-direction. Discrete counters are sold as "Quadrature-Encoder" ICs and

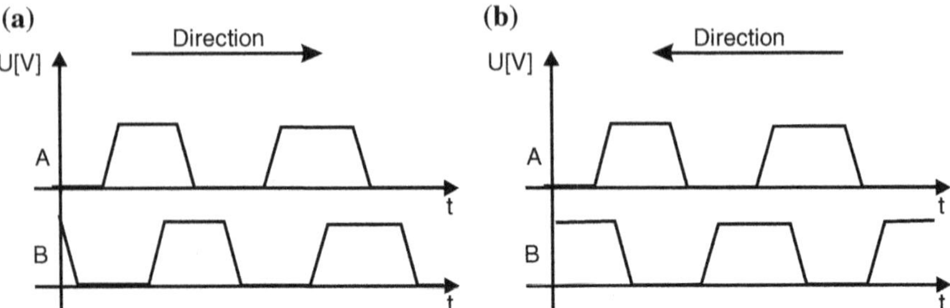

Fig. 10.42 Principle of direction detection with two digital signals with a 90° phase-lag

frequently include actuator drive electronics, which can be applied for positioning tasks. Latter prevents them from being useful for typical haptic applications.

10.2.1.2 Absolute Measurement Principle

Absolute measurement systems acquire a position- or angle-proportional value directly. They are usually analog. A reference position for these systems is not necessary. They have advantages with reference to their measurement frequency, as they are not required to measure with dynamics defined by the maximum movement velocity. The acquisition dynamics of incremental principles is given by the necessity not to miss any events. In case of absolute measurement principles, the measurement frequency can be adjusted to the process-dynamics afterward, which is usually less demanding. However by the analog measurement technology the efforts are quite high for the circuit, the compensation of disturbances, and the almost obligate digitization of the analog signal.

An alternative for the pure absolute measurement with analog technology is given by a discrete absolute measurement of defined states. In Sect. 9.2.2.1, Fig. 9.15, a commutation of EC-drives with a discrete, position coding of magnet-angles with field plates was already shown. This approach is based on the assumption to achieve a discrete resolution of ΔD from m measurement points with n states by

$$n^m = \Delta D. \tag{10.40}$$

In case of the commutated EC-drive $m = 3$ measurement points, which are able to have $n = 2$ states, could encode eight positions on the circumference, but only six were actually used. But there are other more complex code discs with several lanes for one sensor each. These sensors are usually able to code two states. However, e.g., by the use of different colors on the disc many more states would be imaginable. A resolution of, e.g., 1degree (360 discrete steps) would need the number of

$$m = \frac{\log(\Delta D)}{\log(n)} = 8.49 \tag{10.41}$$

at least nine lanes for encoding.

10.2.2 Requirements in the Context of Haptics

Position measurement systems are primarily characterized by their achievable resolution and dynamics. For haptic devices, in dependence on the measurement basis for computer mice and scanners, position resolutions are frequently defined as dots-per-inch ΔR_{inch}. Consequently, the resolution ΔR_{mm} in metric millimeters is given as:

$$\Delta R_{\text{mm}} = \frac{25,4\,\text{mm} \times \text{dpi}}{\Delta R_{\text{inch}}}. \tag{10.42}$$

A system with 300 dpi resolution achieves an actual resolution of 84 μm. In dependency on the measurement principle used, different actions have to be taken to achieve this measurement quality. With incremental measurement systems, the sensors for the acquisition of single steps (e.g., holes on a mask) are frequently less resolutive, requiring a transformation of the user's movement to larger displacements at the sensor. This is typically achieved by larger diameters of code discs and measurement at their edge. These discs are mounted on an axis, e.g., of an actuator. With analog absolute systems, an option for improving the signal is conditioning. It is aimed at reducing the noise component in the signal relative to the wanted signal. This is usually done by a suppression of the noise source (e.g., ambient light), the modulation, and filtering of the signal (e.g., lock-in amplifier, compare Sect. 10.2.6) or the improvement of secondary electronics of the sensors (high-resolution A/D-transformer, constant reference sources).

Beside the position measurement itself, its dynamic has to be considered during the design process. This requirement is relevant for incremental measurement systems only. Absolute measurement systems need a bandwidth equal to the bandwidth provided by the interface and the transmission chain (Chap. 11) for positioning information. Incremental measurement systems, however, have to be capable of detecting any movement event, independent from the actual movement velocity. The protocol format, usually given by counters part of the microcontrollers, has to be dimensioned to cover the maximum incremental frequency. This requires some assumptions for the maximum movement velocity v_{max}. If a system with 300 dpi position resolution moves at a maximum velocity of 100 mm/s, the dynamic f_{ink} for detecting the increments is given as

$$\frac{1}{f_{\text{ink}}} = \frac{\Delta R_{\text{mm}}}{v_{\text{max}}} \tag{10.43}$$

For the example, the necessary measurement frequency is given with $f_{\text{ink}} = 1{,}190\,\text{Hz}$. The effective counting frequency is usually chosen with factor two to four higher than that to have a security margin for counting errors and direction detection.

10.2.3 Optical Sensors

Optical sensors for position measuring are gladly and frequently used. They excel by their mechanical robustness and good signal-to-noise ratios. They are cheap and in case of direct position measurement quite simple to read out.[12]

Code Discs

Code discs represent the most frequently used type of position measurement systems with haptic devices, especially within the class of low-cost devices. They are based on transmission (Fig. 10.43a) or reflection of an optical radiation, which is interrupted in discrete events. The necessary baffle is located near to the receiver. It is manufactured by stamping, or printed on a transparent substrate (glass, plastic material) via thick-film technology or laser printers. For high requirements on resolution, they are made of metal, either self-supportive or on a substrate again. In these cases, the openings are generated by a photolithographic etching process. The receivers can be realized in different designs. Figure 10.43 shows a discrete design with two senders in form of diodes and two receivers (photodiode, phototransistor). The placement of sender/receiver units have to allow the phase shift for directional detection (see Sect. 10.2.1.1). An alternative is given by fork light barriers already including a compact sender/receiver unit. Additionally opto-encoders (e.g., HLC2705) exist including the signal conditioning for direction detection from the two incremental signals. The output pins of these elements provide a frequency and one signal for the direction information.

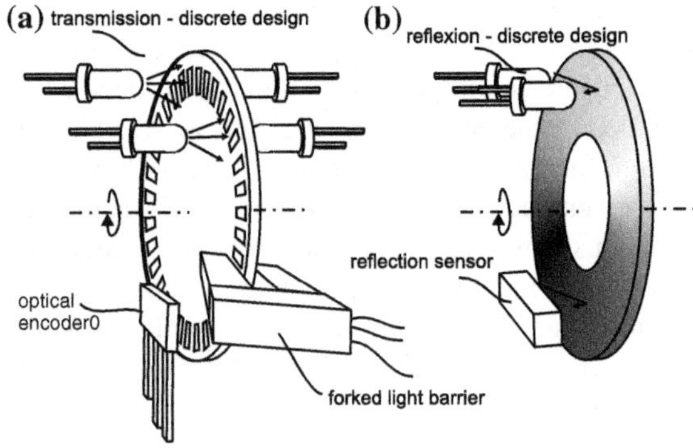

Fig. 10.43 Incremental optical position measurement (**a**), and absolute position measurement via grayscale values (**b**)

[12] The examples presented here are discussed either for translatory and rotatory applications. But all principles may be applied to both, as a translation is just a rotation on a circle with infinite diameter.

Gray Scale Values

With similar components, but for absolute measurement a grayscale disc or grayscale sensor can be built. Once again there are transmission and reflection (Fig. 10.43b) variants of this sensor. In any case the reflection/transmission of the radiation varies depending on the angle or position of a code disc. The amplitude of the reflection gives absolute position information of the disc. For measurement, once again, either a discrete design or the usage of integrated circuits in the form of the so-called reflection sensors is possible. Although such sensors are frequently used as pure distance switches only, they show very interesting proportional characteristics between the received numbers of photons and their output signal. They are composed of a light emitting diode as sender and a phototransistor as receiver. In some limits, the output is typically given by a linear proportional photoelectric current.

Reflection Light Switches

Reflection light switches show useful characteristics for a direct position measurement too. In the range of several millimeters, they have a piecewise linear dependency between photocurrent and the distance from the sensor to the reflecting surface. Consequently, they are useful as sensors for absolute position measurement of translatory movements (Fig. 10.44a). By this method, e.g., with the SFH900 or its SMD successor SFH3201 within a near field up to ≈ 1 mm measurement inaccuracies of some micrometers can be achieved. In a more distant field up to 5 mm, the sensor is suitable for measurement inaccuracies of $\frac{1}{10}$ mm still.

Mice-Sensor

The invention of optical mice without ball resulted in a new sensor type interesting for other applications too. The optical mice sensors are based on an IC measuring an illuminated surface through an optic element (Fig. 10.44b). The resolution of the CMOS sensors typically used range from 16×16 to 32×32 pixels. By the image

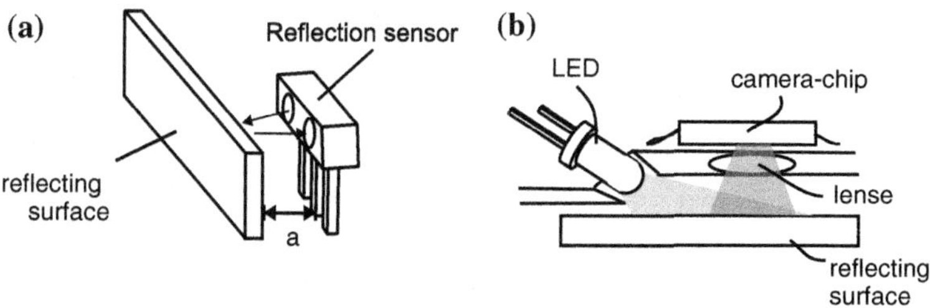

Fig. 10.44 Distance measurement with reflection light switches (**a**), and via the movement of an reflective surface in two DoFs "mouse-sensor" (**b**)

acquired the chip identifies the movement of contrast difference in their direction and velocity. The interface of the calculated values varies from sensor to sensor. The very early types provided an incremental signal for movements in X- and Y-direction identical to approaches with code discs described above. They additionally had a serial protocol included to read the complete pixel information. Modern sensors (e.g., ADNB-3532) provide serial protocols for a direct communication with a microcontroller only. This allowed a further miniaturization of the IC and a minimization of the number of contact pins necessary. The resolution of state-of-the-art sensor is in between 500 and 1000 dpi and is usually sufficient for haptic applications. Only the velocity of position output varies a lot with the sensor types available at the market, and has to be considered carefully for the individual device design. The frequency is usually below 50 Hz. Additionally, early sensor designs had some problems with drift and made counting errors, which could be compensated only by frequent referencing.

The sensors are usually sold for computer-mouse-similar applications and corresponding optics. Besides that it is also possible to make measurements of moving surfaces with an adapted optic design at a distance of several centimeters.

Triangulation

Optical triangulation is an additional principle for contactless distance measurement; however, it is seldom used for haptic devices. A radiation source, usually a laser, illuminates the surface to be measured, and the reflected radiation is directed on different positions along a sensor array (Fig. 10.45). The sensor array may be made of discrete photodiodes. Frequently, it is a CCD or CMOS row with the corresponding

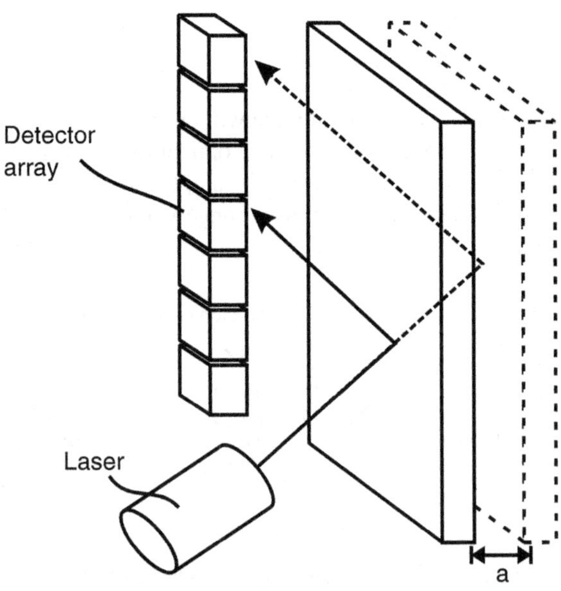

Fig. 10.45 Triangulation of a distance with laser-diode and detector array

high resolution. By focal point identification weighting of several detectors, a further reduction of measurement inaccuracy can be achieved. Compared to other optical sensors, triangulation sensors are expensive as the detection row with a sufficient resolution is a high-cost factor. Their border frequency ($\gg$1 kHz) and their measurement inaccuracy ($<$10 μm) leave nothing to be desired. It is one of the very few principles which can hardly be used for measuring rotating systems.

10.2.4 Magnetic Sensors

Beside the optical measurement principles, especially the group of magnetic measurement principles is relevant for haptic devices. This is a consequence from the fact that electrodynamic and electromagnetic actuators already require magnetic fields to generate forces. For systematization, sensors for static fields, field plates, and hall-sensors, and sensors for induced currents and time-dependent fields can be distinguished.

Field Plates or Magnetic-Dependent Resistors

Field plates or magnetic-dependent resistors (MDR) are two pole elements with the resistance being controlled by the presence of a magnetic field. They make use of the GAUSS effect, which is based on charge carriers being displaced by the LORENTZ force when crossing a magnetic field. The resulting increase of the path length [61] requires an increase of the ohmic resistance of the material. The parameter characterizing this dependency is dependent on the electron mobility and the path length in the magnetic field. A frequently used material is InSb with very high electron mobility. For an additional increase of the effect, the conductor is formed like in the shape of a meander similar to strain gages. MDRs are not sensitive to the polarity of the magnetic field. They are detecting the absolute value only. The increase of resistance is nonlinear and similar to a characteristic curve of a diode or transistor. A magnetic bias is recommended when using the plates to make sure they are in their linear working point.

Hall-Sensors

Hall-sensors are based on the GAUSS effect too. In contrast to field plates, they are not measuring the resistance increase of the current within the semiconductor, but the voltage orthogonal to the current. This voltage is a direct result of the displacement of the electrodes along the path within the material. The resulting signal is linear and bipolar in dependency on the field-direction. ICs with an integrated amplifier electronics and digital or analog output signals can be bought off the shelf. A frequent use can be found with sensors being located at a phase angle α with diametral magnetized rotational magnets (Fig. 10.46). In this application, rotation and rotation-direction are measured.

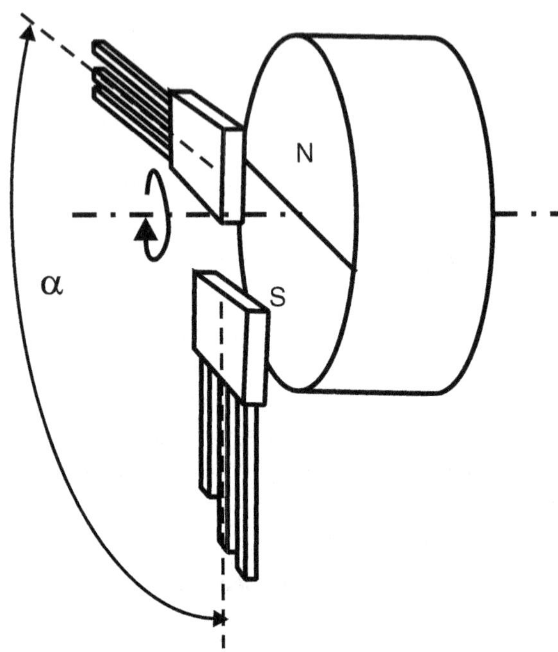

Fig. 10.46 Measurement of the rotation angle of a magnet via field plates or hall-sensors

Inductance Systems

An often forgotten alternative for position measurement is the measurement of changing inductances. The inductance of a system is dependent on many parameters, for example, the magnetic permeability of a material in a coil. Using a differential measurement in between two coils (Fig. 10.47b), a displacement measurement can be made, if a ferromagnetic material moves in between both coils as a position-depending core. As alternatives, the geometry of the magnetic circuit may be changed or its saturation may influence the inductance of the coils. Latter approach is used in systems, where grooves on a ferromagnetic material trigger events in a nearby coil (Fig. 10.47a).

A simple electronic method for measuring inductance is the use of a LR-serial circuit, which—for example with a microcontroller—is triggered with a voltage step. The measurement value is given by the time the voltage at the resistor needs to trigger a comparator voltage. The duration encodes the inductance, assuming a constant resistance. For the actual design, it has to be considered that the winded coil has an own resistance which cannot be neglected. As an alternative a frequency nearby the resonance $\frac{L}{R}$ of the LR-circuit can be applied. The voltage amplitude measured varies dependent on the inductance detuned by the movement of the ferromagnetic core.

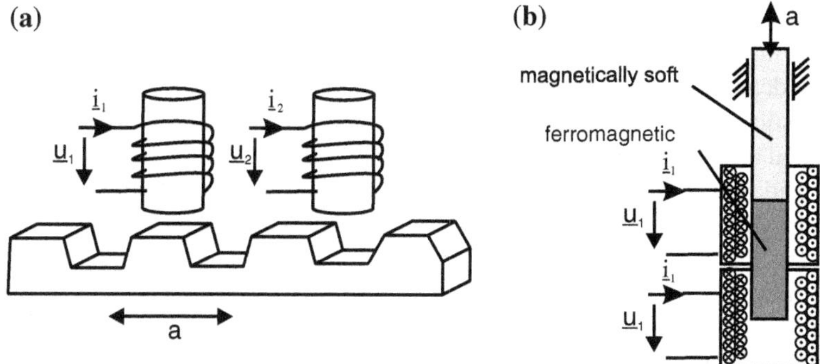

Fig. 10.47 Incremental measurement of a movement via induced currents (**a**) and differential measurement of the position of a ferromagnetic core (**b**)

10.2.5 Other Displacement Sensors

Beside the displacement measurement principles discussed above, there are some rarely used principles still worth to be mentioned here.

Ultrasonic Sensors

Ultrasonic sensors (Fig. 10.48) are based on the running time measurement in between the emission of acoustic oscillations and the moment of the acquisition of their reflection. The frequency chosen is dependent on the requirements on measurement accuracy and the medium for propagation of the wave. As a rough rule of thumb, the denser a material is, the less the damping becomes for acoustic waves. For measurement in tissue frequencies between 1 and 40 MHz are applied. In water frequencies between 100 and 500 kHz and in the atmosphere frequencies well below 30 kHz are used.

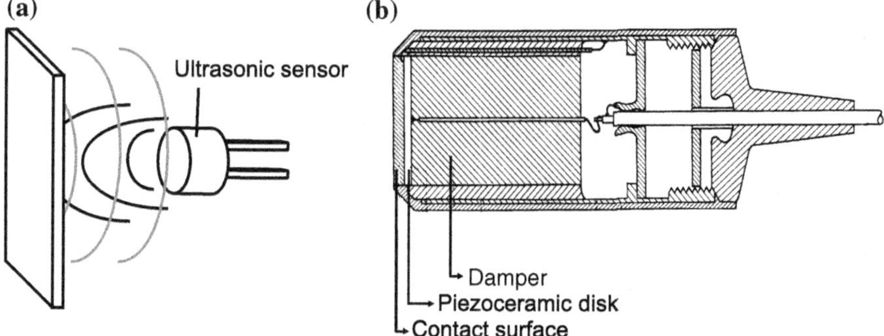

Fig. 10.48 Distance measurement via ultrasonic sensors (**a**) and cross section through a medical ultrasonic head with fixed focus (**b**)

Whereas with medical applications in tissues the medium shows a damping quite linear in the range of 1 dB/Mhz/cm, the measurement within the atmosphere is strongly dependent on the frequency chosen and usually nonlinear. Additionally, the acoustic velocity is dependent on the acoustic density of the medium. For the transversal direction—typically used for measurement—velocities between 340 m/s for air and 1,500 m/s for water can be achieved. According to the wave theory, the minimum measurement accuracy possible in transversal direction is $\frac{\lambda}{2}$, which is coupled to both factors mentioned above. It is a natural border of minimum resolution to be achieved.

The most frequently used source and receiver for the mechanical oscillation are piezoelectric materials (Fig. 10.48b), whose step response oscillations are sharpened by a coupled mass.

Capacitive Sensors

In Sect. 9.5, the equations for the calculation of capacities between plates of electrostatic actuators (Eq. 9.75) were introduced. Of course the measurement of a variable capacity, especially with the linear effect of a transversal plate displacement, can be used for position measurement. This is especially interesting if there are conductive components in the mechanical design, which already move relative to each other. As the capacity is very much dependent on the permittivity of the medium between the plates, which can be strongly influenced by oil or humidity, such a measurement can be done on insusceptible or other well-housed actuators only. Additionally, leakage fields of conductors or geometries nearby are usually of the same size as the capacity to be measured. But capacitive sensors for haptic devices can be found in the context of another interesting application. The measurement of the capacity of the handle, even when isolated by a nonconductive layer, allows identifying a human touch very securely.

10.2.6 Electronics for Absolute Positions Sensors

The absolute measurement of a position requires, as mentioned earlier, some additional effort in the electronic design compared to discrete sensors. Two aspects shall be discussed in the context of this chapter.

Constant Current Supply and Voltage References

For the generation of a constant radiation or the measurement of a bridge circuit, the use of constant currents is necessary. There is always the possibility to wire an operational amplifier as a constant current source, or use transistor circuits. Nevertheless for designs with low quantities, there are ICs which can be used as current sources directly. The LM234, for example, is a voltage-controlled three-pin IC, providing a current with a maximum error of 0.1 %/V change in the supply voltage. The maximum current provided is 10 mA, which is usually sufficient for the supply of optical or resistive sensors.

The change of the signal is usually measured in relation to a voltage in the system. In this case, it is necessary to provide a voltage which is very well known and independent from temperature effects or changes of the supply voltage. Common voltage regulators as used for electronic supply are not precise enough to fulfill these requirements. An alternative is given by Zener diodes operated in reverse direction. Such diodes, however, are not applicable to high loads and are of course only available in the steps of Zener voltages. Alternatively, reference voltage sources are available in many voltage steps on the market. The REF02, for example, is a six-pin IC, providing a temperature-stable voltage of 5 V with an error of 0.3 %. The drivable load of such voltage sources is limited, in case of the REF02 it is 10 mA, but this is usually not a relevant limit as they are not thought as a supply to a complex circuit but only as a reference.

Compensation of Noise

The obvious solution for the compensation of noise in a measurement signal is given by the usage of a carrier frequency for modulating the signal. A prerequisite of course is given by the sensor showing no damping at the modulating frequency. This is usually no problem for optical sensors in the range of several kilohertz. At the receiver, the signal is bandpass-filtered and equalized or otherwise averaged. This suppresses disturbance frequencies or otherwise superimposed offsets.

A simple but very effective circuit for noise compensation is the use of so-called lock-in amplifiers. On the side of the sender, a signal is switched between the states *on* and *off* at a frequency f. In the receiver, the wanted signal such as, e.g., the offset and other disturbing frequencies are received. A following amplifier is switched with the same frequency between $+1$ and -1 in such a way, that with the receipt of the wanted signal including the disturbance the positive amplification happens. During the period without the wanted signal, when the receiver measures the disturbance only, the signal is inverted with -1. The resulting signal is low-pass filtered afterward, resulting in a subtraction of the noise signal and providing a voltage proportional to the wanted signal only.

10.2.7 Acceleration and Velocity Measurement

Beside the direct position measurement, haptic systems sometimes demand some knowledge about the first or second derivative of position in form of velocity or acceleration. Such a necessity may be given with stability issues for closed-loop systems or impedance behavior of users or manipulated objects. The acquisition can either be done by direct measurement or by differentiation of the position-signal with digital or analog circuits. Additionally, it can be imagined to, e.g., measure velocity and calculate the position by integration. The capabilities of integration and differentiation and their limits, such as typical direct measurement principles, are sketched in this section.

10.2.7.1 Integration and Differentiation of Signals

The integration and differentiation of signals can either be done analog or digital. Both variants have different advantages and disadvantages.

Analog Differentiation

The basic circuit for an active analog integrator is shown in Fig. 10.49a. It is a high-pass filter, which already gives hints on the challenges connected with differentiation. The high-pass behavior is limited in its bandwidth. The upper border frequency is given by the resonance frequency $f_R = \frac{1}{2\pi RC}$ and by the bandwidth of the operational amplifier. As these components are sufficiently dynamic for haptic applications, this should be no problem in practical realization. Due to the negative feedback, however, the natural bandwidth limit of the operational amplifier at high frequencies has a phase of 90° adding to the phase of 90° from the differentiation. This makes the circuit sensitive to become electrically instable and oscillate [96].

This effect can be compensated by a serial resistance with a capacity C, which is identical to a linear amplification with the operational amplifier. This diminishes the phase for high frequencies by 45° resulting in a phase margin to the instable border condition. Analog differentiation is an adequate method for the derivation of velocities from positioning signals. A double analog differentiation needs a careful design of the corresponding circuit, as a number of capacitive inputs are placed in series. Additionally, it should be considered that the amplitude of the operational amplifier is limited by the supply voltage. Accordingly, the amplitude's dynamic has to be adjusted to the maximum signal change expected.

Analog Integration

The basic circuit of an active analog integrator is given in Fig. 10.49b. Analog integration is a reliable method from analog calculation technique, but has limited use for haptic applications. The circuit has an upper border frequency given by the resonance $f_R = \frac{1}{2\pi RC}$, and for a nonideal operational amplifier, it has a lower border frequency

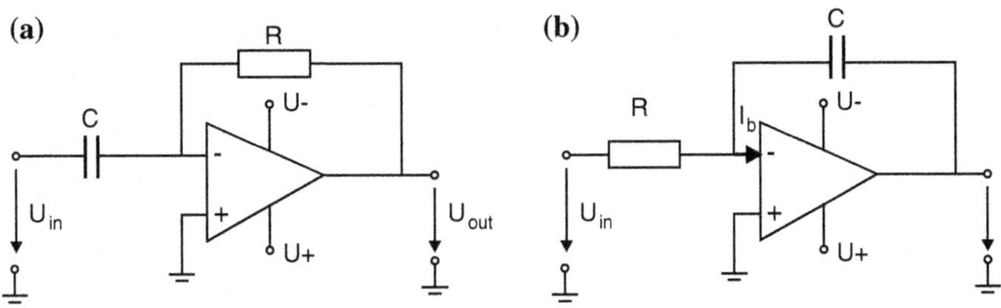

Fig. 10.49 Analog circuits for **a** differentiation and **b** integration [96]

too. This is a result of the current I_b at the input of the OP-amplifier charging the capacitor with $U_{\text{in}} = 0$ V continuously. If $C = 10\ \mu\text{F}$ and $I_b = 1\ \mu\text{A}$, the voltage increases by 0.1 V per second. Whereas in signal processing applications, this can be compensated by high-passes in series, for haptic applications covering a bandwidth from several seconds to 10 kHz this behavior is usually not acceptable.

Digital Differentiation

Digital differentiation is realized by a subtraction of two consecutive measurement values. It is very applicable, especially when the signal is measured at high frequencies. The quality of the signal is dependent on the noise on the input. Frequently, the least-significant bit of, e.g., an AD-conversion is rejected before the differentiation is performed, as it is oscillating with the noise of the AD-conversion (quantization-noise). To derive velocity from position measurements, COLGATE recommends a high-position measurement resolution as well as a low-pass filtering of the generated velocity signal to improve the quality [17].

Digital Integration

Digital integration is the summation of continuous measurement values and the division of the sum by the number of values. Alternatively, it can be the sum of discrete changes of a measurement value. The incremental measurement of a digital encoder is also a form of integration on the basis of change information. The procedure is robust at high frequencies beyond the actual upper border frequency of the signal. Beside a sufficient dimension of the register size for the measurement values to prevent an overflow, there is nothing else to worry about.

10.2.7.2 Induction as a Velocity Measure

The most frequent variant to gain information about velocity is given by the digital signal processing of a position measurement. Nevertheless to be able to measure velocity directly, the use of a velocity-proportional physical effect is mandatory. Beside Doppler-ultrasonic measurement, which is seldom applicable to haptic systems due to the wavelengths (compare Sect. 10.2.6), the use of electrical induction is the most frequently used direct effect. Accordingly, an electrical-induced voltage U is generated in a conductor of the length l, moving orthogonal in a magnetic field B with the velocity v:

$$U = v\,B\,l. \tag{10.44}$$

Special geometric designs as given with electrodynamic actuators (Sect. 9.2) can be used for velocity measurements with inducing voltages in their coils. In contrast to electrodynamic actuators, the design requires a maximization of conductor length, to generate a pronounced voltage signal. The inductivity of the winding generates a low-pass characteristic in combination with its own resistance. This limits the dynamic of the signal. The biggest error made with these kinds of sensors is given by a bad homogeneity of the winding due to dislocation of single turns. This manufacturing error results in different winding lengths moving in the B-field at different positions of the sensor, which is directly affecting the quality of the measured signal.

10.2.7.3 Force Sensors as Acceleration Sensors

In contrast to velocity measurement, to measure accelerations a wide variety of sensors exists. Ignoring some exceptions, most of them are based on the relation

$$\mathbf{a} = \frac{\mathbf{F}}{m}. \tag{10.45}$$

In fact, the force measurement principles given in Sect. 10.1 are added by a known mass m only, resulting in a mechanical strain of a bending element or generating another acceleration-proportional signal. In professional measurement technology especially piezoelectric sensors for high-dynamic measurements, but also piezoresistive sensors for low-frequency accelerations are established. In mechatronic systems with high quantities, micromechanical acceleration sensors with comb-like structures in silicon according to the capacitive measurement principle are used. The requirements of automotive industry for airbags and drive stability programs to measure acceleration in many directions made low-price and robust ICs available at the free market, e.g., the ADXL series of Analog Devices. The bandwidth of these sensors ranges from 400 Hz to 2.5 kHz with maximum accelerations >100 g in up to three spatial directions. Only a wide variance of their characteristic values, e.g., the output voltage at 0 g, requires a calibration of the individual sensor.

10.2.8 Conclusion on Position Measurement

With haptic devices, position measurement is a subordinated problem. In the range of physiological perceived displacements resolutions, there are enough sensor principles which are sufficiently precise and dynamic for position measurement. The calculation or measurement of accelerations or velocities is easily possible too. Without doubt, the optical measurement technology is the most frequently used technical solution. Nevertheless especially for the design of specific actuators it is indicated to ask the questions, whether there are other sensor principles applicable for a direct integration into the actuator.

If there are specific requirements for measurement in the range of a few μm positioning resolution, the proposed principles should be treated with reserve. Measurements in the range of μm require specific optical or capacitive measurement technology. With the exception of special psychophysical questions, it is unlikely that such requirements are formulated for haptic devices.

10.3 Touch Sensors

Christian Hatzfeld

With the increasing number of systems using touch-sensitive surfaces for ↪HCI, touch sensors have become more prevalent. They detect whether a human user touches a sensitive two- or three-dimensional surface of an object or system. One can differentiate between sensors that detect the contact position and ones that detect different types of touch or contact pose.

When analyzing this kind of systems, one can identify several functional principles. Because of robustness, low costs, and high sensitivity, resistive and capacitive principles are among the most used in ↪HCI. DAHIYA AND VALLE provide a thorough analysis of different measurement principles in [19] for the usage in robotic applications. In the following, the function of resistive and capacitive systems is described in more detail.

10.3.1 Resistive Touch Sensors

Resistive touch sensors to detect contact positions are based on two flexible, conductive layers that are normally separated from each other. If a user touches one of the layers, a connection is made between both layers and the position of the connection point can be calculated from the different resistances as shown in Fig. 10.50 based on Eqs. 10.46 and 10.47.

$$u_{\text{x,out}} = \left. \frac{R_2}{R_1 + R_2} U_{x1} \right|_{U_{x2} = 0\text{V}, \quad U_{y3}, U_{y4} \text{ in Hi-Z state}} \tag{10.46}$$

$$u_{\text{y,out}} = \left. \frac{R_4}{R_3 + R_4} U_{y3} \right|_{U_{y4} = 0\text{V}, \quad U_{x1}, U_{x2} \text{ in Hi-Z state}} \tag{10.47}$$

Resistive touch sensors exhibit a high resolution of up to 4,096 dpi in both dimensions and a high response speed (<10 ms). With additional wiring, the pressure on the screen also can be recorded. This principle does not support multitouch detection, i.e., the simultaneously contact in more than one position, with the setup shown in Fig. 10.50. A simple mean to measure multitouch interaction, too, is the segmentation of one of the conductive layers in several (n) conductive strips called *hybrid analog resistive touch sensing*. This increases the number of calculations to obtain a position reading from 2 to $2n$, but this is still less than the calculation of a whole matrix with at least n^2 calculations.

For the usage of resistive touch sensors, there are a couple of commercially available integrated circuits (as for example MAX 11800 with a footprint as low as 1.6×2.1 mm^2) that will alleviate the integration of such a sensor in a new system.

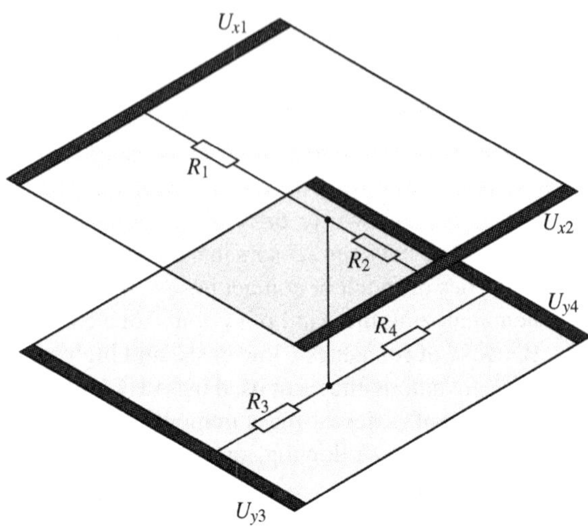

Fig. 10.50 Principle of resistive touch sensors

10.3.2 Capacitive Touch Sensors

For capacitive touch sensors' detecting positions, two general approaches are known. *Self-capacitance* or *surface-capacitance* sensors are built up from a single electrode. The system measures the capacitance to the environment that is altered when a user touches the surface. Based on the measurement of the current that is used to load the changed capacitance, a position measure can be deducted similar to the calculation in the case of resistive sensors. This sensor type is prone to errors from parasitic capacitive coupling, the calculation of multiple touch positions is possible but requires some effort.

When more than one capacitor is integrated in a surface, one can use the mutual capacitance type sensor. In that case, the capacitors are arranged in a matrix and the capacitance of each capacitor is changed by approaching conductive materials like fingers or special styluses. This matrix is read out consecutively by the sensor controller. Because of the matrix arrangement, the detection of multipoint touch is possible. As with resistive sensor systems, there are several commercially available integrated circuits for the readout of such capacitive matrices.

For the identification of contact poses, for example, the touch with a single finger or a whole hand, the self-capacitance approach can be used as well. In that case, the changed capacitance is considered as an indicator for the touch pose with regard to an arbitrary-shaped electrode. Since the realization of such a function is quite simple in terms of the required electronics, this procedure is incorporated in standard components under several brand names, like, for example, ATMEL QTOUCH.

Sometimes, capacitive sensor systems are combined with inductive sensor systems that track the position of a coil with respect to the reference surface. This is for example used to make use of stylus' on touchscreens and graphic tablets. Because of the different sensor principles, one can weigh the tool equipped with the coil more and avoid misreadings by the capacitive effect of the user's hand.

Table 10.6 Advantages and disadvantages of different touch-sensing principles according to [19]

Sensing principle	Advantages	Disadvantages
Capacitive	Sensitive, low cost, commercial readout circuits available	Cross-talk, hysteresis, complex electronic
Conductive composites	Mechanically flexible, easy fabrication, low cost	Hysteresis, nonlinear response, slow reaction time
Magnetic	High sensitivity, good dynamic range, no mechanical hysteresis, robust	Restricted to nonmagnetic medium, complex readout, size, high power consumption
Optical	Sensitive, immune to electromagnetical interference, no electrically conductive parts, flexible, fast response	Bulky, sensitive to bending, power consumption, complex readout
Piezoelectric	High sensitivity, good dynamic response, high bandwidth	Temperature sensitive, difficult electrical connection, no static measurement
Piezoresistive	High sensitivity, low cost, low noise, simple electronics	Mechanically stiff, nonlinear, hysteresis, signal drift
Resistive	High sensitivity, low cost, commercial readout circuits available	Power consumption, no multitouch, no contact-force measurement
Ultrasonic	Fast dynamic response, good force resolution	Limited utility at low frequency, temperature sensitive, complex electronics

10.3.3 Other Principles

However, a lot of other principles are known as well, that are often based on the change of a position and a detection of this change with different sensing principles. Examples include optical and magnetic measuring principles as described above in Sect. 10.2. They are often investigated in the context of robot tactile sensing, where not only touch, but also pose, handling, and collision are of interest [19].

In these cases, the use of flexible materials and microtechnology is of interest, which makes this kind of sensor to be the most welcomed examples for microsystem engineers. For the use in haptic systems, the current state of development of such systems as, for example, shown in [19, 27, 95] has to be critically checked. As a general classification, Table 10.6 gives some advantages and disadvantages for possible sensing principles.

A last example for advanced touch sensors shall be given: Based on an impedance spectroscopy measurement, the TOUCHÉ system by SATO ET AL. allows to differentiate different grips and body poses from an impedance measurement. Examples include the identification of the finger positions on a door knob, the discrimination of arm poses when sitting on a table or even the detection of someone touching a water surface. Applications include gesture interfaces for worn and integrated computers as well as the possibility of touch passwords, that consist out of a predefined sequence of touch poses [80].

10.4 Imaging Sensors

Within the contents of this book, the focus is laid on device-based sensors, like the above-described sensors for force, deflection, and touch. Pure input sensors, such as the imaging sensors, will be not discussed further as per definition no haptic feedback can be given without a real physical contact. Nevertheless, they can be used to build a complex HMI when combined with body-worn tactile devices and should be kept in mind for such applications. For that purpose, one can differentiate two general classes of imaging sensors:

1. *Direct imaging sensors* are usually based on camera systems observing the user or body parts, deriving commands from gestures performed with extremities or facial expressions. Such imaging systems are subject to application in controlled environments (such as inside a car), for direct computer or gameplay interaction (*Microsoft* XBOX KINECT), or even touch interaction on surfaces
2. *Marker-based imaging senors* are again based on camera systems observing special markers worn or held by the user. Such systems can be highly accurate due to the knowledge of relations between markers and their sizes. They are used in professional navigation application such as surgical applications or motion-capture technologies and can also be found in gameplay situation such as the *Sony* PLAYSTATION MOVE CONTROLLER.

10.5 Conclusion

With the exception of force/torque sensors, commercially available sensors for position, velocity, acceleration, touch, and images exhibit sufficient properties for the usage in haptic systems. Within this section, the necessary knowledge of the underlying sensing principles for sound selection from available sensors is reported. For the use of force/torque sensors, the relevant principle and design steps for the development of a customized sensor are given and can be deepened in the below-mentioned background readings.

Recommended Background Reading

[6] Barlian, A. & Park, W. & Mallon, J. & Rastegar, A. & Pruitt, B.: **Review: Semiconductor Piezoresistance for Microsystems**. Proceedings of the IEEE 97, 2009.
Review on piezoresistive silicon sensors. Physics, examples, sensor characteristics.

[11] Bray, A. & Barbato, G. & Levi, R.: **Theory and practice of force measurement**. Monographs in physical measurement, Academic Press Inc., 1990.
Discussion of examples of force sensors. Basic mechanics and hints for designing deformation elements.

[19] Dahiya, R. S. & Valle, M.: **Robotic Tactile Sensing**, Springer, 2013.
Overview about sensing principles used for robotic applications that are also usable for haptic systems.

[35] Keil, S.: **Beanspruchungsermittlung mit Dehnungsmessstreifen**. Cuneus Verlag, 1995.
All about strain gages. History, materials and technology, selection and application.

[47] Lenk, A. & Ballas, R.G. & Werthschützky, R.& Pfeifer, G.: **Electrical, Mechanical and Acoustic Networks, their Interactions and Applications**. Springer, 2011.
Introduction in modeling dynamics of electromechanical systems using network theory. Contains plenty of useful examples.

[72] Rausch, J.: **Entwicklung und Anwendung miniaturisierter piezoresistiver Dehnungsmesselemente**. Dr-Hut-Verlag, München, 2012.
Comparison of sensing principles for strain sensing with focus on piezoresistive silicon elements. Design of a tri-axial force sensor using semiconducting strain gages.

[98] Tränkler H.R.& Obermeier, E.:**Sensortechnik: Handbuch für Praxis und Wissenschaft**. Springer, 1998.
Extensive overview of sensors and sensor electronics.

[107] Young, W.C., Budynas, R.G.:**Roark's formulas for stress and strain. Bd. 6.** McGraw-Hill, New York, 2002
Mechanics Handbook for calculation of stress and strain fields in complex deformation bodies.

References

1. Amor A, Budde T, Gatzen H (2006) A magnetoelastic microtransformer-based microstrain gauge. Sens Actuators A Phys 129(1–2):41–44. doi:10.1016/j.sna.2005.09.043
2. Arshak K, McDonagh D, Durcan M (2000) Development of new capacitive strain sensors based on thick film polymer and cermet technologies. Sens Actuat A Phys 79(2):102–114. doi:10.1016/S0924-4247(99)00275-7
3. Arshak K et al (2006) Development of high sensitivity oxide based strain gauges and pressure sensors. J Mater Sci Mater Electron 17(9):767–778. doi:10.1007/s10854-006-0013-4
4. Ballas R (2007) Piezoelectric multilayer beam bending actuators: static and dynamic behavior and aspects of sensor integration. Springer, Berlin, pp XIV, 358. ISBN: 978-3-540-32641-0
5. Bao M-H (2004) Micro mechanical transducers: pressure sensors, accelerometers and gyroscopes, vol 8, 2nd edn. Handbook of sensors and actuators 8. Elsevier, Amsterdam, pp XIV, 378. ISBN: 9780080524030
6. Barlian A et al (2009) Review: semiconductor piezoresistance for microsystems. Proc IEEE 97(3):513–552. doi:10.1109/JPROC.2009.2013612
7. Berkelman P et al (2003) A miniature microsurgical instrument tip force sensor for enhanced force feedback during robot-assisted manipulation. IEEE Trans Rob Autom 19(5):917–921. doi:10.1109/TRA.2003.817526
8. Beyeler F et al (2007) Design and calibration of a MEMS sensor for measuring the force and torque acting on a magnetic microrobot. J Micromech Microeng IOP 18:025004 (2007). doi:10.1088/0960-1317/18/2/025004

9. Botsis J et al (2004) Embedded fiber Bragg grating sensor for internal strain measurements in polymeric materials. Opt Lasers Eng 43. doi:10.1016/j.optlaseng.2004.04.009
10. Brand U, Büttgenbach S (2002) Taktile dimensionelle Messtechnik für Komponenten der Mikrosystemtechnik. In: tm-Technisches Messen/Plattform für Methoden. Systeme und Anwendungen der Messtechnik 12. doi:10.1524/teme.2002.69.12.542
11. Bray A, Barbato G, Levi R (1990) Theory and practice of force measurement. Monographs in physical measurement. Academic Press Inc., Waltham. ISBN: 978-0121284534
12. Burdea GC (1996) Force and touch feedback for virtual reality. Wiley-Interscience, New York
13. Caldwell DG, Lawther S, Wardle A (1996) Multi-modal cutaneous tactile feedback. In: Proceedings of the IEEE international conference on intelligent robots and systems. Diss, pp 465–472. doi:10.1109/IROS.1996.570820
14. Calvert P et al (2007) Piezoresistive sensors for smart textiles. Electroact Polym Actuators Devices (EAPAD) 2007:65241i-8. doi:10.1117/12.715740
15. Chase T, Luo R (1995) A thin-film flexible capacitive tactile normal/shear force array sensor. In: Proceedings of the 1995 IEEE IECON 21st international conference on industrial electronics, control, and instrumentation, vol 2. Orlando, FL, pp 1196–1201. doi:10.1109/IECON.1995.483967
16. Cochrane C et al (2007) Design and development of a flexible strain sensor for textile structures based on a conductive polymer composite. Sensors 7(4):473–492. doi:10.3390/s7040473
17. Colgate J, Brown J (1994) Factors affecting the Z-Width of a haptic display. In: Proceedings of IEEE international conference on robotics and automation, vol 4, pp 3205–3210. doi:10.1109/ROBOT.1994.351077
18. Cranny A et al (2005) Thick-film force, slip and temperature sensors for a prosthetic hand. Measur Sci Technol IOP 16:931–941. doi:10.1016/j.sna.2005.02.015
19. Dahiya RS, Valle M (2013) Robotic Tactile Sensing. Springer, Berlin. doi:10.1007/978-94-007-0579-1
20. Kurtz AD (1962) Adjusting crystal characteristics to minimaze temperature dependency. In: Dean M, Douglas RD (eds) Semiconductor and conventional strain gages. Academic Press Inc. New York, ISBN: 978-1114789906
21. Dölle M (2006) Field effect transistor based CMOS stress sensors. PhD thesis. IMTEK, University of Freiburg. ISBN: 978-3899594584
22. Ferdinand P et al (1997) Applications of bragg grating sensors in Europe. In: International conference on optical fiber sensors OFS. Wiliamsburg, Virginia. ISBN: 1-55752-485-8
23. First Sensor Technology GmbH (2009) T-Brücke. Technical report. First Sensor Technology GmbH. http://www.first-sensor.com/
24. Fung Y-C (1993) Biomechanics: mechanical properties of living tissues, 2nd edn. Springer, New York, pp XVIII, 568. ISBN: 0-387-97947-6; 3-540-97947-6
25. Gall M, Thielicke B, Poizat C (2005) Experimentelle Untersuchungen und FE-Simulation zum Sensor-und Aktuatoreinsatz von flächigen PZT-Funktionsmodulen. In: Deutsche Gesellschaft fürMaterialkunde e.V. http://publica.fraunhofer.de/dokumente/N-28844.html
26. Gehin C, Barthod C, Teissyre Y (2000) Design and characterization of new force resonant sensor. Sens Actuat A Phys 84: 65–69. doi:10.1016/S0924-4247(99)00359-3
27. Goethals P (2008) Tactile feedback for robot assisted minimally invasive surgery: an overview. Technical report workshop, Eurohaptics conference. Department of Mechanical Engineering, K.U. Leuven
28. Gross D et al (2009) Technische Mechanik: Band 2: Elastostatik. Springer, Berlin. ISBN: 978- 3540243120
29. Hagedorn P et al (1998) The importance of rotor flexibility inultrasonic traveling wave motors. Smart MaterStruct 7:352–368. doi:10.1088/0964-1726/7/3/010
30. Hasegawa Y, Shikida M et al (2006) An active tactile sensor for detecting mechanical characteristics of contacted objects. J Micromech Microeng IOP 16:1625–1632. doi:10.1088/0960-1317/16/8/026
31. Hasser CJ, Daniels MW (1996) Tactile feedback with adaptive controller for a force-reflecting haptic display—Part I: design. In: Proceedings of the fifteenth southern biomedical engineering conference, pp 526–529. doi:10.1109/SBEC.1996.493294

32. Hoffmann K (1985) Eine Einführung in die Technik des Messens mit Dehnungsmessstreifen. Hottinger Baldwin Messtechnik GmbH (HBM)
33. Hou L (1999) Erfassung und Kompensation von Fehlereffekten bei der statischen Kraftmessung mit monolithischen Nd:YAG-Laserkristallen. PhD thesis, Universität Kassel. https://kobra.bibliothek.uni-kassel.de/handle/urn:nbn:de:hebis:34-159?mode=full
34. Inaudi D (2004) SOFO sensors for static and dynamic measurements. In: 1st FIG international symposium on engineering surveys for construction works and structural engineering. Nottingham. http://cordis.europa.eu/result/report/rcn/41123_en.html
35. Keil S (1995) Beanspruchungsermittlung mit Dehnungsmessstreifen. Cuneus Verlag. ISBN: 978-3980418805
36. Kerdok A (2006) Characterizing the nonlinear mechanical response of liver to surgical manipulation. PhD thesis, Harvard University, Cambridge. URL: http://biorobotics.harvard.edu/pubs/akthesis.pdf
37. Kern T (2006) Haptisches Assistenzsystem für diagnostische und therapeutische Katheterisierungen. PhD thesis, Techische Universität Darmstadt, Institut für Elektromechanische Konstruktionen. http://tuprints.ulb.tu-darmstadt.de/761/
38. Kersey A et al. (1997) Fiber grating sensors. J Lightwave Technol 15(8). doi:10.1109/50.618377
39. Kim K et al (2007) Calibration of multi-axis MEMS force sensors using the shape-from-motion method. IEEE Sens J 7(3):344–351. doi:10.1109/JSEN.2006.890141
40. Kizilirmak G (2007) Frei applizierbare MOSFET-Sensorfolie zur Dehnungsmessung". PhD thesis, RWTH Aachen. http://darwin.bth.rwth-aachen.de/opus3/volltexte/2007/1973/pdf/Kizilirmak_Goekhan.pdf
41. Klages S (2004) Neue Sensorkonzepte zur Zungendruckmessung. Diplomarbeit, Darmstadt: Technische Universität Darmstadt, Institut für Elektromechanische Konstruktionen. http://tubiblio.ulb.tu-darmstadt.de/53658/
42. Kon S, Oldham K, Horowitz R (2007) Piezoresistive and piezoelectric MEMS strain sensors for vibration detection. In: Proceedings of SPIE, vol 6529, p 65292V-1. doi:10.1117/12.715814
43. Kunadt A et al (2010) Messtechnische Eigenschaften von Dehnungssensoren aus Kohlenstoff-Filamentgarn in einem Verbundwerkstoff. In: tm-TechnischesMessen/Plattform fürMethoden, Systeme und Anwendungen der Messtechnik 77(2):113–120. doi:10.1524/teme.2010.0014
44. Lang U et al (2009) Piezoresistive properties of PEDOT: PSS. Microelectron Eng 86(3):330–334. doi:10.1016/j.mee.2008.10.024
45. LaserComponents Group (2008) Faseroptische Sensoren. http://www.lasercomponents.com/de/1134.html
46. Latessa G et al (2009) Piezoresistive behaviour of flexible PEDOT.PSS based sensors. Sens Actuat B Chem 139(2):304–309. DOI:http://dx.doi.org/10.1016/j.snb.2009.03.063
47. Lenk A et al (ed) (2011) Electromechanical systems in microtechnology and mechatronics: electrical, mechanical and acoustic networks, their interactions and applications. Springer, Heidelberg
48. Luo F et al (1999) A fiber optic microbend sensor for distributed sensing application in the structural strain monitoring". Sens Actuat A Phys 75:41–44. doi:10.1016/S0924-4247(99)00043-6
49. Maiwald M et al (2010) INKtelligent printed strain gauges. Sens Actuat A Phys 162(2):198–201. doi:10.1016/j.sna.2010.02.019
50. Matsuzaki R, Todoroki A (2007) Wireless flexible capacitive sensor based on ultra-flexible epoxy resin for strain measurement of automobile tires. Sens Actuat A Phys 140(1):32–42. doi:10.1016/j.sna.2007.06.014
51. Mehner J (2000) Entwurf in der Mikrosystemtechnik, vol 9. Dresdner Beiträge zur Sensorik. Dresden University Press. ISBN 9783931828479
52. Meiss T et al (2007) Fertigung eines Miniaturkraftsensors mit asymmetrischem Grundkrper zur Anwendung bei Katheterisierungen. In: MikroSystemTechnik. https://www.vde-verlag.de/proceedings-de/563061014.html

53. Meiss T (2012) Silizium-Mikro-Kraftsensoren für haptische Katheterisierungen: Entwurf, Musterbau und Signalverarbeitung sowie erste Validierung des Assistenzsystems HapCath. PhD thesis, Technische Universität Darmstadt, Institut für Elektromechanische Konstruktionen. http://tuprints.ulb.tu-darmstadt.de/2952/
54. Meißner P (2007) Optische Nachrichtentechnik I—Skriptum zur Vorlesung. Skriptum
55. Meißner P (2006) Seminar zu speziellen Themen der optischen Nachrichtentechnik—Simulatorische und experimentelle Untersuchungen optischer WDM Übertragungssysteme. Skriptum
56. Meschede D (2007) Optics, light and lasers: the practical approach to modern aspects of photonics and laser physics. 2nd edn. Optik, Licht und Laser engl. Weinheim: Wiley-VCH-Verlag, pp IX, 560. ISBN: 978-3-527-40628-9
57. Michel J (1995) Drehmomentmessung auf Basis funkabfragbarer Oberflächenwellen-Resonatoren. PhD thesis, Technische Universität München. http://www.mst.ei.tum.de/forschung/veroeffentlichungen/abgeschlossene-dissertationen.html
58. Micron Instruments (2012) U-shaped semiconductor gage. http://www.microninstruments.com/
59. Müller M (2009) Untersuchungen zu Kraft-Momenten-Sensoren auf Basis von Faser-Bragg-Gittern. PhD thesis, Technische Universität München. https://mediatum.ub.tum.de/doc/956469/956469.pdf
60. Nakatani M, Howe R, Tacji S (2006) The fishbone tactile illusion. In: Eurohaptics. IEEE. Paris. http://lsc.univ-evry.fr/eurohaptics/upload/cd/papers/f46.pdf
61. Nührmann D (1998) Das große Werkbuch Elektronik, vol 7. Franzis' Verlag, Poing. ISBN: 3772365477
62. Oddo C et al (2007) Investigation on calibration methods for multi-axis, linear and redundant force sensors. Measur Sci Technol 18:623. doi:10.1088/0957-0233/18/3/011
63. Palais JC (2005) Fiber optic communications, 5th edn. Pearson Prentice-Hall, Upper Saddle River, pp XIII, 441. ISBN: 0130085103
64. Pandey N, Yadav B (2006) Embedded fibre optic microbend sensor for measurement of high pressure and crack detection. Sens Actuat A Phys 128:33–36. doi:10.1016/j.sna.2006.01.010
65. Partsch U (2002) LTCC-kompatible Sensorschichten und deren Applikation in LTCC-Drucksensoren, 1st edn, vol 9. Elektronik-Technologie in Forschung und Praxis 9. Templin/Uckermark: Detert, pp III, 163. ISBN: 3934142117
66. Pasquale M (2003) Mechanical sensors and actuators. Sens Actuat A Phys 106(1–3):142–148. doi:10.1016/S0924-4247(03)00153-5
67. Peirs J et al (2004) A micro optical force sensor for force feedback during minimally invasive robotic surgery. Sens Actuat A Phys 115(2–3):447–455. doi:10.1016/j.sna.2004.04.057
68. Pfeifer G, Werthschützky R (1989) Drucksensoren. Verlag Technik. ISBN: 978-3341006603
69. Physik Instrumente (PI) GmbH (2010) Flächenwandlermodul. http://www.physikinstrumente.de/de/produkte/prdetail.php?sortnr=101790
70. Ping L, Yumei W (1996) An arbitrarily distributed tactile sensor array using piezoelectric resonator. In: Instrumentation and measurement technology conference IMTC-96. IEEE conference proceedings, vol 1. Brussels, pp 502–505. doi:10.1109/IMTC.1996.507433
71. Puers R (1993) Capacitive sensors: when and how to use them. Sens Actuat A Phys 37–38:93–105. doi:10.1016/0924-4247(93)80019-D
72. Rausch J (2012) Entwicklung und Anwendung miniaturisierter piezoresistiver Dehnungsmesselemente. Dissertation, Technische Universität Darmstadt. http://tuprints.ulb.tudarmstadt.de/3003/1/Dissertation-Rausch-online.pdf
73. Rausch J et al (2006) INKOMAN-Analysis of mechanical behaviour of liver tissue during intracorporal interaction. In: Gemeinsame Jahrestagung der Deutschen, Österreichischen und Schweizerischen Gesellschaften für Biomedizinische Technik 6(9)
74. Reindl L et al (2001) Passive funkauslesbare sensoren (wireless passive radio sensors). In: TM—Technisches messen/plattform für methoden, Systeme und anwendungen der Messtechnik 68(5):240. doi:10.1524/teme.2001.68.5.240

75. Rey P, Charvet P et al (1997) A high density capacitive pressure sensor array for fingerprint sensor application. In: International conference on solid-state sensors and actuators, IEEE. Chicago. doi:10.1109/SENSOR.1997.635738
76. Rosen J, Solazzo M, Hannaford M (2002) Task decomposition of laparoscopic surgery for objective evaluation of surgical residents learning curve using hidden markov model. In: computer aided surgery 7. http://www.ncbi.nlm.nih.gov/pubmed/12173880
77. Roths J, Kratzer P (2008) Vergleich zwischen optischen Faser-Bragg-Gitter-Dehnungssensoren und elektrischen Dehnungsmessstreifen. In: TM-Technisches Messen 75(12), pp 647–654. doi:10.1524/teme.2008.0903
78. Russell R (1992) A tactile sensory skin for measuring surface contours. In: Tencon 1992 region 10 conference. IEEE, Melbourne. doi:10.1109/TENCON.1992.271943
79. Saraf R, Maheshwari V (2006) High-resolution thin-film device to sense texture by touch. Technical report 5779, pp 1501–1504. http://www.sciencemag.org/content/312/5779/1501.full
80. Sato M, Poupyrev I, Harrison C (2012) Touché: enhancing touch interaction on humans, screens, liquids, and everyday objects. In: Proceedings of the 2012 ACMannual conference on human factors in computing systems, pp 483–492. doi:10.1145/2207676.2207743
81. Schaumburg H (1992) Sensoren, vol 3. Werkstoffe und Bauelemente der Elektrotechnik 3. Stuttgart: Teubner Verlag, p 517. ISBN: 3519061252
82. Schlüter V (2010) Entwicklung eines experimentell gestützten Bewertungsverfahrens zur Optimierung und Charakterisierung der Dehnungsübertragung oberflächenapplizierter Faser-Bragg-Gitter- Sensoren. PhD thesis, Bundesanstalt für Materialforschung und -prüfung - BAM. http://www.bam.de/de/service/publikationen/publikationen_medien/dissertationen/diss_56_vt.pdf
83. Schreier-Alt T (2006) Polymerverkapselung mechatronischer Systeme—Charakterisierung durch eingebettete Faser Bragg Gitter Sensoren". PhD thesis, Technische Universität Berlin, Fakultät IV - Elektrotechnik und Informatik. http://opus4.kobv.de/opus4-tuberlin/frontdoor/index/index/docId/1494
84. Sektion für UT (1999) Minimal Invasive Chirurgie Tübingen. Verbundprojekt TAMIC—Entwicklung eines taktilen Mikrosensors für die Minimal Invasive Chirurgie - Schlussbericht. Technical report. https://www.yumpu.com/de/document/view/6617211/verbundprojekttamic-entwicklung-eines-taktilen-experimentelle-
85. Sergio M et al (2003) A dynamically reconfigurable monolithic CMOS pressure sensor for smart fabric. IEEE J Solid-State Circ 38(6):966–968. doi:10.1109/JSSC.2003.811977
86. Simone C (2002) Modelling of neddle insertion forces of percutaneous therapies. Johns Hopkins University, Diplomarbeit. Baltimore. doi:10.1109/ROBOT.2002.1014848, http://ieeexplore.ieee.org/xpls/abs_all.jsp?arnumber=1014848&tag=1
87. Sirohi J, Chopra I (2000) Fundamental understanding of piezoelectric strain sensors. J Int Mater Syst Struct 11(4):246–257. doi:10.1106/8BFB-GC8P-XQ47-YCQ0
88. Silicon Microstructures Incooperated (SMI) (2008) Pressure Sensors Products. http://www.si-micro.com/pressure-sensor-products.html
89. Smith C (1954) Piezoresistance effect in germanium and silicon. Phys. Rev. Am Phys Soc 94(1):42–49. doi:10.1103/PhysRev.94.42
90. Stavroulis S (2004) Rechnergestützter Entwurf von piezoresistiven Silizium-Drucksensoren mit realem mechanischem Wandler. Dissertation, Technische Universität Darmstadt, Institut für Elektromechanische Konstruktionen. http://tuprints.ulb.tu-darmstadt.de/473/
91. Stockmann M (2000) Mikromechanische Analyse der Wirkungsmechanismen elektrischer Dehnungsmessstreifen. Institut für Mechanik der technischen Universiät Chemnitz. http://www.qucosa.de/recherche/frontdoor/?tx_slubopus4frontend[id]=urn:nbn:de:bsz:ch1-200000494
92. Su L, Chiang K, Lu C (2005) Microbend-induced mode coupling in a graded-index multimode fiber. Appl Opt 44(34). doi:10.1364/AO.44.007394
93. Sun Y et al (2002) A bulkmicrofabricated multi-axis capacitive cellular force sensor using transverse comb drives. J Micromech Microeng IOP 12:832–840. doi:10.1088/0960-1317/12/6/314

94. Suster M et al (2006) A high-performance MEMS capacitive strain sensing system. J Microelectromech Syst 15:1069–1077. doi:10.1109/JMEMS.2006.881489
95. Tegin J, Wikander J (2005) Tactile sensing in intelligent robotic manipulation—a Review. Ind Rob 32(1):64–70. doi:10.1108/01439910510573318
96. Tietze U, Schenk C (2002) Halbleiter-Schaltungstechnik, 12th edn. Springer, Berlin, pp XXV, 1606. ISBN: 3-540-42849-6
97. Toda K (1994) Characteristics of interdigital transducers for mechanical sensing and nondestructive testing. Sens Actuat A Phys 44(3):241–247. doi:10.1016/0924-4247(94)00809-4
98. Tränkler H, Obermeier E (1998) Sensortechnik: Handbuch für Praxis undWissenschaft. Springer, Berlin. ISBN: 978-3-642-29941-4
99. Unterhofer K et al (2009) CMOS Stressmesssystem zur Charakterisierung von Belastungen auf MEMS Bauteile. In: MikroSystemTechnik KONGRESS 2009. VDE VERLAG GmbH. https://www.vde-verlag.de/proceedings-en/453183009.html
100. Valdastri P et al (2005) Characterization of a novel hybrid silicon three-axial force sensor. Sens Actuat A Phys 123–124. In: Eurosensors XVIII 2004—The 18th European conference on Solid-State Transducers, pp 249–257. doi:10.1016/j.sna.2005.01.006
101. Vasarhelyi G, Adama M et al (2006) Effects of the elastic cover on tactile sensor arrays. Sens Actuat A Phys 132:245–251. doi:10.1016/j.sna.2006.01.009
102. Vasarhelyi G, Fodor B, Roska T (2007) Tactile sensing-processing—interface-cover geometry and the inverse-elastic problem. Sens Actuat A Phys 140:8–18. doi:10.1016/j.sna.2007.05.028
103. Vazsonyi E et al (2005) Three-dimensional force sensor by novel alkaline etching technique. Sens Actuat A Phys 123–124:620–626. doi:10.1016/j.sna.2005.04.035
104. Voyles R, Morrow J, Khosla P (1997) The shape from motion approach to rapid and precise force/- torque sensor calibration. Trans ASME-G-J Dyn Syst Measur Control 119(2):229–235. doi:10.1115/1.2801238
105. Werthschützky R (2007) Mess- und Sensortechnik - Band II: Sensorprinzipien. Vorlesungsskriptum
106. Werthschützky R, Zahout C (2003) Angepasste Signalverarbeitung für piezoresistive Drucksensoren. In: tm-Technisches Messen/Plattform für Methoden, Systeme und Anwendungen der Messtechnik, pp 258–264. doi:10.1524/teme.70.5.258.20043
107. Young W, Budynas R (2002) Roark's formulas for stress and strain, vol 6. McGraw-Hill, New York. ISBN: 007072542X
108. Ziemann O (2007) POF-Handbuch: optische Kurzstrecken-Übertragungssysteme, 2nd edn. Springer, Berlin, pp XXX, 884. ISBN: 978-3540490937
109. Zwicker TU ((1989) Strain sensor with commercial SAWR. Sens Actuat A Phys 17(1–2):235–239. doi:10.1016/0250-6874(89)80085-X

Chapter 11
Interface Design

Thorsten A. Kern

Abstract This chapter deals with different interface technologies that can be used to connect task-specific haptic systems to an IT system. Based on an analysis of the relevant bandwidth for haptic interaction depending on the intended application and an introduction to several concepts to reduce the bandwidth for these applications (local haptic models, event-based haptics, movement extrapolation, etc.), several standard interfaces are evaluated for use in haptic systems.

After the decision for the actuator (Chap. 9) used to generate the haptic feedback, and after the measurement of forces (Sect. 10.1) or positions (Sect. 10.2), it becomes necessary to focus on the IT interface. This interface has to be capable of providing data to the actuation unit and catch and transmit all data from the sensors. Its requirements result—such as with any interface—from the amplitude resolution of the information and the speed at which they have to be transmitted. The focus of this chapter lies on the speed of transmission, as this aspect is the most relevant bottleneck when designing haptic devices. Haptic applications are frequently located on the borderline, be it with regard to the delay acceptable in the transmission, or the maximum data rate in the sense of a border frequency.

With regard to the interface, two typical situations may be distinguished: spatially distributed tactile displays with a reasonable number of actuators and primarily kinaesthetic systems with a smaller number of actuators. In case of tactile systems, pin arrays, vibrators, or tactors, the challenge is given by the application of bus systems for the reduction of cable lengths, and the decentralization of control. Although there are still some questions of timing left, for example, to provide tactile signals in the right order despite a decentralized control, the data rates transmitted are usually not a challenge for common bus systems. VAN ERP points out [6] that a 30-ms time delay between impulses generated by two vibrators at the limbs may not be distinguished any more. For the data interface, this observation implies for this application that any

T.A. Kern (✉)
Continental Automotive GmbH, VDO-Straße 1, 64832 Babenhausen, Germany
e-mail: t.kern@hapticdevices.eu

C. Hatzfeld and T.A. Kern (eds.), *Engineering Haptic Devices*,
Springer Series on Touch and Haptic Systems, DOI 10.1007/978-1-4471-6518-7_11

time delay below 30 ms may be uncritical for transmitting information haptically. This is a requirement that can be fulfilled by serial automation technology network protocols like CAN or the time-triggered version TTCAN, without any problems. Accordingly, this section concentrates on requirements of haptic kinaesthetic devices with a small number of actuators only, whereas these devices usually have to satisfy tactile requirements according to their dynamic responses.

11.1 Border Frequency of the Transmission Chain

Section 1.2.2 stated that it is necessary to distinguish two frequency areas when talking about haptic systems. The lower frequency range up to $\approx$30 Hz includes a bidirectional information flow, whereas the high-frequency area >30 Hz transmits information only unidirectionally from the technical system to the user. Although the user himself influences the quality of this transmission by altering the mechanical coupling, this change itself happens at lower frequencies only, and is—from the perspective of bandwidth—not relevant for the transmission. If this knowledge is applied to the typical structures of haptic devices from Chap. 6, some fascinating results can be found. For the following analysis, it is assumed that the transmission and signal conditioning of information happens digital. According to NYQUIST, the maximum signal frequency has to be sampled at least two times faster. In practical application, this factor two is a purely theoretical concept, and it is strongly recommended to sample an analog system around 10 times faster than its maximum frequency. The values within figures and texts are based on this assumption.

11.1.1 Bandwidth in a Telemanipulation System

For a telemanipulation system (Fig. 11.1), the knowledge of the differing asymmetric dynamics during interaction gives the opportunity to benefit directly for the technical design. In theory, it is possible to transmit the haptic information measured at the

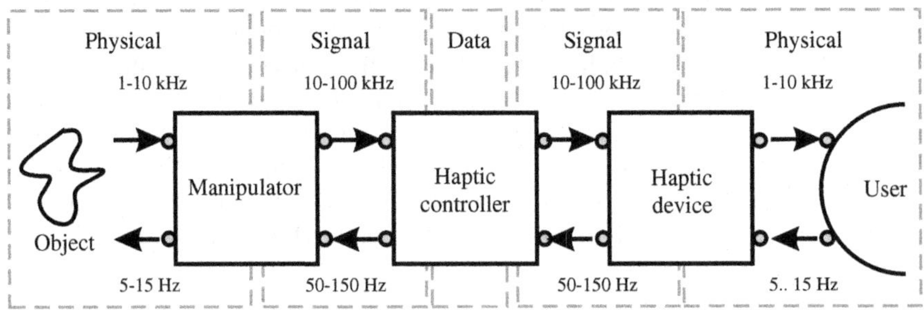

Fig. 11.1 Block diagram of a telemanipulator with haptic feedback

object within the bandwidth of 1 Hz–10 kHz, and replay it as forces or positions to the user. The user's reactions may in this case be measured at a bandwidth from static to 5 or 15 Hz only, and may be transmitted via controller and manipulator to the object. Although this approach would be functional indeed, the simplicity of position measurement and the necessity to process them, e.g., passivity control, result in movements being sampled and transmitted similar in dynamic as in the opposite transmission direction for haptic feedback.

11.1.2 Bandwidth in a Simulator System

For a simulation system with haptic feedback, the different dynamics result in slightly different findings. Nevertheless, it is still true that the movement information may be sampled at a lower rate. However, the simulator (Fig. 11.2) has to provide the force output at a frequency of 1–10 kHz. Due to this simple reason, the simulator has to be aware of the actual position data for every simulation step. Consequently, with simulators, the haptic output and the measurement of user reaction have to happen at high frequency (exceptions, see Sect. 11.2).

There are two approaches to integrate the haptic controller in the simulator. In many devices, it is designed as an external hardware component (Fig. 11.2), which reduces the computing load for the main simulator, and helps reducing the data rate significantly in special data processing concepts with parametrizable models (Sect. 11.2). As an alternative, the controller may be realized in software as a driver computed by the simulation main computing unit (Fig. 11.3). This is a concept used especially for high-power permanently installed simulation machines, or those used in cost-effective haptic devices for gaming industry with little requirements in dynamics and haptic output.

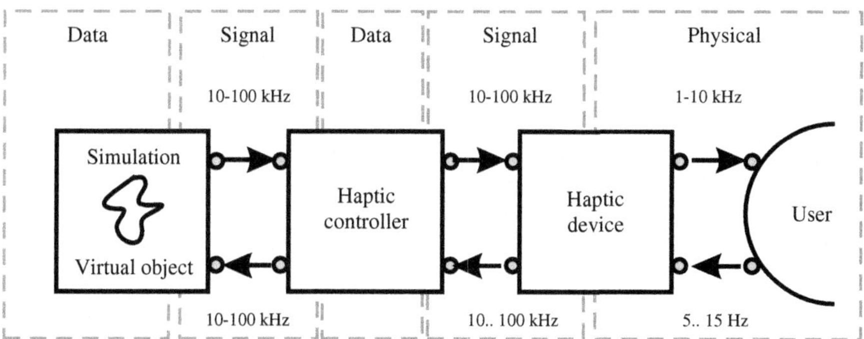

Fig. 11.2 Block diagram of a simulator with haptic feedback and an external controller

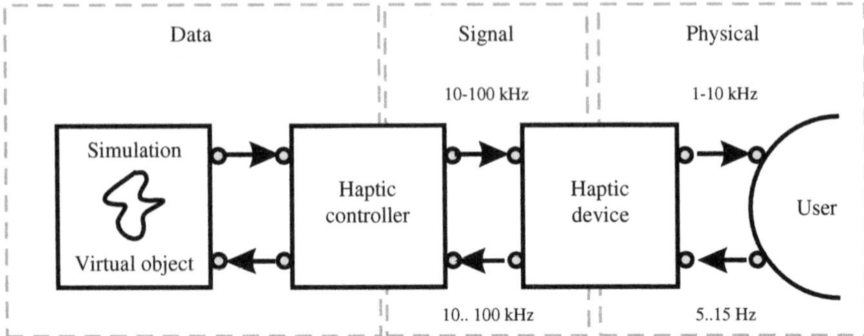

Fig. 11.3 Block diagram of a simulator with haptic feedback and a controller as part of the driver software

11.1.3 Data Rates and Latencies

Table 11.1 summarizes the data rates necessary for kinaesthetic applications in some typical examples. The data rates range from 200 kbit/s for simple applications up to 50 Mbit/s for more complex systems. Such rates for the information payload—still excluding the overhead necessary for the protocol and the device control—are achieved by several standard interface types today (Sect. 11.3).

Besides the requirements for the data rate, there is another requirement considering the smallest possible latency. In particular, interfaces using packets for transmission, with an uncertainty about the exact time of the transmission (e.g., USB), have to be analyzed critically concerning this effect. Variable latencies between several packets are a problem in any case. If there are constant latencies, the reference to other senses with their transmission channel becomes important: A collision is not allowed to happen significantly earlier or later haptically than visually or acoustically. The range possible for latency is largely dependent on the way to present the other sensual impressions. This interdependencies are subject to current research and are analyzed by the group around BUSS at the Technische Universität München.

Table 11.1 Example calculating the required unidirectional data rates for typical haptic devices

DoFs	Resolution (bits)	0.1 kHz	1 kHz (kbit/s)	10 kHz
11 DoF	8	800 bit/s	8	80 kbit/s
	16	1,600 bit/s	16	160 kbit/s
13 DoF	8	24 kbit/s	240	2.4 Mbit/s
	16	48 kbit/s	480	4.8 Mbit/s
16 DoF	8	48 kbit/s	480	4.8 Mbit/s
	16	96 kbit/s	960	9.6 Mbit/s

11.2 Concepts for Bandwidth Reduction

Anyone who has tried to process a continuous data flow of several megabit with a PC, and in parallel make this PC do some other tasks too, will have noticed that the management of the data flow binds immense computing power. With this problem in mind and as a result of the questions of telemanipulation with remotely located systems, several solutions for bandwidth reduction of haptic data transmission have been found.

11.2.1 Analysis of the Required Dynamics

The conscious analysis of the dynamics of the situation at hand should be ahead of every method to reduce bandwidth. The limiting cases to be analyzed are given by the initial contact or collision with the objects. If the objects are soft, the border frequencies are in the range of <100 Hz. If there are stiff objects part of interaction and if there is the wish to feedback these collisions, frequencies up to a border >1 kHz will have to be transmitted. Additionally, it has to be considered that the user is limited concerning its own dynamics, or may even be further limited artificially. The DAVINCI System (Fig. 1.10) as a unidirectional telemanipulator filters the high frequencies of the human movements to prevent a trembling of the surgical instruments.

11.2.2 Local Haptic Model in the Controller

A frequently used strategy being part of many haptic libraries is the use of local haptic models. These models allow faster reaction on the user's input compared to the simulation of a complete object interaction (Fig. 11.4). Such models are typically linearized functions dependent on one or more parameters. These parameters are actualized by the simulation at a lower frequency. For example, each degree of freedom of the haptic system may be equipped with a model of spring, mass, and damper, whose stiffness, mass, and friction coefficients are updated to the actual value at each simulation step, e.g., every $\approx\frac{1}{30}$ s. This approach does not permit the simulation of nonlinear effects in this simple form. The most frequent nonlinear effect when interacting with virtual worlds is the liftoff of a tool from a surface. Dependent on the delay of the actualization of the local model, the liftoff is perceived as "sticking," as the tools are held to the simulated surface by the local model in one simulation step, whereas it is suddenly released within the next. Concepts that model nonlinear stiffnesses compensate this effect satisfactorily. By making the additional calculations necessary for the local model, a significant data reduction between simulation and haptic controllers is achieved. Distantly related concepts are used in automotive

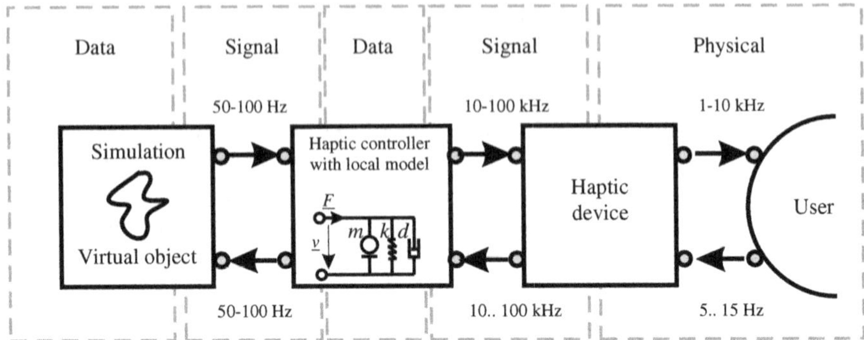

Fig. 11.4 Block diagram of a simulator with haptic feedback and a local haptic model inside the controller

applications, where CAN bus systems are configured in their haptic characteristics by a host, and report selection events in return only.

11.2.3 Event-Based Haptics

KUCHENBECKER presented in 2005 the concept of "event-based haptics" [2] and brought it into perfection since. It is based on the idea to split low-frequency interaction and high-frequency unidirectional presentation, especially of tactile information (Fig. 11.5). These tactile events are stored in the controller and are activated by the simulation. They are combined with the low-frequency signal synthesized from the simulation and are presented to the user as a sum. In an improved version, a monitoring of the coupling between haptic device and user is added, and the events' intensities are scaled accordingly. The design generates impressively realistic collisions with comparably soft haptic devices. As any other highly dynamic system, it nevertheless requires a specialized driver electronics and actuator selection to achieve full performance.

A variant of the concept of event-based haptics is the overlay of measured high-frequency components on a low-frequency interaction. This concept can be found in the case of VERROTOUCH (Sect. 2.4.4) or in the application of an assistive system like HAPCATH (Sect. 14.3). The overall concept of all these systems follows Fig. 11.6. A highly dynamic sensor (piezoelectric or piezoresistive) is implemented in a coupled mechanical manipulation system. The interaction forces or vibrations induced by collisions between tool and object are then transmitted to an actuating unit attached near to the handle of the device. In case of these systems, it is then just a variant whether the interaction path is also decoupled or sticks to the normal mechanical connection.

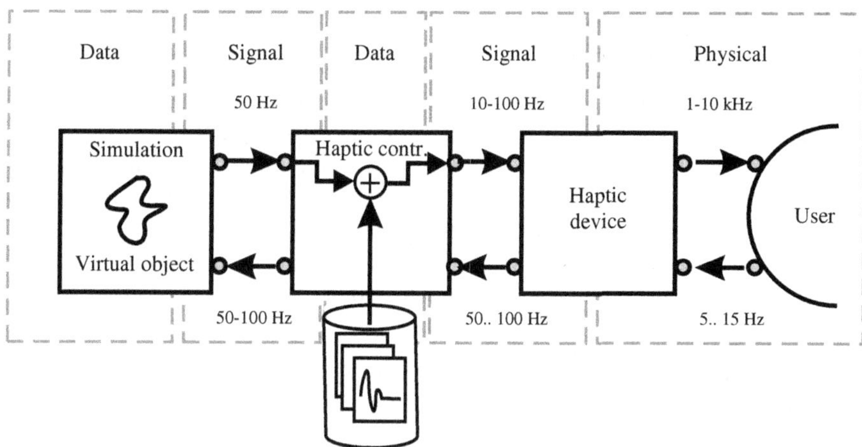

Fig. 11.5 Block diagram of a simulator with haptic feedback and with events of high dynamic being held inside the controlling structure

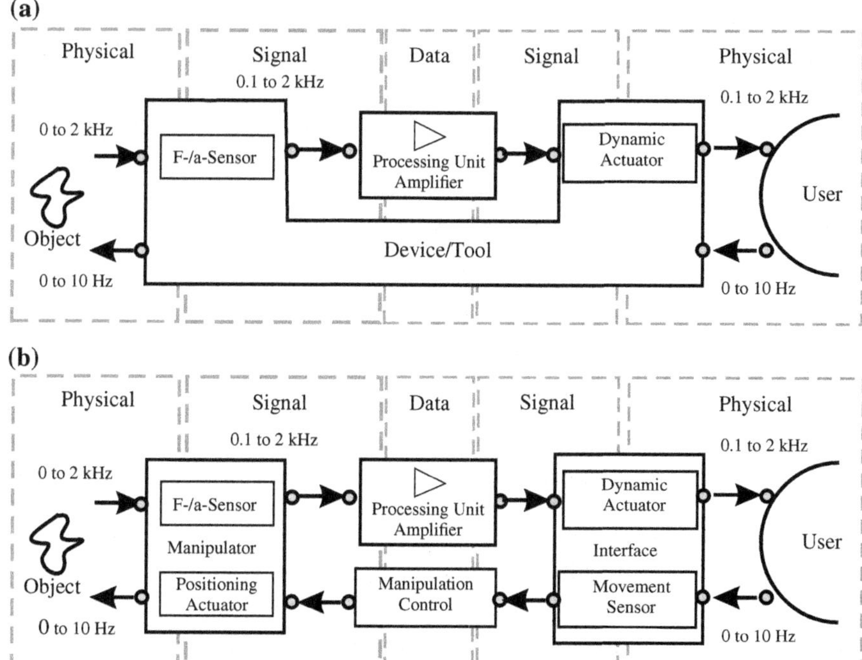

Fig. 11.6 Concept of an event-based haptic overlay of tactile relevant data with (**a**) and without (**b**) mechanical coupling of interface and manipulator

11.2.4 Movement Extrapolation

Another very frequently used method for bandwidth reduction on the path to measure user reaction is given by extrapolation of the movement. In particular, with simulators using local models, it is often necessary to have some information about steps in between two complete measurement sets, as the duration of a single simulation step varies strongly, and the available computing power has to be used most efficiently. The extrapolation becomes a prediction with increased latency and a further reduced transfer rate. Prediction is used for haptic interaction with extreme dead times.

11.2.5 Compensation of Extreme Dead Times

The working group of NIEMEYER from Telerobotics Laboratory at the Stanford University works on the compensation of extreme dead times of several seconds by prediction [4]. The dead time affects both paths: the user's reaction and the information to the user, such as the haptic feedback generated. The underlying principle is an extension of the telemanipulation system, which is added with a controller of the manipulator and a powerful controller for the haptic feedback (Fig. 11.7). The latter can be understood as an own simulator of the manipulated environment. During movement, a model of the environment is generated in parallel. If a collision happens in the real world, the collision is placed as a wall in the model, and its simulation provides a haptic feedback. Due to the time lag, the collision does not happen at the position where it happened in reality. During the following simulation, the collision point is relocated slowly within the model back to its correct position. By successive exploration of the environment, a more detailed haptic model is generated. The method has the status of a research project.

11.2.6 Compression

As any data stream, haptic data can be compressed for reducing their bandwidth. This may happen based on numerical methods on each individual packet; however, it may

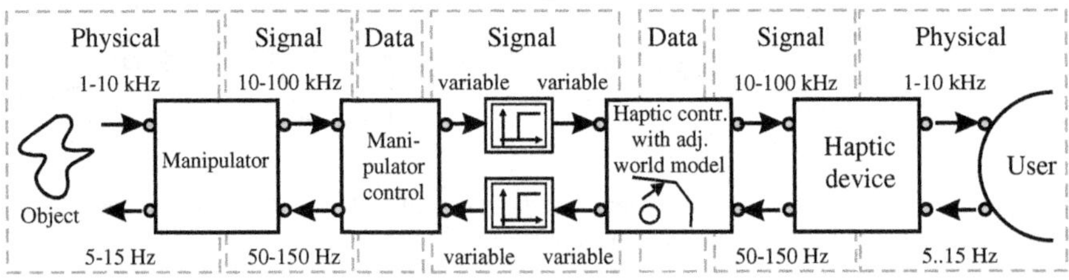

Fig. 11.7 Block diagram of a telemanipulator with compensation of long dead times by an adaptable world model

also be possible to make use of the special properties of the haptic human–machine interaction and haptic perception. The following list shall give a short overview about common approaches:

- A first approach for compressing haptic data is given in the situational adaption of digitalization on the path for measuring user reaction. SHAHABI ET AL. [5] compare different digitalization methods adapting their time and amplitude discretization on the actual movement velocity.
- Since several years now the working group around BUSS does intensive research on the perceptional impact of loss of resolution and bandwidth in haptic data streams [3]. They are coupling their research with the analysis of user reactions and basing their algorithm on the psychophysical perception and a benefit–effort analysis.
- The working group around EL SADDIK wants to achieve data reduction by standardizing the haptic interaction in a descriptive data format "HAML". It models the environment in a comparably little set of parameters, which gives advantages in teleinteraction applications with a larger group of participants. In particular, varying data transmission paths can be compensated more easily on this abstract description, in comparison with classic telemanipulation approaches with a transmission of explicit forces and positions. As a by-product of this work, concepts for the unidirectional replay of haptic data in the form of a "haptic player" are developed [1].
- Another obvious approach for compression is the usage of limitations given by haptic perception. The working group of KUBICA demonstrates [7] an analysis of an interaction with a virtual environment at different velocities. The identified dependency of the force perception threshold on the velocity was successfully used as a basis for data reduction.

11.3 Technical Standard Interfaces

Most haptic devices are operated with personal computer or related systems. They offer a high flexibility in configuration as well as for research projects as for gaming and design applications. Within this section, different standard interfaces typical to PC hardware architecture are highlighted and discussed with respect to haptic device applications.

11.3.1 Serial Port

The serial port is an interface, which is—dependent on the operating system used[1]—quite simple to be addressed. The serial interface of a home computer is based on the

[1] As a rule of thumb: with Linux or DOS systems, a direct communication using the serial interface is lot simpler when compared to Windows operating systems.

RS232 standard. This standard defines, besides several timing aspects, the bits being encoded in between ± 3 to ± 15 V for low and high levels. A connection to digital circuits with different logic has to happen via a converter, such as the MAX232. Usually, only two lines (RxD and TxD) are necessary for data transmission. The maximum specified data rate is 20.000 bit/s, whereas data rates far off specifications with 56 or 128 kbit/s are also possible. Both the data rate and the number of bytes carrying the actual data payload can be configured in a wide range. Two systems communicating with each other have to be adjusted in these parameters. A simple parity control with one bit is also integrated. With respect to its data rate, the serial port is absolutely suitable for the control of simple haptic devices. However, due to its master–slave architecture, a bidirectional data communication is connected with a large data overhead as a result of the coordination of both units. As modern PC is not necessarily equipped with serial ports only, several USB-to-serial converters are offered at the market. They emulate a COM port in software and transmit the data through the USB connector and the typical 9-pole socket. The data rates achievable with these connectors are usually sufficient, but they show some unpredictable delays and time loss due to software emulation of the interface. This makes them hardly usable for time-critical haptic applications.

11.3.2 Parallel Port

The parallel port is, if it is still part of modern PCs, a 25-pole double-row SUB-D-socket on the computer back plane. Similar to the serial interface, its ease of use is largely dependent on the operating system. In an ideal case (with several Linux distributions and with DOS), the address of the port can be directly written with three consecutive bytes. In this case, the first byte represents the eight data lines and the two following bytes are used for setting and reading of the control lines. The parallel port is set to work with 5 V logic levels with a maximum source current of $\approx$5 mA and sinking currents of maximum $\approx$20 mA. An overload current or leveling the pins to wrong voltages should be strictly avoided, as typical PCs do not show any protective circuitry. The data transmission of modern parallel ports is usually bidirectional, allowing read and write operations on the same eight data lines. But as the change between data transmission directions needs some time, the control lines are frequently used as input, whereas the data lines operate as output. Writing to and reading from the port can be made with frequencies up to 100 kbit/s without much effort. An extension of the parallel port is given by the "Enhanced Parallel Port" (EPP) and the "Enhanced Capabilities Port" (ECP). Whereas the ECP excels by Plug and Play functionalities mainly, the EPP had been designed for an increase in bidirectional data transmission rates up to 2 Mbit/s. This increase was achieved by a hardware implementation of the data protocol. For a slave using the EPP—such as a haptic device—this additional hardware requires of course some more hardware on the device's end, as the protocol has to be realized near to the interface. From the perspective of data rates, the parallel port is highly suitable for haptic devices, especially in EPP mode. In particular, low latencies of $<100\,\mu s$ between writing command and the availability of the data make it very attractive. Only the dwindling availability of this port to standard PCs makes it necessary to use other type of interfaces.

By their flexibility in software drivers, the serial and the parallel port makes it possible to be interfaced as debugging ports to microcontroller circuits or to bus systems such as I2C or CAN.

11.3.3 USB

The USB port is a serial port with a predefined data transmission protocol. It contains two data lines, an electrical ground, and a 5-V supply, which can be drained with up to 100 mA per device attached. According to the USB specifications, this load can be increased up to 1 A, if the host accepts it. An extension of the standard named "power USB" considers additional lines for higher currents and even larger voltages. The USB clients receive an identifier when being connected to the bus, which marks the data packages, sent to or from them. Each USB component has "device descriptor" uniquely identifying the manufacturer and the product. Additionally, each device is classified into several standard classes. The "human interface device class" (HID) is reserved for input systems. Devices with active, haptic feedback are grouped in an own class of "physical interface devices" (PID). Each manufacturer of a USB driver circuit has to apply for a unique product id. Due to the requirements and the complexity connected to the implementation of a USB conform protocol, it is recommended to use USB interface circuits off the shelf for product designs with little quantities manufactured. Such interface circuits offer parallel or serial data lines to be interfaced by an own microcontroller, which itself does not have to bother about the USB interface any more. The USB interface can be operated in different modes. For the transmission of larger, time-critical data volumes, the "isochrone" transfer is the most suitable. Its theoretical limit is given by the data packets, which are transmitted in microframes according to the USB 2.0 standard. The duration of a microframe is given by 125 μs. So-called full-speed systems are able to transmit up to 1,023 bytes with each microframe. High-speed devices are able to transmit even three times more bytes per microframe. According to the specifications, transmission rates of up to 40 MByte/s are possible with devices combining several isochronous endpoints in one unit. The data rate of isochronous transfer is optimized for unidirectional transmission only. In case of bidirectional communication, the data rate is reduced accordingly. Nevertheless, the speed of the USB port covers any requirements given by USB devices. However, two special aspects have to be checked in the context of the individual application:

- A microframe of 125 μs duration (8 kHz) is the upper limit of the available bandwidth. Without compression and decoding, this gives the natural bandwidth limit according to NYQUIST of 4 kHz.
- The data rate has a tolerance of 0.1 %.

11.3.4 FireWire: IEEE 1394

FireWire, Apple's brand name, according to the IEEE 1394 standard is a serial transmission format similar to USB. In fact, it is a lot older than the USB specification. The six-pole FireWire Connector includes a ground and a supply line. The voltage is not controlled and may take any value between 8 and 33 V. FireWire 400 defines up to 48 W power to be transmitted. The data rates are—dependent on the port design—100, 200, 400, or 1,600 kbit/s. This is completely sufficient for any haptic application. Even fiber optics transmission over 100 m distance with up to 3,200 kbit/s is specified in the standard. The bus hardware additionally includes a concept to share memory areas between host and client, enabling low-latency transmissions. Even networks without an explicit host can be established. The interface according to IEEE 1394 is the preferred design for applications with high data transmission rates. Only the little propagation of this interface in personal computers hinders a wide application.

11.3.5 Ethernet

The capabilities of the Ethernet interface available with any PC are enormous but largely dependent on the protocol used. Whereas the naked interface enables transmission rates of 10 or 100 Mbit or even gigabit, the available data payload within the transmission is largely dependent on the interlacing of the underlying protocols. The very well-known TCP/IP protocol has a header portion of 40 bytes. The Ethernet protocol adds another 18 bytes for the Ethernet frame, and some more 8 bytes for the whole packet, resulting in an overhead per packet of 66 bytes. This packet may contain up to 1,460 bytes of data. This is sufficient for typical haptic applications with respect to the available space per packet. Assuming a six-DOF kinematics with 16 bit (2 byte) resolution in their sensors and actuators, each packet has to carry only 12 bytes of data, with one packet for force output and one for position input. The number of bytes carried in each packet has a lower limit depending on the physical design of the network. In the area of home networks, it is 50 bytes, making it necessary to add arbitrary data on the example from above. A cycle of the haptic device example would transmit 232 bytes, which is 1.856 kbits. With a 10-Mbit network, theoretical bandwidth of 8 kHz would be available. Even when considering that the data have to be extended with some additional overhead (address negotiations, status information), this is still sufficient for many haptic applications. A disadvantage in using the Ethernet is given by the high efforts necessary for packet confection and protocol formulation, which would usually overload the computation power of standard microcontrollers. Additionally, a high number of clients reduce the data rate within a network significantly. Using switches compensate this reduction to some extend. But the method of choice is usually given by an exclusive network for the haptic application.

11.3.6 Measurement Equipment and Multifunctional Interface Cards

Measurement and multifunction interface cards are a simple approach to interface to hardware designs. They are available for internal and external standard interfaces, such as PCMCIA, USB, or even LAN. They are usually equipped with several standard software-drivers optimized for their hardware capabilities. When considering a prototype design, they should be considered in any case. Their biggest disadvantage is given by the data processing happening inside the hosting PC and within the restrictions of the operating system. In particular, in combination with non-real-time operating systems like Windows, the dynamics of controllers necessary for haptic applications may become not fast enough.

11.3.7 HIL Systems

"Hardware-in-the-loop" (HIL) systems were first used in control engineering and to compensate the disadvantages from multifunctional cards for rapid prototyping and interfaces to haptic systems. HILs include a powerful controller with proprietary or open real-time operating system. The programs operated on these controllers have to be built on standard PCs and are transmitted as with any other microcontroller system. Frequently, the compilers allow programming with graphical programming language such as MATLAB/Simlink or LabView. The processors of the HILs are connected via specialized bus systems with variable peripheral components. Ranging from analog and digital output over special bus and actuator interfaces, a wide range of components is covered. HIL systems are predestined for the always time-critical applications of haptics in design phase. But compared to other solutions, they have a high price.

11.4 Final Remarks on Interface Technology

The interface subordinates to the requirements of the system. Any realistic application and its required data rate can be covered with today's standard components. Only commercial or company interests may prevent the choice of a suitable interface for a haptic device. This is a complete difference to the situation at the beginning of the twenty-first century. At this time, highly specialized interfaces were designed for haptic devices, to cover the high requirements on data transmission rates. Accordingly, even today, commercial products with own ISA or PCI interface cards can be found on the market. Other solutions require an EPP parallel port still. Nevertheless, the design of controller circuit suitable for the USB protocol should not be underestimated. In particular, its layout and programming for the still high data rates of haptic devices offer enough room for errors. Although the technical specifications are sufficient to fulfill the requirements, the first design and operation is far from being trivial.

References

1. Eid M et al (2007) A device independent haptic player. In: Virtual environments, human-computer interfaces and measurement systems (VECIMS 2007). IEEE symposium on University of Ottawa, Ottawa, Ostuni, pp 7–12. ISBN: 978-1-4244-0820-7
2. Kuchenbecker K et al (2010) VerroTouch: high-frequency acceleration feedback for telerobotic surgery. In: Kappers AML, Bergmann-Tiest WM, van der Helm FC (eds) Haptics: generating and perceiving tangible sensations. Proceedings of the eurohaptics conference. Springer, Amsterdam, NL, Heidelberg, pp 189–196. doi:10.1007/978-3-642-14064-8_28
3. Kuschel M et al (2006) Lossy data reduction methods for haptic telepresence systems. In: Proceedings of the IEEE international conference on robotics and automation (ICRA 2006), pp 2933–2938. doi:10.1109/ROBOT.2006.1642147
4. Mitra P, Gentry D, Niemeyer G (2007) User perception and preference in model mediated telemanipulation. In: EuroHaptics conference and symposium on haptic interfaces for virtual environment and teleoperator systems. World haptics 2007. Second Joint Telerobotics Lab. Stanford University, Palo Alto, CA, Tsukaba, pp 268–273
5. Shahabi C, Ortega A, Kolahdouzan MR (2002) A comparison of different haptic compression techniques. In: Proceedings of the IEEE international conference on multimedia and expo (ICME'02), p 1. doi:10.1109/ICME.2002.1035867
6. van Erp J (2005) Vibrotactile spatial acuity on the torso: effects of location and timing parameters. In: First joint eurohaptics conference and symposium on haptic interfaces for virtual environment and teleoperator systems. WHC 2005, pp 80–85. doi:10.1109/WHC.2005.144
7. Zadeh MH, Wang D, Kubica E (2008) Perception-based lossy haptic compression considerations for velocity-based interactions. Multimedia Syst 13(4):275–282. doi:10.1007/s00530-007-0106-9

Chapter 12
Software Design for Virtual Reality Applications

Alexander Rettig

Abstract This chapter addresses the main steps in the development of software for virtual reality applications. After a definition of *virtual reality*, the general design and architecture of VR systems are presented. Several algorithms widely used in haptic applications such as the definition of virtual walls, penalty- and constraint-based methods, collision detection, and the Voxmap-PointShell Algorithm for 6 DoF interaction are presented. The chapter further includes a brief overview about existing software packages that can be used to develop virtual reality applications with haptic interaction. The concepts of *event-based haptics* and *pseudo-haptics* are introduced as perception-based approaches that have to be considered in the software design.

> The ultimate display would, of course, be a room within which the computer can control the existence of matter. A chair displayed in such a room would be good enough to sit in. Handcuffs displayed in such a room would be confining, and a bullet displayed in such a room would be fatal. With appropriate programming such a display could literally be the Wonderland into which Alice walked.
> IVAN E. SUTHERLAND, 1965 [17]

A central application area for haptic systems is the so-called virtual reality. This names a concept of human–computer interaction (HCI), which has developed rapidly during the last 20–30 years. The vision itself has been formulated already at the beginning of the development of computer graphics in the 1960s of the last century by SUTHERLAND. The focus of this chapter about software design is on this very application area of computer haptics in virtual worlds as, for example, training applications or haptic guidance in assistive systems. Other application areas of haptic systems as the modification of haptic properties of manual control elements or telemanipulation need a system structure, which is mainly defined by control engineering aspects as discussed in other chapters of this book (Chaps. 7 and 11).

Following a short overview to motivate the subject and provide some terminology and concepts, different topics will be discussed from a software developer's point of view. According to the goal of this book, the chapter at hand does not provide an

A. Rettig (✉)
ask—Innovative Visualisierungslösungen GmbH,
Heinrich-Hertz-Straße 1, 64295 Darmstadt, Germany
e-mail: a.rettig@hapticdevices.eu

C. Hatzfeld and T.A. Kern (eds.), *Engineering Haptic Devices*,
Springer Series on Touch and Haptic Systems, DOI 10.1007/978-1-4471-6518-7_12

exhaustive depiction but gives a basic understanding by creating interest for further activity with the topic in mind.

12.1 Overview About the Subject "Virtual Reality"

The term *virtual reality* (VR)[1] is defined as follows for the course of this chapter.

Definition *Virtual Reality* Virtual reality or VR, terms a technology, which among others should fulfill the demand to recreate the natural environment as close to real as possible inside a computer simulation or to let unreal things become real and present them to the user quasi as if they were true. The displayed virtual environment should behave in the way the user would expect because of his experiences in the natural world. Ideally, VR allows a perfectly intuitive handling with the computer which needs not to be learned.

Three main criteria have been identified during the research about virtual worlds which should be fulfilled as possible to achieve this goal: a quality of the display and form of presentation, which enables *immersion*, natural or intuitive *interaction*, and realistic or at least plausible *behavior* of the displayed environment.

12.1.1 Immersion

Immersion, "diving" into the virtual world, happens when the user quasi forgets that he is interacting with a virtual world and the real world largely steps into the background. To achieve this, it is crucial to activate as many senses as possible with the best quality as possible, mainly the dominant visual sense as well as the auditory sense, but also—as soon as manipulative interaction becomes relevant for the application—the haptic perception and finally the olfactory or even the sense of taste may be stimulated. The latter ones typically are only dealt with in a kind of exotic scenarios though. At least a variety of sense modalities should be incorporated, and this is called *multimodal* presentation.

12.1.2 Natural Interaction

Experienced computer users are easily apt to consider operating the computer by a mouse to be natural and intuitive. However, actually so-called more or less abstract interaction metaphors find use which must be understood and trained first: Is it

[1] There are concepts "augmented reality" (AR), "mixed reality" (MR), or even "augmented virtuality" as well, which name different kinds of mixture and embedding of real and virtual objects into a real or virtual environment. This differentiation is not needed in our context, though.

"natural" to "grab" a virtual object by pressing a mouse button after an arrow symbol was moved onto the visual representation of the object on the screen by means of a movement spatially separate of it of a small plastic box (the mouse)? Really, natural grabbing is certainly something else: The hand is moved to where the object is seen, although in a fraction of seconds, contact is taken up with the sense of touch before finally the fingers are closed around the object. In addition, that the meaning associated with an interaction metaphor often is only partially obvious—one *opens* a document by dropping its icon onto the icon of a text processor program? Those doubtless are useful and reasonable concepts, but their semantics have to be learnt and understood explicitly. By using different input devices as data gloves up to full body tracking systems, which track the movement of all extremities, the VR research tries to provide actually natural interaction with complex data worlds.

12.1.3 Natural Object Behavior

The third criterion, plausible behavior, inter alia includes, that dropped things fall, liquids swash and flow, that you cannot go through walls, and objects in the virtual environment cannot penetrate each other. These behavioral characteristics have to be provided by different simulations.

Particularly, the latter aspect of natural object behavior, the impenetrability of solid objects, is of importance: Although it could be implemented using collision detection and more or less physically correct simulation in a purely graphical presentation, then there is a possible discrepancy between the real movement of the user and the visual echo of this movement. It occurs, for example, if a virtual hammer stops on the surface of the virtual wall, while the real hand of the user has already deeply banged into the virtual brickwork. *Haptic rendering*[2] with force-reflecting robots makes it possible to overcome this shortcoming.

This reflection is not only of academic significance but motivated by a practical benefit for the user (or his employer). One benefit which by no means should be underestimated in this respect is the fun factor. Apparently, this is directly commercially relevant in the computer games industry. In the wake of the rapid development of computer technology, which is very much driven by computer gaming, the personal computers available on the mainstream market by now are capable to display VR scenes in a complexity and quality, which was achievable solely with specialized graphics workstations from SGI, SUN, and IBM still a few years ago. Thus, VR technologies are about to enter the mass market: A lot of computer games in particular from the genre of so-called first-person shooters or motorsport simulations with their excellent graphics and realistic physics simulations can—albeit with some

---

[2] The term *rendering* generally denotes the presentation process done by a computer system. Without further definition, often the image generation process for the graphical representation is meant, but analogously, the production of structured, information-carrying stimuli for other modalities is called acoustic, olfactory, or just haptic rendering.

conceptual concessions—be called "virtual reality." Today, the wealth of detail of these virtual worlds is enormous; immersion happens very quickly; you forget that you play; almost everything behaves plausibly. The interaction, however, is hardly natural or intuitive as long as the game is controlled by keyboard and mouse. Suitable input devices fill this gap partly: The motorsport simulation makes much more fun immediately when you have a vibrating steering wheel in your hands and throttles below your feet—and the driving performance gets better.

This factor, better user performance, is also a driving force for the use of virtual reality in industrial or medical applications. Another major aspect is the possibility to save costs, while increasing the quality of products at the same time by using virtual prototypes. The automotive industry as a pioneer in this area demands, for example, that vehicles could be presented ahead of the construction of any physical prototype in life size in photo quality and accurate, verifiable, and interactively adjustable lighting, enabling designers to decide on form details, the car paint, and exact color of the interior. In the best case, the software should be that intuitive that nobody needs an explanation on how to use it.

In the same way, virtual prototypes are used to examine whether the vehicle can be assembled as planned. Since the worker himself is subject to this analysis, also it suggests itself to let a real person do the virtual assembly. This approach is superior in many aspects to a noninteractive assembly simulation relying only on a human model.

To get applicable results from this kind of investigation, the virtual assembly scenario has to be as realistic as possible. Particularly in situations with narrow construction spaces that are difficult to access, haptic feedback for collisions must be provided, because the mechanic needs and uses it more or less consciously for finding his way and accurately placing either the tool or the component.

Quite similarly, haptic feedback vastly improves interactive simulations of medical interventions for the training of certain surgical procedures. Obviously, the manipulative fine motor skills of the surgeon are of central importance during operations. Particularly in minimal-invasive surgery, much practice is needed for achieving these skills, since the physician cannot interact in the operation field with his fingers, but has to do it in an entirely unfamiliar manner: The interaction is indirect using long thin instruments, which are inserted into the situs through small incisions. These punctures act as pivot points, so that some of the movement directions are mirrored relatively to the movement of the hands. Additionally, also the view is not direct, but via a screen showing an image taken by an endoscopic camera. The spatial orientation and hand–eye coordination in this situation have to be learned and trained intensively. Traditionally, models or animal cadavers are used, but more and more VR training simulators are establishing in this domain. Providing realistic visualization, simulation of soft tissue deformation, haptic feedback, and tools for training analysis and evaluation, they are much more flexible concerning different scenarios and much better suited for tracing the learning progress.

12.2 Design and Architecture of VR Systems

12.2.1 Hardware Components

Generally, the central hardware component of virtual reality systems consists of high-performance computers or clusters of computers which communicate via network. Connected to them are input devices and display systems using different interfaces.

The input devices mouse and keyboard as known from desktop computers are only of minor importance in VR applications. To fulfill the claim of natural interaction, many other input apparatuses have been developed. Central are so-called tracking systems, which offer absolute three-dimensional measuring of positions and mostly also of orientations in space, i.e., these provide six degrees of freedom (DoF). Using a tracking system, it is possible to locate an interaction device exactly as well as to measure the (head) position and viewing direction of the user. The latter is among others necessary to achieve a visual rendering with correct perspective of the virtual environment or an appropriate spatial acoustics simulation. Several technologies are used for 6 DoF tracking: Magnetic tracking systems consist of sensors measuring the magnetic field, which is emitted by a field generator. Analogously, ultrasound tracking systems have ultrasound emitters and sensors. Another class are camera-based systems, where image processing methods are used to detect selected features of the real world (so-called marker) in video streams. Their relative location is then calculated by taking the camera parameters into account. Usually, such tracking systems operate with an update rate of 30–50 Hz. Next to the update rate, also the latency between measuring and provision of the data in the VR system is important: If it gets too large, noticeable delays between user action and the visual result arise, which destroy immersion: When, for example, the image follows with some time offset when the head is moved, immediately the impression gets lost that the perceived visual stimuli come from objects with fixed location in space.

On the side of output devices, several technical approaches are used for generating specific stimuli for the different sense modalities. For the auditory perception, it is the task of speaker systems, from headphones to surround sound systems and finally to speaker arrays by which precisely located sound events could be generated.

However, by far, the most attention is given to the most dominant distance sense of man, the visual sense. Based on the physiological model of the trichromatic vision, almost all optical display systems generate image frames at frequencies between 50 and 120 Hz, which are built of square picture elements (*pixels*) mixing the additive primary colors red, green, and blue. The impression of continuity of motion is created by means of that the change of images is fast enough to trick the human vision system, which commonly happens from about 15 Hz [5].

To generate the impression of spatial depth, stereoscopic displays are of exceeding importance in VR, because the presentation of a separate image with a different perspective for each eye is the most effective way to stimulate the depth impression at least for people with normal seeing capabilities.

The sizes of graphical displays used in VR applications range from small optical units in data glasses or *head-mounted displays* (HMD) to large tiled projections beamed by a bunch of projectors or projections onto the sides of a big cube which one enters into a so-called CAVE (recursive acronym for *cave automatic virtual environment*). Standard monitors or autostereoscopic monitors are also used, but they do not meet the demand for immersion very well.

Force-reflecting haptic devices take an exceptional position among the hardware components of VR systems: They are both input and output devices for they provide the input functionality of (mechanical) tracking systems in their *passive* degrees of freedom and also the (stimulus) output functionality by force transmission in their *active* DoF.

12.2.2 Device Integration and Device Abstraction

Many modern VR systems implement the concept of *device servers* and *logical devices* for the integration of various input and output devices. A server is a computer or a running software process, respectively, which offers a service to other programs. The service a device server provides is to supply input data on request or receive output data following a specified protocol.[3] The device server abstracts from the special characteristics of the physical device as far as reasonable. For example, all 6 DoF tracking devices provide position and orientation data, but they differ from manufacturer to manufacturer with respect to the control sequences which are used to initialize and calibrate them and to trigger the data retrieval. These technical details could be hidden by the device server, which is adapted to the particular tracking system. It then as a service provides the logical device "position and orientation" comprising a standardized protocol for the transfer of control and payload data. This service furthermore can occur *network transparently*: The input or output device may be connected to another computer in the network, if the communication protocol between VR system and device server is designed accordingly and suitable configuration methods enable the VR systems to gain access to the remote server.

The concept of logical devices disburdens the application developer who designs a virtual scenario from thinking about the exact type, e.g., of the tracking system to be used. Later on, it is even possible to replace an absolute by a relative 6 DoF input device like the so-called SpaceMouse[4] without the need to change anything of the scenario, as long as the device driver of the SpaceMouse also implements the logical device "position and orientation." By the way, this concept is quite common in the graphical "desktop" user interfaces of modern PCs: Computer mice which are

[3] The term *protocol* refers to a scheme which describes the semantics of an otherwise abstract data stream and determines the order of its elements.

[4] The SpaceMouse consists of a knob roughly of the size of the palm of the hand. It includes sensors, which measure forces and torques the user exerts on it. The internal software converts these to relative movements.

connected via the USB port register themselves as *human interface device* (HID). Included device drivers take care that the operating system can use the device as a "pointing device."

Quite similarly, at least on a single computer, the VR system developer[5] need not to worry about servers for the output of audio and image data as the operating system and standard libraries provide software interfaces to the respective device drivers, which control the hardware, i.e., the sound and graphics card. The latter on their own are responsible for sending appropriate signals to loudspeakers or monitors via their hardware interfaces. As soon as more complex output systems on distributed systems get involved though, the standard driver interfaces are not sufficient any more and VR systems have to implement the needed functionality itself.

This holds, e.g., for graphical display systems as tiled projection screens or CAVEs, which often are driven by clusters of computers, each one of it controlling one or more projectors. Server processes running on all of these computers collectively provide a service for the output of the composed image.[6] Similar concepts are used, when the acoustic output should be carried out by a separate audio workstation.

The device technologies in the area of VR are utterly varied, the development of concepts is still very dynamic, and thus, no common standards have been established yet. Every VR system defines its own protocols for device connections.[7] It would be desirable, if there were cross-system classes for haptic devices similar to the USB device class HID in the future. They should be based on popular interfaces which are capable of real-time data transmission as, e.g., IEEE 1394[8] (Sect. 11.3).

Whereas small latencies in the data transmission via networks are mostly unproblematic for pure input devices, they become critical if they occur in the control loop of haptic devices. Therefore, a network transparent device abstraction is not suitable for haptics without further ado. We go into that in more detail for the special case of the software renderer in Sect. 12.2.5 and relating to the latencies in Chap. 11.

Another aspect of device abstraction is of major relevance for haptic devices: The application developer is used to model the virtual environment in a Cartesian coordinate system, and in the same way, simulations are typically performed in Cartesian coordinates. The necessary conversion of the input data from device coordinates to Cartesian coordinates should be "hidden" from the application by the device driver or server as well as the transformation of the Cartesian output forces and moments into motor forces and moments and the appropriate control currents. Both conversions

---

[5] Very often, one distinguishes between the system developer, who creates the software infrastructure (in our case, the VR system libraries and executables), and the application developer, who uses this infrastructure in order to model a virtual environment with object data and behavioral descriptions.

[6] On UNIX-like systems with graphical user interface based on the X-Windows system, this concept is well established. Any computer with a running X-Server can display graphical output for all computers in the network which use the X protocol.

[7] There are some efforts to develop software libraries which define a standard server interface for various input devices, e.g., the free open source VRPN (*virtual reality peripheral network*) system.

[8] Apple's *FireWire*, SONYs *i.Link*.

need internal knowledge about the kinematics of the device and the mechanical and electronic design, which almost without exception are irrelevant for the design of the virtual scenario.

12.2.3 Software Components

The central software instance of a VR system is often called object manager or VR kernel. It manages the scene, i.e., all objects of the virtual environment, and controls the data flow between all hardware and software components.

In many cases, the central data structure is a so-called scene graph. It structures objects into groups in a transformation hierarchy and defines their relations. The whole virtual environment shares a global world coordinate system. Nodes that hold transformations define the relative location of their subordinate objects in the virtual world, and parent–child relations in the graph represent spatial and kinematic dependencies.

For example, the four geometry objects, which represent the wheels of a vehicle, are connected via transform nodes to the node, which contains the car body, which itself is linked to the world coordinate system via a transformation node (Fig. 12.1). By changing the parameters of the transform nodes over time, it is easily possible to describe the dependent movements of the individual parts of the car, i.e., the rotation of the wheels around their axes relative to the car body as well as the superimposition of the forward movement of the whole vehicle.

The object manager furthermore manages the behavioral descriptions of the scene. These define the effects of an "event" on the objects and which subsequent events are initiated by it.

According to the implementation concepts of the particular VR system, these dependencies are coded directly in scripts or parameterizable modules or they are expressed in so-called behavior graphs or a combination of both is used.

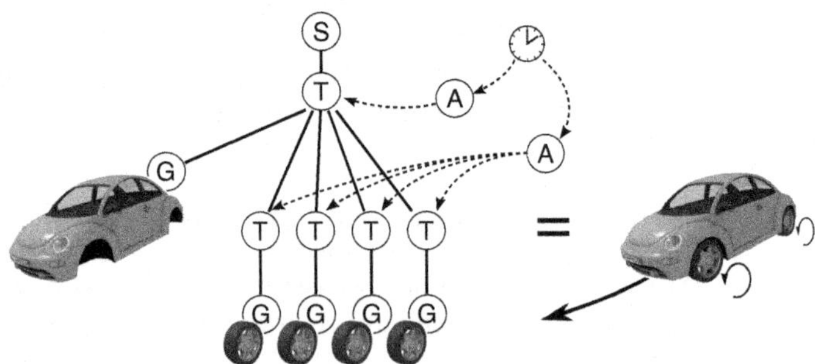

Fig. 12.1 Simple scene and behavior graph of a driving vehicle: *S* root node of the scene, *T* transformation node, *G* geometry node, *A* animation node, *clock symbol* timer. Routes are *dotted*

After loading the scene description and setup of all data structures, the VR kernel starts a cycle repeating upon system termination, which can be phased into three steps[9]:

1. Collecting all events
2. Propagation of the events through the behavior graph or triggering the event processing in the appropriate modules, respectively, and
3. Display of the final state of the cycle on all output devices

The first phase among others comprises of requesting the data of the input devices via the device servers depicted above. In the second phase, the event data are passed to the various simulations and these calculate the next step. Since simulations may run concurrently as described below, this phase ends with a synchronization leading to a consistent state, which is interpreted by the different renderers for the individual sense modalities to create the respective display data.

Events, Event Propagation, and Behavior Graph

There are several sources for events: Obviously, user actions via input devices are events, which occur either at discrete points in time (finally you only click a mouse button now and then) or when using a tracking system at every time step, since tracking systems usually send data continuously.

Less obvious are those events, which originate from internal changes as the stepping forward of the simulation time, which triggers the calculation of a new state of animations or simulations. Moreover, there are events, which result indirectly from user input or simulations as, for example, the collision of two moved objects.

On the implementation side, events are represented by state changes of objects, i.e., by changes of the values of internal variables of them, which define their current properties.

The task of the object manager is to propagate event changes according to the scene description. This means that it has to send messages to all concerned objects to inform them about relevant events, which they should react onto. To illustrate the underlying principles, we will utilize the terminology of the ISO standard X3D [3]. As a rule, even VR systems that do not implement this standard work with similar concepts.

Besides the software object instances that correlate with those objects of the virtual environment, which are arranged in the scene graph and rendered by a renderer (visible objects by the graphic renderer, sound sources by the acoustic renderer, and tangible objects by the haptic renderer), there are software object instances which exclusively provide behavioral functionality. Included among others, these are simulations or animations, which, for example, store a predefined motion to be played back, controlled by a timer.

Both object types—scene and behavioral objects—are stored as nodes in the behavior graph. Therefore, the scene graph could be considered as a subgraph

---

9 Basically, this process is the same for every interactive software system including all widely used programs that have graphical user interfaces—from word processors up to computer games.

embedded into the behavior graph. In addition to the parent–child relations of the transformation hierarchy, the behavior graph contains directed edges which model reaction chains of events. These edges are called *routes* (in X3D).

Routes deliver messages between nodes. In the simplest case, a message is the transfer of a new value: An animation may, for example, generate a new position which should be transferred to the transform node belonging to the object to be animated. Within other concepts, the message transfer correlates with the call of a function of the receiving node which may take several parameters. The state change initiated by the message in the receiving node very often results in the generation of new events, which again have to be transferred to linked nodes. This process of distributing or *propagating* messages is called *event cascade*.

We will give a simple scenario to exemplify this concept: For the replay of a time-controlled animation, a timer is instantiated in the scene. It generates a time signal for each run of the VR application loop (the event is "a new loop has been started"). The event is transferred to an animation node by a message which contains the new time stamp. The animation node evaluates a stored function which maps the time value to a new position value (this evaluation is the subsequent event). The event is propagated to a transform node, which sets the new position value in the transformation it holds internally. When the next rendering is done (Sect. 12.2.5), all objects that are descendants to the transform node in the hierarchy will appear at the new position.

This composition of behavioral and scene objects makes it possible to construct networks of dependencies which model very complex behavior inside a virtual world. Messages about events commonly are not only sent to one but many objects (by creating several outgoing routes). In the same way, an object may receive many types of messages and react differently to them.

The control of this data flow is the responsibility of the object manager. In some constellations of the event cascade, cycles could happen which have to be broken to avoid infinite loops, which otherwise would lead to a seemingly frozen system. The events that could not be processed for that reason in one time step but should not get lost are fed into the event cascade of the next time step.

12.2.4 Simulation

Despite the diversity of behavior that could be created by the composition of simple behavioral elements, it is not possible to describe everything what eventually is of interest in a virtual environment. Many desired behavioral properties of a scenario cannot be defined by simple timer or event-controlled processes but need more complex simulations. For example, it is not feasible to predefine animations for every single activity during a surgical operation—taking hold of tissue, clamping, cutting at arbitrary spots, and suturing—which would properly produce local deformations or even topological changes due to incisions. Rather, this behavior of organs must be simulated in real time, and depending upon the use case, different levels of physical correctness are necessary: If, e.g., the focus is to train dexterity (motor skills), the reactions of the virtual tissue only need to be plausible and a kind of close to reality. If on the other hand, a lesson is about improving diagnostic abilities (sensor skills and interpretation), where specific properties of anatomical structures should be examined for pathological alterations, these have to be simulated with much higher accuracy and realism.

To implement the relations that have been identified during simulation modeling, usually data structures and calculation methods are needed, which are not part of a general VR system. Very often, moreover, there already exists an implementation of the physical model in a specialized library or system for physical simulation, which at the best could be reused and integrated into a given VR system. To do so, a concept is needed how the application loop of the VR system exchanges data with the simulation component and synchronizes with it.

Simple simulations possibly may be implemented directly as nodes of the behavior graph to be processed within the application loop. The node encapsulates the interface of the simulation library, delivers the events it receives as input data to the simulation, and feeds the results of it as subsequent events into the event cascade.

However, this approach is only suitable if the simulation does not need too much computing power. Inside the application loop, only a restricted amount of computation time can be allocated for performing a simulation step, since the various output systems need to be updated at regular intervals in order to guarantee a minimal graphical frame rate for example. Otherwise, the virtual environment may feel sluggish or start to judder. Even if the simulation itself fulfills real-time demands, it may happen that the necessary computing resources are not available in the context of the complete system which also has to manage input and output devices and process the event cascade. In this case, it is inevitable to decouple the simulation from the cycle of the application main loop.

To support the clarification of the problem, we will at first discuss the term "real time", which is used more sloppy in the domain of computer graphics as, e.g., in robotics: One essential temporal requirement of a graphical display is obviously to nearly always achieve update rates above a frequency of 15 Hz to create the visual impression of motion and avoid perceivable delays between user interaction and the reaction of the system. This often is already taken to be enough to refer to it as real-time computer graphics. More precisely, though it would be to say that an application provides *interactive computer graphics* to emphasize that the system is fast enough for fluid interaction.

When speaking of real time in the simulation domain, it is meant that the simulated time or model time proceeds as fast as the real-world time. Simulations can stretch or compress time (slow motion or fast motion) for to, for example, show processes, which are too fast or too slow in reality to observe them. Simulations of this kind do not run in real time. The underlying interpretation of "real time," however, does not define any constraint on the time needed to calculate one single simulation step.

On the other hand, real-time capability of a control system especially in robotics denotes that the result of a calculation is available within a given period for use as control signal. If it is enough to fulfill this requirement on average and exceptions are largely harmless as long as they are not too frequent, this is called "soft" real time. Within "hard" real-time constraints, the system must complete each calculation before a deadline in any case. The time to deadline may differ largely though: Some systems only require a guaranteed result in a period on the timescale of seconds, whereas others need to be clocked in microseconds.

To meet the requirement of hard real time with multitasking (see below), control computers have to run under a real-time operating systems such as LynxOS, QNX, and VxWorks. These take care that processes with respective privileges get the processor within a guaranteed time span for a defined period. Based upon this guarantee, the processes running on the system can give real-time guarantees themselves.

The problem touched above—that not enough resources can be allocated for a simulation—now presents itself as follows: A simulation may, for example, not be able to perform a simulation step during the duration of one cycle of the application main loop, which may have to run with 15 Hz. Thus, the simulation does not satisfy the soft real-time requirements of the VR system. But it possibly meets the real-time condition for simulations if it, for example, can perform a step of 0.2 seconds of model time with a frequency of 5 Hz and a processor load of 50 %. Now, the integration of the simulation into the VR system becomes possible if it is separated from the main loop into its own loop: About every three main application loop cycles, the simulation can feed new data into the behavior graph via the node associated with it. Then, the next simulation step can be started with new input data coming from other parts of the virtual world.

The concurrency of the VR main loop and the simulation is realized by outsourcing the simulation into its own so-called *thread*. Threads are a technology provided by the operating system, which allow it to let parts of a program run temporally independent from the rest of the program. This is similar to the concept of processes, which is the basis of *multitasking*. Simply speaking, the processor switches the program it processes in very short time slices. For the user, it appears as if the programs would run at the same time. On modern processors with multiple processor cores or computer systems with multiple processors, the execution of some of the programs (in case of processes) or program parts (in case of threads) actually takes place at the same time. For complex scenarios, this concept is developed even further: It is also possible to outsource the simulation onto a separate computer or a computer cluster[10] as well. This is called a *distributed* system.

The next issue is the correct synchronization between the simulation and the rest of the system: Each time data have to be exchanged between the two loops, both systems have to be in a consistent state. If a rotation matrix was only filled with new data to the half when it is transferred, this would result in data corruption. Most likely, the inconsistent copy of the rotation matrix would not have certain essential mathematical properties. Problems of this kind are avoided by defining dedicated spots in the program where the involved threads wait for each other, and thus, only "approved" data are exchanged. We do not go into the problems of the synchronization of processes and threads in more detail here but as to be expected

[10] *Cluster* in this context denotes a network of computers which collectively perform a task.

the effort to implement it becomes the more, the more complex the distribution of the system is: If threads run on *one* computer, common memory in the RAM can be used for data exchange, whereas in the case of cluster systems, network communication including inherent latency problems gets involved.

12.2.5 Subsystems for Rendering

The main purpose of a VR system is of course to display the simulated virtual environment. To do so, various rendering subsystems have to interpret the current state of the scenario regularly and create an appropriate stimulus (a rendering) for every intended sense modality.

For the most obvious one, the visual or graphical presentation, this means that about every 10–50 ms, the graphical rendering has to be triggered. Since it is only sensible to create a new image when a new state of the virtual scene is available, usually the graphical rendering is directly integrated into the main application loop of the VR system. After the event cascade has been processed and, if necessary, data from concurrent simulations have been synchronized into the main thread, the system starts a so-called traversal of the scene graph: The renderer recursively marches through the transformation hierarchy, and thus, by and by, all nodes that contain information relevant for visualization are visited. The respective data are transferred to the basic graphics library[11]: Affine transformation matrices are accumulated from data for translation, rotation, and scaling in transform nodes, material nodes provide color settings and textures, light nodes direction, color, brightness and attenuation of light sources and last but not least the triangles and other polygons stored in geometry nodes are “pumped”[12] to the graphics library. The library is responsible for transferring the data efficiently into the graphics card via appropriate drivers. The graphics card finally performs the computation of a color for each pixel autonomously and creates the output signal for the connected monitor or projector.

It is worth to notice that the graphics card controls the output device with a constant frequency [e.g., 60 Hz for TFT flat screens and preferably higher for cathode ray tube (CRT) displays] which is completely asynchronous to and independent of the rate in which images are computed. For CRT displays, a frequency of 85 Hz and above is recommendable for the reason that this avoids the perception of flicker by the receptors in the periphery of the retina, which are specialized in the detection of movement: Due to the functional principle of a CRT, the brightness of each pixel alternates, because the phosphor on the screen gets activated only for a short moment when the electron beam passes it on its zigzag way and luminesces decreasingly until activated again. On the one hand, the afterglow duration should not be too long to enable the sharp display of fast movements without smear artifacts, but on the other

[11] This basic graphics library usually is *DirectX* on Windows systems and on others the platform-independent *OpenGL* (open graphics library).

[12] It is spoken of “polygon pumps” in this context as well.

hand, a short afterglow duration abets the distracting perception of flicker. Flat screens or LCD or DLP projectors do not have this problem, because their backlight provides quasi-constant brightness and pixel colors are created by filtering.

Hence, for the graphical display, one has to distinguish between image update rate and display refresh rate. The first one is determined by the cycle time of the VR application loop and has to be high enough that single images fuse into motions, whereas the latter higher one is determined by the graphics card and drivers and adjusted to the frequency, at which flicker gets imperceptible dependent on properties of the display hardware. Obviously, there is no benefit of an image update rate of the VR system above the refresh rate of the monitor or projection system and only would wastefully occupy computing power.

Some of the concepts of acoustic rendering are similar to those of graphic rendering. To take the location of a virtual sound source into account for sound synthesis, the transformation hierarchy is evaluated and the orientation and position of the sound source as well as of the user are determined. These parameters for the generation of stereoeffects[13] are input into the audio library[14] together with *audio chunks*—the audio data fragmented into short pieces—from each sound source. The audio library forwards these data to the audio card, which generates the final output signal for the loudspeakers. The VR system has to take care that enough audio data are available for the audio card until the next update cycle. Otherwise, distracting dropouts become audible.

Humans perceive frequencies above more than 10 kHz to achieve a clear audio experience free of artifact dynamics of about 20 kHz are preferable. Due to the enormous computing power, it would need that the generation of sounds by simulating the vibration behavior of the objects in the virtual scenario in real time is impracticable. Most VR systems therefore use the trick to combine the sounds from a set of prepared samples, even if that does not allow for the full diversity of sounds which is possible when interacting with real oscillatory objects. To compensate this, the audio renderer may also interpolate between samples associated with different locations of the object the user interacts with [15].

Thus, the requirements for the graphical and acoustic rendering are determined essentially by the biological properties of the senses. Neither the visual nor the auditive sense retroact to the technical device, there is no coupling loop[15]

This may be different with acoustical rendering. There it indeed could happen that the user's interaction and a slow synthesis of the sounds lead to resonance effects, which—if the sound system is accordingly powerful—may cause acoustic

---

[13] For realistic surround sound simulation, geometry information has to be included additionally.

[14] For example, platform-independent *OpenAL* (open audio library) or specific for Microsoft *Direct-Sound3D*.

[15] Strictly speaking, even in this case, there is a feedback loop which is closed via the interaction of the user with the virtual world. The instabilities, which occur in practice, usually do not result from specific properties of the graphical or acoustical rendering, but from unstable physics simulations. Though resonances occur *in* the presentation, but not *because of* the representation. An object may jitter on the monitor, but the luminous flux from the monitor does not get chaotic thereby or overdrives and endangers the user either.

overdrive. But this problem is mostly uncritical due to the limited capacity of the sound hardware, which possibly could cause instabilities.

In contrast, the haptic rendering has besides the *physiological*, a *control theoretical* component too, because within haptic systems, there is bilateral energy exchange between haptic device and user. Therefore, as in general, for controlling coupled systems, hard real-time requirements have to be fulfilled when driving haptic devices, in particular force feedback devices: Too large latencies in the control loop—i.e., control information is not updated for too long—destabilize the system. Because typical haptic devices are tuned for low damping, this could happen in a resonance frequency and carries the danger of self-destruction of the system or hurting the user. This leads to the postulation that high control frequencies of above 1,000 Hz should be achieved for the most frequently used devices with low impedance or when high force gradients are wanted. Besides this technical reason, the high frequency has implications for the quality and fidelity of the haptic impression. Since the haptic sense perceives vibrations of above one kilohertz (Sect. 2.1), people find haptic systems which are able to display information in this range to be more realistic. This especially applies for the extreme cases, free space movement and stiff contacts.

12.2.6 Decoupling of the Haptic Renderer from Other Sense Modalities

The visual presentation needs update rates, which are oriented to the temporal resolution of the visual sense. Even for the perception in the peripheral field of view and for extra fast movements, 70 Hz is quite enough. For professional applications, even lower frequencies of about 20 Hz suffice, because their virtual objects as well as the users usually move much slower than in computer games. Thus, the requirements to a graphical renderer with respect to the update rate are about the factor 10–100 lower than to a haptic renderer. It follows that the haptic renderer has to be decoupled from the cycle of the main application loop of the VR system and the other slower renderers, for which such high update rates are neither feasible nor fruitful.[16]

The decoupling of the haptic rendering loop is implemented using the same principles as the concurrency of simulations (see Sects. 11.1 and 11.2): Following the initialization phase of the VR application, the system starts a thread, which concurrently to the application main loop executes the haptic rendering loop with high priority and frequency. It consists of the query of position data from the haptic device, collision detection, contact classification with force simulation, and finally actuating the device driver with the calculated forces and moments. At the beginning of its cycle, the main thread reads position data of the haptic device from the haptic thread

[16] Although the limit frequencies for acoustic rendering are above those of haptic rendering, it could stay in the main application loop, when the sample-based approach is used. Only a physically based sound simulation would have to be decoupled, but this should not be discussed further here.

to feed it into the event cascade. The data flow in the opposite direction is more intricate:

In complex scenes, the collision detection is the time-critical part of haptic rendering. Therefore, it is an efficient approach to reduce the amount of data on which collision detection has to be performed in the haptic thread: If the main thread only supplies the haptic thread with a local extract of the scene, the *global* collision detection could run with the lower frequency of the main loop, whereas the high-frequency collision detection only "sees" object parts, which currently are close to the haptic probe. After processing the event cascade, the main thread estimates an area from the current position and movement of the haptic probe,[17] which includes the haptic probe and the predicted path of it at least up to the next collision cycle of the main loop. The estimated area should have the smallest volume possible, but it is better to overestimate a little than to underestimate. All polygons of the scene that intersect the area are collected and pumped to the haptic thread and rendered by it. Collecting the relevant polygons itself can be implemented using collision detection algorithms by approximating the area by a polygonal hull object, which then is tested against the polygons. Obviously, the collision detection in the main thread has to be done regularly and in not too long intervals, because otherwise the prediction of the motion of the haptic probe gets too inaccurate and the portion of the scene that has to be provided to the haptic thread gets too large. This again would make the local collision detection in the haptic thread more costly.

The data flow between main application thread and haptic thread has to be designed carefully; thus, that the haptic thread never has to be stopped longer to wait for new data from the main thread. During the data exchange between threads, it is inevitable to protect the memory areas which currently are written by one thread from any, even reading access by other threads, because the data are not consistent during the write operation. To achieve this, the memory area is locked by the writing thread using operation system functionality. Any other thread running into a statement during the program execution, which needs access to the said memory area, has to be stopped until the writing thread unlocks it again. To prevent that the haptic thread gets thwarted by the main thread when it writes the local scene portion, the locking phase has to be kept short.

A good approach for this is to use *double buffering*: While the haptic thread is rendering a local copy of the scene in one memory area, the main thread fills a second memory buffer with the next local scene copy. Not before this copy is transferred completely to the buffer, the haptic thread is notified that it should start using the data in the second buffer for rendering and release the first one to be refilled by the main thread. Only for the transfer of the notification and the buffer switch, both threads must be synchronized. This should be done that way the main thread has to wait for the haptic thread if necessary but not vice versa—on the one hand because of the higher importance of the haptic rendering, on the other hand because the haptic loop reaches the synchronization point in the program more often due to the higher loop

[17] The velocity of the haptic probe can be calculated from the position and duration of the last cycle in the main thread of course, but it is more accurate, if this is done in the haptic thread also.

frequency anyways and therefore the average waiting period of the main thread is much shorter than it would be for the haptic thread conversely.

The concept of the local scene copy is also used with success, if the haptic rendering loop runs on another computer than the main application loop. For example, there are haptic devices such as the HAPTICMASTER (Fig. 6.9), which are sold with a separate driver computer on which the haptic loop is executed. In this case, it is not necessary any more to have a dedicated high-frequency haptic thread in the VR system itself. Apart from the differences in the interprocess communication on operation system level, the same principles are used for data exchange and synchronization.

12.2.7 Haptic Interaction Metaphors

As aforementioned shortly, the term *interaction metaphor* calls a concept in the context of "HCI", which defines how a user action is linked to the reaction of the computer system. The way people work with a computer today by moving the mouse to control a little arrow and to click on mouse buttons to trigger actions is based on a complex interaction metaphor, which consists of many smaller interaction metaphors. One of them is, e.g., the metaphor "drag and drop": pointing to an object, click, hold the button moving the mouse and thereby dragging the object onto another one as the icon of the tray, let it go. This interaction metaphor is patterned after the natural grasping, moving, and releasing of an item.

For the haptically enriched interaction with a virtual environment on a very basic level, two interaction metaphors can be discerned.

Ideally, if a perfect haptic device was available, the user could directly interact with all objects of the virtual environment including the sense of touch: When grasping in the very moment of the contact between fingers and hand to the virtual object, a haptic perception would occur. The fingers of the hand as well as any other part of the body would be totally free in movement without resistances except for those which are defined by the virtual world. Everywhere at the body, haptic stimuli could be provided. This shall be called *direct* haptic interaction.

Partially, it is possible: The end-effector of the first versions of the haptic device PHANTOM, for example, is a small plastic cone like a thimble where the user puts his fingertip into to move the TCP. The force feedback of the system leads to the impression that you (almost) directly touch the surface of virtual objects.

To extend this to the whole hand, it would be necessary to have many effectors, one for each finger or even better one for each phalanx and several for the palm and the back of the hand as well. Some developments of exoskeletons lead to this direction, but there are no devices yet, which could generate an even only roughly realistic impression. The software side seems to be the less demanding part of this challenge: The hand can be modeled in sufficient granularity using today's software technologies. The collision detection and force simulation between hand model and virtual environment is an ambitious, but not a fundamentally impossible task. On the side of the hardware development, even the exact tracking of all possible hand movements is extremely difficult: Each finger can be moved in four DoF and the thumb in three, and additionally, multiple motions of the metacarpus are possible. Conversely, for each of these degrees of freedom, actuators of best quality have to be plugged together in a very confined space. Even if only the fingertips are considered, the technical complexity is enormous.

Because of that, another interaction metaphor comes in handy which is used with all widespread haptic devices: a kind of indirect interaction with the virtual environment called *tool-handle metaphor*. Within this concept, one does not directly touch the virtual environment with the fingers, but indirectly via the item one holds in the hand. It is as one would touch the environment with the tip of a pen instead of the fingertip.

This admittedly is a severe restriction, but it still allows utterly valuable haptic interaction, while requiring much less effort with regard to the device technology. First of all, in many applications in reality, the haptic perception is indirect also. That is obvious for minimal-invasive surgery, but also eating with a knife and fork, feeling out the notch of a screw with the screwdriver, writing, or drawing with a pen or brush, modeling clay with tools, and doing woodwork with a chisel are examples. When simulating these applications in virtual reality, it is feasible to equate the end-effector with the tool. If the end-effector of the haptic device could be exchanged, then it is possible that the effector can be a clone of the grip of the real tool. It is noticeable that the applications mentioned above are of less explorative nature but are more related to the modification of objects.

But the tool-handle metaphor is also useful in different problems: Whenever it is suitable to equate the end-effector with a (substantially stiff) virtual object which should be moved around in the virtual environment, it is a good choice to use a real representative of it as end-effector. In combination with rapid prototyping, this approach is of special interest for assembly simulation as aforementioned in the introduction. The part that should be tested for assembly ease in VR can be produced by stereolithography and mounted as end-effector. The mechanic simply takes and moves it as in the real assembly situation.

The choice of the interaction metaphor plays an essential role for the design of the device as well as for the planning of the software application and the algorithms to be used.

12.3 Algorithms

Depending on the purpose of a visual presentation of virtual objects, different properties can be used, and others can be ignored. Many aspects such as contour, color, reflective properties, transparency, and local color variations (texture) contribute to a realistic visual impression. The most important properties, however, are those which allow us to recognize the shape of the object. Consequently, first efforts were aimed at contours and shape-defining edges during the historical development of computer graphics. With increasing computing power of the hardware and improved capabilities of the software, the generation of brightness variations, color, and other visual object properties could be realized.

A very similar approach can be found with haptic rendering, enabling the user to perceive shapes first and material properties second. Such impressions are generated when body parts of the user—usually finger or hand—hit a mechanical resistance during explorative movements. This resistance has to show certain properties: With regard to haptic rendering, a spatial and temporal coherence has to exist, it has to depend on the exerted force, and there must be a spatial relation between user and the resistance. Consequently, a haptic device has to be able to generate varying forces at different positions in space.

To get a three-dimensional representation of the involved objects, they must be described in a mathematical model first. In general, the same representations can be used as in 3D computer graphics applications. Instead of those attributes necessary for a visual representation, properties such as stiffness, elasticity (hardness), roughness, or stickiness stand in.

The dominating representation to define spatial structures within the so-called real-time computer graphics are sets of triangles approximating the surface of an object. Even CAD systems providing surface representations of higher order[18] for modeling generate a triangle-based intermediate representation to render on the screen. This is a concession to the fact that today's graphic hardware is highly optimized for the handling of triangles. In special application areas (e.g., image acquisition technology in medical context like computed tomography or magnet resonance tomography), direct volume models are visualized via *direct volume rendering* (DVR), making use of regular 3D grids of scalar or vector data (voxels). Both representations may be used for haptic simulations too; however, additional data structures for collision detection (Sect. 12.3.5) have to be generated.

Besides the virtual objects, the interacting user has to be represented within the virtual environment too. In the graphical rendering, the corresponding counterpart of the eye would be the camera, which is characterized by position, orientation, aperture angle, and other parameters. For the haptic rendering, at least the hand—for the most frequent type of interaction in virtual worlds, which is manual interaction—is represented by a point in space. Its coordinates are derived from the position of the handle of the haptic device, the ↪TCP. However, this point in the virtual environment has to be distinguished from the TCP in the real world. In general, coordinate transformations are necessary to match the position data from the device's TCP to the position of the point in the virtual environment. A simple example would be a molecular construction kit with haptic feedback, making it necessary to scale movements by several orders of magnitude. The representation of the TCP in the virtual scene is called *haptic interaction point*, or short HIP, in the following.

For some haptic rendering algorithms, another object besides the HIP exists to represent the user interaction. This object may be pointlike or more complex. It is called *haptic probe*. Whereas the HIP always represents the position of the TCP within the virtual scene, the haptic probe is the object used to figure out the contact

---

[18] E.g., nonuniform rational b-spline tensor surfaces (NURBS).

situation. In so-called penalty-based approaches (Sect. 12.3.2), the position of HIP and haptic probe is identical, whereas they may differ in other methods.

The most basic component of a haptic rendering algorithm is a collision detection tuned to the object representation. It calculates the localization of the contacts between user and virtual object. The general concepts of collision detection are subject to Sect. 12.3.5.

The collision detection is strongly integrated within the total structure of haptic rendering algorithm. Put in a scheme, the cycle of haptic rendering can be systematized as follows:

- Position/orientation detection of the user's interaction
- Collision detection and evaluation
- Simulation
- Force or position output via the haptic interaction device

The following sections present several methods to render virtual objects haptically.

12.3.1 Virtual Wall

The simplest haptic model enables the virtual touch of a smooth plain surface with a dotlike probe, as it would happen with, e.g., a hard pencil's tip. For this purpose, a haptic device is actuated that way that no resistance is exerted as long as the user moves in free space. For simplification, we assume the free space to be the half space with positive x coordinate. Then, any movement of the HIP through the yz-plane would get immediately slowed down by a counterforce $\mathbf{F}$. The user perceives this constraint as contact with a virtual wall.

The calculation of the counterforce in the simplest format is as follows:

$$\mathbf{F} = \begin{cases} K \times (x, 0, 0) & \text{for } x < 0 \\ (0, 0, 0) & \text{else} \end{cases} \tag{12.1}$$

with K being the proportional factor which controls the stiffness of the wall. Conceptional a spring according to HOOK's law is simulated, being effective in the normal direction of the yz-plane only (Fig. 12.2). Movement components are not constrained in parallel with the plane. This makes the plane feel perfectly smooth. A large spring constant K results in a fast force increase, making the virtual wall feel stiff, whereas a small K generates the impression of a soft wall, feeling like made of foam plastic.

This model of the virtual wall also could be interpreted as force field. In front of the wall (the yz plane), the force field is zero. Behind the wall surface, the potential of the field increases proportionally to the distance to the surface.

This model can be easily generalized to a plane of arbitrary orientation and position. If a plane is defined in the HESSE's normal form with a normalized normal vector $\mathbf{n}$ and a distance d of the plane to the origin according to

Fig. 12.2 Virtual wall

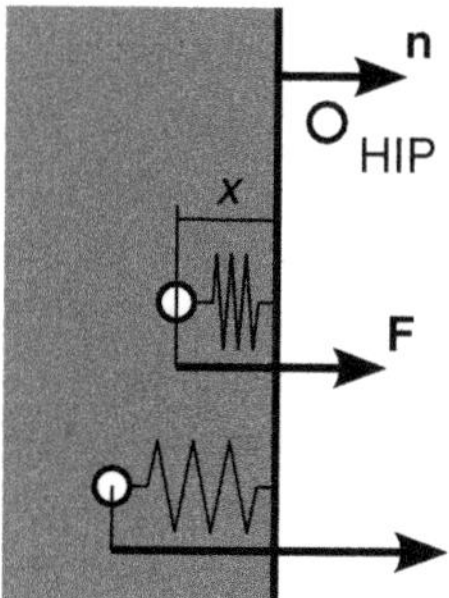

$$d = \mathbf{xn} = \begin{pmatrix} x_0 \\ x_1 \\ x_2 \end{pmatrix} \begin{pmatrix} n_0 \\ n_1 \\ n_2 \end{pmatrix} \tag{12.2}$$

the calculation of the force **F** for a position **p** of the HIP is given as follows:

$$\mathbf{F} = \begin{cases} K(d - \mathbf{pn})\mathbf{n} & \text{for } \mathbf{pn} < d \\ 0 & \text{else} \end{cases} \tag{12.3}$$

The formulation of the control loop of the haptic rendering for the virtual wall is given as follows:

```
repeat:
  read HIP position p
  s = d - DotProduct(p, n)
  if s > 0:
    F = ScalarMultiplication(K * s, n)
  else:
    F = 0
  actuate HIP with F
```

If K increases, the control loop has to be processed with an increased frequency to guarantee stability of the whole system. As a result of dead times from discretization and other system components, the system tends to generate more energy with increasing product of K and δt. This makes the virtual wall to feel not as passive anymore as it is "expected" from a real wall. If the damping is not sufficient, the resulting energy makes the system oscillate in its resonance, which can be perceived easily (Sect. 7.2.2.3).

If an additional damping hardware or some closed-loop control in an external unit is not possible or reasonable (Chap. 7), some additional damping in the control loop of the haptic rendering may increase the stability of the overall system. One option is to add some damping force to the spring force, actuating against the movement of the HIP and being proportional to its velocity **v** with a proportional factor D, the damping constant. Consequently, a spring-damper system is

$$\mathbf{F} = \begin{cases} K\mathbf{q} - D\mathbf{v}, & \text{for } s > 0 \\ 0 & \text{else} \end{cases} \tag{12.4}$$

with $s = d - \mathbf{pn}$ and $\mathbf{q} = s \times \mathbf{n}$ results. The vector $\mathbf{q}$ describes the distance between the HIP $\mathbf{p}$ from the plane in the reverse normal direction, being the penetration depth in the virtual wall. To still render an ideally smooth surface, it has to be made sure to use the normal component of the velocity for the damping only: $\mathbf{v} = \dot{\mathbf{q}}$. If otherwise $\mathbf{v} = \dot{\mathbf{p}}$ is used, the impression of friction on the surface is generated. This impression, however, is not realistic, as it is independent of the contact force and shows no transition from static to dynamic friction.

In the implementation[19] of the control loop, the velocity $\mathbf{v}$ is approximated by the differential quotient $\frac{\mathbf{q}_i - \mathbf{q}_{i-1}}{t_i - t_{i-1}}$ of the values from the actual time step (i) and the time step before $(i-1)$:

```
initialize q_old, t_old
repeat:
  read HIP position p
  read current time t
  s = d - DotProduct(p, n)
  q = ScalarMultiplication(s, n)
  if s > 0:
    v = ScalarMultiplication(1/(t_old - t),(q_old - q))
    F = ScalarMultiplication(K, q) -
        ScalarMultiplication(D, v)
  else:
    F = 0
  display F at HIP
  q_old = q
  t_old = t
```

It is obvious that even with damping included, the duration of a control cycle should not get too long, as the quality of the velocity approximation degrades. Nevertheless, a clever tuning of the parameters K and D may achieve some good haptic quality even with slow sampling rates. If oneself is in the lucky situation to cooperate with the hardware engineers and to be able to influence the hardware architecture, it is recommended to examine the whole signal processing chain including hardware

[19] In the implementation of time-critical systems such as haptic control loops, care should be taken not to make repetitive calculations of similar or unnecessary values. Modern compilers are able to optimize such calculations out of the machine code. Some manual "optimizations" may even hinder the compiler to generate an optimized code. This makes it necessary to optimize the algorithm first by some complexity analysis. Afterward, it is implemented, and only after careful considerations, detail optimizations should be made to the code. Anyway tests should be made to prove that the optimized code is performing better than the nonoptimized version. Only very experienced programmers are able to compete against good compilers in this discipline.

and control architecture according to Chaps. 6, 7, and 11. Usually, this makes it possible to find a much more optimized solution than tuning the rendering algorithm only.

12.3.2 "Penalty" Methods

The haptic rendering method of a virtual wall in the previous section follows a so-called penalty-based approach. These methods are characterized by a cycle starting with a collision detection to check whether the HIP is within the boundaries of the object—whether it is "in contact".[20] Afterward the contact is classified, a repulsion or "penalty" force is calculated. This force is directed against the impact direction and tries to push the HIP out of the boundaries of the object.

The "classification of contact" characterizes the penetration depth and direction as a basis for the calculation of the penalty force. This step is trivial for the virtual wall and resembles the calculation of the vector $\mathbf{q}$. The function of the collision detection in this simple model is covered by the query whether $s > 0$. Collision detection and classification with complex objects requires much more efforts. In fact, the most time-critical part of real applications is given by the implementation of the collision detection.

As written before, penalty-based methods with dotlike haptic probes do not have to distinguish between their own position and the position of the HIP. An obvious definition of the penetration depth and direction is given by the minimum distance of the HIP to the surface of the virtual object (Fig. 12.3). The repulsive force shall try to move the HIP to the nearest surface point. This allows the calculation of the penetration depth and direction for every point within the object in advance, as it does not change during interaction. In this case, the classification of the contact can be done very efficient in a preprocess, which will be described for the case of the polygonal representation of virtual objects as polygonal surfaces in the following.

Besides some negligible exceptions, every point of the object's interior has a minimum distance to exactly one point of the object's surface. Polygonal surface models have the features vertices, edges, and surfaces. All points of the object's

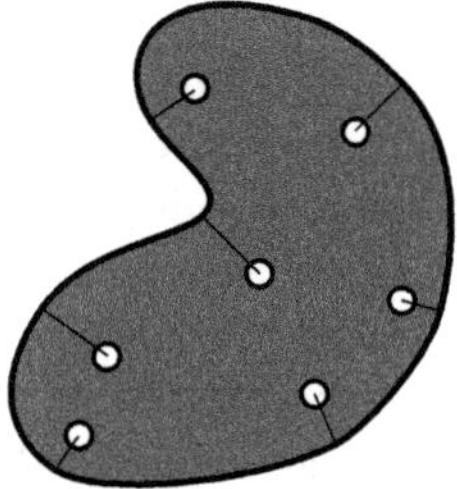

Fig. 12.3 Minimum distance between several points and object surface

[20] The exceptional case of the HIP being exactly on the virtual surface at the moment of measure is very unlikely due to the time discretization. Nevertheless, formally, it is part of the contact situation.

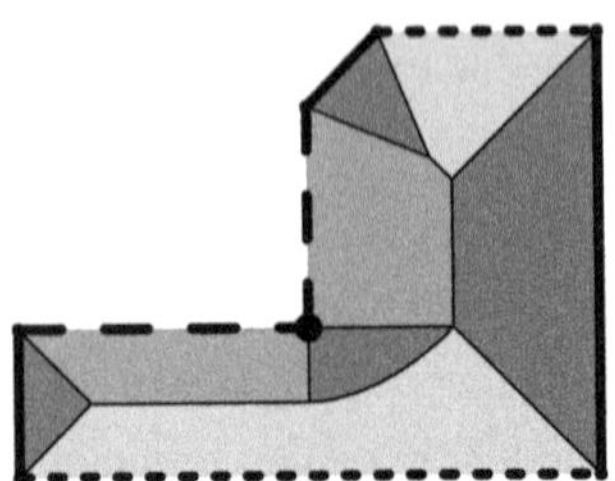

Fig. 12.4 Inner VORONOI regions of an object. Areas in *light gray* belong to *dotted* features, *medium gray* to *dashed* features, and *dark gray* to drawn through features (and the *thick dot* representing a convex vertex)

interior that have an equal distance to the same feature—the same edge, surface, or vertex—can be summarized in a common region, called VORONOI *region* [8] of the attribute. The VORONOI regions of all features partition the interior of the object (Fig. 12.4). The partition is computed via so-called *medial axes transformation* (MAT), which will not be detailed further in the context of the book. The term origins from the name of the boundary surfaces of the said regions which are called *medial surfaces* (and in the two-dimensional case *medial axes*). They consist of the set of points which have minimum distance to more than one single surface point.

Each Voroni region of a feature of the object's surface is bounded by the feature and medial surfaces. Convex polyhedrons have flat medial surfaces, whereas in general, VORONOI regions of concave edges or vertices show curved medial surfaces. To represent VORONOI regions, they usually are approximated by polyhedrons themselves.

During runtime of the haptic rendering algorithm, the VORONOI region has to be identified, which contains the HIP. If the HIP is exactly on a medial surface during one cycle (a usually purely theoretical situation), the above definition of the penetration direction is not unique. In this case, one neighboring area has to be chosen, e.g., that one lying into the direction of the actual movement. The search for the VORONOI region can be done with a classic hierarchical bounding volume algorithm (Sect. 12.3.5).

Once the VORONOI region has been identified, the rest of the classification algorithm is simple: The vector from the HIP to the closest point of the adjacent feature is calculated, which is a standard procedure in any geometry calculation.

Penalty-based approaches have some fundamental difficulties though. The classification of the contact and the identification of the penetration depth and direction consider the actual situation only. This results in instabilities at the borders of the VORONOI regions with perceivable and frequently annoying changes in the force direction and intensity. Particularly with thin structures and relatively small spring constants, this effect becomes very dominant, when the HIP switches into the region corresponding to an object surface opposite to the original penetration point. As a result, an unexpected change in force direction happens; the original resistance is replaced by a force acting in the direction of movement pushing the HIP out of the object on the opposite surface. Figure 12.5 visualizes this situation.

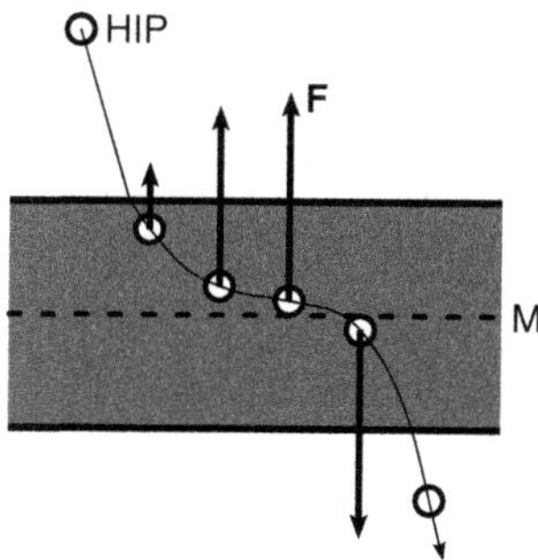

Fig. 12.5 "Tunneling through" thin structures: sudden inversion of the force direction when crossing the medial axis (M)

This conceptional disadvantage—not to cover the "history" of the contact—and its consequence lead to the constraint-based methods presented in the following.

12.3.3 Constraint-Based Methods

The principle of constraint-based methods is based on the approach to model the virtual objects of the scene as borders of the free space impenetrable for the haptic probe. Whereas the user-guided HIP is able to penetrate an object, the haptic probe is kept on the surface of the object and additionally constrained to keep the minimal distance to the HIP. From the relative position between HIP and haptic probe, a force is calculated pulling the HIP in direction to the haptic probe. This force is made perceivable for the user when displayed by the haptic device.

In 1995, Zilles and Salisbury [19] proposed one representative of this algorithmic class called *god-object*[21] algorithm (also named *surface contact point* algorithm). This algorithm is one of the most central algorithms for haptic rendering and is implemented in many haptic rendering libraries. It solves the major problems of the penalty-based methods described before and enables a one-point interaction with polygonal surface models. The algorithm is suitable for the control of haptic devices with three active translatory DoF.

The method can be described best when following the trajectory of the HIP and a virtual probe in contact with a virtual object (Fig. 12.6). Initially, the HIP is in free space. The location of the haptic probe $\mathbf{p}_0$ is identical to the HIP. Now, the user moves the handle of the haptic device. The algorithm checks whether the haptic probe is able to follow the HIP by performing a ray intersection test against the scene. If the movement does not cross an object's surface, the haptic probe is set to the new HIP position and the next cycle begins. Otherwise, the crossed polygon's supporting

[21] The haptic probe is called "god-object" by the authors, and another typical term is *virtual proxy*.

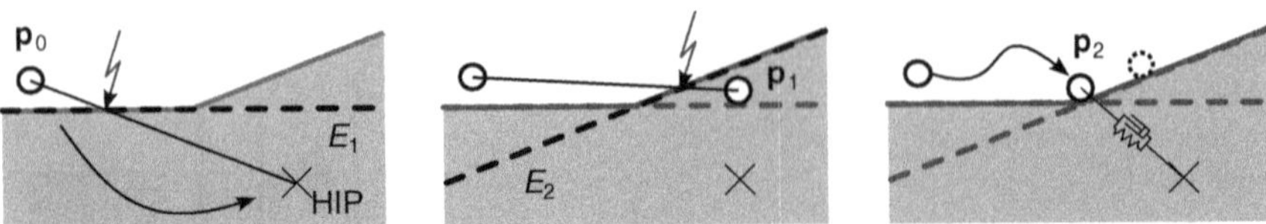

Fig. 12.6 Procedure of the god-object algorithm (*diagram* for 2D case). The optimum position of the haptic probe is *dotted*

plane[22] E_1 is set as a constraint to the haptic probe and limits its movement: The point $\mathbf{p}_1$ on the plane has to be identified, which has the minimum distance to the HIP. This is the new candidate for the position of the haptic probe. But before setting this point, it has to be checked whether the passage way between $\mathbf{p}_0$ and $\mathbf{p}_1$ is not blocked. With nonconvex objects, it may happen that a second polygon has to be considered, resulting in its supporting plane E_2 becoming another constraint to the position of the haptic probe. In a three-dimensional case, the position of the probe is limited to the intersection line of both supporting planes. The new candidate $\mathbf{p}_2$ for the position of the haptic probe is the point nearest to the HIP on this line. A third time the passage way between the points $\mathbf{p}_0$ and $\mathbf{p}_2$ has to be checked. It may be that the supporting plane of another polygon limits the possible positions of the haptic probe on the collective intersection point of all planes. Otherwise, $\mathbf{p}_2$ is the new location of the haptic probe. As three supporting planes constrain the possible position of the haptic probe to one point already, another test in the same cycle is not necessary.

After the new position of the haptic probe was found, a spring-damper system (Eq. 12.4) between HIP and haptic probe is used to calculate the force to be displayed haptically.

It should be noted that the algorithm does not guarantee that the haptic probe reaches a local minimum distance to the HIP within one time step. Such a situation is displayed in Fig. 12.6, where the optimum point lies on the second supporting plane and not on the first one. Nevertheless, the algorithm limits the haptic probe to the intersection lines of both supporting planes. This is not critical, as already within the next time step, the haptic probe moves into the direction of the optimum point. Practically due to the high update rates, no artifacts can be perceived, as the method converges in a few time steps to the optimum position of the haptic probe. The theoretical disadvantage that the algorithm in some rare cases may need several time steps to converge is compensated by the advantage that it terminates for one time step after three tests at the latest: A result is guaranteed to be computed within a reliable time frame.

[22] The *supporting plane* is the infinitely extended plane the points of a flat polygon lie on. If the vertices of a polygon are not in one plane, it has to be subdivided into flat polygons before, usually into triangles. This step should go into a preprocess and will not be discussed in further detail in this book.

In this algorithm, the largest part of the runtime is usually consumed by the ray intersection tests for the identification of constraint planes. This makes it necessary to use sophisticated collision detection methods. On the contrary, the minimization of the distance while satisfying the plane constraints is less demanding. It is done as follows: The distance between haptic probe **p** and HIP **q** is minimized by searching the minimum of the energy function (Eq. 12.5) with one to three constraints given by the plane equations (Eq. 12.6).

$$f(\mathbf{p}) = \frac{1}{2}(\mathbf{p} - \mathbf{q})^2 \tag{12.5}$$

$$E_i\colon \mathbf{n_i p} - d_i = 0 \tag{12.6}$$

By the method of LAGRANGE multipliers, the function to be minimized can be set up as

$$h(\mathbf{p}, \lambda_1, \lambda_2, \lambda_3) = \frac{1}{2}(\mathbf{p} - \mathbf{q})^2 + \sum_{i=1}^{3} \lambda_i (\mathbf{n_i p} - d_i) \tag{12.7}$$

The minimum is found by setting the partial derivatives $\frac{\delta h}{\delta p_i}$ and $\frac{\delta h}{\delta p_i}$ to zero and solving the resulting system of equations:

$$\begin{pmatrix} 1 & 0 & 0 & n_{10} & n_{20} & n_{30} \\ 0 & 1 & 0 & n_{11} & n_{21} & n_{31} \\ 0 & 0 & 1 & n_{12} & n_{22} & n_{32} \\ n_{10} & n_{11} & n_{12} & 0 & 0 & 0 \\ n_{20} & n_{21} & n_{22} & 0 & 0 & 0 \\ n_{30} & n_{31} & n_{32} & 0 & 0 & 0 \end{pmatrix} \begin{pmatrix} p_0 \\ p_1 \\ p_2 \\ \lambda_1 \\ \lambda_2 \\ \lambda_3 \end{pmatrix} = \begin{pmatrix} q_0 \\ q_1 \\ q_2 \\ d_1 \\ d_2 \\ d_3 \end{pmatrix} \tag{12.8}$$

The matrix is symmetric and always contains the identity matrix within the upper left 3×3 submatrix and the zero matrix within the lower right corner. In the first two passes per time step, only one respectively two constraints are defined. Consequently, the system of equations is reduced to the upper left 4×4, respectively, 5×5 matrices. In the third pass, it is sufficient to solve the system of equations according to

$$\begin{pmatrix} n_{10} & n_{11} & n_{12} \\ n_{20} & n_{21} & n_{22} \\ n_{30} & n_{31} & n_{32} \end{pmatrix} \begin{pmatrix} p_0 \\ p_1 \\ p_2 \end{pmatrix} = \begin{pmatrix} d_1 \\ d_2 \\ d_3 \end{pmatrix} \tag{12.9}$$

which represents the calculation of the intersection points of the three supporting planes, as the solution for the Lagrange multipliers is of no importance.

Due to the very simple structure, all systems of equations can be inverted analytically in all three cases so that just the coefficients have to be set at runtime. It is recommended to derive manually optimized program code out of this method, minimizing the number of necessary mathematical operations. As an example, the solution of the 3×3 system of equations (Eq. 12.9) can be implemented most efficiently with the CRAMER's rule.[23]

An extension of the god-object algorithm from a one-point interaction to an interaction with a spherical probe was suggested by RUSPINI ET AL. [16]. It solves one problem of the god-object algorithm that due to numerical errors, the haptic probe may drop at edges of polygons into the inner volume of the object. By replacing the dimensionless point by an object with volume, this effect of numerical gaps can be prevented.

Additionally, the algorithm, which in some haptic libraries implementing it is called RUSPINI algorithm, contains a concept for *force-shading*. It is used for smoothing the sensibly edged surface of polygonal objects resulting from the discontinuities of the normal directions at vertices and edges.

In graphical rendering, e.g., PHONG *shading* is used to cover the edges between polygons. For lighting calculation for each surface point, a normal vector is used, which is interpolated from the normal vectors stored at each vertex of the polygon. This idea is adapted for force-shading, but has to be extended by an additional step. As the direction of the force is defined by the direction of the spring-damper system between the haptic probe and the HIP, a new position for the haptic probe has to be derived from the interpolated normal first. The authors of the algorithm propose an adaptation in two phases: After calculating a first position $\mathbf{p}$ according to the constraint method of the god-object algorithm, a plane is put through $\mathbf{p}$ whose normal vector is collinear to the interpolated normal. Onto this force-shading plane, the HIP $\mathbf{q}$ is projected. In the second phase, the final position $\mathbf{p}'$ of the haptic probe is calculated by using the newly found point $\mathbf{q}'$ in a second pass of the constraint method. By this means the direction of the spring-damper system approximates the direction of the interpolated normals.

An interesting fact about this approach for smoothing the direction of forces is the modification of the displayed force, which is achieved by altering the position of the haptic probe and not by directly adding forces to the HIP. Avoiding a conceptional break in doing so is highly valuable for the stability of the method. Additionally, this method is suitable for the integration of friction, the display of locally different stiffness of the same object, and the generation of fine surface structures by *haptic textures*.

12.3.4 6 DoF Interaction: Voxmap-PointShell Algorithm

Another fundamental algorithm was developed by MCNEELY ET AL. [13] from Boeing in 1997 to solve questions in the area of virtual prototyping. The VPS algorithm combines a penalty-based approach including hybrid collision detection with the

[23] For larger systems of equations, this approach leads to inefficient code though.

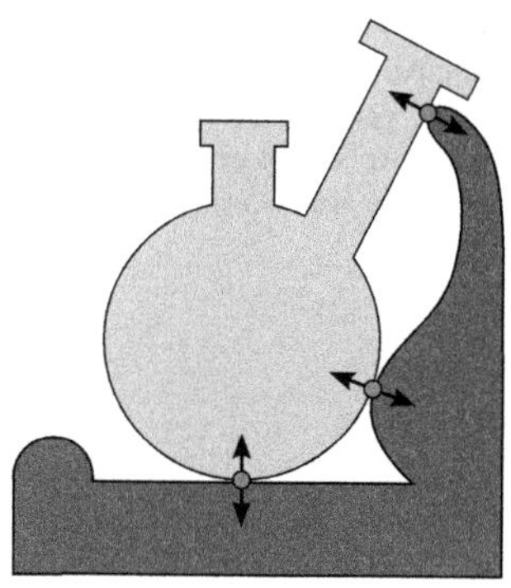

Fig. 12.7 Collision between dynamic object (*light gray*) and static part of scene (*dark gray*) with antiparallel surface normals at the contact points

simulation of rigid body dynamics and the concept of a *virtual coupling* as suggested in 1993 by Colgate et al. [2] (see below).

In contrast to the methods discussed above, the algorithm allows the output of a force vector and additionally of torques in three DoF. This is a basic requirement for the interactive planning of assembly paths with virtual prototypes. It makes use of the tool-handle metaphor and equates the virtual part with the handle of the haptic device.

One challenge in the development of the algorithm was to enable haptic rendering for highly complex scenes with huge polygon counts as they are generated from airplane CAD data. A clever representation of the scene was the right solution: The collision detection is not performed on the polygonal data but in the form of hybrid collision detection between a so-called *Voxmap* representing the static part of the scene and a so-called *PointShell* which represents the "dynamic object," i.e., the part to be assembled (Fig. 12.7).

Hybrid Collision Detection

The Voxmap is a global data structure representing the whole static scene. It consists of a three-dimensional grid of cubes and is generated in a preprocess in several steps. In each grid element or *voxel* (abbreviation for volume element in analogy to pixel), the information is stored, whether it is part of an object, whether it contains surfaces, or whether it represents empty space of the scene (Fig. 12.8). In the following step, the voxels are classified in more detail, by, e.g., storing the distance to the next surface point in each empty space voxel.[24] Afterward, the Voxmap is available as three-dimensional array of scalar data in a coherent memory area.

The rasterization process of the first generation step for the volume generation can be performed highly efficient on the graphics card by simply rendering the scene layer for layer. The volume model is the stack of the generated layers. The resolution of the layers should be chosen according to the detail level necessary for the application. As an example, a resolution of 1 mm is usually sufficient for assembly in automotive applications.

[24] The calculation of this distance field could be done via an *Euclidian distance transformation* (EDT).

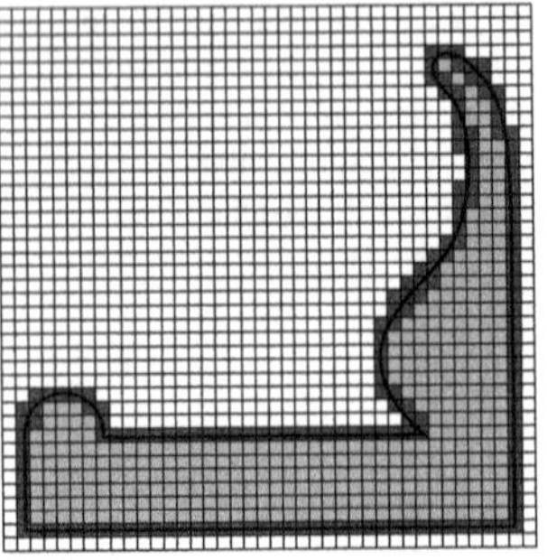

Fig. 12.8 Voxmap of the static part of the scene: surface voxels are *dark gray*, inner voxels *light gray*, and free space voxels *white*

The dynamic part, the one to be moved by the user through the scene, is represented by the PointShell (Fig. 12.9), which can be calculated in advance too. It consists of a multitude of points located at the surface of the dynamic object comparable to the data a laser scanner would generate. Also, the digital generation process of the point cloud is similar to a scan. The points are sampled by ray tests in a regular raster, which can be realized with the help of the graphics card, too. Once again, the density of the sampling should be chosen to resemble the main details of the object and has to be tuned to the resolution of the Voxmap. For each surface point, the corresponding surface normal is calculated and stored negated. The reason for this is given a little further below.

Within each cycle of the haptic rendering, a simple collision detection algorithm is applied on the Voxmap-PointShell representation of the scene: According to the actual transformation of the dynamic object, each point of the PointShell is transformed into the coordinate system of the Voxmap. Afterward, it is checked, in which voxel the point comes to rest: If the hidden voxel represents empty space, no local collision happens. Otherwise, the point must be taken into account in the subsequent

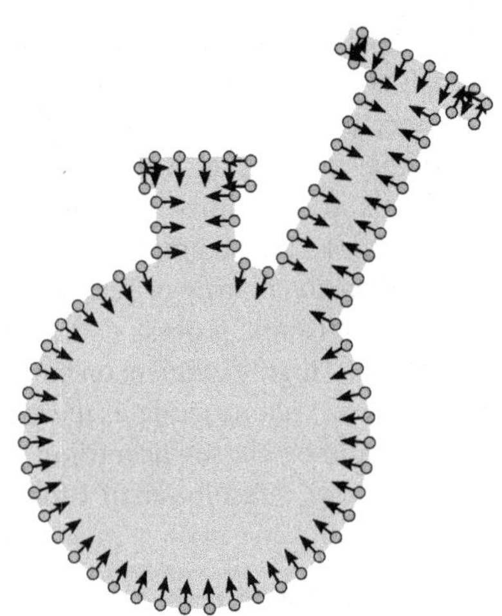

Fig. 12.9 PointShell of the dynamic object with negated surface normals

classification and force simulation. The advantage of this hybrid representation in the collision detection is given by the fact that for each PointShell point, only one access to the Voxmap has to be done. In addition, this access could be calculated highly efficiently directly from the transformed coordinates of the PointShell point: The three-dimensional index into the Voxmap can be achieved by rounding the point coordinates to integer values.

As a result, the runtime behavior of the collision detection is completely independent of the complexity of the static model part of the scene. This advantageous runtime complexity, however, is paid with relatively high memory footprint,[25] as the memory requirements scale cubically with the resolution of the Voxmap.

The runtime complexity of the total collision detection step with one test for each PointShell point scales linearly with the number of points in the point cloud. The latter in itself depends quadratically on the size of the dynamic object in relation to the sampling density. Consequently, dynamic objects with a large surface may be problematical.[26]

Local Penalty Method

After collecting all colliding points of the PointShell, local penalty forces $\mathbf{F}_i$ are calculated and summed up. For the identification of the individual forces, the surface normals $\mathbf{n}_i$ stored in the PointShell will be used: During contact with two real objects, their surface normals at the contact point are antiparallel (Fig. 12.7). On a microscopic size, this holds also for edges and vertices. As a consequence, the contact normal for each PointShell point can be used as surface normal of the dynamic object too.

The next challenge is to identify the penetration depth of the PointShell point $\mathbf{p}_i$. With a clever trick, this problem is transferred to the model of the virtual wall. A plane is put into the center $\mathbf{m}_j$ of the surface voxel containing $\mathbf{p}_i$ with the normal direction $\mathbf{n}_i$. Then, a distance vector $\mathbf{p}_i$ to this plane is calculated. Multiplied with a spring constant, the local penalty force $\mathbf{F}_i$ (Fig. 12.10 and Eq. 12.3) is calculated:

$$\mathbf{F}_i = \begin{cases} K(\mathbf{m}_j\mathbf{n}_i - \mathbf{p}_i\mathbf{n}_i)\mathbf{n}_i & \text{for } \mathbf{p}_i\mathbf{n}_i < \mathbf{m}_j\mathbf{n}_i \\ 0 & \text{else} \end{cases} \tag{12.10}$$

If the local forces $\mathbf{F}_i$ of all n detected contacts have been calculated, the total collision force $\mathbf{F}_K$ and the total collision torque $\mathbf{M}_K$ are summed up:

$$\mathbf{F}_K = \frac{1}{n}\sum_i \mathbf{F}_i$$
$$\mathbf{M}_K = \frac{1}{n}\sum_i \mathbf{F}_i \times \mathbf{l}_i \tag{12.11}$$

---

[25] A quite common correlation: The more the memory can be used, the better the runtime performance will be.

[26] In these cases, hierarchic bounding volume methods (Sect. 12.3.5) provide some room for optimization, which, however, are not discussed further in the context of this book.

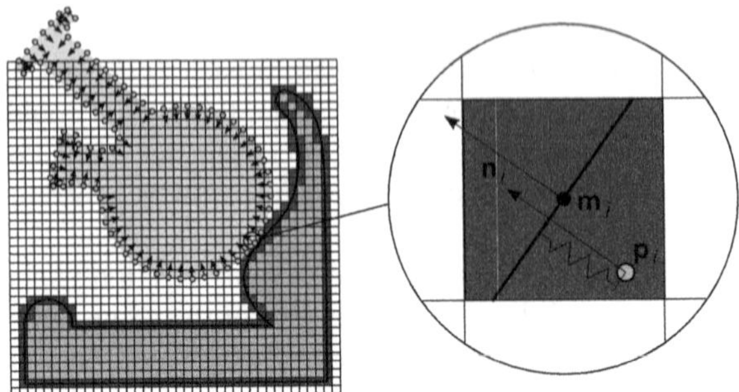

Fig. 12.10 Determination of the local penalty force

with $\mathbf{l}_i = \mathbf{p}_i - \mathbf{S}$ and $\mathbf{S}$ being the actual position of the reference point (preferably the center of gravity) of the dynamic object.

Scaling by the factor $1/n$ avoids the emergence of too strong forces and smoothes the effects of contact points being added and falling away between the cycles.

Now, an obvious idea would be to actuate the 6 DoF haptic device directly with F_k and M_k. This would be identical to a hard coupling between HIP and the dynamic object resembling the haptic probe. Practically, this approach frequently results in instable system behavior and a bad quality of the haptic impression. The dynamics of the haptic device is usually not sufficient to display the simulated forces.

Virtual Coupling

Similar to the god-object algorithm, which derives the force to be displayed by the device from the force of the spring-damper system between haptic probe and HIP, a so-called *virtual coupling* couples the dynamic object with the handle of the haptic device. The position of the handle in the virtual scene is given by the HIP. In the VPS algorithm, the coupling is designed as a six DoF spring-damper system: Translatory springs with corresponding dampers represent three translatory DoF, and rotatory springs with corresponding dampers represent the other three rotatory DoF. The simulation of the torques generated by the virtual coupling is implemented analogously to the force simulation of the translatory spring-damper system:

$$\begin{aligned} \mathbf{F}_V &= K_t\mathbf{q} - D_t\mathbf{v} \\ \mathbf{M}_V &= K_r\theta - D_r\omega \end{aligned} \tag{12.12}$$

with K_t and K_r being translatory and rotatory spring constants and D_t and D_r representing the corresponding damping constants, respectively. The values $\mathbf{q}$ and $\mathbf{v}$ are the vector distance between dynamic object and virtual handle (representing the displacement of the virtual spring in idle position 0) and their *relative* velocity. The vector θ represents the relative rotation between the dynamic object and the virtual handle, and ω is the relative angular velocity.

Rigid Body Dynamics

Up to now, it is open how the position and motion of the dynamic object are identified within the virtual scenario. In the steps described till here, each haptic rendering cycle calculates forces and torques of the virtual collisions and inside the virtual coupling. Now, still their impact on the dynamic object has to be simulated.

One approach is given by the integration of motion equations according to NEWTON and EULER[27]:

$$\mathbf{F} = \mathbf{F}_K + \mathbf{F}_V = m\mathbf{a}$$
$$\mathbf{M} = \mathbf{M}_K + \mathbf{M}_V = \mathbf{I}\alpha + \omega \times \mathbf{I}\omega \tag{12.13}$$

with the translatory acceleration $\mathbf{a} = \dot{\mathbf{v}}$ and the angular acceleration $\alpha = \dot{\omega}$. As additional necessary value, the mass m of the dynamic object and its inertial tensor **I** describing the mass distribution of the object have to be provided.

Both values can be derived from the surface model of the object with the methods already implemented: A volume model of he dynamic object is generated with the same voxelization method known from the generation of Voxmaps. To each center $\mathbf{w}_k$ of an occupied voxel, a point mass m_k is assigned. The total mass m is given by the sum of all m_k, and the components I_{ik} of the inertial tensor

$$I = \begin{pmatrix} I_{00} & I_{01} & I_{02} \\ I_{10} & I_{11} & I_{12} \\ I_{20} & I_{21} & I_{22} \end{pmatrix}$$

are calculated with

$$I_{ij} = \sum_{k=1}^{n} m_k(\mathbf{r}_k^2 \delta_{ij} - \mathbf{r}_{ki}\mathbf{r}_{kj}) \tag{12.14}$$

with the KRONECKER symbol δ_{ij} and $\mathbf{r} = \mathbf{v} - \mathbf{S}$. **S** denotes the point of action of the virtual coupling at the object. Optimally, the object is modeled, so that **S** coincides with the center of mass of the object. The latter is important to prevent the emergence of torques during translatory movements in collision-free space, which otherwise would disturb the haptic impression.

Including these components, the VPS algorithm is complete: In the virtual scene, the collision detection and identification of collision forces are done. Together with the forces and torques from the virtual coupling, they act upon the dynamic object and result in its movement. Finally, the torques and forces of the virtual coupling are displayed by the haptic device. The coupling therefore builds the bridge between real handle and virtual part: It is like the user would move the dynamic object through the world by pulling a rubber band attached to it. According to the application, the

---

[27] For a very helpful introduction into the implementation of rigid body dynamics, simulation based on NEWTON–EULER equations [1] is recommended.

haptic impression can be controlled by configuring the mass of the dynamic object adapted to the spring stiffness and particularly to the damping of the virtual coupling.

This overview about the VPS algorithm shows how object representations, collision detection, physical modeling, and finally the simulation of the forces to be displayed are interconnected and have to be harmonized with each other.

At the same time, the limits of a certain modeling concept for a usage beyond the intended application become obvious. A volumetric description of the static part of the scene cannot be used without major efforts to represent deformable models. Movements of single objects of the static scene or interaction between objects except for the dynamic object cannot be realized directly with the hybrid collision detection. Nevertheless, it should be noted that the algorithm proved very valuable for the interactive verification of assembly steps of rigid parts in rigid installation spaces. For this purpose, it is one of the best methods at hand.

12.3.5 Collision Detection

The core element of a haptic rendering algorithm is efficient collision detection. At a frequency of close to 1,000 Hz, it must be checked whether there is contact between the haptic probe and the virtual object. Collision detection is also relevant for applications in the area of virtual reality which not necessarily include haptics. Fundamentally, any kind of physics simulation, where virtual objects interact with each other, should be mentioned. In scientific as well as in industrial applications and in computer games, the collision detection is of central importance. Consequently, it was focused on this topic since the very beginning of virtual reality.

Besides the obvious usages of collision detection algorithms—to find out whether virtual objects collide—there are other less obvious ones: within so-called raytracers, the paths of single light rays are simulated physically to create a photorealistic image. For one image, millions of collision tests have to be made between rays and the virtual scene (using ray intersection tests).

Based on the following considerations, the most fundamental questions and concepts of collision detection methods as well as principles for their optimization shall be discussed. This will provide a good basic understanding of the challenges related to it. For a more comprehensive discussion, specialized literature for collision detection is recommended [4, 18].

For all problems where large amounts of data have to be processed efficiently, algorithms with low *complexity* have to be designed. Complexity is a measure how processing time, memory requirements, and the need for other resources for the algorithm depend on the problem size, which is usually measured by the number of elements to be processed.

If, for example, a collision between two objects shall be tested, this can be done "brute force" by testing all triangles of the one object against all triangles of the other object. The effort increases quadratically with the size of the problem: If the number of triangles of both objects doubles, the runtime of the algorithm quadruples. Thus, the algorithm has quadratic order, and the (runtime) complexity is $O(n^2)$.

The size of the models that can be displayed graphically on state-of-the-art VR systems and even on off-the-shelf PCs in real time (Sect. 12.2.4) is several hundred thousand or even more than a million triangles. Even in such complex scenarios, efficient collision detection is desired, quadratic complexity, however, is completely impracticable.

The total complexity of a collision detection method can be improved with a plausible idea: In several steps, the size of the problem is reduced by performing simple and more "economic" tests, excluding as many elements as possible from the more complex tests. The necessary steps are usually arranged in a *collision pipeline*, which is processed step by step.

For a fast exclusion of many elements, their spatial coherence is analyzed and exploited. For example, neighboring triangles can be collected into a bounding object. If a test of the bounding object shows no intersection, all contained elements can be excluded from all further analysis.

Different geometry types are suitable as bounding objects: spheres, cubes, and convex polyhedrons with defined orientation of surfaces [so-called *discrete-oriented polytopes* (DOP)], to name the most frequently used ones. Usually, only one bounding geometry type is used within the same collision detection step. A "good" bounding geometry can be tested efficiently for intersections and approximates the included elements closely, because this reduces the number of false-positive tests. The quality of the approximation achievable with a bounding geometry depends highly on the shape of the bounded object. Figure 12.11 shows this effect for different bounding geometry types: Spheres, for example, approximate elongated objects only badly. Consequently, flexible bounding geometry types are advantageous for the approximation quality. However, they require more effort for intersection tests. A good compromise are DOPs which reach a good approximation quality in combination with relatively simple intersection tests. But there is no universal optimum choice of a bounding geometry type for every application, as the runtime performance at collision detection is dependent on many factors.

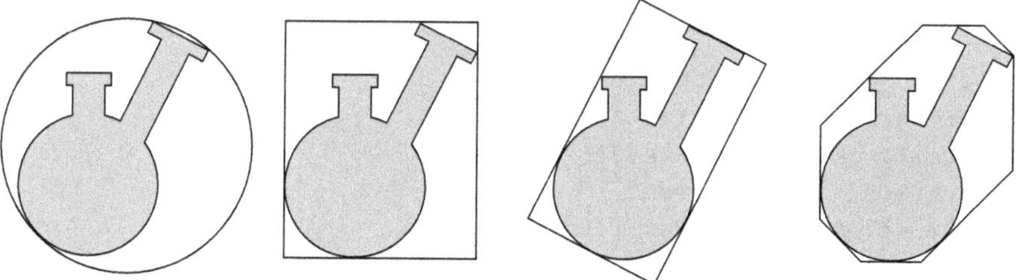

Fig. 12.11 Types of hull geometries in 2D (from *left* to *right*): minimal sphere, axis-aligned bounding box (*AABB*), object-oriented bounding box (*OOBB*), and discrete-oriented polytope (*DOP*)

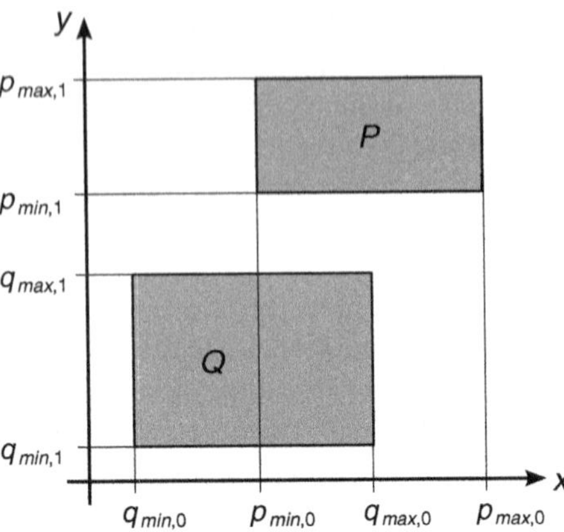

Fig. 12.12 Intersection test of boxes: no intersection of the intervals on the y-axis guarantees the absence of collisions

The test for the intersection of two spheres is simple: May m_0 and m_1 be the centers of two spheres and r_0 and r_1 their radii. They do not intersect, if $|m_0 - m_1| > r_0 + r_1$. For an efficient implementation, the monotony of the square in the first quadrant should be used to avoid the extraction of the root for the calculation of the distance between m_0 and m_1. The test condition then is $(m_0 - m_1)^2 > (r_0 + r_1)^2$. This saves one root for the costs of an additional multiplication for squaring the sum of the radii, which still requires considerable less processor cycles. The optimization of these tests is relevant, as they are frequently used within the collision detection process as depicted below. The calculation of the bounding sphere (which exists and is unique) is not trivial, but can be done using the algorithm of HOPP [6].

In average, fewer effort is required by the test for intersection of two-axis-aligned boxes. One-axis-aligned box Q can be characterized by two vertices $q_{\min}$ and $q_{\max}$. The intersection of both boxes is empty—no collision happens—if at least one coordinate of the minimum of one box is larger than the maximum of the other box (Fig. 12.12).

This makes it possible to formulate a test which has to be processed in the average of all collision-free cases only to the half of its code, as it already excludes a collision before:

```
if      qMin[0] > pMax[0]:
  return "no collision"
else if qMax[0] < pMin[0]:
  return "no collision"
else if qMin[1] > pMax[1]:
  return "no collision"
else if qMax[1] < pMin[1]:
  return "no collision"
else if qMin[2] > pMax[2]:
  return "no collision"
```

```
  else if qMax[2] < pMin[2]:
    return "no collision"
  else:
    return "boxes intersect"
```

The full potential of tests via bounding volumes is used in hierarchic collision detection algorithms working with trees of nested bounding volumes.[28]

The root of such a tree resembles the total object, which is nothing but "all triangles." During generation of the bounding volume hierarchy (BVH), all triangles of a node are distributed onto its child nodes recursively until the leaves of the tree contain exactly one triangle (Fig. 12.13).

An advantageous split is essential for the efficiency of the following collision detection. Quality criteria are as follows:

- The tree should be *balanced* according to the number of its child nodes and therefore be optimized in its depth. This is identical to the requirement that the geometric primitives of a node should be distributed equally among its child nodes.
- The distribution of the geometrical primitives should be chosen to minimize the volumes of the bounding geometries and reduce overlapping on the same level of the tree.

How good both criteria can be fulfilled at the same time depends on the structure of the individual models. For models made of triangles with large variations in their size, no optimum distribution can be found covering both criteria equally good.

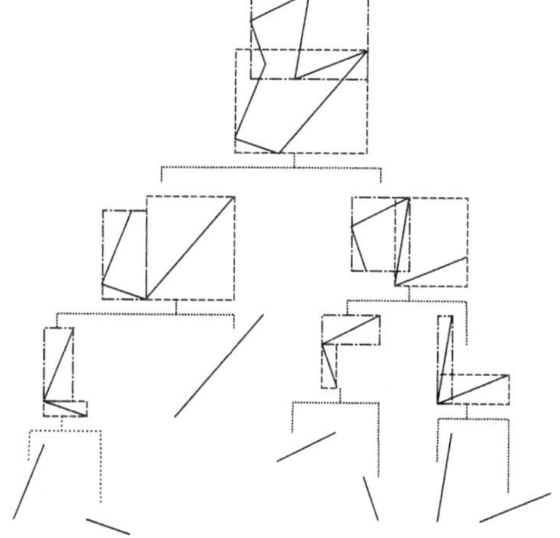

Fig. 12.13 Schematic representation of the split of a geometry inside a bounding volume hierarchy

[28] A tree is a cycle-free interconnected *graph*, consisting of *nodes* and connecting *edges*. *Directed trees* have one *root*. *Leaves* are nodes which do not have any "child nodes" and no outgoing *branches*. Usually, hierarchic collision detection algorithms use binary trees, i.e., trees where all nodes except for the leaves have two child nodes. *Octrees*, trees with eight child nodes, are another frequent variation.

After the generation of the bounding volume hierarchy in a preprocess, the data structures are available, which enable the application to quickly filter out large parts of the objects during the collision tests. If the bounding volumes do not intersect in a node of the tree, all child nodes and the included triangles can be neglected for the following tests, whereas nodes, which could not be excluded, are recursively processed in more detail during the subsequent tests. Very fast, only branches close to the actual contact point remain.

If the question is about the existence of collision only, the collision detection can be interrupted as soon as a collision was found for one branch. For this question, therefore, a near-contact leads to the worst average performance: The hierarchy has to be traversed deeply up to the demanding triangle tests to just return "no collision" as result finally.

However, for haptic applications, exact information about the contact is necessary to be able to classify it. This requires all contact points to be identified, i.e., all triangles that intersect another object. In this *exact* collision detection, the near-collision does not differ from collision concerning its runtime performance.

In many applications, it is not sufficient to regard the collision between a geometrically simple probe (e.g., a point or sphere) with objects of the virtual scene but the contact between complex geometrical models. In this situation, both tested objects are represented by bounding volume hierarchies, which are processed in parallel recursive traversal.

```
function CollisionTest( node0, node1 ):
  if bounding volumes of node0 and node1 are disjoint:
    return
  else if node0 and node1 are leafs:
    if TriangleTest( node0, node1 ) is positive:
      add triangle pair to collision list
  else if node0 "larger than" node1:
    for each child node cn of node0:
      CollisionTest( cn, node1 )
  else:
    for each child node cn of node1:
      CollisionTest( node0, cn )
```

The function `CollisionTest` of the pseudo-code listing above is called with the root nodes of both object's bounding volume hierarchies. During the recursive processing, the list is filled with all pairs of triangles of the one or the other object which collide (it is an algorithm of the class of exact collision detections). If the list is empty at its end, in the current time step, there are no collisions. As metric (for the test "larger than") for the choice of the path to descend, the number of children or the size of the bounding volume may be chosen.

Besides spatial coherence, also temporal coherence is exploited often: In a system simulated with small time steps, usually only small changes happen from step to step. Information from the last collision detection cycle can be recycled as a starting point

Fig. 12.14 Separating plane E

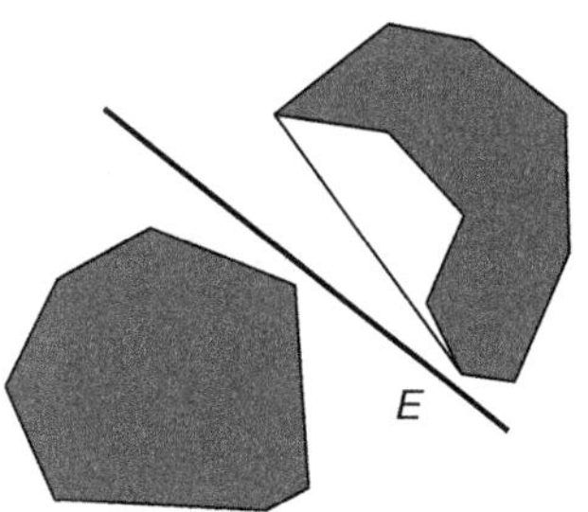

for the search of a collision in the current time step. Collision detection algorithms according to this principle are called *incremental.*

A good example for this is an algorithm which often very quickly proves that no collision exists and thus used in early phases of collision pipelines. The convex hulls of two objects given it tries to identify a plane, which lies in between them. If so, the objects are in different of the both disjunctive half spaces generated by the plane. The existence of such a *separating plane* guarantees that no collision happens.

The first test within each cycle of the collision detection checks whether a separating plane existed within the prior cycle. This plane is tested for being valid in the current cycle too. If this is the case, the further processing of the collision pipeline for the object pair can be stopped. If this is not the case, a new separating plane is searched nearby the old plane until one is found or a maximum number of trials were made. In the latter case, the processing of the collision pipeline is continued, as collision cannot be excluded (Fig. 12.14).

Another incremental procedure is given by the *closest feature tracking* algorithm of LIN and CANNY [12]. It searches for the pair of surface features (points, edges, faces) of two polygonal objects which constitute the minimum distance between both (Fig. 12.15). The method starts with an arbitrary feature pair and analyzes specifically[29] pairs of neighboring features whether they are closer to each other. The search gives a fast result if a good starting pair was chosen. Such a starting pair is often available in the form of the feature pair of the last collision test.

Fig. 12.15 Pairwise closest features of a set of objects

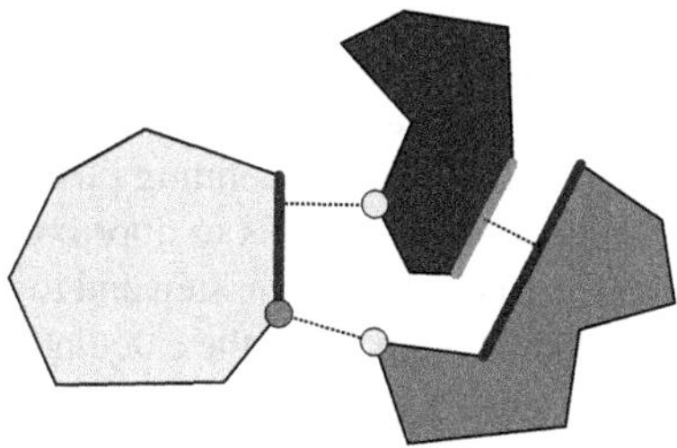

[29] For the choice of the next feature pair, the VORONOI regions of the objects are analyzed.

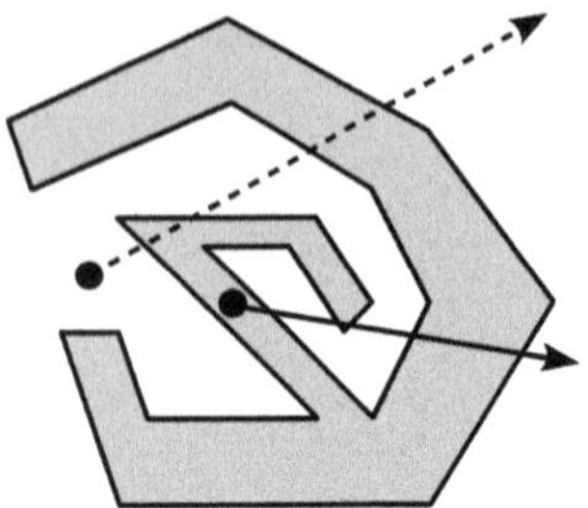

Fig. 12.16 Ray intersection tests with odd (drawn through ray) and even (*dashed* ray) number of intersection points for an inner and an outer point, respectively

For the haptic rendering with a pointlike probe, the question for the collision detection has to be formulated different from the collision with complex models: It has to be identified whether the point is within the object or whether the movement of the point passed a surface of an object within the last time step.

A point is within a closed polyhedron[30] if a ray cast to infinity starting from this point in an arbitrary direction hits an odd number of surfaces. With convex polyhedrons, the ray hits exactly once. With concave objects, numerous hits may occur (Fig. 12.16). When implementing this test, some exceptional cases have to be considered: the ray may hit an edge or point within the numerical precision. In this case, the corresponding surfaces have to be counted very carefully. Grazing touches of the corresponding surfaces (only relevant for the concave case) should not be counted. A penetration should be counted only once even when there are two or more surfaces touching the ray.

The question about a point shape probe having crossed a surface within the last time step is answered by a ray intersection test, too, but the test is performed with the line between the last and the current position of the probe instead with the ray.

In both cases, the ray intersection test can be optimized using the efficient collision detection methods described above: the ray, respectively the line, may be tested recursively against the bounding volume hierarchy of the object in question.

The analysis of the two collision queries for the one-point interaction leads to another basic distinction. The collision tests between complex objects as described so far only consider static states of the objects within one time step. As a consequence of the time discretization, collisions can be missed if a relative small or fast object is in collision only in between two simulation steps. These cases are addressed by the methods of *dynamic collision detection* (in contrast to *static collision detection*), which additionally are suitable to calculate the moment of collision back in time and interpolate the corresponding position of the involved objects.

The basic concept is to approximate the volume occupied by the moving object since the last simulation step and to test this volume for collisions with other objects. It is important to avoid the calculation of the envelopes of *both* objects. If the relative movement of the objects is expressed within the coordinate system of one of the objects, it is sufficient to calculate the envelope of the other "dynamic" object.

---

[30] An object modeled of triangles representing a solid body is such a closed polyhedron. For nonclosed polyhedrons with a "hole" in their surface, an inner region cannot be defined so easily.

The methods for hierarchic collision detection can be applied to dynamic collision detection too. We start from the assumption that for both involved objects, boundary hierarchies exist. During the recursive collision detection algorithm, an envelope for the bounding volume of the current node in the bounding volume hierarchy of the dynamic object is approximated and tested for intersection. Once again, the principle is to exclude large parts of the geometry as effortless and early as possible from the more performance-critical collision tests.

The approximation of the envelope of the bounding geometry as well as of the individual triangles is simple for translations. In these cases, the movement volume itself can be characterized by a polygonal object. The bounding surfaces of rotating objects become so-called *ruled surfaces* though which are curved and need a more complex representation. Therefore, once again, a rough outer approximation of the envelope by an bounding object of the chosen bounding geometry type (sphere, box, DOP) is used. Only, in the leaves of the hierarchy, exact tests are conducted for the triangles, which are necessary when the contact moments are calculated back in time.

12.4 Software Packages for Haptic Applications

Several software packages supply frameworks for the development of haptic and multimodal interactions. They can be helpful to quickly develop test setups based on ↪COTS devices. On the other side, the development of an own device-handling software component will enable the use of this framework with newly developed haptic interfaces. In the following, a couple of frameworks are discussed. Please note that the following is primarily based on secondary references like [7] and the individual project Web sites.

CHAI3D—www.chai3d.org This framework is open source and maintained by several companies (including *ForceDimension*, *Hansen Medical*, and *Microsoft Research*) and universities (*EPFL*, *Stanford*, and *Siena*). It supports all major haptic interfaces (ForceDimension omega and delta devices, Novint Falcon, Phantom devices, and MPB Freedom 6) and uses OPENGL for graphics rendering. It is written in C++ and supports all operating system platforms, and the current version is 2.0 released in 2009.
The framework incorporates potential field models, finger proxy models, several static and dynamic friction models as well as stick-slip, viscous, vibration, and magnetic effects. AABB and spherical models for collision detection are included, and several additions, including an audio library and a deformable model engine, are available.

H3D API—www.h3dapi.org This framework is maintained by the hardware and system supplier *SenseGraphics* and historically closely related to the *Geomagic* haptic interfaces from the PHANTOM series. There is an open source as well as an closed source version of the framework that uses OPENGL for graphics as well and the open source HAPI (Haptics ↪API, as well developed by *SenseGraphics*)

for the rendering of haptics. This API can use several actual renderers, including the haptic renderer of CHAI3D, the *geomagic* renderer OPENHAPTICS, and two additional renderers. It is written in C++ as well and supports all operating systems, but provides additionally high-level abstraction with interfaces to X3D and python. The current version is 2.2.0, which was released in 2013.
All major haptic interfaces are supported as well (Phantom, omni, delta, Falcon, Moog FCS HapticMaster), and several force and surface effects are available. Furthermore, several stereographic displays are supported for the development of multimodal applications.

Geomagic OPENHAPTICS Toolkit As mentioned above, this toolkit incorporates function for haptic rendering with hardware from the Phantom series. The framework is written in C++, and the source is closed.

Quanser QUARC Quarc is an real-time control software that also includes drivers for haptic systems, i.e., the high-fidelity systems by *Quanser*, but also the Falcon and the Phantom series. The toolkit is closed source and only commercially available, but supports other hardware such as robot manipulators, image-based tracking, cameras, and GPS to build complex mechatronic and also autonomous systems.

Frameworks for the Novint Falcon Because of the low price, the Novint Falcon is a very attractive device for teaching and do-it-yourself projects. The hardware manufacturer Novint supplies HDAL, a closed source low-level communication layer between the Falcon and the application, and F-Gen, as a higher level component intended for the integration of haptic effects into gaming applications. HDAL is based on C++, but only available for Windows distributions. These packages are maintained by the manufacturer. Additionally, there is an open source interface library called LIBNIFALCON for the Novint Falcon. While the last version is dated in 2010, the library was integrated in the H3D API according to the project Web site.

Additional Toolkits for Interactions with Surfaces There are several toolkits for the interaction with surfaces, i.e., toolkits designed for two-dimensional interaction. These incorporate a tabletop touch input display and some kind of additional hardware to generate signals to display haptic surface properties. Examples are the HAPTICTOUCH toolkit [11] that is using the Haptic Tabletop Puck (HTP) as an additional haptic device capable of displaying friction, height, and surface properties. The REFLECTIVE HAPTICS TOOLKIT by *Nuilab* (see Fig. 5.1) modulates the friction between a stylus and the touch-sensitive surface.

Further toolkits are or were available for the development of touch applications based on Java (JTouch), but are not considered sufficiently developed for productive use. Based on the above-mentioned properties of the individual toolkits and the current development state, the H3D API seems to be most feasible for new developments.

12.5 Perception-Based Concepts for VR software

This section briefly introduces two concepts that originate in properties of human perception. These concepts can reduce the requirements on a systems' hardware, but have to be incorporated into the systems' software accordingly.

12.5.1 Event-Based Haptics

The concept of event-based haptics was already introduced in Sects. 2.4.4 and 11.2.3. Briefly, it is based on two properties of human haptic interaction, i.e.,

- Humans cannot exert haptic interactions with frequencies higher than about 20 Hz (see Fig. 1.7)
- Humans cannot detect the direction of a high-frequency haptic stimulus

As shown in Sect. 11.2.3 and Fig. 11.5, the software of a haptic system has to incorporate means to store prerecorded contact transients.

12.5.2 Pseudo-haptic Feedback

LÉCUYER ET AL. introduced the concept of pseudo-haptic feedback in 2000 to reduce system and device complexity while still offering a haptic feedback to the user [10]. The authors of [14] define pseudo-haptics as follows:

Definition *Pseudo-haptics* "Pseudo-haptics corresponds to a haptic percept that is different from what the real haptic sensory supply would suggest by 'playing' with the multimodal—mainly the visual—feedback of a system." (*Quote from* [14, p. 1]).

More pictographic, one achieves the impression of a haptic feedback using a stiff input device that does not change its haptic characteristics and a changing image of the environment interacted with. The input device measures the force exerted by the user and changes the visual characteristics of the object or material interacted with respect to an underlying model. A stiffer material will deform less as a more compliant material on the visual display when the same force is applied. The perceptual basis of this effect is not analyzed fully yet [14]. It is assumed that the sensory input from haptic and visual channels will be integrated in an inner model. Because of the experience of the user and the visual impression, the contradictory haptic impression will be overruled and creates a realistic haptic feedback.

Examples for the use of haptic feedback include the simulation of friction, stiffness, mass, and texture [9]. If one wants to incorporate pseudo-haptic feedback in an application, PUSCH ET AL. give some design guidelines in [14].

12.6 Conclusion

For the integration of haptic rendering in existing virtual reality systems, a modification of their architecture is usually necessary. In case of a new development of a multimodal system which should include the haptic sense, an appropriate architecture should be chosen: It should provide functionalities for multithreading and/or distribution including mechanisms for synchronization. This is needed to decouple the higher frequency haptic rendering of the application behavior and graphic rendering cycle.

Additionally, the VR system has to include data structures and algorithms for exact collision detection between haptic probes and virtual objects as well as physical models for the calculation of collision or frictional forces. If the forces calculated inside the haptic simulation are intended to affect the virtual scene, instances for physical simulation, rigid body dynamics, deformable objects, etc., have to be implemented in the scenario and linked to the objects of the scene. The resulting data flow requires an open eye on the synchronization between the different modules.

Due to the hybrid functionality of haptic devices—they are input and output devices at the same time—an adequate concept of device abstraction has to be realized too. At least the communication between high-frequency haptic rendering thread and haptic device has to be designed with little latency. With "dangerous" devices, which are extremely strong or fast, it is strongly recommended to give hard real-time guarantees to the haptic rendering loop. Among other things, this requires to build on top of real-time operating systems. Softer real-time requirements should be fulfilled for smaller haptic devices in any case, as the fundamental system stability and an acceptable quality of the haptic feedback can be achieved easier.

References

1. Baraff D, Witkin A, Kass M (1999) Physically based modelling, course notes 36. In: ACM SIGGRAPH '99, Aug 1999. https://graphics.stanford.edu/courses/cs448b-00-winter/papers/phys_model.pdf
2. Colgate JE et al (1993) Implementation of stiff virtual walls in force-reflecting interfaces. In: Proceedings of the IEEE virtual reality annual international symposium (VRAIS). Seattle, Sept 1993, pp 202–2008. doi:10.1109/VRAIS.1993.380777
3. Consortium W X3D, H-Anim, and VRML 97 (2014) specifications. Website. http://www.web3d.org/x3d/specifications/
4. Ericson C (2004) Real-time collision detection. Morgan Kaufmann, Burlington. ISBN: 1558607323
5. Goldstein EB (2006) Sensation and perception, 7th edn. Wadsworth Publishing Co., Belmont

6. Hopp TH, Reeve CP (1996) An algorithm for computing the minimum covering sphere in any dimension. In: Jan 1996. http://www.mel.nist.gov/msidlibrary/doc/hopp95.pdf
7. Kadlecek P (2010) A practical survey of haptic APIs. Bachelor thesis, Charles University, Prague. http://www.ms.mff.cuni.cz/~kadlp7am/kadlecek_petr_bachelor_thesis.pdf
8. Langetepe E, Zachmann G (2005) Geometric data structures for computer graphics. AK Peters, Natick. ISBN: 9781568812359. http://akpeters.com/product.asp?ProdCode=2353
9. Lécuyer A (2009) Simulating haptic feedback using vision: a survey of research and applications of pseudo-haptic feedback. Presence: Teleoperators Virtual Environ 18(1):39–53. doi:10.1162/pres.18.1.39
10. Lecuyer A et al (2000) Pseudo-haptic feedback: can isometric input devices simulate force feedback? In: Proceedings of virtual reality, 2000. IEEE, pp 83–90. doi:10.1109/VR.2000.840369
11. Ledo D et al (2012) The haptictouch toolkit: enabling exploration of haptic interactions. In: Proceedings of the 6th international conference on tangible, embedded and embodied interaction, Kingston, Ontario, Canada, pp 115–122. doi:10.1145/2148131.2148157
12. Lin MC, Canny JF (1991) Efficient algorithms for incremental distance computation. In: IEEE conference on robotics and automation, pp 1008–1014. doi:10.1109/ROBOT.1991.131723
13. McNeely WA, Puterbaugh KD, Troy JJ (1999) Six degree-of-freedom haptic rendering using voxel sampling. In: SIGGRAPH, pp 401–408. doi:10.1145/1198555.1198605
14. Pusch A, Lécuyer A (2011) Pseudo-haptics: from the theoretical foundations to practical system design guidelines. In: Proceedings of the 13th international conference on multimodal interfaces. ACM, pp 57–64. doi:10.1145/2070481.2070494
15. Reissell L-M, Pai DK (2007) High resolution analysis of impact sounds and forces. In: EuroHaptics conference, 2007 and symposium on haptic interfaces for virtual environment and teleoperator systems. World haptics 2007. Second joint. Tsukaba, pp 255–260. doi:10.1109/WHC.2007.70
16. Ruspini DC, Kolarov K, Khatib O (1997) The haptic display of complex graphical environments. In: Proceedings of the 24th annual conference on computer graphics and interactive techniques, SIGGRAPH '97. ACM Press/Addison-Wesley Publishing Co., New York, NY, USA, pp 345–352. ISBN: 0-89791-896-7. http://doi.acm.org/10.1145/258734.258878
17. Sutherland IE (1965) The ultimate display. In: Proceedings of IFIP congress, pp 506–508. http://www.cs.utah.edu/classes/cs6360/Readings/UltimateDisplay.pdf
18. Zachmann G (2000) Virtual reality in assembly simulation—collision detection. Dissertation, Department of Computer Science, Darmstadt University of Technology, Germany. http://citeseer.ist.psu.edu/zachmann00virtual.html
19. Zilles CB, Salisbury JK (1995) A constraint-based god-object method for haptic display. In: Proceedings of the international conference on intelligent robots and systems, IROS '95, vol 03. IEEE Computer Society, Washington, DC, USA, p 3146. ISBN: 0-8186-7108-4. doi:10.1109/IROS.1995.525876

Chapter 13
Evaluation of Haptic Systems

Carsten Neupert and Christian Hatzfeld

Abstract In this chapter, a number of measurement methods and tests are presented that can be used either for verification or validation or—sometimes—both. We therefore refrain from an ordering based on these steps, but will present the methods based on the focus of the evaluation method. In that sense, one can identify three main foci of evaluation methods: system-centered methods that will test system properties (and are mostly used for verification, Sect. 13.1), task-centered methods that will test the task performance of a user working with the haptic system (such tests are mainly used for validation, but they can also verify system properties depending on the test design, Sect. 13.2), and user-centered methods that will measure the impact of the haptic system on the user. The latter are almost exclusively used for system validation and further described in Sect. 13.3.

As stated in Chap. 4, the evaluation of a haptic system will test whether a haptic system fulfills all requirements defined in the development process (verification) and whether it is conformed with the intended usage of the haptic system (validation). In the following, we will focus on the evaluation method itself, the selection of a proper method and the test design is left to the reader. For the design of task-specific haptic interfaces, we consider this to be the most practicable approach because of the uniqueness of the designs. This chapter therefore will not deal with optimized evaluation processes as for example established by SAMUR for the class of haptic interfaces [32]. This work shows that a more standardized testing of haptic system will exhibit advantages in terms of testing time and comparison of different systems.

Figure 13.1 gives an overview of the application of task-centered and user-centered evaluation methods with regard to the intended application as described in Sect. 1.3 for the selection of a proper method described in the following sections.

C. Neupert (✉)
Institute of Electromechanical Design, Technische Universität Darmstadt,
Merckstr. 25, 64283 Darmstadt, Germany
e-mail: c.neupert@hapticdevices.eu

C. Hatzfeld
Institute of Electromechanical Design, Technische Universität Darmstadt,
Merckstr. 25, 64283 Darmstadt, Germany
e-mail: c.hatzfeld@hapticdevices.eu

C. Hatzfeld and T.A. Kern (eds.), *Engineering Haptic Devices*,
Springer Series on Touch and Haptic Systems, DOI 10.1007/978-1-4471-6518-7_13

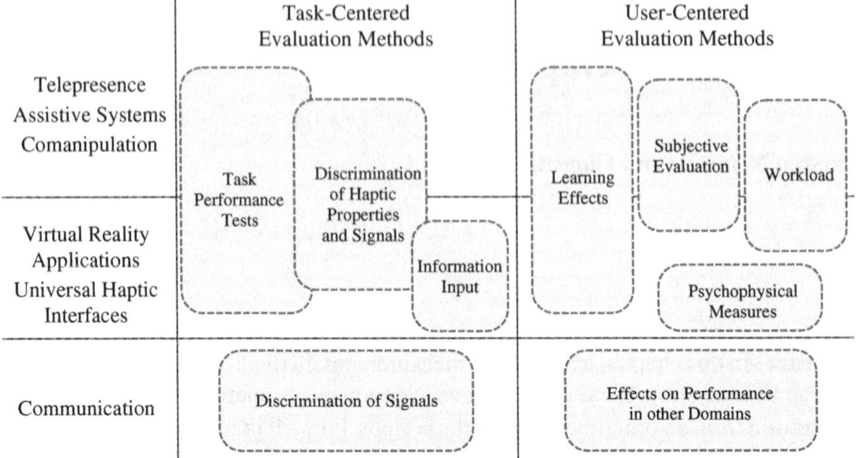

Fig. 13.1 Assignment of evaluation methods to different applications of haptic systems. System-centered evaluation methods are not included for clarity, since they can be applied to almost every system

13.1 System-Centered Evaluation Methods

The prevalent goal of system-centered evaluation is the generation of comparable technical ratings and values. These are used to verify the developed system against the requirements defined in the development process and to compare different systems with each other. The latter is especially relevant for haptic displays and haptic interfaces, since these systems are intended to be used universally in a variety of applications. Because of that, system-centered evaluation methods are not dependent on a certain type of task or a user. From this, it follows that normally there are no different experiment conditions to be considered in the interpretation of such acquired evaluation values.

When designing a haptic system, one defines properties of the haptic system based on requirements derived from the application. The assumed properties are usually calculated precisely and are well-known in theory. But to confirm the promised characteristics at the final haptic system, it is necessary to perform measurements, at least of the most crucial parts. With respect to standard measurement hardware, in the following parts, some hints for performing performance measurements are given.

The main focus of this evaluation part is the characterization of force reflecting haptic system. In the different sections, values of interest are figured out and some hints are given to implement and conduct the measurements. So, the measurement of a haptic system's workspace, output force-, and motion-depending values are discussed as well as the measurement of dynamic behavior of mechanical parts and the displayable impedance of different systems. Secondary, special properties of admittance systems and teleoperation systems are discussed.

13.1.1 Workspace

The workspace, respectively, the number and nature of the ↪DoF, is the most prominent characteristic of a haptic system. To analyze the workspace, the measurement of distances and angles can simply be done with a ruler or a measuring tape and an angle meter, thereby one can differ between active and passive degrees of freedom with corresponding workspaces. Active degrees of freedom are actuated and crucial for the haptic feedback. Passive degrees of freedom show the workspace that is reachable and driven by the user.

Further, the characterization can consider the ability to reach every point of the workspace, the independence of the end effectors orientation from the position in the workspace, constraints of the workspace because of singularities, the conditioning number κ at each point in the workspace, and the global conditioning index ν (see Sect. 8.4).

13.1.2 Output Force-Depending Values

Since haptic systems are bidirectional in energy flux, the verification of output force-depending values describes the energy flux directed to the user. Hence, the user is seen as passive in this case. The verification of the force-depending values can be done in two steps. The first is to investigate static or quasi-static output force signals; the second is to investigate the dynamic output force behavior such as frequency response, step response, and impulse response.

At first, the verification of static force values is shown. Therefore, a force sensor, attached to the end effector of the device, with a resolution higher than the minimal desired force output resolution of the haptic system, is needed. The measurement is done by supplying an input to the haptic system and simultaneous inspection of measured output forces.

13.1.2.1 Static Analysis

Peak force is the highest displayable force with very low duration, limited in time due to heat dissipation. When testing the maximum peak force, it might be wise to measure the time till the actuator brakes down.

Maximum continuous force is the highest continuous force that can be displayed without concerns of thermal destroying of the actuator and is sometimes called as saturation level or saturation force.

Minimal force (offset) is the lowest force that can be displayed. It is caused by force bias, slip-stick effects, backdrive friction, or stiction.

Hysteresis describes the difference of displayed forces during increasing and decreasing device input signals.

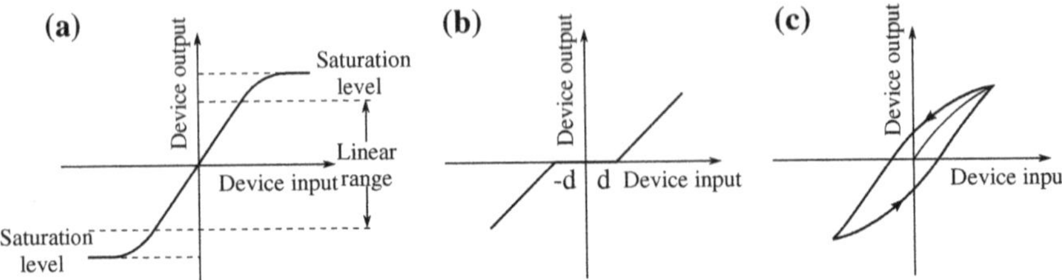

Fig. 13.2 Common nonlinearities in the characterization of haptic devices: **a** saturation, **b** dead zone, **c** hysteresis, figure based on [31, 37]

Sensitivity is the change of output signal for a change of the device input signal in the linear region. It can also be described as the slope of the graph.

Output force resolution is the minimal displayable force step resulting due to minimal change of the device input signal, thereby the minimal D/A resolution should be considered as well.

Cross talk describes the influence of an output forces in one direction, produced while displaying a force in a different direction of a multidimensional haptic system.

Dynamic range is defined as the difference of maximum continuous force and minimum force in dB.

While inspecting the measurement signals, one may find different characteristic graphs showing the output force signals with respect to device input signals displayed in Fig. 13.2. The previously described characteristic nonlinearities can be derived out of these graphs, called calibration curves [32].

13.1.2.2 Dynamic Analysis

The verification of static force signals addresses only the actuator and gear capabilities of the haptic system. To maintain the overall mechanical properties of the haptic system, including inertia and damping, it is necessary to characterize the dynamical behavior of the system. One has to note that the movement of the system will induce a unwanted output signal of dynamic force sensors because of acceleration of the mass of the sensor. This systematic error can be dealt with an additional calibration or all measurements have to be conducted in a mechanical fixed condition.

Frequency response shows the force transfer function between the device input and the measured force response value with respect to frequency. The transfer function is represented by a bode plot and includes amplitude and phase values. With the bode plot, one can get information about the useful frequency range of the system, represented as the flat region of the plot without a significant drop, as well as the half-power bandwidth represented by the −3-dB gain drop of the transfer function where the stop band of the system begins. The frequency response is most relevant in control design. Due to the comparison of the waveform of an sinusoidal excitation and the displayed force, a measure called force fidelity can be performed to measure the distortion of a haptic signal [17, 32].

Step response represents a time-domain analysis method to get quantified information about rise time, percentage overshoot or settling time. The rise time can be derived by measuring the time between the excitation step and the pass of the desired steady-state value of the response and is a rate for the maximal displayable bandwidth. Percentage overshoot and settling time are rates for damping in the system and stability. Also information about the force output accuracy with respect to the device input excitation can be derived, represented by the closeness of the output force value to the desired value, as well as information about force precision represented by the repeatability of the output force value and commonly indicated as $\pm$ standard deviation.

Impulse response is also a time-domain-based verification method. With the excitation of a short impulse with a magnitude of, for example, peak force, it is possible to derive maximum speed and maximum acceleration of the system.

13.1.2.3 Measuring Conditions

In the most cases, it is not possible to provide a universally valid test scenario. Hence, before execution of the measurements, one should create defined measurement conditions. Therefore, it might be crucial to use special constrains to get best-fit measurement values for the specific case. The four major conditions to be mentioned are constrains of the end effectors motion and can be figured out as fixed end, open end, handheld, and user phantom.

Fixed end means to constrain the user interface to zero velocity. This case might be used for static tests and frequency response without having the influence mechanical properties in the signal. The output force is measured between the haptic system and a rigid clamp.

Open end means a free movement of the haptic system without any constrains. This means zero force at the end effector of the haptic system. In this case, velocity or acceleration at the end effector should get taken into account. Therefore, an accelerometer could be attached instead or in addition to the force sensor. To derive velocity, the acceleration could be integrated.

Handheld stands for the users hand and arm gets coupled to the haptic system. The goal is to create mechanical confinements of the haptic system in a realistic manner. Hence, it is possible to get measurement result involving all realistic occurring parameters. One should be aware that experiments with human users in the measurement loop vary from trial to trial. For low frequency ranges, humans can adjust their mechanical impedance quite freely, so that such measurements are probably most suited to evaluate the interference liability of the haptic system.

User phantom should overcome the disadvantage of undefined conditions by using a human in the measurement loop. With a user phantom, designed to provide almost equal characteristics than a human, it is possible to create acceptable realistic conditions and repeatable results. As a user phantom, one can use a rubber with known mechanical properties or in the easiest case a spring and damper combination [17].

To provide meaningful results and a complete set of measurement values, it might be necessary to repeat all measurements for different points and orientations of the end effector in the workspace of a multidimensional system.

13.1.3 Output Motion-Depending Values

Most common for measuring motion values of a haptic system, also multidimensional, is the use of accelerometers. Besides the verification of maximum accelerations, due to integration of the acceleration signals, velocities, and due to double integration, the position of a specimens end effector can be determined almost without effecting the measurement. For measurements with necessity of high accuracy, it might be meaningful to use special sensors for measuring velocity or position. In this case, the measurement should be reduced to at least one degree of freedom for the same time.

Measuring motion capabilities can be combined with the measurement of frequency-, step-, or impulse-response measurements. Due to the characterization of phase differences of force and motion values, one can derive information about inertia and damping of the system [32].

While force-depending values are most important for systems based on the impedance structure, for admittance-based haptic systems, the motion-depending values are in the main focus. For a large number of haptic admittance systems, the accuracy of displayable velocities or positions is relevant. For example, during the characterization of braille displays, the rise time and stationary accuracy of the different pins are of note. To measure the position of small structures, for example, optical measurement systems like laser triangulators for small unidirectional deflections, tracking systems for large multidimensional movements or vibrometers for highly dynamic movements can be used.

13.1.4 Mechanical Properties

The achievable haptic quality is direct depending to the mechanical properties of a haptic system. Mechanical properties can vary depending on the position and orientation in the workspace; hence, these factors should be evaluated as well.

Stiffness of the mechanical structure is an indicator for maximal displayable stiffness, thereby the displayable bandwidth is directly limited by the stiffness of the mechanical system. When measuring the stiffness, the displacement of the end effector under external force excitation is of interest, while the actuator or the actuators are fixed. The measurement value can be calculated due to the division of the acting force at the end effector and the measured displacement. If the position sensors for the end effectors displacement are integrated to the actuators, there is the possibility to measure the stiffness of the system the other way round. While fixing the end effector with a force sensor, one can excite the mechanical structure with forces produced by the actuators of the system. The stiffness can be calculated due to the division of the forces, measured at the end effector, and the virtually measured displacement by the sensors in the actuators. This calculation is equivalent to the slop of the graph till the force excitation limit is reached by the maximum continuous force value of the system.

Backlash can be measured by comparing the externally measured deflection of an end effector and the signals of the system integrated sensory. The difference of the signals is a measure for backlash and is usually a problem arising due to the reversal of force directions.

13.1.5 Impedance Measurements

One important measure of haptic displays is the ability to provide a wide range of mechanical impedance. The perceptible haptic impression of a haptic system is given due to a combination of force reflection and velocity. In this case, the mechanical impedance means the force reflection of the haptic system with respect to the users inserted velocity. The impedance is a frequency-dependent value and can be terminated by $Z = F/v$. Besides the mechanical impedance for one operating point, the so-called *Z-width* of the system is of interest [8]. The Z-width means the value of impedances that can be displayed by the haptic device. Mostly the development objective of a haptic system is to design the lowest impedance to be like transparent and the largest impedance to be as high as possible to display stiff walls. Since, a haptic systems tend to get instable while displaying very stiff signals, the upper limit of displayable impedances is given by the maintainable stability of the system.

Usually, the mechanical impedance gets displayed in a bode plot showing amplitude and phase of a signal. The so-generated signals can be used for system identification to create models of the system, showing damping, stiffness, and inertia of the system.

For measuring the mechanical impedance of a haptic system, one may use a special setup with external force excitation to input a defined periodical force sweep to the haptic system. To calculate the impedance, a force sensor and a velocity sensor need to be attached between the external force source and the specimen. In this case, the impedance can directly be calculated as the quotient of the measured force and velocity.

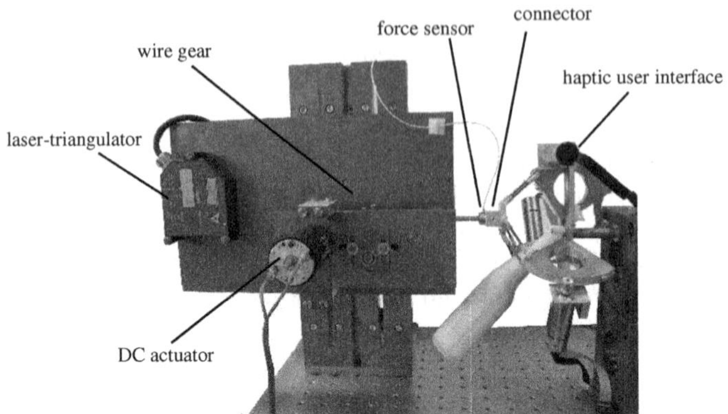

Fig. 13.3 Measurement setup for the evaluation of mechanical impedances [22]

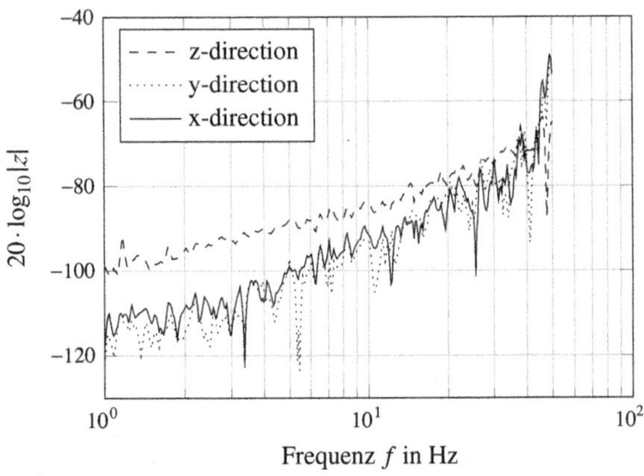

Fig. 13.4 Measurement of the mechanical impedance of the passive delta structure of the haptic interface of the INKOMAN system from Fig. 13.3

An exemplary measurement setup for measuring mechanical impedance is shown in Fig. 13.3. The main parts are a force source consisting of a DC-actuator coupled to linearly guided rod, an attached force sensor as coupling element between the actuated rod and the specimen and a velocity measurement system. To measure the velocity of the system, a laser triangulator with additional analog differentiator is attached.

The shown setup is attached to a handheld haptic controller, based on a delta robot. In this case, the impedance measurement is done for the passive system in different directions at the toll center point of the delta robot to examine its mechanical properties. The resulting impedance amplitude is shown in Fig. 13.4 [22].

To determine the Z-width, one has to measure the impedance of the haptic system when moving in free space to get the lowest displayable impedance of the system. To measure the maximum displayable impedance of the system, one can repeat the measurement while rendering a stiff wall with the haptic display by fixing the tool center point with the haptic systems actuators at its maximum continuous force.

If there is no special measurement setup available to excite the system, it is possible to jiggle the systems end effector with the humans hand while measuring the force and velocity at the end effector. In this case, one is restricted to the dynamic range of the frequency output capability of a human's hand. Also the calculation of impedances out of excitation values and responding signals for force or velocity independently is possible [32].

Due to the impedance measurement of a system at different points and directions in the working space of the haptic system on can measure the homogeneity or dexterity of the system. Therefore, the quotient of the generally smallest and largest measured impedance of the passive haptic system can be calculated.

13.1.6 Special Properties

Besides the characterization of output capabilities of haptic interfaces, one can find different values in complex haptic systems that might be of interest regarding the haptic quality of haptic system.

One of these is, for example, the input signal of a haptic system that can affect the haptic quality. Hence, the bandwidth and accuracy of the provided signals should be larger and more precisely, then the set requirements for the displayable haptic feedback. Another point might be the slaves sensing capability of a haptic teleoperation system. The dynamics of the signal can affect the displayable haptic feedback as well as the accuracy of the sensor signals. Also the latency of (force) signals might be of interest and can heavily affect the transparency of the haptic system [18]. Even the time shift between haptic feedback, visual and acoustic feedback might be of interest.

Further properties that have to evaluate depending on system structure and application are transparency and the transparency error (see Sect. 7.4.2), latencies in the control loop (especially when the system contains packet-based information transfer, for example, the Internet), control stability, and energy consumption.

13.1.7 Measurement of Psychophysical Parameters

Psychophysical parameters such as absolute and differential thresholds are fairly well-investigated (see Sect. 2.1). Based on the suggestions of WEISENBERGER ET AL., it can be nevertheless useful to measure a psychophysical parameter, since deviations to (well-known) thresholds can be attributed to the fidelity of the device [42]. This procedure is implemented in different evaluation test beds by SAMUR [32], but can also be applied individually. To assess the fidelity of a haptic system, absolute and differential thresholds are useful measures, evaluating resolution and reproducibility of a device. This method is therefore preferable for the evaluation of haptic interfaces. Similarly, the discrimination of haptic properties and signals can be used to assess haptic systems for communication means as described in the next section.

13.2 Task-Centered Evaluation Methods

The above-described methods of system-centered evaluation are used to determine concrete values for the system's properties. However, to validate whether a system performs correctly, further evaluations of the task performed with the system are needed. Such task-centered methods investigate the feasibility of a system to perform beneficial in the intended usage.

Typical for this purpose is a simple test task that is performed by several test subjects under different boundary conditions. A typical boundary condition is, for example, the kind of feedback such as no feedback, haptic feedback, visual feedback, and haptic and visual feedback under which the test is performed or different properties of the haptic system used (e.g., amplitudes, frequency, speed). Because of the dependence on test persons and different boundary conditions, one should carefully design the experiment to prevent errors because of inter- and intra-personal effects as well as learning.

Results of task-centered method can be used to compare different systems in the same task or to assess the effectiveness of haptic systems in a given application. Furthermore, such tests can be useful to identify promising parameters to optimize a given design or interaction. Starting with these methods, an engineer will take large steps toward the sometimes odd-looking (from an engineers point of view) approaches of human factors and ergonomics. However, the authors find it advisable to consider such approaches as early as possible in the design process (e.g., by defining evaluation tests in the requirement derivation process) to incorporate noteworthy aspects that will actually contribute to a good product.

13.2.1 Task Performance Tests

Typical task performance tests are conducted with haptic interfaces using ↪VR test setups or with teleoperation system. One should identify a task that is very close to the intended usage of the system to use all interaction primitives involved, but that is simple enough to be understood quickly by the test persons and to be completed in time.

Typical embodiments of such tests are given in the following list.

Pick and Place Tasks Subjects have to pick up an item or object, move it to another location, and place the object in a predefined way. Depending on the application, objects, and placing rules differ, simple setups only require to place a round peg into a hole, while more complicated setups require to place micro- or nanoscale objects in predefined orientations. Because of the simple implementation, this type of test is often used to evaluate universal haptic interfaces and—of course—systems designed for microassembly [2]. Typical parameters in these tests are the kind of feedback (kinaesthetic/tactile feedback, combinations with visual and auditory feedback, etc.).

Tracing In these kind of tests, subjects have to follow a given contour in 2D or 3D space or move in given constraints. These kind of tests are especially useful to evaluate the effect of haptic guidance techniques and new rendering techniques in multimodal applications as, for example, shown in [29], where three-dimensional perceptual deadband coding (see Sect. 2.4.4) was evaluated.

Application Task Primitives The third major group of task performance tests are derived from typical tasks in applications. For surgery, this could be suturing or the handing over of a needle or an electrocautery electrode, typical assembly task could include the fastening of a nut or the selection of an object with a certain property. Such tasks can be derived from generalized assessment tests as, for example, reported in [9] for minimal invasive surgery.

To gain quantified measures from these tests, several outcomes are commonly used. They can be used in combination with almost every test type and have to be chosen according to the goal of the evaluation and the technical capabilities of the test environment.

Completion Time The most basic measure is the time needed to complete a given task. Variations of this test outcome include the number of fulfilled tasks in a given time, sometimes also known as the completion rate.

Errors Another outcome is the number of errors made by the test subject. The type of error depends on the task and the intended outcome of the test, commonly used as errors are collisions with other tools or boundaries, the dropping of objects, deviation of the intended placing locations, or errors in task performance like the ripping of tissue. Of course, the failing of the task itself can be an error, too.

Handling Forces Since one of the main goals for assistive and teleoperation systems is the reduction of handling forces, the evaluation of average, maximum, or contact forces is a common outcome of task performance tests. One has to note that this outcome as well as some of the above-mentioned error definitions will require additional sensory equipment such as tracking systems or reliable force sensors.

Examples for practical realizations of such tests can be found in a vast number of studies, for example [27], where haptic feedback for robotic surgery is evaluated. Figure 13.5 shows an example using a DAVINCI surgical robot. The work of PONGRAC gives some general guidelines for the evaluation of virtual reality and teleoperation systems [30]. In the studies included in the meta-analysis about the effects of haptic feedback by NITSCH [28], further task performance tests can be found for various applications.

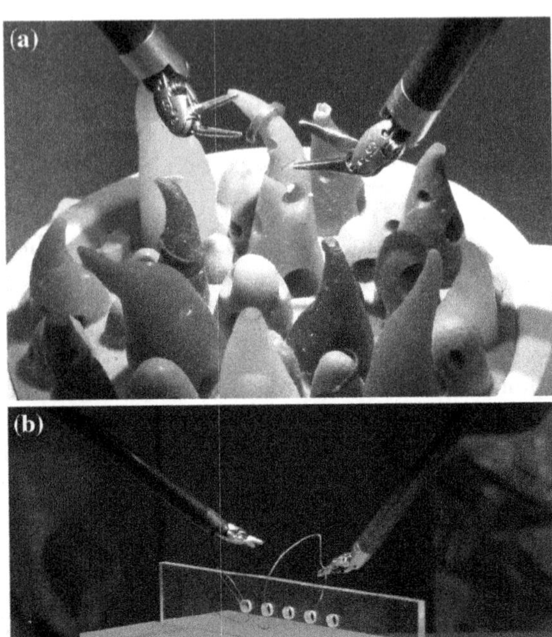

Fig. 13.5 Task performance test setups for minimal invasive surgery: **a** pick and place setup of a DAVINCI surgical robot at the University Medical Centre Mannheim, Germany, **b** needle transfer setup as reported in [27]. Pictures courtesy of Peter P. Pott, Technische Universität Darmstadt and Katherine J. Kuchenbecker, University of Pennsylvania

13.2.2 Identification of Haptic Properties and Signals

One of the main goals of haptic systems intended for communication is to transfer information from the system to the user. Similar is the necessity of a teleoperation to convey enough information to the user to differentiate between relevant components and materials, for example between tissue and vessels in a surgical application. For such kind of evaluations, TAN ET AL. proposed the evaluation of the ↪information transfer (IT) of a haptic application, a measure that is used widely in the evaluation of haptic systems [32]. JONES AND TAN give a detailed explanation in [20], on which this section is predominantly based.

This approach is orientated on a information theoretical framework and is normally based on a absolute identification experiment. A user is presented one of K stimuli S_i and has to choose a response from a set of K responses R_j with $i, j = 1, \ldots, K$. Based on the answers, an confusion matrix is constructed, that denotes how often each response R_j was chosen, when a certain stimulus S_i was presented. Stimuli are represented in rows while responses are given in columns. Based on this matrix, an estimate for the information transfer IT_{est} can be calculated according to Eq. (13.1).

$$IT_{\text{est}} = \sum_{j=1}^{K}\sum_{i=1}^{K} \frac{n_{ij}}{n} \log_2 \left(\frac{n_{ij} \times n}{n_i \times n_j}\right) \text{bit} \tag{13.1}$$

with n total number of trails

$n_{i,j}$ number of occurrences of (S_i, R_j)

n_i row sum

n_j column sum

Based on the estimation IT_{est}, the number of correctly identifiable stimulus levels n_C can be calculated according to Eq. (13.2).

$$n_C = 2^{IT_{\text{est}}} \tag{13.2}$$

The upper limit of n_C for a given information channel is also called *channel capacity*. For haptics, typical values of n_C are in the range of 3...4 for unidimensional information transfer [6, 11]. Specialized systems like the TACTUATOR achieve higher values up to 12 bit [38]. For auditory and visual examples, values of n_C tend to be higher in the range of 5...7 identifiable levels of, for example, force or compliance. To correctly measure the channel capacity, an sufficient high number of stimulus alternatives have to be incorporated in the study. JONES AND TAN give a rule of thumb to evaluate $K = 2^{IT'_{\text{est}}+1\ldots2}$ stimulus alternatives in $n > 5K^2$ trials to minimize statistical bias and to exceed the maximum channel capacity.

The specification of the information transfer in terms of IT is preferable to the description in terms of percentage correct that can be found in many studies. The measure of IT is sensitive to chance performance as well as confusions of the response alternatives by the test subject. One has to note that the information transfer capacity of a channel is considerably lower than the number of ↪Just Noticeable Differences (JND) in the same range lets assume. According to DURLACH ET AL., this is due to the fact that information transfer also involves the subjects memory and not only the sensory system [11]. In terms of the general perception primitives described in Sect. 1.2.1, the ↪JND measures the discrimination ability of a subject, whereas IT describes the ability to identify a certain stimulus. The latter is much more difficult.

Test setups evaluating the identification of haptic properties and signals are based on real and virtual samples with different object properties (see e.g., the vast number of experiments with real objects by the group of KAPPERS [21]) for the evaluation of teleoperation systems and haptic interfaces. A typical experiment could assess the number of different compliances that can be discriminated when using a haptic teleoperation system or the different amounts of damping, that can be rendered and perceived when using ↪VR interfaces and corresponding applications. An example is presented by SCILINGO ET AL. that investigate the ability of a magnetorheologic haptic display to render compliances [35]. For the evaluation of haptic displays intended for communications, stimuli are constructed from the different attributes the system can display (e.g., speed, distance, frequency, amplitude) and the informa-

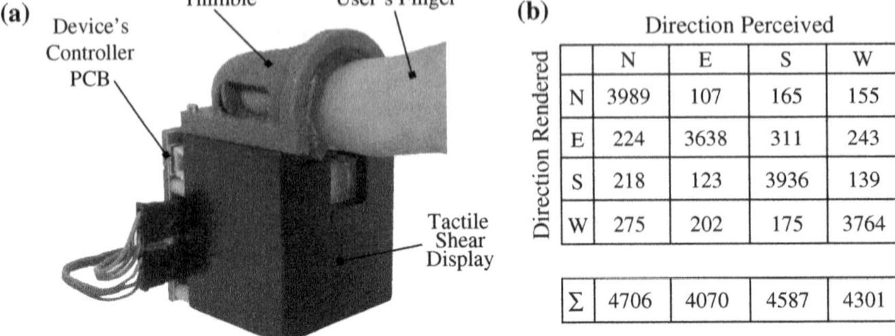

	N	E	S	W
N	3989	107	165	155
E	224	3638	311	243
S	218	123	3936	139
W	275	202	175	3764
Σ	4706	4070	4587	4301

Fig. 13.6 Evaluation result of a shear display (**a**) with regard to the displayed directions (**b**). Values in the confusion matrix are taken from [14] and report all stimuli presentations of the study without differentiating further experiment variables like moving distance and speed

tion transfer is considered with respect to these attributes. This is illustrated in the following example.

Example: Tactile Shear Display by GLEESON ET AL. [14]

GLEESON ET AL. developed a tactile shear display to convey information in mobile applications. Figure 13.6 gives a picture of the completed system that is based on a pin that can be moved in different directions with different kinematic properties (see [13, 19] for further information). A confusion matrix of several tests with the tactile shear interface, that displayed stimuli in north, south, east, and west directions, is also given in Fig. 13.6.

Based on the values in the confusion matrix, the calculations according to Eqs. (13.1) and (13.2) give $IT_{\mathrm{est}} = 1.23\,\mathrm{bit}$ and $n_C = 2.35$. With regard to the above rules-of-thumb, one could argue that the test setup contained a too little number of stimulus alternatives K to evaluate the maximum channel capacity, but that was not the intention of the original study the data where taken from. One has to note also that GLEESON ET AL. conducted several studies with different hypothesis and the confusion matrix in Fig. 13.6 contains all trails from these different studies. It is only used as an example for the usage and interpretation of ↪IT here.

13.2.3 Information Input Capacity (Fitts' Law)

The evaluation of information input capacity is somewhat the opposite side of the assessment of information transfer as described in the last section. While this quantity measures the amount of information transferred from system to user (interaction path **P'** in Fig. 2.24), the information input capacity measures the amount of information that can be transferred from user to system (interaction path **I'**). The concept was developed by FITTS [12] and is often referred to as *Fitts' law*. It describes the accuracy of movement with regard to the size of the movement. It was proven originally for unidirectional movement tasks such as tapping (see Fig. 13.7), peg-in-hole, and item transfer and extended to two-dimensional tasks by ACCOT AND ZHAI [1].

To measure the information input capacity, FITTS defined the index of performance I_p according to Eq. (13.3)

$$I_\mathrm{p} = -\frac{1}{t} \log_2 \left(\frac{w}{2d} \right) \text{ in bits/s} \tag{13.3}$$

with W as measure of the target size and A as distance between targets, as shown exemplary in Fig. 13.7. The logarithmic term is also called the index of difficulty I_D. To incorporate the temporal dimension, the average time for a single movement t is used.

This approach is, however, difficult to use, when one wants to compare different systems or interaction setups. Therefore, the usage of the relation given in Eq. (13.4) is used more often. It relates the moving time t_m with the index of difficulty I_D and two device and test-dependent constants c_a and c_b.

$$t_\mathrm{m} = c_a + c_b \log_2 \left(\frac{d}{w} + 1 \right) \tag{13.4}$$

Equation (13.4) can be derived from Eq. (13.3) directly, but contains a slightly modified formulation of the index of difficulty I_D. According to MACKENZIE AND BUXTON, this so-called *Shannon formulation* provides slightly better fits to given data, is more consistent with the underlying information theorem, and ensures an always positive rating for the index of difficulty [26]. Given that, it is obvious, that the device and test-dependent constant c_b directly relates to the index of performance I_p. In an evaluation, one will conduct tests with different indices of difficulty of performance and record the movement time needed to move to and select a target. The values of c_a and c_b are determined by fitting Eq. (13.4) to the data. The such acquired indices

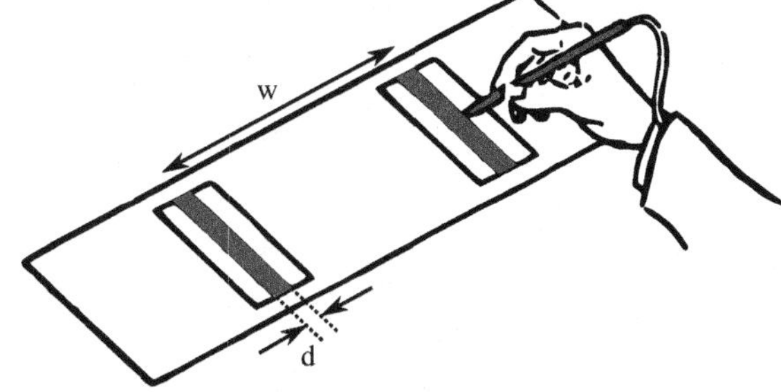

Fig. 13.7 Tapping experiment used to evaluate the information input capacity. Test persons were told to tap both target regions with width d at distance w from each other as fast as possible without missing the targets. Figure is based on [12]

of performance allow to measure the input capability of a user with a given system and the comparison of different haptic systems.

For haptics, one can find studies employing *Fitts' Law* to investigate the effect of haptic feedback on task performance with different interface configurations [15, 32, 41] and multimodal interfaces [7].

13.3 User-Centered Evaluation Methods

The usage of haptic systems will not only have an effect on the task performance of a user, but also on the user himself. User-centered evaluation methods will give some insights into these processes in order to assess advantages and disadvantages of haptic systems in the investigated application.

Compared to task- and system-centered evaluation methods, there is no prevailing test form for user-centered methods. Depending on the intended informative value of the test, one will find comparative tests as well as tests producing single-test values.

13.3.1 Workload

Ergonomics and occupational sciences know different forms of workloads as defined by the standard ISO 10075. For the evaluation of haptic systems, the influence of a haptic system on the physical and mental workload of a user is probably the most interesting problem. For the evaluation of haptic systems, the assessment of workload is mainly of interest for the following intentions of haptic feedback:

Lowering the workload If a haptic system is designed to lower the workload by adding haptic feedback to an existing application, one should verify if this goal is achieved. In that case, one will probably compare workload assessments in test conditions with and without haptic feedback. It might be possible to integrate this workload assessment in other evaluation tests, for example, task-centered evaluations described above. In most of these cases, mental workload is of interest.

Assessment of new kinds of haptic interaction When a haptic interaction replaces other forms of interaction, it might be beneficial to evaluate the effects on physical (sometimes also mental) workload to gain knowledge about the usefulness of the new interaction scheme.

As an overview, Fig. 13.8 shows some influence factors on the workload of a human operator that probably can be influenced by the designer of a haptic system and the usage definition. Based on this considerations, one should also note that a medium workload is preferable for a safe usage of a haptic system—too low or too excessive demands on the user will result in errors.

In the following, a quite simplified view on workload evaluation is presented to have applicable methods for the comparison of different systems or different

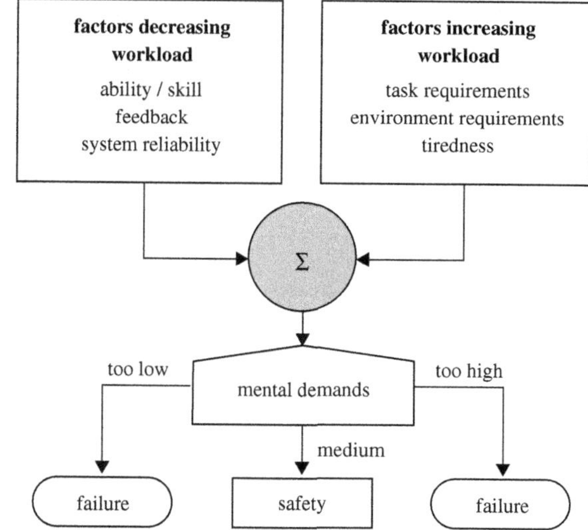

Fig. 13.8 Influences on the workload of a human user and possible outcomes. Figure based on [34]

conditions of a single system. Aspects like emotional workload are neglected, since they mainly depend on the conditions, a task is executed in and not on the system used [34]. The application of the methods presented here therefore does not fulfill all requirements of a workload analysis as formulated by ISO 10075-3.

One of the easiest ways to assess workload is the usage of standardized questionnaires. One of the widespread is the *NASA Task Load Index (NASA-TLX)* [16]. The resulting scale is a workload measure based on six different areas: mental demand, physical demand, temporal demand, performance, effort, and frustration. Test persons first evaluate the amount of contribution of each of these areas to the overall workload, and in the second step, the areas are evaluated on bipolar scales [34]. Despite the NASA-TLX, there are quite a few of other workload assessment tools. An overview is given in [25], the German Federal Institute for Occupational Safety and Health (BAuA) also provides an online database and toolbox for this purpose [4].

Another possibility to assess workload is the analysis of biometric measures. Pulse rate, blood pressure, respiratory frequency, brain activity, ↪electromyography (EMG), pupil diameter, blinking frequency, skin conductivity, and eye movement are indicators for a persons' attention and concentration. Advantages of these biometric measures are the possibility of continuously data acquisition without major interference with the actual task. Disadvantages of these measures are the dependence on the individual user (therefore, one often has to record a baseline before actual performing the experiment), and the lacking of a definite assignment of biometric measures to a defined measure of workload. Table 13.1 gives some biometric indications that can be used for a selection of suitable measurements.

For haptic systems designed to assist a human operator, for example, in handling tasks with complex trajectories, the assessment of muscular activity and fatigue could be of further interest. While muscular activity is monitored by EMG, the assessment

Table 13.1 Biometric indications to newly arriving information and evidence for information processing. Table based on [34]

Arrival of information	Information processing
• Body orientates toward stimulus	• Blood pressure increases
• Current activity halts	• Pulse rate increases
• Increasing EMG activity	• Pulse rate variability
• Pupil diameter increases	• Skin conductivity increases
• Skin conductivity increases	• Blood glucose level increases
• Respiratory frequency decreases	• Respiratory frequency increases
• Pulse rate decreases	
• Blood pressure decreases	

of muscular fatigue is somewhat more complicated. According to [40], EMG signals do not correlate well with muscular fatigue. The proposed evaluation based on maximum voluntary contraction (MVC) of the muscles is probably challenging to implement in a test of a haptic system. Further possibilities to assess muscular fatigue can be adapted from sport science literature [44], if the presented possibilities are not applicable or insufficient for the intended test.

13.3.2 Subjective Evaluation

While subjective evaluations can be done with regard to almost every aspect of the usage of a haptic system, the measure of the amount of immersion into a virtual or teleoperated environment is of interest for these kinds of applications. PONGRAC reports several standardized questionnaires for this purpose [30], for example, the *Witmer-Singer-Presence Questionnaire* or the *ITC-Sense of Presence Inventory*. More recently, CHERTOFF ET AL. presented an approach for the evaluation of multimodal virtual environments claiming to improve the integration of sensory, cognitive, affective, personal, and social elements of experience [5]. Although primarily developed for the assessment of computer games, this could be a valid alternative for the evaluation of haptic telepresence systems and ↪VR interactions as well.

For the design of training systems, a subjective evaluation, i.e., a self-report of the test subjects, that compares the virtual or teleoperated training condition to the real condition without system-mediated haptic feedback is advisable to investigate acceptance processes of the users. KRON ET AL. investigate preferences of users for certain kinematic structures of a telepresence system for disposal of explosive ordnances by using a subjective evaluation after usage [23]. Further evaluations could cover emotional and social aspects of a system. This may seem to be exaggerated for the majority of haptic systems, but is of importance in application areas like ↪Ambient Assisted Living (AAL) or other assistive systems for elderly and disabled users. In that case, one has also to consider impacts on other people not directly interacting with the haptic system that are only involved as a relative or assistant.

Self-reports can also be used to evaluate the subjective performance of users, when some kind of feedback about the task performance or criteria for good performance are given [43]. From the evaluation of usability, aspects like the *joy of use* could be assessed by self-reports or even the assessment of verbal statements during the evaluation test [33, Chap. 6].

13.3.3 Learning Effects

In particular for training applications, the results of regular performance tests can be compared over time and learning effects can be quantified in terms of these test outcomes. Another approach to measure learning effects is the comparison of a trained group of subjects with an untrained control group in the intended application, for example, shown by AHLBERG ET AL. for the effect of ↪VR training on the error rate in cholecystectomies by novice surgeons [3].

13.3.4 Effects on Performance in Other Domains

Another approach to measure the effect of a haptic system on the user is to assess the performance in another domain. Predominantly one will find such kind of tests in areas, where a haptic system is intended to assist in a secondary task, while the primary task is not to be affected. Obviously, this is true for communication applications in vehicles. In that case, the standardized ↪Lane Change Test (LCT) is performed with interaction tasks using the haptic communication system as a secondary task to the driving on real roads or in a simulator, for example. Despite the standardized outcomes like the medium lane deviance, other parameters as for example the viewing direction and duration or the task performance of the secondary tasks can be evaluated. Examples for such kinds of evaluation of touch devices can be found in [10, 24, 36].

There are also cases, where such tests aim at the primary goal of the intended application: VÁRHELYI ET AL. investigate the effect of an active accelerator pedal that would create a counterforce when passing the speed limit. The evaluation conducted in 284 vehicles showed an improvement of the driver's compliance with the speed limit as well as reduced average speeds, speed variability, and emission volumes [39].

13.4 Conclusion

As it can be seen from this section, the evaluation of haptic systems is complex and exhibits a large number of different facets. For each newly developed task-specific haptic device, evaluation methods and goals have to be selected from the above-mentioned (and other applicable) measures. Depending on the kind of application, existing studies can give hints about the selection of applicable methods. The works of WILDENBEEST ET AL. evaluating teleoperated assembly tasks [43] and the evaluation of an assistive system for minimal invasive surgery by MCMAHAN ET AL. [27] should be recommended here because of the wide scope and the thorough methodology for this kind of systems. For new kinds of universal haptic interfaces, the work of SAMUR is naturally a must-read [32].

Recommended Background Reading

[17] Hayward, V.& Astley, O.R.: **Performance measures for haptic interfaces**. In: Robotics research, 1996.
Extensive list of possible physical measures for the evaluation of haptic interfaces.

[31] E. Samur: **Performance metrics for haptic interfaces**. Springer, 2012.
The probably most advanced work about evaluation techniques for haptic interfaces with a strong focus on the interaction with virtual environments.

References

1. Accot J, Zhai S (1997) Beyond Fitts' law: models for trajectory-based HCI tasks. In: Proceedings of the ACM SIGCHI conference on human factors in computing systems. ACM, pp 295–302. doi:10.1145/258549.258760
2. Acker A (2011) Anwendungspotential von Telepräsenz-und Teleaktionssystemen für die Präzisionsmontage. Dissertation, Technische Universität München, http://mediatum.ub.tum.de/doc/1007163/1007163.pdf
3. Ahlberg G et al (2007) Proficiency-based virtual reality training significantly reduces the error rate for residents during their first 10 laparoscopic cholecystectomies. Am J Surg 193(6):797–804. doi:10.1016/j.amjsurg.2006.06.050
4. Bundesanstalt für Arbeitsschutz und Arbeitsmedizin (BAuA). Toolbox: Instrumente zur Erfassung psychischer Belastungen. last accessed 2014-Feb-4. 2014. http://www.baua.de/de/Informationen-fuer-die-Praxis/Handlungshilfen-und-Praxisbeispiele/Toolbox/Toolbox.html
5. Chertoff D, Goldiez B, LaViola J (2010) Virtual experience test: a virtual environment evaluation questionnaire. In: 2010 IEEE virtual reality conference (VR), Mar 2010, pp 103–110. doi:10.1109/VR.2010.5444804
6. Cholewiak SA, Tan HZ, Ebert DS (2008) Haptic identification of stiffness and force magnitude. In: Symposium on haptic interfaces for virtual environments and teleoperator systems. Reno, NE, USA. doi:10.1109/HAPTICS.2008.4479918
7. Chun K et al (2004) Evaluating haptics and 3D stereo displays using Fitts' law. In: IEEE proceedings of the 3rd IEEE international workshop on haptic, audio and visual environments and their applications, 2004, HAVE 2004, pp 53–58. doi:10.1109/HAVE.2004.1391881

8. Colgate J, Brown J (1994) Factors affecting the Z-Width of a haptic display. In: IEEE proceedings international conference on robotics and automation, 1994, vol 4. May 1994, pp 3205–3210. doi:10.1109/ROBOT.1994.351077
9. Derossis AM et al (1998) Development of a model for training and evaluation of laparoscopic skills. Am J Surg 175(6):482–487. doi:10.1016/S0002-9610(98)00080-4
10. Domhardt et al M (2013) Evaluation eines haptischen Touchpads für die Fahrer-Fahrzeug-Interaktion. In: Brandenburg E et al (eds) Grundlagen und Anwendungen der Mensch-Maschine-Interaktion: 10. Berliner Werkstatt Mensch- Maschine-Systeme (Berlin 2013). Fortschritt-Berichte VDI, Reihe 22, Mensch-Maschine-Systeme. VDI-Verlag, Düsseldorf, pp 9–18. https://www.tu-berlin.de/zentrum_mensch-maschine-systeme/menue/veranstaltungen/berliner_werkstaetten_mms/10_berliner_werkstatt_mms/
11. Durlach N et al (1989) Notes and comment resolution in one dimension with random variations in background dimensions. Attention Percept Psychophys 46(3):293–296. doi:10.3758/BF03208094
12. Fitts P (1954) The information capacity of the human motor system in controlling the amplitude of movement. J Exper Psychol 47(6):381. doi:10.1037/h0055392
13. Gleeson BT, Horschel SK, Provancher WR (2009) Communication of direction through lateral skin stretch at the fingertip. In: Third joint EuroHaptics conference and symposium on haptic interfaces for virtual environment and teleoperator systems (WorldHaptics Conference). Salt Lake City, UT, USA. doi:10.1109/WHC.2009.4810804
14. Gleeson BT, Horschel SK, Provancher WR (2010) Perception of direction for applied tangential skin displacement: effects of speed, displacement and repetition. In: IEEE transactions on haptics 3.3 (2010), pp 177–188. ISSN, pp 1939–1412. doi:10.1109/TOH.2010.20
15. Hannaford B et al (1991) Performance evaluation of a six-axis generalized force-reflecting teleoperator. IEEE Trans Syst Man Cybern 21(3):620–633. doi:10.1109/21.97455
16. Hart S, Staveland L (1988) Development of NASA-TLX (task load index): results of empirical and theoretical research. Hum Mental Workload 1:139–183. doi:10.1016/S0166-4115(08)62386-9
17. Hayward V, Astley OR (1996) Performance measures for haptic interfaces. Robot Res 1:195–207. doi:10.1007/978-1-4471-0765-1_22
18. Hirche S, Buss M (2007) Human perceived transparency with time delay. In: Advances in telerobotics (2007), pp 191–209. doi:10.1007/978-3-540-71364-7_13
19. Horschel SK, Gleeson BT, Provancher WR (2009) A fingertip shear tactile display for communicating direction cues. In: Third joint EuroHaptics conference and symposium on haptic interfaces for virtual environment and teleoperator systems (WorldHaptics Conference), pp. 611–612. doi:10.1109/WHC.2009.4810906
20. Jones L, Tan H (2013) Application of psychophysical techniques to haptic research. IEEE Trans Haptics 6:268–284. doi:10.1109/TOH.2012.74
21. Kappers AM, Bergmann Tiest WM (2013) Haptic perception. Wiley Interdisc Rev: Cogn Sci 4(4):357–374. doi:10.1002/wcs.1238
22. Kassner S (2013) Haptische Mensch-Maschine-Schnittstelle für ein laparoskopisches Chirurgie-System. Dissertation, Technische Universität Darmstadt. http://tubiblio.ulb.tudarmstadt.de/63334/
23. Kron A, Schmidt G (2005) Haptisches Telepräsenzsystem zur Unterstützung bei Entschärfungstätigkeiten: Systemgestaltung, Regelung und Evaluation. In: at- Automatisierungstechnik/Methoden und Anwendungen der Steuerungs-, Regelungs-und Infor- mationstechnik 53.3/2005, pp 101–113. doi:10.1524/auto.53.3.101.60272
24. Liedecke C, Baumann G, Reuss H-C (2014) Potential of the foot as a haptic interface for future communication and vehicle controlling. In: Proceedings of 10th ITS European congress. Helsinki, FIN
25. Lysaght RJ et al (1989) Operator workload: comprehensive review and evaluation of operator work-load methodologies. Technical report DTIC document. http://www.dtic.mil/dtic/tr/fulltext/u2/a212879.pdf

26. MacKenzie IS, Buxton W (1991) Extending Fitts' law to two-dimensional tasks. In: Proceedings of the SIGCHI conference on human factors in computing systems. ACM, pp 219–226. doi:10.1145/142750.142794
27. McMahan W et al (2011) Tool contact acceleration feedback for telerobotic surgery. IEEE Trans Haptics 4(3):210–220. doi:10.1109/TOH.2011.31
28. Nitsch V, Färber B (2012) A meta-analysis of the effects of haptic interfaces on task performance with teleoperation systems. IEEE Trans Haptics 6:387–398. doi:10.1109/ToH.2012.62
29. Nitsch V et al (2010) On the impact of haptic data reduction and feedback modality on quality and task performance in a telepresence and teleaction system. In: Haptics: generating and perceiving tangible sensations (2010), pp 169–176. doi:10.1007/978-3-642-14064-8_25
30. Pongrac H (2008) Gestaltung und evaluation von virtuellen und Telepräsenzsystemen an hand von Aufgabenleistung und Präsenzempfinden. PhD thesis, Universität der Bundeswehr, München. http://athene.bibl.unibw-muenchen.de:8081/node?id=86166
31. Samur E (2010) Systematic evaluation methodology and performance metrcis for haptic interfaces. Dissertation, École Polytechnique Fédérale de Lausanne. doi:10.5075/epfl-thesis-4648, http://infoscience.epfl.ch/record/145888
32. Samur E (2012) Performance metrics for haptic interfaces. Springer, Berlin. doi:10.1007/978-1-4471-4225-6, ISBN:978-1447142249
33. Sarodnick F, Brau H (2006) Methoden der usability evaluation. Huber, Bern
34. Schlick C, Luczak H, Bruder R (2010) Arbeitswissenschaft. Springer, DE. ISBN:978-3-540-78333-6
35. Scilingo EP et al (2003) Haptic displays based on magnetorheological fluids: design, realization and psychophysical validation. In: IEEE proceedings of 11th symposium on haptic interfaces for virtual environment and teleoperator systems, 2003, HAPTICS 2003, pp 10–15. doi:10.1109/HAPTIC.2003.1191217
36. Serafin C et al (2009) International product user research: concurrent studies comparing touch screen feedback in Europe and North America. In: SAE world congress and exhibition, SAE International. doi:10.4271/2009-01-0779
37. Silva CWD (2007) Sensors and actuators: control system instrumentation. CRC Press, Boca Raton. ISBN:978- 1420044836
38. Tan H, Rabinowitz W (1996) A new multi-finger tactual display. J Acoust Soc Am 99(4):2477–2500. doi:10.1121/1.415560
39. Várhelyi A et al (2004) Effects of an active accelerator pedal on driver behaviour and traffic safety after long-term use in urban areas. Accid Anal Prev 36(5):729–737. http://dx.doi.org/10.1016/j.aap.2003.06.001
40. Vollestad NK (1997) Measurement of human muscle fatigue. J Neurosci Methods 74(2):219–227. http://dx.doi.org/10.1016/S0165-0270(97)02251-6
41. Wall S, Harwin W (2000) Quantification of the effects of haptic feedback during a motor skills task in a simulated environment. In: Proceedings of second PHANToM users research symposium, 2000. http://www.personal.reading.ac.uk/shshawin/pubs/wall_purs2000.pdf
42. Weisenberger JM, Krier MJ, Rinker MA (2000) Judging the orientation of sinusoidal and squarewave virtual gratings presented via 2-DOF and 3-DOF haptic interfaces. Haptics-e 1(4):1–20. http://www.haptics-e.org/Vol_01/he-v1n4.pdf
43. Wildenbeest JG et al (2013) The impact of haptic feedback quality on the performance of teleoperated assembly tasks. IEEE Trans Haptics 6(2):242–252. doi:10.1109/TOH.2012.19
44. Williams C, Ratel S (2009) Human muscle fatigue. Taylor & Francis. ISBN:9781134053520

Chapter 14
Examples of Haptic System Development

Limin Zeng, Gerhard Weber, Ingo Zoller, Peter Lotz, Thorsten A. Kern, Jörg Reisinger, Thorsten Meiss, Thomas Opitz, Tim Rossner and Nataliya Stefanova

Abstract In this section, several examples of task-specific haptic systems are given. They give an insight into the process of defining haptic interactions for a given purpose and illustrate the development and evaluation process outlined in this book

L. Zeng · G. Weber
Human–Computer Interaction Research Group, Technische Universität Dresden, Dresden, Germany
e-mail: limin.zeng@tu-dresden.de

G. Weber
e-mail: gerhard.weber@tu-dresden.de

I. Zoller · P. Lotz
Continental Automotive GmbH, Babenhausen, Germany
e-mail: ingo.zoller@continental-corporation.com

P. Lotz
e-mail: peter.lotz@continental-corporation.com

T. A. Kern (✉)
Continental Automotive GmbH, VDO-Straße 1, 64832 Babenhausen, Germany
e-mail: t.kern@hapticdevices.eu

J. Reisinger
Daimler AG, Sindelfingen, Germany
e-mail: joerg.reisinger@daimler.com

T. Meiss · T. Opitz · T. Rossner · N. Stefanova
Institute of Electromechanical Design, Technische Universität Darmstadt, Darmstadt, Germany
e-mail: t.meiss@emk.tu-darmstadt.de

T. Opitz
e-mail: t.opitz@emk.tu-darmstadt.de

T. Rossner
e-mail: t.rossner@emk.tu-darmstadt.de

N. Stefanova
e-mail: n.stefanova@emk.tu-darmstadt.de

C. Hatzfeld and T.A. Kern (eds.), *Engineering Haptic Devices*,
Springer Series on Touch and Haptic Systems, DOI 10.1007/978-1-4471-6518-7_14

so far. Examples were chosen by the editors to cover different basic system structures. Section 14.1—*Tactile You-Are-Here-Maps* illustrates the usage of a tactile display in an assistive manner, enabling a more autonomous movement of people with visual impairments. Section 14.2—*User Interface for Automotive Applications* presents the development of a haptic interface for a new kind of user interaction in a car. It incorporates touch input and is able to simulate different key characteristics for intuitive haptic feedback. Section 14.3—*HapCath* describes a comanipulation system to provide additional haptic feedback in cardiovascular interventions. The feedback is intended to reduce exposure for both patient and physician and to permit new kinds of diagnosis during an intervention.

14.1 Tactile You-Are-Here Maps

Limin Zeng and Gerhard Weber

14.1.1 Introduction

You-are-here (YAH) maps as one of the truly ubiquitous tools are installed on walls or wooden boards to allow sighted persons quickly acquire the information about where they are within their surroundings [6]. In addition to various geographic features rendered on a YAH map, a YAH symbol indicating readers' current position on the map must be represented. In recent years, a large number of location-based maps on handheld devices become available and provide digital YAH maps, which represent dynamic data about position and orientation.

However, for visually impaired people, these visual YAH maps are inaccessible, neither YAH maps installed on walls nor ubiquitous electronic YAH maps. Although people having visual impairments are able to use global positioning system (GPS)-based navigation devices to query their current location and follow turn-by-turn guidance to reach unfamiliar destinations, they cannot explore the surroundings and acquire more related spatial geographic information, such as the layout of street networks and the shape of a complex crossing, which might help them understand an unfamiliar point of interest better.

To improve accessibility features of map exploration applications for the visually impaired, a number of haptic devices have been employed in previous studies. The TouchOver Map application indicates that visually impaired users find it was challenging to find out the details of short streets, e.g., direction, connection on regular touchscreen displays, due to lack of explicit haptic feedback [18]. Even if visually impaired users could distinguish several simple tactile patterns (e.g., circles, squares) on electrostatic haptic displays [22, 25] that offer better haptic sensation than common touchscreen displays, it is still unclear whether they can read a complex city map with various geographic features, because at present, there is no map-rendering system developed for such an electrostatic haptic display. Recently, ZENG AND WEBER implemented a city map system on a pin-matrix display with 7,200 pins, which is suitable for desktop use [26]. Haptic sensation of raised and lowered pins on the pin-matrix display allows individuals with visually impairments to explore streets and points of interest (POIs) interactively. However, the desktop-based tactile map system does not render the current users' location.

Considering the good haptic sensation of a pin-matrix display, we develop a ubiquitous tactile YAH map system (named TacYAH map). A touch-enabled pin-matrix display with 960 pins allows mobile usage if a map can be rendered accordingly. In addition to a series of tactile map symbols rendered on the TacYAH map, we design a set of tactile YAH symbols consisting of raised and lowered pins to indicate users' updated location and heading orientation. Moreover, the visually impaired users can interact with the TacYAH map by panning and zooming through a haptic controller.

14.1.2 The TacYAH Map Prototype

(a) Mobile HyperBraille Display

The employed tactile display is a mobile version of the HyperBraille display [23], consisting of a matrix of 30×32 piezoelectric refreshable pins (see Fig. 14.1). The innovative array cells reduce the size of the display considerably compared to existing Braille displays consisting of one row of Braille cells. The weight of the display is about 600 g, and its pin matrix area covers 7.0×8.1 cm. Each pin has only two statuses, that is, either raised or lowered. The height of a raised pin is about 0.7 mm, and the space between each pin reaches 2.5 mm; thus, the raised pins can be distinguished by fingertips. Tactile force strength of each pin is more than 30 cN, which ensures the raised pins would stand up always even if under high finger pressure. The whole screen refreshes rapidly with the help of its high-speed data bus and piezoelectric actuators (up to 5 Hz refresh rate). There is a capacitive layer on the top of the pin-matrix area that makes the display touch enabled. Furthermore, due to its low power consumption, the tactile display makes use of a common USB cable to supply power and data.

(b) System Components

In addition to the portable pin-matrix display, the prototype consists of several additional components (see Fig. 14.2). A smartphone equipped with a GPS sensor and a digital compass provides users' updated location and orientation accordingly. A WiiCane whose handle is enhanced by mounting a Wii remote control is used to input commands (e.g., panning, zooming) via its built-in keys, such as the key "A" for querying where users are, and the keys "–" and "+" for zooming out and

Fig. 14.1 The mobile HyperBraille display (a matrix of 30 × 32 pins)

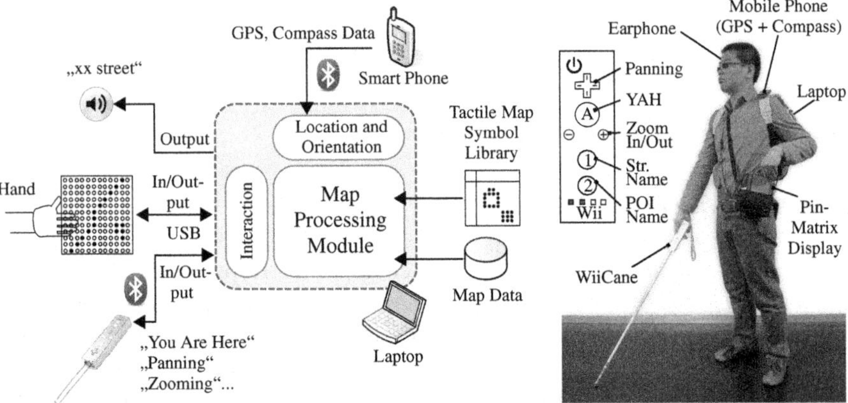

Fig. 14.2 The overview of the TacYAH Map prototype (*left* the system architecture; *right* the system worn by a user)

zooming in, respectively. Note that the WiiCane offers tactile feedback via a vibration to inform on the completion of processing the input commands. Users acquire additional auditory information through an earphone, like the street name or POI name. Besides, a light laptop is required to run the main program, and it is connected to all other devices by Bluetooth or wired connections.

(c) Map Data Processing

At present, most current commercial map providers (e.g., Google Maps, OpenStreetMap) offer tile-based map data as images to various clients, like Web browsers or map applications on mobile phones. Considering the challenging tasks to generate tactile maps from image processing methods [24], the TacYAH map parses the vector-based map data in the format of Geography Markup Language (GML,

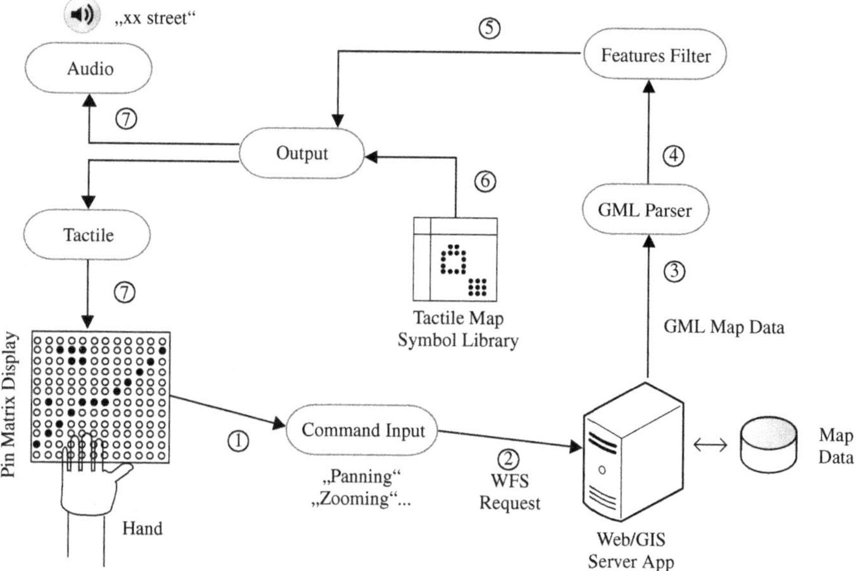

Fig. 14.3 Flowchart of map data processing

ISO 19136:2007). To generate GML map data according to users' input commands, an open-source Geographic Information System (GIS) software, namely GeoServer, has been employed in the prototype. As illustrated in Fig. 14.3, there are seven main steps to prepare a response to users' input (e.g., panning, zooming, and acquiring auditory information). After pressing a key in the second step, users' commands are translated into Web Feature Service (WFS) requests. The WFS utilizes HTTP to express and transport detailed geographic features. Visually impaired people touch the tactile representation and listen to auditory information within the output modules. The TacYAH map prototype can easily import worldwide city map data from the OpenStreetMap repository.

(d) Tactile Map Symbols

Generally, on a visual map, various geographic features can be rendered by different icons, styles, and colors, even on overlapping layers. However, it is challenging to render a city map on such a low-resolution pin-matrix display (10 pins per inch). Thus, to represent various geographic features, a set of distinguishable tactile map symbols has been designed in the TacYAH map system, as shown in Fig. 14.4.

Furthermore, in addition to the set of geographic features, a set of YAH symbols has been designed for rendering not only users' location but also users' heading orientation. During the design process, 7 visually impaired individuals (3 female and 4 male, mean age 30 years) were invited to choose the best set among 3 candidate

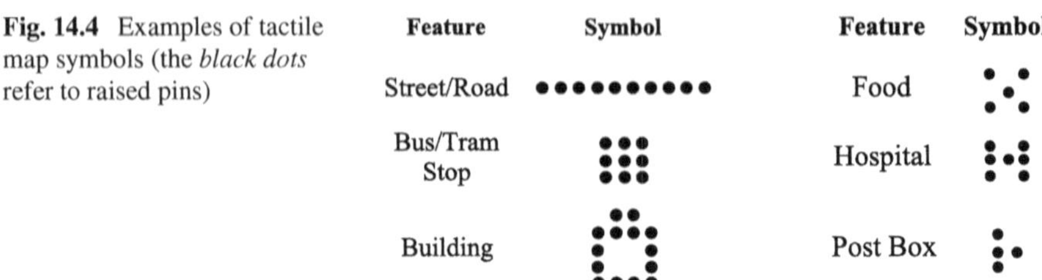

Fig. 14.4 Examples of tactile map symbols (the *black dots* refer to raised pins)

sets (see Fig. 14.5). Five of them preferred the second set, and 2 participants chose the third one. Therefore, the TacYAH map system employs the second set. It is important that the tactile symbols should not be changed (like their size) when the map is zoomed in or zoomed out. Otherwise, the visually impaired might be confused with the modified symbols that are different to the ones they learned before.

The tactile map symbols and the YAH symbols make up a city map on a pin-matrix display as shown in Fig. 14.6. Visually impaired individuals learn the layout of streets and complex crossings, as well as the surrounding buildings and POIs. Note, it is necessary to keep enough space between each symbol (e.g., about 2 pins), to ensure the symbols can be identified correctly. However, the bus stop symbol is an exception, as it can be recognized even if close to a street [26].

(e) Map Interaction

Convenient exploration of maps requires support for a variety of user strategies. Regardless of the specific information, the WiiCane vibrates for a short amount of time (the vibration duration: 300 ms) to inform users on the completion of rendering output, and then, the users can touch the updated screen. Note that, the north is always at the top of the map (north-up map).

"You Are Here" It informs users about where they are, which renders the surrounding map on the tactile display, and the updated YAH symbol in the center of

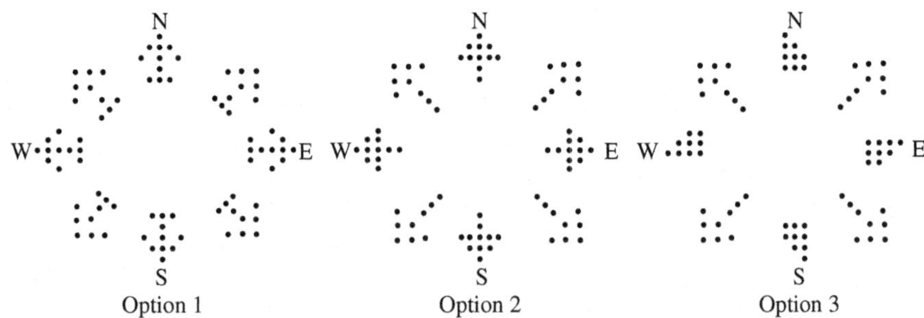

Fig. 14.5 The three candidate sets of tactile YAH symbols [27]

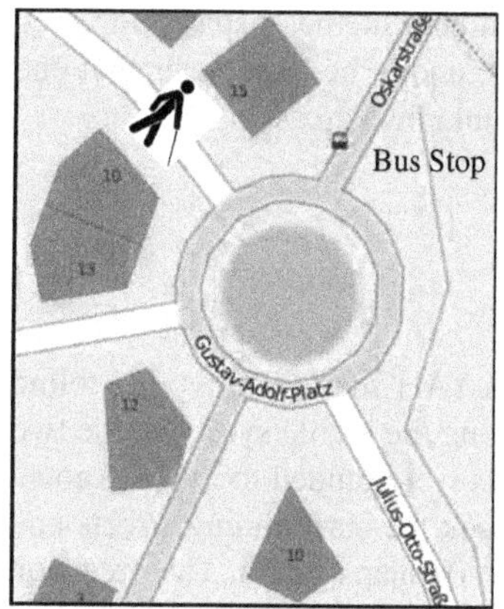

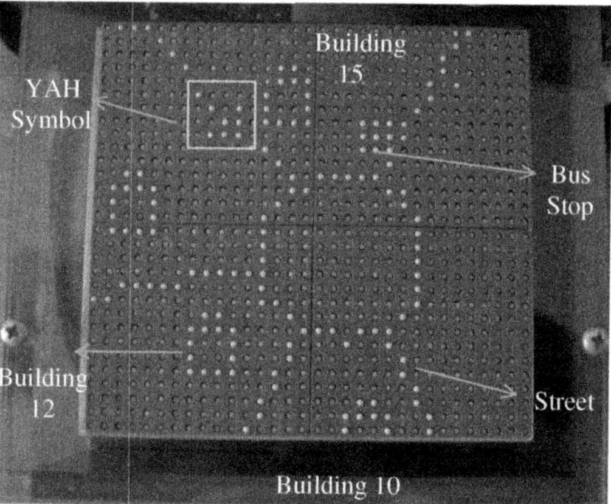

Fig. 14.6 A circle crossing rendered on the pin-matrix display (*left* a screenshot of the crossing from OpenStreetMap; *right* a representation on the pin-matrix display)

the display. When a user presses the key "A" on the WiiCane, the function will be triggered.

Panning Users can pan a map to the left/right/up/down side by pressing the appropriate button on the WiiCane. Each panning operation will update one-third of the screen vertically or horizontally, rather than the whole screen. Otherwise, the visually impaired users might become confused after panning due to losing the previous context.

Zooming In the TacYAH map, two zoom levels are supported. At the first level, only street networks and YAH symbols are rendered (i.e., Street View), while at the second level, it renders streets and POIs at the same time (i.e., POI View). The whole screen represents an area covering 150×160 m (5 m/pin) in Street View, and covers an area of 100×106 m (3.3 m/pin) in POI View. Certainly, many more zoom levels can be designed and implemented, such as a building view to represent the exact shape of the building.

Auditory geographic information To acquire the related auditory information about the surrounding geographic features, like the name of a street or a building, and the bus lines offered at a bus station, a visually impaired user needs to touch the corresponding map symbol with one finger on the tactile display and press the key "1" or "2" on the WiiCane simultaneously. The auditory information is generated by Text-to-Speech (TTS) software and based on the information stored in the GIS, such as names of streets or POIs and bus lines at bus stations.

In summary, the TacYAH map prototype provides not only audible geographic information, but also tactile representation of street maps via the predesigned tactile map symbols and YAH symbols. It is necessary to validate whether visually impaired people are able to benefit from the proposed system, under the hypothesis that end users can locate themselves independently, even in unknown areas.

14.1.3 Evaluation

In order to investigate the performance of the TacYAH map prototype, 5 blind participants were invited in a pilot study. After training the tactile symbols, the two map views, and usage of the system, the participants were guided to two unknown outdoor test sites and asked to locate themselves, seek the surrounding streets and nearby POIs within 100 m with the help of the TacYAH map system. The precision of identifying the YAH symbols is 95 % after a short amount of training time (mean: 75.0 s). In the 10 outdoor field tests, all the participants can locate themselves successfully and find out the surrounding streets and the targeted POIs (mean recall: 100 %, mean precision: 80 %). Specifically, they would estimate the distance (mean error: 9.1 m; ST. Dev.: 3.82) and the orientation (mean error: 18°; ST. Dev.: 0.25) between themselves and the nearby POIs in 100 m.

14.1.4 Conclusion and Outlook

The results of the conducted pilot study with blind users indicate through the proposed haptic YAH map system individuals having visually impairments are able to locate themselves on the move and explore the surroundings in unknown areas. In addition to many more evaluations with end users, the functionalities of the TacYAH map as a proof of concept prototype should be improved in the future toward a final product, like including a head-up map. It will be helpful to extend the TacYAH map to render indoor floor plan, specifically in airports and railway stations. Through this new assistive navigation system, the visually impaired will be able to be guided with turn-by-turn instructions by an appropriate low-cost app in their mobile phones and explore the surroundings easily both.

14.2 Automotive Interface with Tactile Feedback

Ingo Zoller, Peter Lotz, Thorsten A. Kern
Jörg Reisinger

Please note that this Sect. 14.2 *is a reworked version of* [28, 29].

14.2.1 Context

Products with active tactile feedback inside a car cover all elements which are in direct contact with its occupants. The more classic applications are active steering wheels with tactile and/or kinaesthetic support [14] and break or throttle pedals [16]. Research additionally focuses on tactile signals to create situative awareness or give assistive clues [15]. In addition to these feedback loops directly related to the task of driving, comfort functions like multimedia and climate control elements are also subject to active tactile feedback. Numerous solutions were presented for tactile feedback on automotive touchscreens, but lately faceplates (Fig. 14.7a) and remote control touch devices like touchpads (Fig. 14.7b) are also equipped with active haptics. The kind of feedback generated by these devices is usually designed to mimic the impression of real push buttons.

The intention to create a situative and configurable tactile feedback to confirm the activation of a function is not new at all. Different technologies are in application to create such feedback. The type of technology in use is always strongly influenced by the volume available. In mobile application, vibrotactile solutions range from standard rotary motors with eccentric discs over high-current versions of such motors for the creation of shorter and sharper pulses (TOUCHSENSE©by *Immersion*) to voice coil systems which allow a broader frequency response. Lately, other systems can be seen on the market reducing building volume and costs to create vibrotactile systems with broad perceptional bandwidth by an intelligent way to mix just few oscillating frequencies [7]. Besides these classic actuators based on rotary or linear movement, lately mobile devices can be noted based on piezoelectric actuators, or electroactive polymers. Ultrasonic [1] and electrostatic [8] devices gain increased attention, too. Originally, they were not invented for confirmation of a function, but for tactile texture simulation. However, they show some fascinating performance in that point.

Although this bunch of technology exists and some of those already have reached a series level in mobile devices, only few solutions can be applied to automotive applications too. This fact is due to market requirements and technical reasons: Automotive

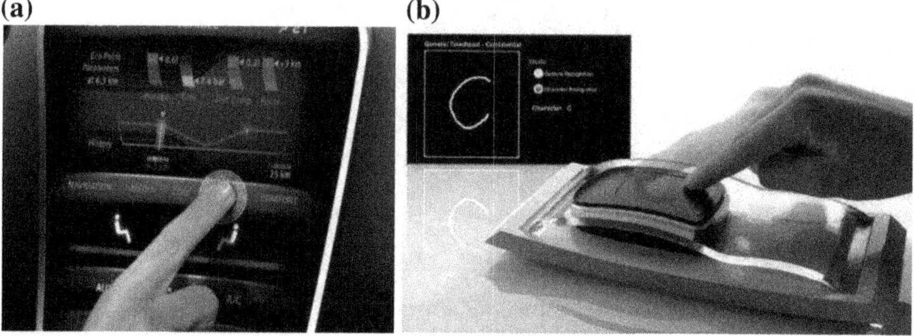

Fig. 14.7 Faceplate with active haptic feedback (**a**); stand-alone touchpad with handwriting recognition and haptic feedback (**b**)

industry has a certain delay in the application of new technologies. Typical design sequences for automotive products last 3 years, which makes them less interesting for young emerging companies which are required to make profitable business within a short period of time. In addition, quality requirements are high, making an entrance into this market almost impossible for anyone but an established Tier 1 or Tier 2 supplier. Besides these general challenges, automotive operating conditions differ in two significant aspects from mobile devices: Their operation is truly one-handed, which cancels any approach to create the haptic feedback in the holding hand. In addition, the feedback has to be very precise and strong to create a good signal-to-noise ratio in the vibrating surrounding of a moving car. This makes actuators for haptic feedback which are still suitable for automotive to become very special.

14.2.2 The Floating TouchPad of Mercedes Benz

The floating touchpad of Mercedes Benz is an example for such an actuated automotive touch control (Fig. 14.8). Comparable to a smartphone, several functions of the headunit can be controlled by finger gestures and handwriting recognition. At all it offers a haptic feedback on the touch surface to confirm input. A vertical movement of the touch area helps to increase feedback quality and reduces working gaps compared to lateral movement. The flat building actuator enables the realization of a raised esthetic shape. Regarding to this exposed position of the touchpad above the rotary COMAND®Controller, the touchpad acts also as a hand rest. This doubled functionality requires a switchable haptic input effect. To achieve the differentiation between valid and invalid touch, a robust sophisticated 3-dimensional algorithm is

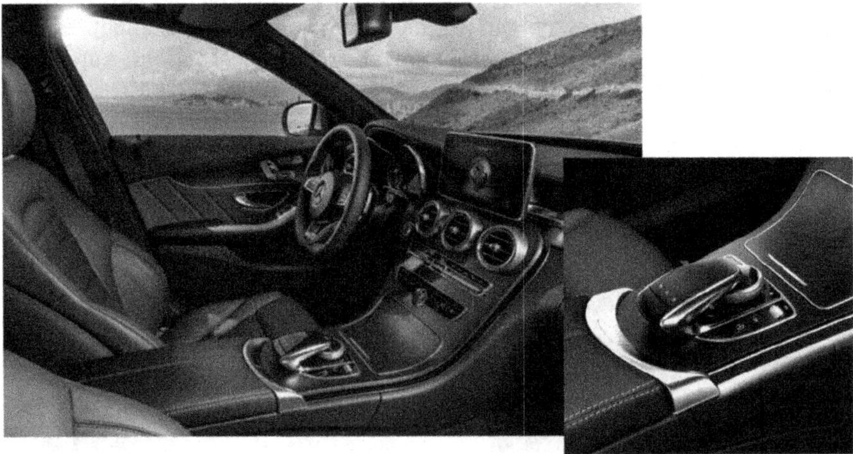

Fig. 14.8 Applied automotive haptic device—touchpad in the Mercedes Benz C-Class, Year 2014

developed (patent pending). Thus, the haptic effect only is released if a valid finger interaction is recognized.

The best hardware and algorithms mean nothing, if user interaction and perceptive quality are not sufficient. Aside of hard environmental requirements, the whole user interaction has its own rules in automotive applications. In addition to a straining primary driving task, strong vertical oscillations take effect on the whole human body, especially arm, hand, and finger. Thus, the serial mechanical linkage holding and controlling the relative position between fingertip and user interface has to be reduced and optimized. In the past, this led to separated user interaction concepts [21]. Anyway—probably regarding to cost effects—touchscreen interfaces are more widely spread, while separated interaction concepts usually are found in upper-class cars.

At all these disturbances require a suiting, not only a stronger, haptic feedback. Automotive devices usually are not handheld, and the remaining contact point between the user and the device is the fingertip. The haptic sensing area of the thenar eminence which often casually is used additionally as a very sensitive human input channel [3] is not available. At all, a clear and distinctive feedback is required. Even standard actuators which are often used in the handheld market do not provide a sufficient haptic feedback, even working under all automotive required environmental conditions.

But how should a perfect automotive haptic feedback look like? Initially, it has to be stated that both interaction certainty and the impression of "liking" are criteria for perceptive quality. Regarding to information, interaction certainty of user interfaces needs to provide a robust signal transfer. For that purpose, the interface has to use a known perceptive alphabet to prevent a not wanted strenuous learning. This simply means that the displayed haptic effect has to show known haptic effects, which are correlating to commonly known haptic experiences. Thus, new effects must be able to differentiate information and have to be understood intuitively. Moreover, the effect has to show a positive impression. The number of such effects seems unfortunately small and ordinary. At all in a context of Mercedes Benz push buttons surrounding the touchpad, which show all described quality aspects, such a push feedback of the touchpad seems to be a valuable and promising feedback. The expectations are exceeded if the user does not realize that the touchpad's feedback is artificial.

But at all how these requirements can be realized? To answer that question, aspects of perceiving a push button have to be taken into account. Maybe a little strange, but it can be supposed that human beings like being lazy. Several, non-published studies show that people like low actuation force combined with highest precision impression (see [19]). Comparing with energy effort, low actuation force means energy saving as well as high precision means high sensing efficiency, because no energy has to be wasted while clarifying a sensual uncertainty.

Our haptic sensing mechanism is extremely efficient, and the difference between vibrating effect and a "snap" is small, and even some strange frequencies denote aberrance from the expected. So besides a well-tuned mechanical system, the electrical control plays an important role. A filtered actuation allows the reduction of disturbing frequencies [4].

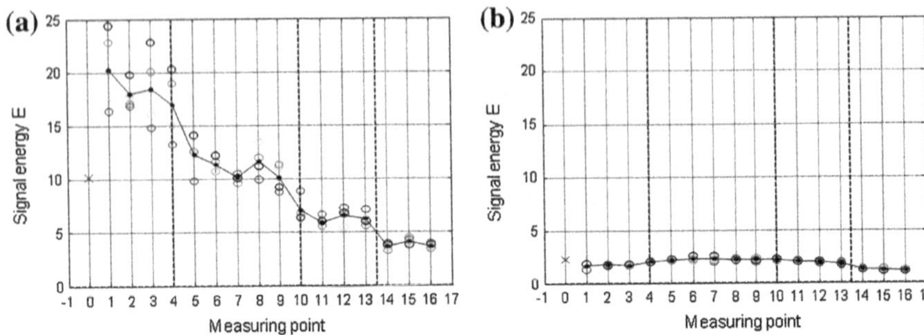

Fig. 14.9 Optimization of the homgeneity of the haptic effect on the surface of the touchpad. Considered measure is the signal energy. The abscissa shows different measuring points distributed over the touch surface. The ordinate shows the parameter values. The different colors show different measurement speed. **a** measurement before optimization, **b** after optimization [28]

A more macrostructural look at haptic quality means the homogeneity of the haptic effect distributed over the whole touch surface as well as different actuation speed. A uniform movement of the surface is required as well as a subjectively identical haptic effect in each point of the touchpad and at different interaction speeds, which is called homogeneity. The interaction-based measurement described later in this chapter plays a key role for an efficient development of the haptic effect and homogeneity reaching a high quality level. Figure 14.9a, b shows the development progress regarding to one dynamic feature. A strong inhomogeneity regarding location and speed can be seen in Fig. 14.9a, reflecting subjective impression. Figure 14.9b shows the effect of the optimization to achieve a homogeneous dispay of haptic effects.

As mentioned in the beginning, besides the perceptive quality, also the passive behavior of the touchpad plays an important role. A so-called phantom feedback, a hard reaching of the end stop that causes unwanted impression of a haptic snap, has to be prevented. Included in the actual COMAND Online®-HMI, the whole system can be controlled easily also while driving. Thus, it is an important step introducing touch holistically to a drivers' environment, to reach best acceptance and comfort.

14.2.3 Actuator Design

The design requirements for an actuator to create precise and strong haptic feedback are not at all obvious. Analysis of the performance of real push buttons indicate that the actuator has to create extremely sharp oscillations (>3 g) which are short in time (damped, see Fig. 14.10 left). In addition, a mechanical switch always operates at a certain pretension in force (e.g., 2–5 N) and after a clearly defined path of travel (e.g., 0.25–0.5 mm, see Fig. 14.10 right). Accordingly, the functional elements for an actuator to create haptic feedback on a decorative surface can be identified as follows:

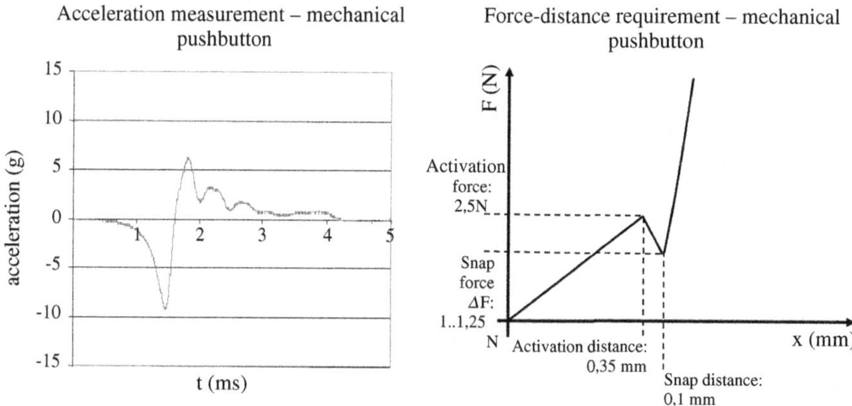

Fig. 14.10 Acceleration measurement and force–distance requirements of a mechanical push button

- Force source to create an acceleration of a mass weighting approximately 100 g
- Parallel guidance to match the force–displacement curve independent from the area touched
- Force-sensing mechanism to guarantee a reproducible switching point

For all components, packaging requirements are formulated to limit the thickness of the device to below 6 mm. For this application, an actuation system compromising the following components has been designed:

- Electromagnetic actuator based on punch-bended pole and anchor with PCB-based coils
- Parallel guidance mechanism based on flexible metallic hinges
- Optical SMD displacement sensors

14.2.3.1 Electromagnetic System

The magnetic system consists basically of two steel plates and one PCB located between the steel plates. The PCB contains the electromagnetic coils; the coil cores are supplied by one of the steel plates. Thus, a simple yet effective electromagnet is formed (Fig. 14.11). This system offers significant advantages compared with a conventional electromagnet. The inductance of the coils is small, thus allowing for very short raise and fall times and very short magnetic pulses. The PCB-based coils are capable of carrying effective current densities in the range of $10\,\text{Å}\,\text{mm}^{-2}$ over the whole PCB thickness, while the direct connection between PCB and supportive metal parts ensures an effective heat transfer and thus prevents overheating of the coil system.

Electromagnetic simulation was used to calculate expected force displacement curves. Measurements showed a good agreement between these calculated curves

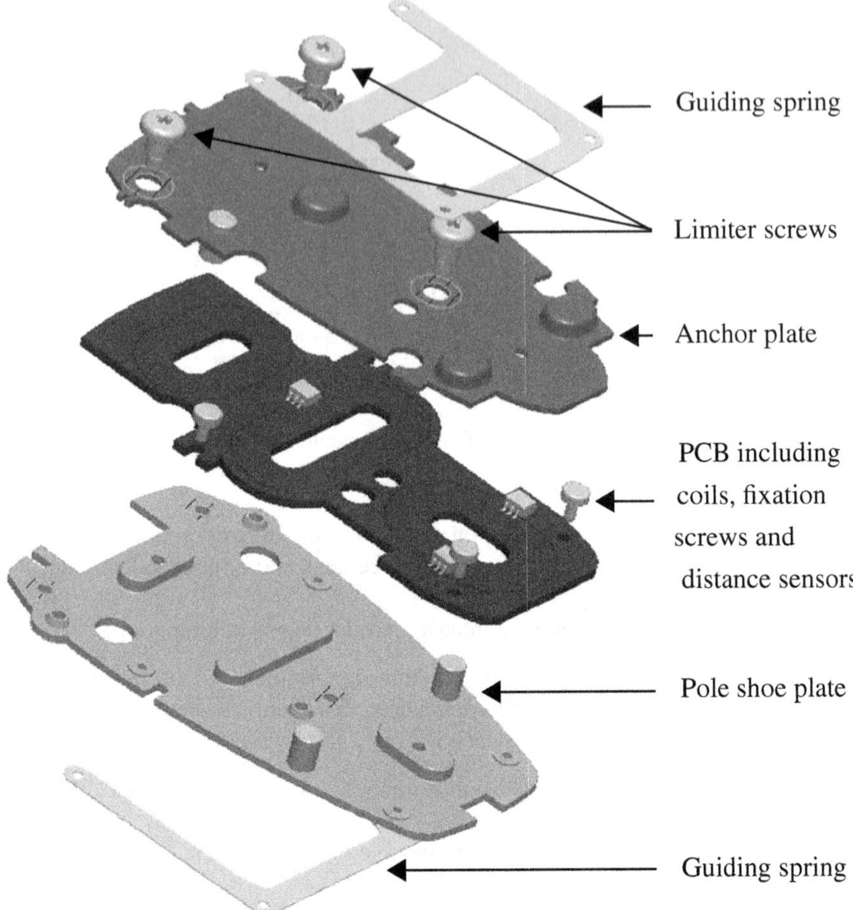

Fig. 14.11 Basic actor design—PCB between two steel plates. The springs represent the elastic component of the module, at the same time they provide the parallel guiding system

and real measurements (Fig. 14.12). The influence of mechanical tolerances and magnetic imperfections were included into the simulation of the system, and after several optimization loops, the magnetic system was finalized to be manufactured by well-established, cost-efficient processes and capable technologies suitable for automotive use. The current density/force displacement curves allow for significant dynamic ranges in feedback both due to force profile and acoustic feedback options.

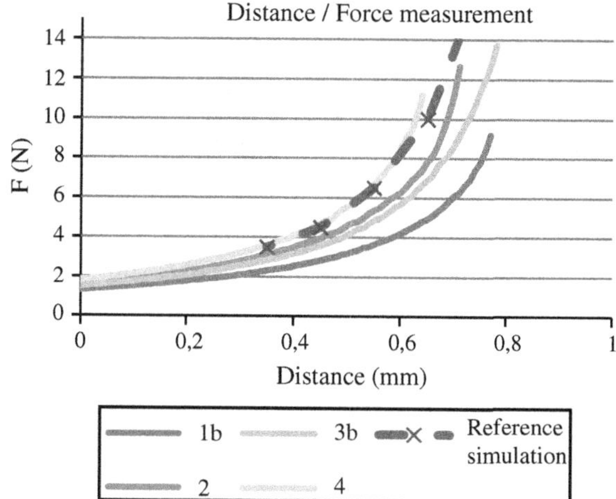

Fig. 14.12 Distance/magnetic force measurement using pole shoe plates out of different manufacturing processes—sample 4 is closest to CAD target. Spring is excluded from measurement; thus, only the magnetic snap force ΔF (see Fig. 14.10b) depending on distance is measured. Distance is measured from mechanical rest position

14.2.3.2 Parallel Guidance

Customers explicitly ask for a homogeneous movement of the surface, with a preferred movement range of less than 0.3 mm. The intention behind this is to visually inhibit actual movement of the input device. For a high-quality feel of the context-sensitive feedback, it is required to activate the haptic pulse at the same user input force anywhere inside the active area. Both aspects are achieved by a parallel guiding system free from play (Figs. 14.11 and 14.13b). Two parallel springs guide the anchor plate with respect to the pole shoe plate. At the same time, they provide the return force for the actuator module. The transfer rod between the two spring arms transfers a share of the force to the other side of the guiding system to prevent a rotation and tilting of the anchor plate and stabilize the parallel guiding system. As a result, homogeneous force–displacement curves are reached almost independent of the location on the actuator surface (Fig. 14.13). The system leads to an almost vertical movement of the anchor plate, and a homogeneous force at the switching point over the active area.

14.2.3.3 Sensors

Sensing the user input is a crucial issue within the whole system. Additionally, the automotive context requires a high robustness against electromagnetic interference, operation in a broad temperature range, and reliability during a lifetime which is much higher than of any consumer device. In contrast, for a valuable haptic impression,

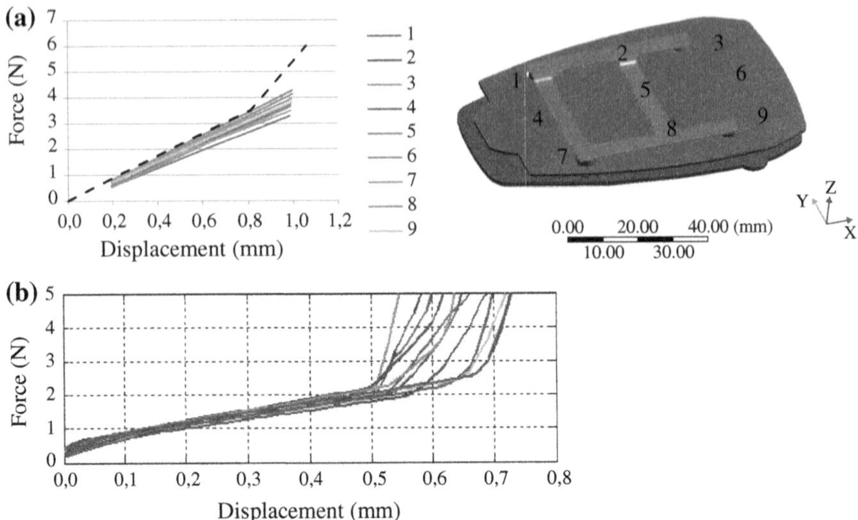

Fig. 14.13 FEM results of simulated force displacement curves at several locations on actuator surface (**a**) compared with real measurement results (**b**). Only mechanical spring force is shown; snap force ΔF (see Fig. 14.10b) is excluded

Table 14.1 Sensor requirements

Parameter	Value
Tolerance of switching point	±10 %
Sensitivity	700 mV/mm
Cutoff frequency	>500 Hz

high sensitivity and precision of the sensors are necessary. Table 14.1 summarizes these requirements. A user gets a constant and valuable impression at repeated inputs if the traveling of the surface varies less than 10 % around the predefined switching point. In combination with the small distance, the surface can travel at all (see Sect. 14.2.3), and the resolution of the ADC of the used microcontroller, the sensitivity of the sensor has to be better than 700 mV/mm. To detect fast movements, the sensor has to provide a cutoff frequency above 500 Hz.

To fulfill all of these different and opposed requirements as much as possible, an integrated reflective interrupter is used in this haptic input device. The most important advantages are as follows: The optical functional principle of the distance measurement does not interfere with any electromagnetic fields; the sensitivity can be adjusted by two resistors R_F and R_L (Fig. 14.14a) and the design of the reflective surface (Fig. 14.14b). The elevation of the reflective surfaces from the anchor plate level controls the operating point of the sensor. Hence, the output of the sensor is almost linear for the whole movement of the input surface (Fig. 14.15).

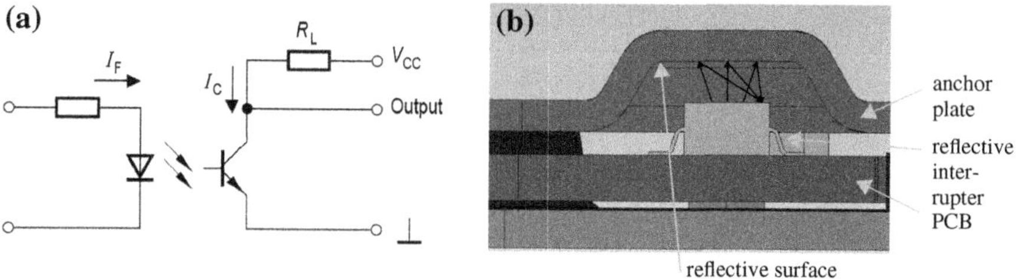

Fig. 14.14 Equivalent circuit of the reflective interrupter (**a**); assembly of sensor and reflective surface (**b**)

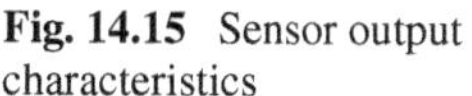

Fig. 14.15 Sensor output characteristics

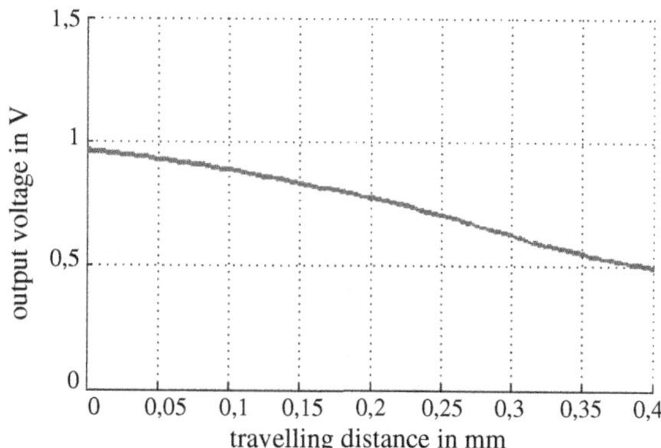

14.2.3.4 Acceleration Amplitude Range

Simulation results lead to the expectation that the maximum acceleration achievable is mostly determined by the air gap remaining at the switching point. Accelerations of more than 10 g are achievable. Smaller accelerations can be controlled by limiting the available current. Initial customer evaluations showed that this performance exceeds the actual needs. A nominal acceleration of 3 G corresponds to user expectations (Fig. 14.16) in the intended context.

14.2.4 Evaluation

With this actuator being used in high-volume series-production devices, extraordinary requirements on its quality and the measurement of its performance are essential.

As described in [28] the existing static measurement principles are not sufficient to describe human haptic perception and even the new, artificially actuated user interfaces cannot be described correctly. Figure 14.17 shows measuring force

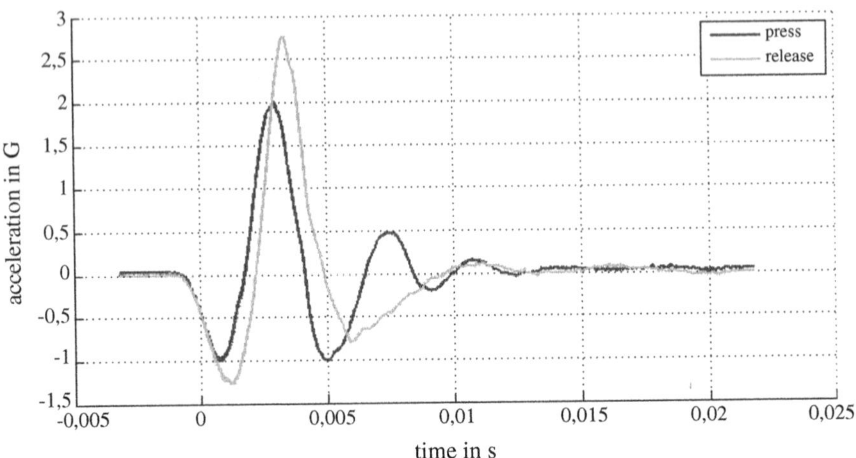

Fig. 14.16 Examples of haptic output characteristic with a real finger as a load [28]

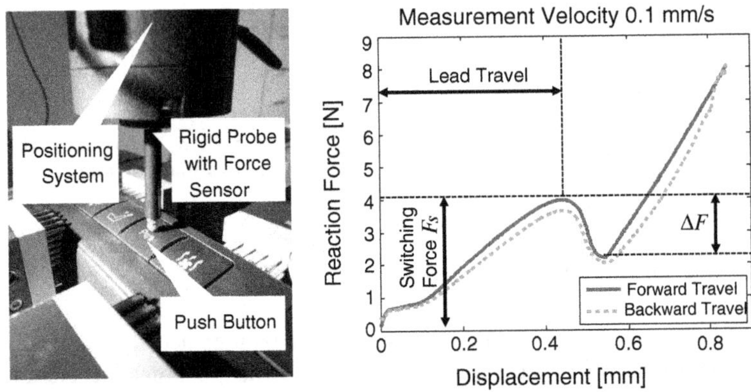

Fig. 14.17 Static measurement for a push button: measurement device (*left*); force–displacement curve and technical features derived from the *curve* (*right*) [28]

versus displacement with a rigid probe and the resulting characteristic curve plotted force versus displacement. Zhou shows several examples where clearly perceived differences cannot be seen in static measurement, such as different inertia by simply adding masses (see Fig. 14.18) or different viscoelasticities by influencing the frictional pairings of the control's bearings. Electromechanically driven switches usually do not show typical snap-behavior as a rapid force reduction, which is one of the key features in the static measurements (see Fig. 14.19).

To solve this problem, [28] show that the switch itself and the human finger influence each other so strongly that the resulting dynamic behavior changes strongly if, for example, the finger's impedance is changed as shown in Fig. 14.20. This

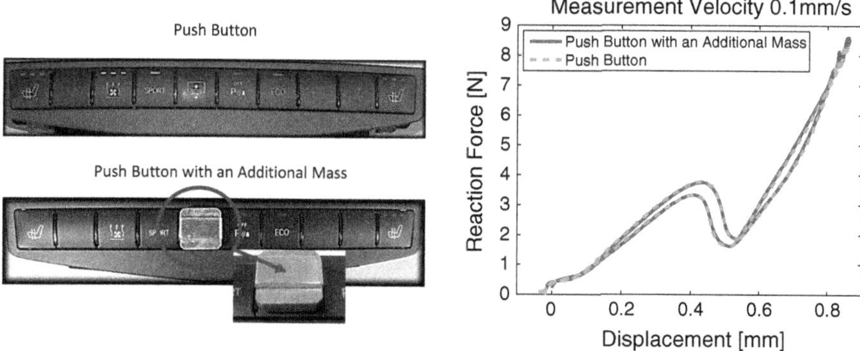

Fig. 14.18 Two push buttons of the same type (*left*), and the corresponding static force-displacement measurements (*right*). One has an added mass of 2.5, Both curves seem the same [28]

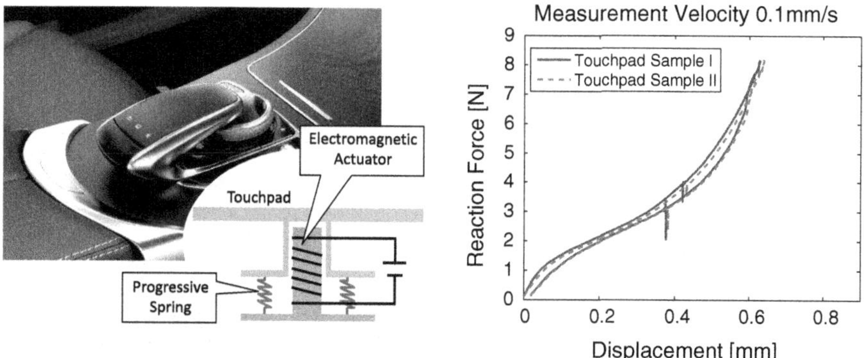

Fig. 14.19 New C-class Touchpad with functional sketch of its electromechanical actuation system (*left*), static force-displacement curve (*right*) [28]

expounds the missing information, or in other words: the reduced bandwidth when the impedance rises to a rigid measurement finger that gets dominant to the switches' impedance.

The described method of "interaction-based-dynamic measurement" as shown in Fig. 14.21 considers both impedances of the switch and the human finger to measure the mechanical interaction-data like acceleration versus time. In comparison to static or also the opposite contactless measurement, the mechanical behavior of a perceived haptic feeling and not the standalone mechanical parameters of the system are measured as perceived. Regarding its interaction-perceiving principle its biggest advantage out of a specifications' view is that these parameters and their tolerance ranges can be transferred to different components, because perception will be constant for all. To get to the point, the perception, and not the device, is specified.

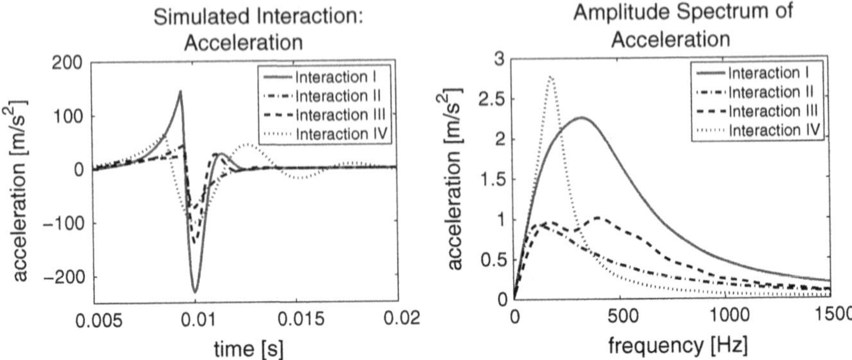

Fig. 14.20 Acceleration in the interaction during pushing a button is simulated (Interaction I). If the probe parameters are changed in the simulation, the interaction will be changed significantly. Interaction II: Probe damping is 3 times greater; Interaction III: Probe stiffness is 10 times greater; Interaction IV: Probe mass is 10 times greater [28]

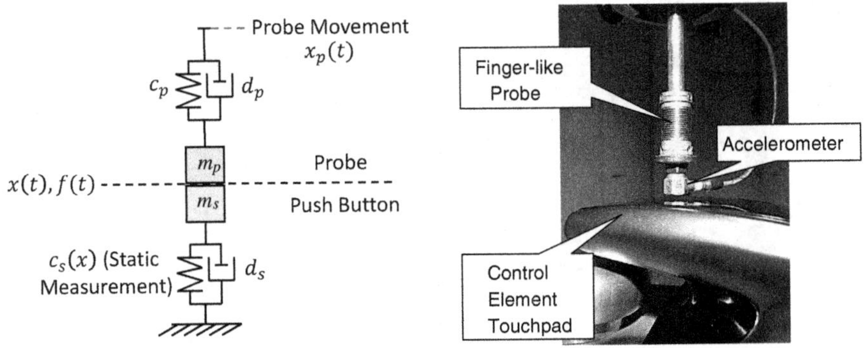

Fig. 14.21 Interaction model when pushing a button (*left*); finger-like measurement device (*right*) [28]

Measurements comparing the finger-like measurement tip with real human fingers in Fig. 14.22 show similarity. Returning to the previously described examples the new interaction-based-measurement shows differences with changed inertia in Fig. 14.23 as well with different active haptic feedback in Fig. 14.24.

The haptic features that connect the objective with subjective perception are the focus of current research.

14.2.5 Discussion and Outlook

Any engineering work focuses on designing a product. A successful industrialization of haptic actuation technology in any industrial context, however, requires always two things:

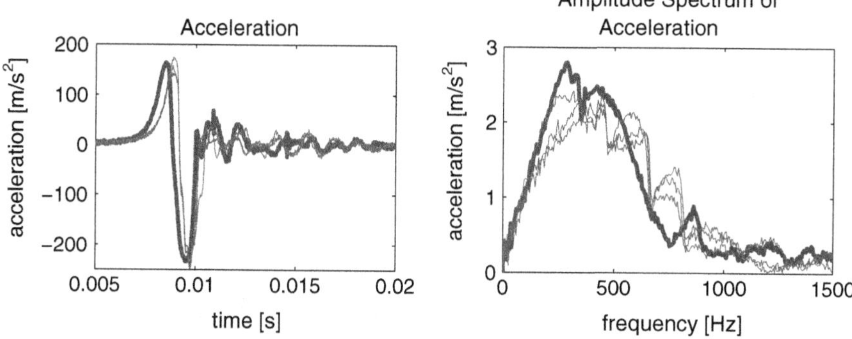

Fig. 14.22 Three measurements of acceleration (*left*, in time domain; *right*, in frequency domain) between three different fingers and a push button (*red curves*) are compared to one measurement of acceleration between the probe and the same button (*blue curve*). The two groups of the *curves* are very similar [28]

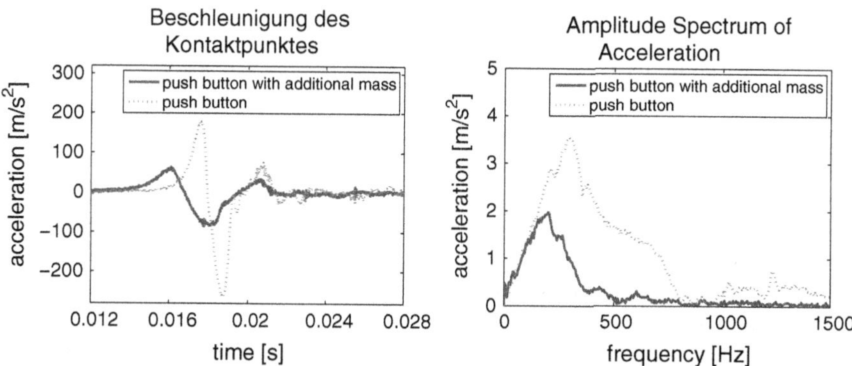

Fig. 14.23 Dynamic measurement of two push buttons with the same force–displacement *curve*, but different masses [28]

- a technology with high performance and optimized quality-to-cost ratio for the intended market
- a measure for this performance and quality

Both items are not to be underestimated. A successful product is nothing without a corresponding quality control. In particular, in the area of haptic technology, this is something which always needs to be developed according to the application at hand. The Mercedes C-class touchpad is an excellent example of this conjunction of research and development and is proposed as an example to extend the real product range of active haptic devices available to the market.

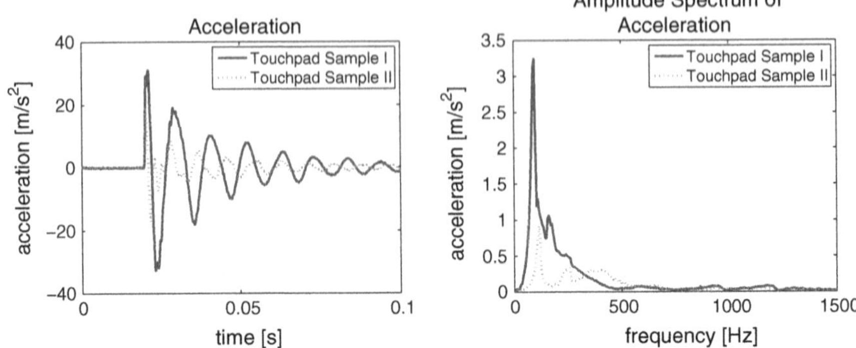

Fig. 14.24 Dynamic measurement of two touchpad samples with the same stiffness, but different active haptic feedback [28]

14.3 HapCath: Haptic Catheter

Thorsten Meiss, Thomas Opitz, Tim Rossner, and Nataliya Stefanova

14.3.1 Introduction

Catheterizations are a conventional process for diagnostic and interventional treatment of blood vessels, which suffer from atherosclerotic depositions, diminishing the blood flow and causing infarcts. In the USA, diagnostic and interventional catheterizations of the heart were performed approximately 1,059,000 times in 2007 (Data base 2011 [20], and about the same amount in Germany in 2009 [2]). In many cases, catheterization is a simple process for well-trained cardiologists: A guide wire is inserted into an artery, usually the arteria fermoralis at the upper leg, and is slid toward the heart. By rotating the proximal end of the wire (the end in the physicians hand), the physician leads the tip at the wire's distal end into the coronary arteries. To visually control the guide wire movement, short-time 2D-X-ray video is used. By sliding a catheter over the guide wire, the physician can lead contrast fluid into the vessels to visualize the course of the arteries for diagnostic purposes. Through this hollow catheter, the physician can lead tools to the upper branches of the coronary vessels or may change and reposition the guide wire very quickly. To reopen totally closed or occluded vessels, the physician can thread a balloon catheter over the proximal end and slide it over the guide wire, through the catheter, into the occluded part of the vessel. Then, he can widen the affected vessel by inflating a balloon (dilatation) and optionally expand a stent to prevent the vessel from contracting again.

However, in many cases, the vessel is totally closed and/or penetrating the obstruction with the guide wire tip becomes very difficult. Additionally, navigation of

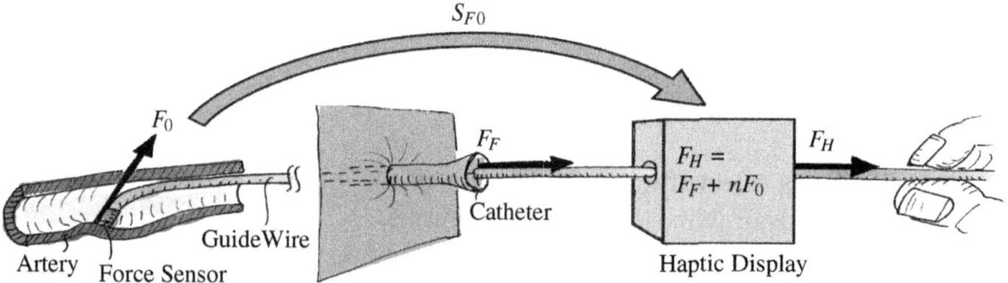

Fig. 14.25 Schematic of the assistive system HapCath: The forces F_0 at the tip of the guide wire are measured by means of a small force sensor. The signal S_{F0} is transmitted out of the patient's body over the wire. Within a haptic display, the signal S_{F0} is reconverted into a scaled force $n \cdot F_0$ by means of amplifiers and actuators, thereby overcoming the friction force F_F within the catheter and vessels. This force is displayed to the surgeon's hand as the amplified force F_H

the wire often turns into a challenging task. Due to the small diameter and the low stiffness of the wire, the physician cannot feel the forces at the wire's tip; the 2D-X-ray imaging linked with the legal limitation of the amount of noxious contrast fluid requires well-trained operators and a long training phase of new cardiologist. To overcome these challenges originating from a lack of intuitive usable information from the guide wire's tip, the HapCath system provides haptic feedback of the forces acting on a guide wire's tip during vascular catheterization [12] (Fig. 14.25). In order to achieve this, force measurement and signal transmission out of the patient's body is realized. Thus, the transmitted signals are used to control actuators within a special haptic display to provide a scaled, amplified force, which is coupled back onto the guide wire. These scaled forces surpass the friction forces, which arise along the length of the guide wire in the vessel and catheter, enabling the user to feel the tip interacting with the walls and obstructions inside the vessel. This shall simplify and accelerate the navigation of the wire and reduce the risk of punctuating the vessel or damaging plaque (depositions). The aim of providing haptic feedback is to enable grasping the right way through the vessels just like with a blind man's cane. For this purpose, very small force sensors have been designed, fabricated, tested, and integrated into guide wires. Special electronics to calculate the 3D force vector acting at the tip have been designed, and a haptic display with a translational and a rotational degree of freedom to couple the forces back onto the guide wire to amplify the measured forces has been constructed and tested.

14.3.2 Deriving Requirements

To our knowledge, the exact forces at the guide wire tip during catheterization have been unknown up to recently. For this project, detailed analysis of the advancing of the guide wire within the vessels [13] with simulation [5] and experimental measurements of the guide wire interactions have been performed.

14.3.3 Design and Development

14.3.3.1 Force Sensor Design

Figure 14.26 shows selected relevant scenarios of the interaction of the guide wire with the vessel walls or with stents inside the vessel.

A force sensor can be integrated at the tip of the guide wire or with some distance to the tip. To allow for the measurement of the interaction forces when the guide wire is advanced with the tip backward (Fig. 14.26d), the sensor needs to be integrated with several centimeters of distance from the tip. This will lead to additional friction forces in the sensor signal and will result in lower frequency response due to higher mass and damping. The most beneficial location to integrate a force sensor therefore is directly into the tip of the wire, due to higher amplitude and frequency resolution of the contact force measurement.

Simulations haven been performed were the guide wire is modeled as distributed elastic elements interacting with viscous elastic walls of the arteries with MATLAB© [5]. Additionally, experiments to determine the buckling load of different types of guide wires where conducted [13]. Both methods reveal a maximum force in axial direction of the wire of around 100–150 mN, e.g., for penetration occlusions, depending on the kind of guide wire used. The forces during advancing and navigation as well as for detecting surface properties, e.g., roughness or softness—for diagnostic purposes—were estimated to be in the range of 1–25 mN.

To allow for force measurement at the tip, two types of microforce sensors have been designed and fabricated [10, 11, 13] (Fig. 14.27). They are built from monocrystalline silicon with implanted boron p-type resistors. This technology was chosen to fulfill the requirements on microscale manufacturing with its high level of integration,

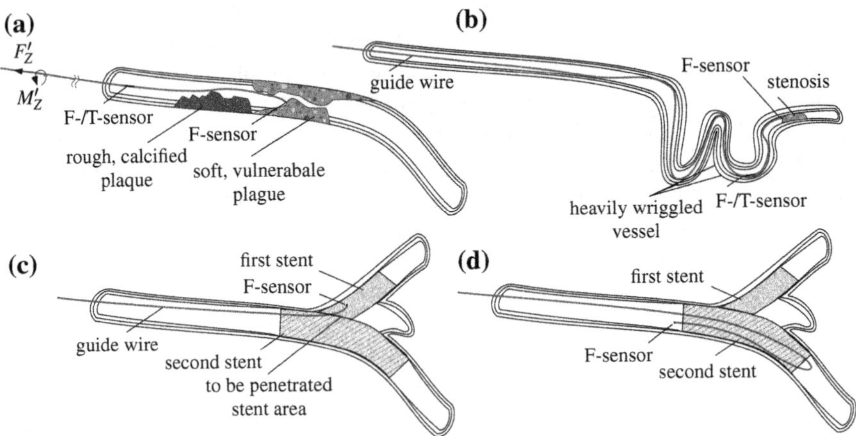

Fig. 14.26 Pictures of different interactions of the guide wire with the vessel, with different kind of plaque (**a**), within a heavily wriggled vessel path (**b**), and with a stent (**c**) and (**d**)

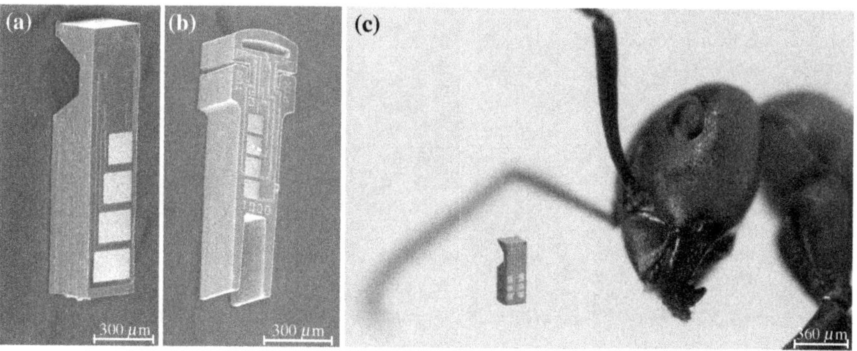

Fig. 14.27 Two types of monocrystalline silicon force sensors: Both are designed to resolve the full force vector in amplitude and angles. Their size is compared to an ant

a relatively high voltage output for robust external readout as well as high mechanical stiffness to fulfill the requirements on high-frequency resolution up to 1000 Hz as well as the need for quasi-static measurements when the guide wire remains in static contact with a constriction.

14.3.3.2 Guide Wire and Sensor Packaging

Guide wires are disposable medical products, manufactured with technologies of precision engineering. The guide wire requires a maximum torsional stiffness and a variable bending stiffness along the wire. To integrate an electrical connection of the sensor over or within the guide wire is a challenging task. A loss in rotational stiffness due to softer materials of the conductors than stainless steel or nickel–titanium will result in less mechanical performance. This is the main reason why the space for the integration of electrical wires is very sparse. The electrical connection is established with four insulated robust copper wires, each with a small diameter of 27 μm.

The sensors are glued onto the tip of the wire with UV curable medical adhesive. They are enclosed into a flexible polyurethane polymer [9] and covered with medically compatible parylene C. Figure 14.28 gives an insight into the assembly.

14.3.3.3 Haptic Display Design

The guide wire is navigated through the vessels with two degrees of freedom: translationally, to advance the guide wire, and rotationally, to choose the relevant wire branch. The haptic display is designed to provide these forces and motions (Fig. 14.29) [17].

The haptic display supports the generation of static forces to display the penetration of occlusions. To give feedback of surface roughness and to reflect the dynamic amplitudes during penetration or when the wire is moved over the grid of stents or rough depositions, the haptic interface is designed with low mechanical inertia to

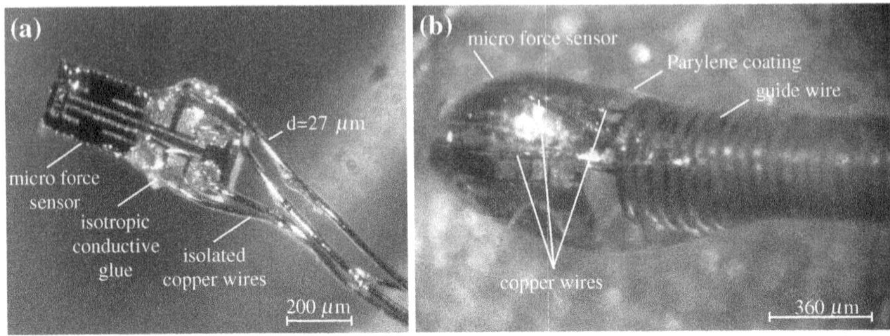

Fig. 14.28 The integration of the sensor into the guide wire tip encompasses several steps of precision mounting, dispensing, and covering with glues and cover polymers

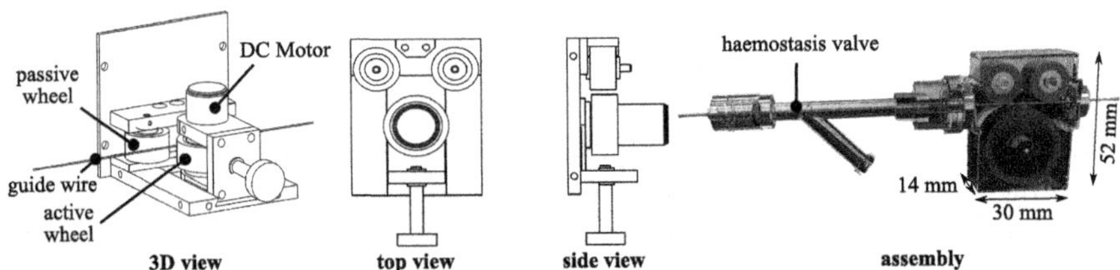

Fig. 14.29 Basic design of the haptic user interface with the translational degree of freedom and side view of the implementation [17]

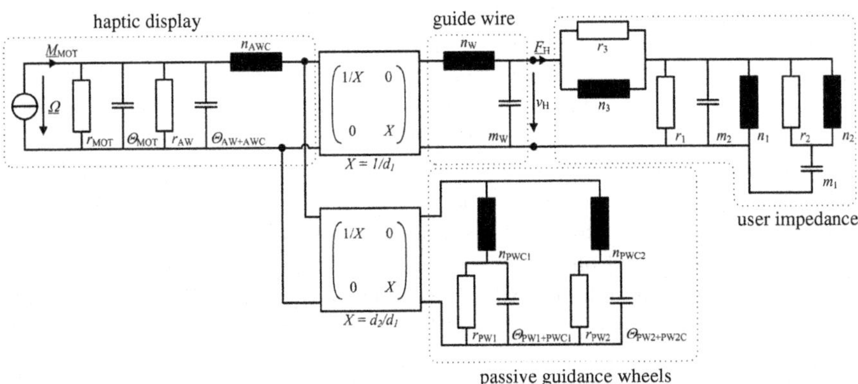

Fig. 14.30 Network model of the haptic user interface including the guide wire and the user's passive mechanical impedance

generate high-frequency feedback as well. To optimize the dynamic performance of the haptic interface, the equivalent circuit representation of the electromechanical setup with guide wire and passive user impedance is used (Fig. 14.30) [17].

14.3.3.4 Electronic Design

The system is powered with one single electronic system. The electronics provide power supply to the force sensor and a unique six channel high-resolution analog front-end. A microcontroller with floating point unit is used to control the sensor readout and to calculate the 3D force vector of the contact forces. It provides angle measurement and PWM control for two brushless DC motors for the haptic feedback interface as well. Force signals are transferred to a PC and to a display for information purposes. The control loop is implemented in the microcontroller itself without the need for time-critical communication with the PC. This allows for a fast control loop with up to 10 kHz s^{-1} control rate for smooth haptic feedback [13].

14.3.4 Verification and Validation

To validate the function of the whole system, the tactile guide wire, the electronics, and the haptic display are connected [12]. The guide wire is advanced into a model of the arteries built from silicone tubes (smooth, healthy arteries) partially filled with epoxy glue mixed with sand to model rough depositions (calcified plaque). Figure 14.31 shows the sensor signal over time when the guide wire is moved in the model and Fig. 14.32 over plates with defined surfaces. The haptic feedback at the interface allows for discrimination of smooth and rough surfaces. When touching different surfaces, the contact force is amplified and different surfaces can be distinguished clearly.

By increasing the amplification factor of the forces, the sensation of soft or elastic surfaces changes to that of much more rigid features due to the higher stiffness emulated by the system. This makes soft, fragile surfaces much easier to detect. They

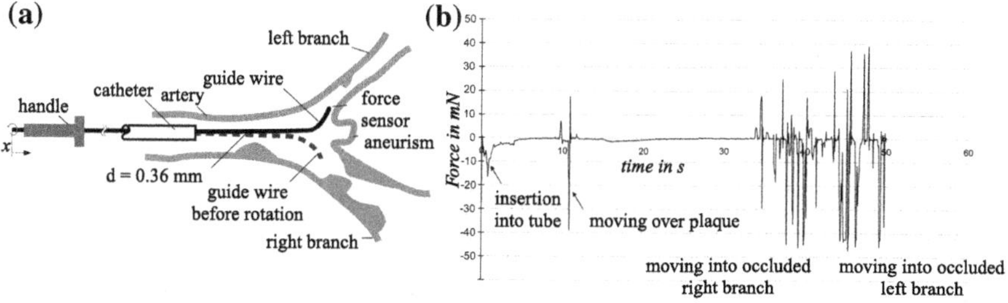

Fig. 14.31 Moving the guide wire with a minimal contact force. The physician maneuvers the guide wire by moving the handle by rotating, pushing, and pulling (**a**). Measurements with a prototype of the tactile guide wire within a model of the arteries with artificial plaque (**b**). The measurement shows the sensor signal during inserting, moving the wire forward, rotating the tip, and going into the right and then into the left vessel branch

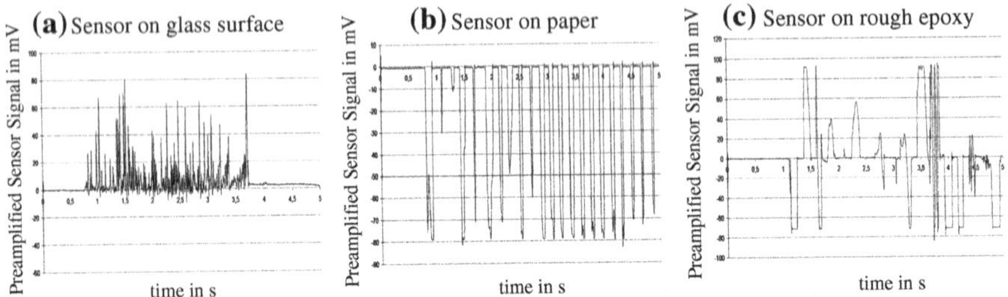

Fig. 14.32 Force signal over time recorded from packaged sensors integrated into a guide wire prototype to evaluate different surface roughness. Glass (**a**), paper (**b**), and sand in epoxy glue (**c**). Notable is the reproducible, nearly periodic output of the packaged sensor on paper (**b**). Increasing roughness of the surfaces leads to increasing output signals (**a**), (**b**) to (**c**)

become virtually harder, whereby elongation due to force is reduced. We assume that this can lead to much less ruptures of vulnerable features, leading to fewer complications during catheterizations in the future, as well as a higher success rate for complicated interventions with wriggled arteries.

14.3.5 Conclusion and Outlook

The project involves several technical challenges, encompassing sensor design and sensor integration, as well as adapted haptic feedback. The current field of research is focused on transferring the results into application by refining the design and the technical implementation of the sensor, electrical wire, and guide wire assembly. The application will benefit from ongoing research regarding optimized filtered signal feedback from touch scenarios of the tip with different surfaces. The project is funded by the German Research Foundation DFG which supports this project under Grant No. WE 2308/3-3.

References

1. Bau O et al (2010) TeslaTouch: electrovibration for touch surfaces. In: Proceedings of the 23nd annual ACM symposium on user interface software and technology (UIST '10). ACM, New York, NY, USA, pp 283–292. doi:10.1145/1866029.1866074
2. Bruckenberger E (2010) Herzbericht 2009. http://www.bruckenberger.de/pdf/hzb22_09auszug.pdf
3. Deetjen P, Speckmann EJ (2004) Physiologie. 4nd Edition. Urban & Fischer Verlag. ISBN: 978-3437413179
4. Hauptle P, Hubinsky P, Gruhler G (2011) Harmonic modulated feedback in control to lower oscillations in mechatronic systems. In: 11th international conference on control, automation and systems (ICCAS). IEEE, pp 273–276.

5. Kern T et al (2006) Force acting on guide wires during vascular navigation calculated with a viscoelastic vessel-model. In: Biomedizinische Technik (BMT). ETH Zürich, Schweiz, Jan 2006. http://tubiblio.ulb.tu-darmstadt.de/28962/
6. Levine M (1982) You-are-here maps psychological considerations. Environ Behav 14(2): 221–237. doi:10.1177/0013916584142006
7. Makino Y, Maeno T (2011) Dual vibratory stimulation for mobile devices. In: World haptics conference (WHC). IEEE. http://www.worldhaptics.org/2011/haptics2011.dekon.com.tr/en/contente84b.html?PID=%7BD5A331DC-CEA7-4968-8D9C-986762A36871%7D
8. Marchuk N, Colgate J, Peshkin M (2010) Friction measurements on a large area TPaD. In: Haptics symposium.IEEE, pp 317–320. doi:10.1109/HAPTIC.2010.5444636
9. Meiss T et al (2008) The influence of the packaging on an in-vivo micro-force sensor. In: Conference proceedings of Eurosensors, Dresden, pp 1557–1560. Aug 2008. http://tubiblio.ulb.tudarmstadt.de/35354/
10. Novel A (2007) Highly miniaturized force sensor for force feedback during catheterization. ISBN:9783981099317
11. Meiss T et al (2007) Fertigung eines Miniaturkraftsensors mit asymmetrischem Grundkörper zur Anwendung bei Katheterisierungen. In: MikroSystemTechnik (2007). https://www.vde-verlag.de/proceedings-de/563061014.html
12. Meiss T et al (2009) Intravascular palpation and haptic feedback during angioplasty. In: Euro-Haptics conference 2009 and symposium on haptic interfaces for virtual environment and teleoperator systems. World Haptics 2009, pp 380–381. Third Joint. Mar 2009. doi:10.1109/WHC.2009.4810904
13. Meiss T (2012) Silizium-Mikro-Kraftsensoren für haptische Katheterisierungen: Entwurf, Musterbau und Signalverarbeitung sowie erste Validierung des Assistenzsystems HapCath. PhD thesis. Technische Universität Darmstadt, Institut für Elektromechanische Konstruktionen, 2012. http://tuprints.ulb.tu-darmstadt.de/2952/
14. Morioka M, Griffin MJ (2007) Frequency dependence of perceived intensity of steering wheel vibration: effect of grip force. In: Proceedings of World Haptics 2007, pp 50–55. doi:10.1109/WHC.2007.58
15. Morrell J, Wasilewski K (2010) Design and evaluation of a vibrotactile seat to improve spatial awareness while driving. In: Haptics symposium, 2010. IEEE, pp 281–288. doi:10.1109/HAPTIC.2010.5444642
16. Mulder M et al (2011) Design of a haptic gas pedal for active car-following support. In: Intelligent transportation systems, IEEE transactions, vol 12, pp 268–279. doi:10.1109/TITS.2010.2091407
17. Opitz T et al (2013) Miniaturized haptic user-interface for heart catheterizations—concept and design. ENG. In: Biomed Tech (Berl). Sept 2013. doi:10.1515/bmt-2013-4405
18. Poppinga B et al (2011) TouchOver map: audio-tactile exploration of interactive maps. In: Proceedings of the 13th international conference on human computer interaction with mobile devices and services. ACM, pp 545–550. doi:10.1145/2037373.2037458
19. Reisinger J (2009) Parametrisierung der Haptik von handbetätigten Stellteilen. Dissertation. Technische UniversitätMünchen. https://mediatum.ub.tum.de/doc/654165/654165.pdf
20. Roger VL et al Heart disease and stroke statistics-2011 update: a report from the American Heart Association. In: Circulation 4, pp e18–e209, Feb 2011. doi:10.1161/CIR.0b013e3182009701
21. Schattenberg K (2002) Fahrzeugführung und gleichzeitige Nutzung von Fahrerassistenz- und Fahrerinformationssystemen:Untersuchungen zur sicherheitsoptimierten Gestaltung und Positionierung von Anzeige-und Bedienkomponenten im Kraftfahrzeug. PhD thesis. RWTH Aachen, 2002. http://sylvester.bth.rwth-aachen.de/dissertationen/2002/181/
22. Tang H, Beebe DJ (1998) A microfabricated electrostatic haptic display for persons with visual impairments. Rehabil Eng IEEE Trans 6(3):241–248. doi:10.1109/86.712216
23. Völkel T, Weber G, Baumann U (2008) Tactile graphics revised: the novel brailledis 9000 pin-matrix device with multitouch input. In: Computers helping people with special needs. Springer, Berlin, pp 835–842. doi:10.1007/978-3-540-70540-6_124

24. Wang Z et al (2009) Instant tactile-audio map: enabling access to digital maps for people with visual impairment. In: Proceedings of the 11th international ACM SIGACCESS conference on Computers and accessibility. ACM, pp 43–50. doi:10.1145/1639642.1639652
25. Xu C et al (2011) Tactile display for the visually impaired using TeslaTouch. In: PART 1–proceedings of the 2011 annual conference extended abstracts on Human factors in computing systems. ACM, pp 317–322. doi:10.1145/1979742.1979705
26. Zeng L, Weber G (2010) Audio-haptic browser for a geographical information system. In: Computers helping people with special needs. Springer, Berlin, pp 466–473. doi:10.1007/978-3-642-14100-3_70
27. Zeng L, Weber G, Baumann U (2012) Audio-haptic you-are-here maps on a mobile touch-enabled pinmatrixdisplay. In: 2012 IEEE international workshop on haptic audio visual environments and games (HAVE). IEEE, pp 95–100. doi:10.1109/HAVE.2012.6374428
28. Zhou W et al (2014) Interaction-based dynamic measurement of haptic characteristics of automotive control elements. In: Eurohaptics 2014 (submitted paper). Final contribution and bibliographic data is available from www.springerlink.com
29. Zoller I, Lotz P, Kern T (2012) Novel thin electromagnetic system for creating pushbutton feedback in automotive applications. In: Isokoski P, Springare J (2012) Haptics: perception, devices, mobility, and communication, vol 7282. Lecture Notes in Computer Science. Springer, Berlin, pp 637–645. doi:10.1007/978-3-642-31401-8_56

Chapter 15
Conclusion

Like any other design process, the design of haptic systems is largely influenced by the optimization of a technical system based on the balancing of a plurality of decisions on separate components, which, as a rule, influence each other. In the beginning, the requirements of the customer, respectively, of the project have to be defined. The methods presented in Chap. 5 are intended to systematically identify the most important aspects of these requirements. However, the engineer should be conscious of the fact that for the design of a sense-related interface, less precise and definite terms are available than he may be used to. Additionally, the knowledge on the part of the customer may result in considerable confusion, as especially haptic terms, e.g. resolution or dynamics, may be used in the wrong context or understood in a wrong way. A better definition of the requirements without major misunderstandings is achieved by, e.g. giving the customer aids, "shows-and-tells" of haptics. It is necessary for the customer and the engineer to come to a common understanding based on references known to both. It seems promising to describe the interactions the user should be able to do with the task-specific haptic system very thoroughly, since they have a large impact on the design of the system and the requirements derived from the capabilities of the haptic sense. For this reason, an understanding of the specialties of haptic perception and interaction on the part of the engineer is necessary. It should not be limited to the technical characteristics described in Chaps. 2 and 3, but also include some knowledge about the "soft," i.e., psychological and social aspects of haptics as described in Sect. 1.1.

Based on the requirements discussed above, the technical design process may begin. An adapted version of the commonly known *V-model* is given in Chap. 4 to do this. This approach tries to integrate all of the above-mentioned aspects in a structured way. One of the very first decisions is the choice of the haptic system's structure (Chap. 6. Although this decision is at the very beginning of the design process, a rough sketch of the favored structure of the device to be developed is necessarily to be made. This demands a considerable knowledge of all the branches

C. Hatzfeld and T.A. Kern (eds.), *Engineering Haptic Devices*,
Springer Series on Touch and Haptic Systems, DOI 10.1007/978-1-4471-6518-7_15

of haptic device design, which later will be needed again during the actual design phase.

Besides the already mentioned decision on the general structure, the basis of the design of kinaesthetic and tactile systems is its kinematic structure (Chap. 8). After the considerations made for kinematics, concerning the transmission and gearing proportions, the working volume, and the resolution to be achieved, suitable actuators are chosen or even designed. In Chap. 9, the basis for this is provided by comparing the different actuation principles. Examples of their realizations, even of unusual solutions for haptic applications, provide a useful collection for any engineer to combine kinematic requirements of maximum forces and translations with impedances and resolutions.

As closed-loop admittance controlled systems with kinaesthetic and tactile applications are gaining in importance, force sensors have to be considered as another component of haptic devices. In Sect. 10.1, this technology is introduced, providing the tools, as well as conveying the chances and also the challenges connected with their application. A frequent application of haptic devices is to be found in the human–machine interface of simulators, be it for games ranging from action to adventure games, or for more serious applications for training surgeons or in the military, respectively, in industrial design. In addition to the output of haptic information, an input of user movements is required. The measurement principles typically used are discussed in Sect. 10.2.

The design steps presented so far will enable the haptic device to provide a tactile or kinaesthetic output to the user, often measuring a reaction too. In particular, with today's computer technology, the data will be almost always interfaced with a standard PC. The requirements derived from this interface are subject to a presentation of standard interface technology given in Chap. 11, whereby the interfaces' performances are compared with each other.

Due to the rather frequent application of haptic devices in simulators, an interface with a simulation engine is required. An insight into the requirements and challenges of suitable haptic algorithms is helpful for the hardware engineer to improve the communication with software engineers and the interfacing with their VR environments. An appropriate introduction is given in Chap. 12.

The cross section given in this book is meant to improve and further speedup the design of haptic devices and to avoid the most critical errors typically made during the design process. The research in the area of haptic devices is making impressive progress. Every few months adapted control engineering concepts appear; the usage of haptic perception for the design is subject to current research. Actuators are being continuously improved; even new principles with haptically interesting properties regularly appear on the market. Closed-loop controlled systems become more and more interesting, due to the slowly increasing availability of highly dynamic high-resolution force sensors. This dynamics of a still young discipline commits developing engineers to monitor the current research attentively. For this purpose, finally, a list enumerating teams active in the haptic area has been compiled in Appendix B.

Appendix A
Impedance Values of Grasps

Tables A.1 and A.2 provide the parameter for the model given by Fig. 3.7 and Eq. (3.7). They parametrize the different grasping situations discussed in Sect. 3.1.6.

Table A.1 Mean values of the mechanical impedance model according to Fig. 3.7 for different grasping situations

Grasp/touch	k_1 (N/m)	m_1 (kg)	d_1 (Ns/m)	d_2 (Ns/m)	k_2 (N/m)	k_3 (N/m)	d_3 (Ns/m)	m_2 (kg)
Power grasps								
Cylinder	412.61	1.577	43.43	33.06	31,271	15,007	182.77	0.13
Sphere	2,500.7	4.32	45.72	31.35	21,033	9,743	150.60	0.098
Ring 17.71	10.0	0.0032	31.35	5,843.7	2,906.7	34.54	0.016	
Precision grasps								
Pen 45°	1,357.1	1.7376	23.38	3.269	36,672	3,544.6	12.22	0.029
Pen vertical	44.73	5.44	4.55	17.92	17,794	1,782.7	12.92	0.029
Pen horizontal	212.49	3.26	7.56	8.15	22,092	3,672.7	13.73	0.043
Finger								
Normal 2 mm	203.21	75.02	1.0854	3.1672	6,656.0	478.73	8.3689	0.0114
Normal 15 mm	0.091	37.28	3.79	3.18	9,273.5	839.92	12.22	0.018
Shear lateral	54.5	10.0	0.323	4.88	12,935	191.62	4.4342	0.0178
Shear distal	77.56	9.892	9.443	3.003	22874	2004	4.0377	0.0195
Shear 45°	1,053.0	90.44	5.47	7.16	26,854	1,090.2	15.26	0.006

C. Hatzfeld and T.A. Kern (eds.), *Engineering Haptic Devices*,
Springer Series on Touch and Haptic Systems, DOI 10.1007/978-1-4471-6518-7

Table A.2 Linear interpolated dependencies of the model's parameters from grasp and touch forces according to Fig. 3.7 for different grasping situations

Grasp/touch		k_1 (N/m)	m_1 (kg)	d_1 (Ns/m)	d_2 (Ns/m)	k_2 (N/m)	k_3 (N/m)	d_3 (Ns/m)	m_2 (kg)
Power grasps									
Cylinder	a	−62.4	−0.216	1.46	−0.409	−365	1,330.0	3.27	0.0043
	b	1,360	4.88	21.20	39.3	36,800	5,300.0	133	0.065
Sphere	a	−49.0	−0.111	−0.0359	−0.788	13.3	109.0	4.94	0.00015
	b	3,250	6.01	46.3	43.4	20,800	8,090.0	75.2	0.096
Ring	a	2.26	-2.26×10^{-16}	−0.0054	0.143	304.0	150.0	1.72	−0.0003
	b	−14.0	10.0	0.107	5.47	1,590.0	811.0	10.5	0.206
Precision grasps									
Pen 45°	a	−74.3	0.0616	−0.776	0.0247	−134.0	363.0	0.551	0.00372
	b	1,840.0	1.34	28.4	3.11	37,500	1,190.0	8.64	0.00447
Pen vertical	a	−21.4	−1.26	1.69	0.428	1,460.0	204.0	0.759	0.00465
	b	584.0	13.6	6.46	15.1	82.80	454.0	7.99	−0.0087
Pen horizontal	a	−27.5	0.56	1.16	−0.619	380.0	229.0	0.409	0.00883
	b	391	−0.371	0.193	12.2	19,600	2,190	11.1	−0.0144
Finger									
Normal 2 mm	a	−124.0	15.2	−0.088	−0.106	−1,350	−36.3	1.84	−0.002
	b	606.0	25.5	2.09	3.2	11,000	361.0	2.39	0.0180
Normal 15 mm	a	0.187	−5.95	0.861	−0.233	−1,940	374.0	1.48	0.000675
	b	0.0311	56.6	0.993	3.94	15,600	375	4.75	0.0159
Shear lateral	a	−14.0	-106^{-10}	0.177	0.509	1,250.0	63.3	0.363	0.00141
	b	100.0	10.0	−0.558	3.23	8,860	14.1	3.25	0.0133
Shear distal	a	−291.0	0.0571	1.61	−0.711	−8,590	367.0	0.266	0.00405
	b	1,720	9.71	4.22	5.31	508,000	811.0	3.17	0.00636
Shear 45°	a	−54.7	−1.94	0.469	1.65	−299.0	74.5	−0.0776	0.000295
	b	1,230.0	96.8	3.95	1.73	27,800	848.0	15.5	0.00538

Interpolation according to $c = a \cdot F + b$

Appendix B
URLs

Tables B.1, B.2 and B.3 are a collection of all names and groups familiar to the authors. Naturally, the lists are not compelling, but they provide a starting point for the research about relevant sources. Additions to this list for future editions of this book may be sent to the editor via e-mail.

Table B.1 URLs of laboratories and individuals working in the area of haptics

Institute	Head
ALAB, Tokyo University www.alab.t.u-tokyo.ac.jp	SHINODA
A Laboratory for Teleoperation and Autonomous Intelligent Robots (ALTAIR) http://metropolis.sci.univr.it/altair	FIORINI
Algemeine Psychologie, Uni Gießen http://www.uni-giessen.de/cms/fbz/fb06/psychologie/abt/allgemeine-psychologie	DREWING
Artificial Intelligence Lab, Robotics http://robotics.stanford.edu	KHATIB
Bioengineering and Robotics Research Center E. Piaggio www.centropiaggio.unipi.it	BICCHI
Department of Bioengineering http://www3.imperial.ac.uk/humanrobotics	BURDET
Bioinstrumentation Lab http://bioinstrumentation.mit.edu	JONES
Biorobotics Laboratory http://brl.ee.washington.edu	HANNAFORD

(continued)

C. Hatzfeld and T.A. Kern (eds.), *Engineering Haptic Devices*,
Springer Series on Touch and Haptic Systems, DOI 10.1007/978-1-4471-6518-7

Table B.1 (continued)

Biomimetics and Dexterous Manipulation Laboratory http://bdml.stanford.edu	CUTKOSKY
Computer Graphics Laboratory ETH Zürich http://graphics.ethz.ch	GROSS
Chair of Information-Oriented Control http://www.itr.ei.tum.de	HIRCHE
Cybernetics Intelligence Research Group www.reading.ac.uk/sse/research/sse-cybernetics.aspx	HARWIN
Delft Haptics Lab www.tudelft.nl	HELM and others
EduHaptics www.eduhaptics.org	PROVANCHER
Faculty of Human Movement Sciences www.fbw.vu.nl/en/	KAPPERS
Fujimoto Lab http://drei.mech.nitech.ac.jp/fujimoto	FUJIMOTO
Group of Robots and Intelligence Machines http://138.100.76.36/en/default.asp	ARACIL, FERRE and others
Haptics and Embedded Mechatronics Lab http://heml.eng.utah.edu/pmwiki.php/Main/HomePage	PROVANCHER
Haptics Grasp Lab http://haptics.grasp.upenn.edu	KUCHENBECKER
Haptic Exploration Lab www.haptics.me.jhu.edu	OKAMURA
Haptic Interface Research Lab http://engineering.purdue.edu/hirl	TAN
Haptics and Virtual Reality Lab http://hvr.postech.ac.kr	CHOI
Haptiklabor www.haptiklabor.de	GRUNWALD
Haptix Laboratory http://www-personal.umich.edu/%7Ebrentg/Web	GILLESPIE
Harvard Biorobotics Laboratory http://biorobotics.harvard.edu	HOWE
Human Sciences Group, Institute of Sound and Vibration Research www.isvr.soton.ac.uk	GRIFFIN
Institut des Systèmes Intelligents et de Robotique, Interaction group www.isir.upmc.fr	HAYWARD

(continued)

Table B.1 (continued)

Institute of Automatic Control Engineering www.lsr.ei.tum.de	BUSS, PEER
Institute of Robotics and Mechatronics www.dlr.de/rm/en	ALBU- SCHÄFFER
Institute of Electromechanical Design (EMK) www.institut-emk.de	SCHLAAK, WERTHSCHÜTZKY
Interactive Graphics and Simulation Group http://informatik.uibk.ac.at/igs	HARDERS
Ishibashi and Sugawara Lab http://nma.web.nitech.ac.jp/index_e.html	ISHIBASHI, SUGAWARA
Kognitive Neurowissenschaften www.uni-bielefeld.de/biologie/cns/index.html	ERNST
Laboratoire de Systemes Robotiques (LSRO) http://lsro.epfl.ch	BLEULER
Mechatronics and Haptics Interfaces (MAHI) http://mahilab.rice.edu	O'MALLEY
Mechatronics Lab www.mechatronics.me.kyoto-u.ac.jp	MATSUNO
Mensch-Computer-Interaktion www.inf.tu-dresden.de/index.php?ln=de\&node_id=937	WEBER
Microdynamic System Laboratory www.msl.ri.cmu.edu	HOLLIS
MIT Touch Lab http://touchlab.mit.edu	SRINIVASAN
Multimedia Communication and Research (MCR) Lab www.mcrlab.uottawa.ca	EL SADDIK
Multimodal Interaction Group www.dcs.gla.ac.uk/~stephen	BREWSTER
Perception, Cognition and Action Group www.kyb.tuebingen.mpg.de/de/forschung/abt/bu.html	BÜLTHOFF
PERCRO www.percro.org	BERGAMASCO
Precision and Intelligence Laboratory www.pi.titech.ac.jp	SATO
Psychology at Hamilton http://academics.hamilton.edu/psychology	GESCHEIDER

(continued)

Table B.1 (continued)

Psychology at Carnegie Mellon, Spatial and Haptic www.psy.cmu.edu	KLATZKY
Laboratories for Intelligent Machine Systems (LIMS) http://lims.mech.northwestern.edu	COLGATE, PESHKIN, LYNCH
Salisbury Research Group http://jks-folks.stanford.edu/home.html	SALISBURY
Sensory Motor Neuroscience www.symon.bham.ac.uk/labs.htm	WING
Sensory Perception and Research Group www.cs.ubc.ca/labs/spin	MACLEAN
System Design Engineering, University of Waterloo http://uwaterloo.ca/systems-design-engineering	
Tachi Lab http://tachilab.org	TACHI
Tele-Rehabilitation Institute www.ti.rutgers.edu	BURDEA
Telerobotics and Control Lab (TCL) http://robot.kaist.ac.kr	KWON
Telerobotics Lab http://telerobotics.stanford.edu	NIEMEYER
TNO, Haptics www.tno.nl	VAN ERP
Tokyo Institute of Technology, Advanced Information Processing www.pi.titech.ac.jp/english/organization/detail_50.html	SATO
Mechanical Engineering, University of Utah http://mech.utah.edu	ABBOT, MASCARO, PROVANCHER
Virgina Touch Laboratory www.sys.virginia.edu/ggerling	GERLING
Virtual Reality Lab www.haptics.buffalo.edu	KESAVADAS
Visualization and Image Analysis Laboratory (VIALAB) www.vialab.org	STETTEN
VR Lab, University of Tsukuba http://intron.kz.tsukuba.ac.jp	IWATA, YANO
Welfen Lab www.welfenlab.de	WOLTER

Table B.2 Established conferences, societies, and journals

Title	URL	Type
Asiahaptics Conference	http://asiahaptics.vrsj.org	Conference
Eurohaptics Conference	www.eurohaptics.org	Conference
Eurohaptics Society	www.eurohaptics.org	Society
Haptics-e	www.haptics-e.org	Journal
Haptics Symposium	www.hapticssymposium.org	Conference
Haptics Technical Committee (IEEE)	www.worldhaptics.com	Society
Human Factors	www.hfes.org	Journal
Intelligent robots and systems (IROS)	www.ieee-ras.org/conferences-workshops/iros	Conference
Conference on robotics and automation (ICRA)	www.ieee-ras.org/conferences-workshops/icra	Conference
Journal of robotics research	www.ijrr.org/	Journal
Presence: teleoperators and virtual environments	www.mitpressjournals.org/loi/pres	Journal
Transactions on haptics (IEEE)	www.computer.org/th	Journal
Transactions on mechatronics (IEEE/ASME)	www.ieee-asme-mechatronics.org	Journal
Worldhaptics conference	www.worldhaptics.com	Conference

Table B.3 Commercial manufacturers of haptic-related products

Manufacturer	URL	Type
Butterfly Haptics	www.butterflyhaptics.com	HW
Chai3D	www.chai3d.org	SW
ForceDimension	www.forcedimension.com	HW
Geomagic (former SensAble)	www.sensable.com	HW and SW
Haptiklibary	www.haptiklibrary.org	SW
Haption	www.haption.com	HW
Immersion	www.immersion.com	HW and SW
Moog	www.fcs-cs.com/robotics	HW
MPB Technologies Inc.	www.mpb-technologies.ca	HW
Novint	www.novint.com	HW
Quanser	www.quanser.com	HW
Reachin	www.reachin.se	SW
SenseGraphics	www.sensegraphics.com	SW and HW
Xitact	www.xitact.com	HW

HW hardware, *SW* software

Glossary

Term	Definition	Page
Braille	Literary language based on raised dots in a surface and therefore used by the blind and visually impaired	60, 100
Comanipulator	Telepresence system with an additional direct mechanical link between user and the environment or object interacted with	39
Cutaneous	Located in the skin	35
Kinaesthetic	Belonging to the kinaesthetic system consisting of muscles, bones and joints, and the mechanoreceptors located therein or influencing the kinaesthetic system, in the course of the book, the notation kinaesthetic is used according to ISO9241-910	35
Mechanoreceptor	Entity consisting of one or more sensory cells, the corresponding nerve fibers, and the connection to the central nervous system	56
Percept	Conscious mental representation of a stimulus that evoked a sensation in the central nervous system	54, 63, 65
Psychometric function	Function to describe perception processes, where detection probability is plotted with regard to stimulus intensity or another changing stimulus parameter	65
Psychophysics	Part of experimental psychology that deals with the analysis of the relations between objectively measurable stimuli and the subjective perception thereof by a person or user	33
Receptive field	Skin area, from that a single nerve fiber can be excited by a mechanical stimulus	57
Stimulus, *pl.* stimuli	Excitation or signal that is used in a psychophysical procedure. It is normally given with the symbol Φ. The term is also used in other contexts, when a (haptic) signal without further specification is presented to a user	33, 54, 65
Tactile	Belonging to the skin and the mechanoreceptors located therein or influencing the skin	35
Taxonomy	Classification of objects based on standardized criteria, a well-known example of a taxonomy is the classification of biological organisms based on species and kind	34

C. Hatzfeld and T.A. Kern (eds.), *Engineering Haptic Devices*,
Springer Series on Touch and Haptic Systems, DOI 10.1007/978-1-4471-6518-7

Index

Symbols
Ψ method, 45
π-coefficients, 391

A
Absolute threshold, 47
Active haptic interaction, 101
Active touch (*Definition*), 70
Actuation principle, 254
Actuator design, 253
Actuator designs, piezoelectric, 294
Actuator, electrostatic, 327
Actuator, piezoelectric, 288
Admittance controlled, closed loop, 173
Admittance controlled, open-loop, 173
Admittance-type system (*Definition*), 77
Aesthetics, 6
Ampere turns, 266
Anatomical terms of the human hand, 53
Anisotropy of haptic perception, 66
Application areas, 13
Application definition, 127
Application(*Definition*), 74
Art, 6

B
Backlash, 509
Basic equations, piezoelectric, 290
Basic piezoelectric actuator designs, 294
Being, physical, 5
Bending actuator, 256, 305
Bending moment, 384
Bernoulli, 385
Bidirectional, 13
Bimorph, piezoelectric, 302
Block-commutation, 280
Bragg, 407
Braille, 21, 305
Brake, electromagnetic, 322

C
C 82, material property, 293
Calibration, 385
Capacitive actuator, 256
Capacitive actuators, 329
Capacitive principle, 254
Capacitive sensors, 428
Capstan drive, 260
Cascade control, 206
Categorized information, 68
Chai3D, 497
Channel, 31
Channel (*Definition*), 32
Channel function, 34
Charge constant, 292
Classification of haptic systems, 73
Closed-loop admittance controlled, 169
Closed-loop impedance controlled, 169
Code disc, 422
Coefficient of strain, piezoelectric, 291
Coefficient of tension, piezoelectric, 291
Coefficients, piezoelectric, 290
Collision detection, 490
    dynamic, 496
    static, 496
Comanipulation system (*Definition*), 80
Comanipulator, 16
Communication, 21
Commutated, electronic, 279
Commutation, 279
Completion time, 513

C. Hatzfeld and T.A. Kern (eds.), *Engineering Haptic Devices*,
Springer Series on Touch and Haptic Systems, DOI 10.1007/978-1-4471-6518-7

Compliance (perception of), 58
Component design, 129
Composites, 384
Comprehensive model, 129
Compression, 450
Concepts of interaction, 69
Conductor, 264
Constraint, 233
Constraint-based method, 481
Consumer electronics, 19
Contact grasp, 106
Control design, 197
Control of linear drive, 206
Control of teleoperation systems, 208
Control stability, 131
Coulomb's law, 327
Coupling factor, piezoelectric, 292
Cross talk, 506
Curie temperature, 293
Current source, analog, 284
Curvature (perception of), 58
Curves of equal intensity, 62
Customer, experiments with, 146

D

D/A-converter, 282
DC-drive, 278
DC-motor, 255
Definition of application, 127
DELTA, 235, 241, 249
Design goals for haptic system, 131
Design piezoelectric actuators, 298
Designs of DEA, 338
Dielectric elastomer actuator (DEA), 335, 336
Differential threshold, 48
Differentiation of signals, 430
Digital to analog converter, 282
Direct current magnetic field, 270
Disturbance compensation, 202
DL (differential limen), 48
Dots-per-inch, 421
DPI, 421
Driver electronics, electrodynamic, 281
Dynamic collision detection, 496
Dynamic range, 506

E

EC-drive, 279
EC-motor, 255
ECP, 452
Effect, piezoelectric, 289
Elastic constant, 291
Elastomechanic, 382, 383
Electric field, 327
Electric motor, 255
Electrical time constant, 274
Electrochemical principle, 254
Electrodynamic principle, 254
Electromagnetic actuators, 314
Electromagnetic brake, 322
Electromagnetic principle, 254
Electromechanical network, 301
Electronic-commutated, 279
Electrorheological fluid, 342
Electrostatic actuators, 327
Energy consumption, 511
Energy density, magnetic, 271
Energy, magnetic, 314
Enhanced parallel port, 452
EPP, 452
Equations traveling wave motor, 297
ERF, 342
Errors, 513
Errors of haptic systems, 121
Ethernet, 454
Euler angle, 138
Evaluation, 130, 503
Evaluation criteria, 151
Event, 154
Event-based haptics, 87, 448
Evolution, 5
Experiments, 146
Exploratory procedures, 69
External effects on perception, 51
External supply rate, 195
Extreme dead times, 450

F

Fabry-Pérot, 405
Fechner's law, 50
Feedforward, 203
Fiber bragg grating, 406
Field driven actuator, 327
Field plates, 425
Field response, 275
Field strength, magnetic, 266
Filling factor, 265
FireWire, 454
Fitts' law, 516
Fixed angle, 138
Flow mode, 345
Flux, 266
Flux density, magnetic, 266

Flux, magnetic, 266
Foil sensor, 395, 399
Fooling the sense of touch, 66
Force feedback device, 78
Force sensing resistor, 395
Force, magnetomotive, 266
Forced-choice paradigm, 46
Formation of the sense of touch, 5
Frequency response, 506
Function of kinaesthetic receptors, 37

G
Gage factor, 386, 387, 414
Gears, 258
Gearwheel, 260
General design guidelines, 148
Gesture, 71
Golgi tendon organ, 37
Grasp, 105
Grayscale sensor, 423
Grip, 105
Guess rate, 41

H
H-bridge, 283
H3D API, 497
Hall-sensors, 425
Hammerstein-model, 185
Handling force, 513
Haptic compression, 87
Haptic controller (*Definition*), 81
Haptic display (*Definition*), 74
Haptic icon, 21, 68
Haptic illusion, 66
Haptic interface (*Definition*), 76
Haptic loop, 165
Haptic quality, 120, 131
Haptic rendering, 459, 497
Haptic system control (*Definition*), 80
Haptic systems (*Definition*), 73
Haptic transparency, 131, 210
Haptic-assistive system (*Definition*), 75
Hapticon, 21
Haptics in cockpits, 7
Hardware in the loop, 455
HID, 453, 463
HIL, 455
Histology, 31
Human interface device, 453
Human movement capabilities, 71
Human–machine interface, 7
Hydraulic, 256
HyperBraille, 305
Hysteresis, 505

I
Icon, 21
Identical condition, 233
IEEE 1394, 454
Impedance controlled, closed loop, 171
Impedance controlled, open-loop, 170
Impedance coupling, 103, 105
Impedance measurement, 107
Impedance-type system (*Definition*), 77
Impulse response, 507
Induction, 274
Industrial design, 18
Influencing factors, 51
Information display, 18
Innervation density of mechanoreceptors, 33
Input strictly passive, 195
Integral criterion, 202
Integration of signals, 430
Intensity, 402
Interaction, 10
Interaction analysis, 127, 149
Interaction concepts, 69
Interaction metaphor, 458, 473
Interaction with haptic systems, 73
Iron-less rotor, 276

J
JND (just noticeable difference), 48
Joint, 228, 234, 236
JTD (just tolerable difference), 48

K
Kinaesthetic, 381
Kinaesthetic (*Definition*), 12
Kinaesthetic receptors, 37
Kinematic structure description, 137
Kinematics (*Definition*), 229

L
Lapse rate, 42
Latency, 511
Linear state space, 183
Linearity of haptic perception, 65
Local haptic model, 447
Longitudinal actuator, 291
Longitudinal effect, magnetic, 316

Longitudinal effect, piezoelectric, 289
Lorentz force, 262
Lossless, 195

M
Magnetic circuit, 266, 317
Magnetic cross section, 318
Magnetic energy, 314, 319
Magnetic field strength, 266
Magnetic field, direct current, 270
Magnetic flux, 266
Magnetic flux density, 266, 268
Magnetic resistance, 266
Magnetic-dependent resistors, 425
Magnetomotive force, 266
Magnetorheological fluid, 347
Magnetorheological principle, 254
Manipulator (*Definition*), 79
Masking, 65
Material properties, 58
Material properties, piezoelectric, 293
Materials, piezoelectric, 292
Measurement of mechanical impedance, 107
Measuring conditions, 507
Mechanical commutation, 278
Mechanical impedance, 52
Mechanically commutated, 279
Mechanoreceptor function, 34
Mechanoreceptor(*Definition*), 31
Mechatronic design, 125
Medical diagnosis, 20
Medical robotics, 17
Medical training, 18
Meissner corpuscle, 35
Merkel disk, 35
Mice-sensor, 423
Micro frame, 453
Microbending sensor, 404
Microneurography, 31
Model time, 467
Model-based psychometric methods, 44
Motor capabilities, 71
Motor control, 10
Moving coils, electrodynamic, 276
Moving magnet, 278
Multimodal displays, 18
Multiple stimulation, 65

N
Natural behavior, 459
Natural interaction, 458
Navigation, 21
Network parameter, 134
Neural processing, 38
Neuromuscular spindle, 37
Nominal load, 375, 415
Novint falcon, 19
NP-I, 35
NP-II, 35
NP-III, 35
Nuclear bag fiber, 37
Nuclear chain fiber, 37
Nyquist criterion, 187

O
Object exploration, 69
Object properties, 58
Observer-based
  state space control, 204
Open-loop admittance controlled, 169
Open-loop impedance controlled, 169
Optical position sensors, 422
Optimization, 130
Output strictly passive, 195
Overshoot, 199

P
Pacinian corpuscle, 35
Pain, 12
Pain receptors, 38
Paradigm, 45
Parallel mechanism, 230
Parallel port, 452
Parallel-plate capacitor, 327
Passive haptic interaction, 102
Passive touch (*Definition*), 70
Passivity, 13, 217
Passivity, control engineering, 103
PC, 35
Peak force, 505
PEDOT:PSS, 388
Penalty method, 479
Perceived quality, 7
Perception, 9, 11, 29
Perception of transparency, 211
Perceptual deadband, 87
Permanent magnet, 269, 320
Permeability, 266
Permeability number, 319
Permittivity, 266
PEST method, 44
Philosophical aspects, 4

Photoelastic effect, 401
Physical interface device, 453
Physiological basis, 30
Pick and place, 512
PID, 453
PID-control, 201
Piezoelectric actuators, 288
Piezoelectric actuators, design, 298
Piezoelectric basic equations, 290
Piezoelectric bimorph, 302
Piezoelectric coefficient of strain, 291
Piezoelectric coefficient of tension, 291
Piezoelectric coefficients, 290
Piezoelectric coupling factor, 292
Piezoelectric effect, 289
Piezoelectric equation, 291
Piezoelectric longitudinal effect, 289
Piezoelectric material properties, 293
Piezoelectric materials, 292
Piezoelectric motor, 256
Piezoelectric principle, 254
Piezoelectric sensors, 408
Piezoelectric shear effect, 290, 297
Piezoelectric special designs, 295
Piezoelectric stack, 256
Piezoelectric stepper motors, 297
Piezoelectric transversal effect, 290
Piezoelectrical Bimorph, 305
Pixel, 461
Plunger-type magnet, 255, 323
Pneumatic, 256
Popov inequality, 191
Popov plot, 191
Power grasp, 106, 110
Power law, 50
Power loss, electrodynamic, 263
Precision grasp, 106, 112
Primitives, 9
Progression rule, 43
Propagation, 466
Properties of tactile channels, 35
Proprioception, 12
Prototyping, 147
PSE (point of subjective equality), 48
Pseudo-haptic feedback, 67
Pseudo-haptics *(Definition)*, 499
Psi method, 45
Psychometric function, 40
Psychometric methods, 42
Psychometric parameters, 47
Psychometric procedures, 42
Psychophysics, 9, 29, 38
Pulse-width modulation, 282
PVDF, material property, 293
PWM, 282
PZT-4, material property, 293
PZT-5a, material property, 293

Q
Quadrant controllers, 281
Quality of haptic systems, 120
Quality of perception studies, 63
Quartz crystal structure, 289
Quartz, material property, 293
Quaternion, 139

R
RA-I, 35
RA-II, 35
Rare earth, 269
Real time, 467
Receptive field, 31
Reflection light switches, 423
Relative resistivity change, 386
Reluctance, 266
Reluctance drives, 322
Reluctance effect, 315
Reluctance effect, magnetic, 316
Remanence flux density, 271
Rendering, 459
Rendering algorithms, 474
Rendering of surfaces, 20
Rendering software, 497
Requirement specification, 127
Requirements, 379, 413
Resistance, magnetic, 266
Resistive, 386
Resistivity change, 390
Resolution of haptic systems, 120
Resonance actuator, 255
Resonance principle, 410
Risk analysis, 160
Root locus method, 187, 201
Rotation, 137
Roughness (perception of), 58
Routh-Hurwitz criterion, 187
Ruffini ending, 35

S
SA-I, 35
SA-II, 35
Safety requirements, 159
Safety standards, 159
SAW sensors, 411

Scaling, 50
Scattering theory, 220
SEA, 354
Self-supportive, 277
Semiconductor-strain gage, 389
Senses, 4
Sensitivity, 506
Sensor-less commutation, 279
Sensory cells, 31
Sensory integration, 38
Sensory physiology, 29
Serial coupled actuators, 354
Serial elastic actuator (SEA), 354
Serial mechanism, 229
Serial port, 451
Serial viscous actuator, 354
Servo-drive, 279
Shaker, 255
Shape, 154
Shape-memory alloy, 254
Shape-memory wire, 256
Shear effect, piezoelectric, 290, 297
Shear mode, 344
Signal detection theory, 45
Silicon sensors, 392
Simulator, 18
Simulator system, 445
Sinus-commutation, 280
SISO, 133
SL (successiveness limen), 48
Slipperiness (perception of), 58
Social aspects, 4
Solution cluster "kinaesthetic", 155
Solution cluster "surface-tactile", 156
Solution cluster "omnidimensional", 158
Solution cluster "vibro-directional", 157
Solution cluster "vibro-tactile", 156
Solution clusters, 153
Spatial distribution of mechanoreceptors, 33
Special designs, piezoelectric, 295
Squeeze mode, 345
Stability, 131, 217, 511
Staircase method, 43
Standardizing organizations, 159
State feedback control, 204
State space vector, 184
State strictly passive, 195
Static collision detection, 496
Step response, 507
Stepper motor, 256, 321
Stepper motors, piezoelectric, 297
Stiffness, 508
Stimulus (*pl.* stimuli) *(Definition)*, 9
Strain gage, 387
Stress, 384
Stress tensor, 382
Strictly passive, 195
Summation, 65
Surface micromachining, 399
Surface-wave actuators, 256
Surgical robotics, 17
System control structures, 169
System design, 128

T

Tactile, 377, 379, 393, 399, 410
Tactile *(Definition)*, 12
Tactile icon, 21
Tactile properties, 35
Tactile receptors, 30
Tactile systems, 340, 348
Tacton, 21
Tangible objects, 6
Task analysis, 149
Task performance test, 512
Taxonomy of haptics, 10
Taxonomy of psychometric procedures, 42
TCP, 229
Technical solution clusters, 153
Telemanipulation system, 444
Teleoperation, 15
Teleoperation systems (*Definition*), 79
Telepresence, 15
Temperature (perception of), 58
Temperature receptors, 38
Temperature, monitor, 285
Texture, 154, 377
Thermal principle, 254
Thermal sensors, 38
Threshold, 47
Threshold values, 53
Time constant, electrical, 274
Time delay, 219
Tool usage, 6
Tool-center point, 229
Tool-mediated contact, 6
Toolkits for haptic prototyping, 147
Total reflection, 402
TPTA system, 15
Tracing, 513
Transistor, 393
Transmission chain, 444
Transmission ratio, 259
Transparency, 210, 511
Transversal actuator, 291

Transversal effect, 315
Transversal effect, electromagnetic, 315
Transversal effect, piezoelectric, 290
Traveling wave, 296
Traveling wave motor, 312
Traveling wave motor, equations, 297
Traveling wave motor, linear, 296
Triangulation, 424
TSA, 260
Twisted-string actuator, 260
Two-point threshold, 48

U
Ubi-Pen, 307
Ultrasonic actuator, 255
Ultrasonic sensors, 427
Universal serial bus, 453
Upper cutoff frequency, 415
Usability, 132
USB, 453
User as a measure of quality, 120
User as mechanical load, 101
User phantom, 508
User(*Definition*), 74

V
V-model, 125
Validation, 131, 503
Verification, 130, 503
Vibrotactile display, 306
Virtual environment, 17
Virtual reality (VR) *(Definition)*, 458
Virtual wall, 476
Viscoelastic material performance, 376
Viscosity (perception of), 58
Voice-coil actuator, 255
Voxmap-pointshell algorithm, 484

W
Wave variable, 221
Wave, traveling, 296
Weber's law, 48
Wheatstone, 386
Wiener-model, 185
Wire, 264
Workspace, 505

Y
You-are-here maps, 526

Z
Z-width, 509

Printed in the United States
By Bookmasters